NC MACHINE PROGRAMMING AND SOFTWARE DESIGN

NC MACHINE PROGRAMMING AND SOFTWARE DESIGN

CHAO-HWA CHANG

MICHEL A. MELKANOFF

University of California, Los Angeles

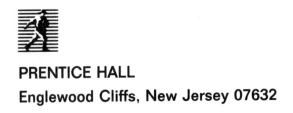

PRENTICE HALL

Englewood Cliffs, New Jersey 07632

Library of Congress Cataloging-in-Publication Data

Chang, Chao-Hwa
    NC machine programming and software design.

    Includes bibliographies and index.
    1. Machine-tools — Numerical control — Programming.
I. Melkanoff, Michel A. II. Title.
TJ1189.C445 1989   621.9'023'028551   88-17874
ISBN 0-13-610809-1

Cover design: Edsal Enterprises
Manufacturing buyer: Mary Noonan

© 1989 by Prentice-Hall, Inc.
A Division of Simon & Schuster
Englewood Cliffs, New Jersey 07632

Printed in the United States of America

10 9 8 7 6 5 4 3 2 1

ISBN 0-13-610809-1

Prentice-Hall International (UK) Limited, *London*
Prentice-Hall of Australia Pty. Limited, *Sydney*
Prentice-Hall Canada Inc., *Toronto*
Prentice-Hall Hispanoamericana, S.A., *Mexico*
Prentice-Hall of India Private Limited, *New Delhi*
Prentice-Hall of Japan, Inc., *Tokyo*
Simon & Schuster Asia Pte. Ltd., *Singapore*
Editora Prentice-Hall do Brasil, Ltda., *Rio de Janeiro*

LIMITS OF LIABILITY AND DISCLAIMER OF WARRANTY

AN IRRESISTIBLE OFFER

Since we realize that, despite our best efforts, a number of typographical and even logical errors may remain in this book, we should like to enlist the help of our readers in locating and expurgating them from future printings and editions. To that effect we shall pay to the first reader who discovers them one dollar per typographical error and five dollars per logical error described to us in writing. Winner among readers who discover the same error will be selected by post marks. The decision of the judges (i.e., the authors) will be final. Readers who wish to participate in this endeavor should write to either author at the following address

Chao-Hwa Chang or M. A. Melkanoff
Manufacturing Engineering Program
3532 Boelter Hall
University of California, Los Angeles
Los Angeles, CA 90024
U.S.A.

The authors will personally send checks to the winners.

To my dear aunts VICKY and SIU-WING

C. -H. C.

To my beloved daughter, FABIENNE, and
in memory of my wife, GENEVIEVE,
my parents, NADINE AND MICHEL, and
my parents-in-law, JEANNE and EUGENE

M. A. M.

contents

PART II/COMPUTER-AIDED NC PROGRAMMING IN THE APT LANGUAGE, 85

preface

Since World War II, the trend in engineering education has been to emphasize fundamentals and design. Manufacturing was given less attention in American universities, because it was considered as knowledge related to technology rather than to engineering science. Manufacturing was also regarded by industry as being of secondary importance with respect to improvement of product performance and even with respect to cost reduction. Such an attitude is reflected by the following typical point of view: "The most significant manufacturing-cost reductions and cost avoidances are those that result from changes in product design rather than from changes in manufacturing methods or systems."* This is true when one design is apparently more rational than another. However, when two designs of a given product are comparable or similar, each product's performance, quality, and cost are determined by the manufacturing operation and by the manufacturer's management. This has been abundantly demonstrated by the achievements of Japanese industries, whose share of the world market has been increasing steadily and remarkably through improvements in manufacturing processes and management over the last 15 years. American industry has learned from painful lessons the vital importance of manufacturing, which can no longer be neglected. An important indication of that change in attitude is that in the American (and foreign) automobile industry, "the whole emphasis" has now "switched to manufacture."[†]

An important factor that has changed manufacturing significantly, and our point of view toward it, is the advent of the computer and of a new generation of production equipment, particularly computer numerical control (CNC) machines.

*Bralla, J. G., *Handbook of Product Design for Manufacturing,* p. xiii. New York: McGraw-Hill, 1986.

[†]See the Preface in *Advanced Manufacturing in the Automotive Industry,* edited by J. Mortimer (Great Neck, N.Y.: IFS Publications, 1987).

With the increasing use of all kinds of programmable devices and systems, including CNC machines, robots, computers, CAD/CAM systems, flexible manufacturing systems (FMS), and computer-integrated manufacturing systems (CIMS), manufacturing has been completely revolutionized. Computer-aided manufacturing (CAM) requires that a process be planned in every detail and that every step in the operation be clearly defined before a part is manufactured. This means that a manufacturing process cannot be loosely defined, as had been done for many years in a process router or general process plan wherein many details were left to machinists or operators. In other words, manufacturing can no longer be treated simply as practice-related knowledge in the modern CAD/CAM environment. Much research work is needed in the field of manufacturing management, process planning, parts classification, automatic parts and materials handling, automatic inspection, automatic monitoring of machine tool status, integration of computer with production equipment, manufacturing data base, manufacturing application software design, and even manufacturing processes. Our new manufacturing engineers must be taught a comprehensive knowledge of manufacturing processes, manufacturing management, and the application of computers and new-generation production equipment, in addition to engineering fundamentals. Such a program of research and education work requires the active participation of all our engineering schools.

Confronting the new situation and responding to urgent industry demand, the University of California at Los Angeles (UCLA) established the Manufacturing Engineering Program in 1980. Since modern computer-aided manufacturing may be considered as "the outgrowth of the marriage of NC machines to on-line computers,"* NC was high on the list of courses to be offered in our program. When we started to prepare our NC courses, we were surprised to discover that there were few NC books in English. Furthermore, such books as were available were mainly concerned with general description of NC technology and manual NC programming. Many important topics, such as computer-aided NC programming, NC programming through CAD/CAM systems, postprocessor design, NC program verification, and the design of NC machine–controller systems, were not covered or were just lightly touched. These books could not possibly serve as university-level textbooks. Lecture notes were then prepared for teaching our two NC courses at UCLA during the past several years. Good NC textbooks are urgently needed by the numerous universities that offer manufacturing engineering programs and NC courses.

This book is offered in response to this urgent need. It is based on our experience in teaching two NC courses, namely, NC Machine Programming and Control Software Design and Numerical Control Manufacturing Machinery Laboratory. This book systematically discusses NC machine programming through different approaches and the design and implementation of postprocessors. Emphasis is given to computer-aided programming in the APT language, NC programming through CAD/CAM systems, and postprocessor design. Although we discuss manual programming of modern CNC machines, this material is included only to the extent necessary to understand the rest of the book.

The book is intended as either a textbook for an upper-division undergraduate NC course or a reference book for NC and APT programmers, manufacturing

---

*Gunn, T. G., *Computer Applications in Manufacturing*. New York: Industrial Press, 1981.

engineers, and NC software designers. It can also be used as a supplementary textbook for a CAM course at the undergraduate or graduate level. We normally present most of the material contained in this book in two courses consisting of lectures and laboratory work.

The book contains 15 chapters and is divided into four parts:

I.   Introduction and Manual NC Programming in Machine-Readable Codes

II. Computer-Aided NC Programming in the APT Language

III. The Generation of NC Programs Through a CAD/CAM System

IV. The Design and Implementation of Postprocessors

The first two parts are written in such a way as to be understood by a reader with little programming background. Readers should preferably have some experience with a CAD system to follow the discussion in Part III (NC programming through a CAD/CAM system). Knowledge of FORTRAN programming is necessary to understand Part IV.

Part I comprises two chapters. Chapter 1 introduces background information on NC and the influence of NC technology on modern manufacturing engineering. Chapter 2 presents NC machine codes, the block formats of an NC program, and the method of manual programming for modern CNC machines. Although FANUC controller systems are selected for illustration, the programming principles can be applied to any system.

In Part II we discuss computer-aided NC programming in APT. The IBM system was selected to describe the APT language and its programming principles because it is the system for which the language was initially designed. Numerous examples are given to explain programming details. This part is divided into six chapters. Chapter 3 introduces the elements of the APT language, including vocabulary words, symbols, numbers and scalars, punctuation and delimiters, and statement labels.

Chapter 4 is concerned with the definition of geometric entities, which are the basic elements required for defining cutter motion. The formats and usages of the statements needed for defining various geometric entities are described in detail.

Chapter 5 describes the definition of cutter motions, including point-to-point motion, contouring motion, and a sequence of motion for a group of points defined as a pattern.

Chapter 6 explains the statements that are used to define machining specifications and machine operations.

Chapter 7 introduces statements and methods that can be used to simplify programming of the cutter path and machining operations. Loop programming, programming with subprograms, and the transformation of cutter path and geometric entities are discussed. The use of the statement defining a pocketing routine is also described.

Chapter 8 explains the general structure and the debugging of an APT program, and the processing of an APT program by the IBM APT-AC Numerical Control processor.

The generation of NC programs through a CAD/CAM system is discussed in Part III. The principles underlying the generation of NC programs based on a CAD model, and the method and procedures for defining cutter paths on the CADAM system, are described in Chapters 9 and 10, respectively.

The postprocessor (i.e., the software used to transform the output from automatic NC programming systems, either an APT or a CAD/CAM system, into codes readable by an NC machine) is necessary for realizing computer-assisted NC programming. Its design and implementation are discussed in detail in Part IV. Although the IBM system was again selected to describe the subject in depth, emphasis is given to the general principles of design and implementation of a postprocessor, with the goal that the reader should be able to design a postprocessor for his own system.

An introduction to the postprocessor and the output from a computer-aided NC programming system (i.e., CLDATA) is given in Chapter 11. In Chapter 12 we describe the method and means of communication within a postprocessor. The general processing routines that are independent of the machine are explained in Chapter 13 and the machine-dependent processing routines in Chapter 14. Chapter 15 describes the procedures for creating a postprocessor, including design, test, and implementation. A postprocessor example is described in detail.

A discussion on the trend in NC postprocessing is also included in the epilogue to this part.

Such an extensive discussion of NC programming and postprocessor design could hardly be given without relating to given NC machine-controller and computer-programming systems. We therefore selected actual systems among the most widely used. However, emphasis is given to the fundamentals and principles of programming and software design so that they can also be applied to other systems.

As a textbook, our work inevitably draws heavily on the works of others. We wish to express our appreciation to the IBM Corporation for permission to use their copyrighted material. Special thanks are due to CADAM INC. for providing the photograph and drawings and to companies that have allowed us to quote their material. Finally, we wish to express our sincere thanks to Prof. Yoram Koren for his constructive suggestions.

C.-H. Chang and M. A. Melkanoff

NC MACHINE PROGRAMMING AND SOFTWARE DESIGN

part I

NC Programming in Machine-Readable Codes

chapter 1

Introduction to Numerical Control Technology

Numerical control (NC) is a technique permitting automatic operation of a machine or process through a series of coded instructions consisting of numbers, letters, and other symbols. Although NC technology was originally developed for automatic control of metal-cutting machine tools, its applications have been extended to a large variety of machines and processes. One of the major contributions of NC technology is that it changes the way machines are automated. Machine automation based on NC can be easily adjusted and applied to different production situations. Combined with the applications of computer technology, NC opens the door to modern computer-aided manufacturing (CAM) and provides the basis for realizing unmanned automation of manufacturing systems and processes in the future.

1.1 A BRIEF HISTORY OF NC TECHNOLOGY

Before the 1950s, there were two different types of production methods in the manufacturing industry: small- or medium-volume production, which was characterized by manual operation, low production speed, and diversified parts or products, and large-volume, automatic production, which used specially designed or adjusted automatic machine tools to make a single type of parts with consistent quality in large quantity and at high speed. The adoption of the latter approach always involved a significant capital investment in machines, tools, fixtures, and auxiliary equipment. It was justified only when the quantity of the parts to be made was large enough to compensate for the large initial capital investment. Production rates by these two approaches were in sharp contrast. For example, an automatic screw machine could produce several thousand screws a day, whereas the manually operated screw machine could make only a few hundred. Because the automatic machine or system was specially designed for a single type of parts, it was very difficult, and often even impossible, to adjust the automatic production system or machine to make other types of parts.

Since World War II, changing demand, technological development, and international competition have led to the appearance of new product designs at a much faster pace than before. A product can no longer exist for a long period without improvements in its quality, property, and performance — in other words, without design changes. For most parts, the old automation production process, which required as little change in part design as possible, became unjustifiable. The old type of automatic machine tool or manufacturing system was characterized by processes controlled by mechanical, electromechanical, pneumatic, or hydraulic hardware systems, which were difficult to change. For example, the classic automatic machine tools made use of cams, drums, mechanical stops, limit switches, and templates to realize a designated sequence of motions and operations on the machine. Any change in the part design required replacement of cams, drums, and templates and readjustment of the mechanical stops and limit switches. In some cases the automatic machine could not be used at all to produce a new part. Therefore a new type of control system, based on a new principle and easy to adapt to variations in part design and production situation, was needed.

The new control system also required that the positions of the cutting tool be controlled with higher accuracy but without human intervention. After World War II, more and more products, such as high-performance aircraft and automobiles, had parts with complicated profiles or shapes that were very difficult and time-consuming to make on hydraulic tracer milling machines. A new type of control system was needed that could process signals at high speed to maintain the cutter profiling motion at a high degree of accuracy. The advent of the first digital electronic computer at the end of World War II, with a processing speed hundreds of times higher than the previous ones, provided the possibility for developing such a new type of control system.

What directly helped to bring about the new control system was the requirements of new high-performance combat aircraft designed after World War II. A crucial problem in the manufacture of these aircraft was that they required precise-profiling work well in excess of the maximum capacity of tracer milling machines. Furthermore, designs were constantly modified and improved. The U.S. Air Force realized that a new technology for precise machining of metal with improved productivity had to be developed.

During World War II the Parsons Corporation had used calculated machining coordinate tables (i.e., tool-position numerical data) to move the milling machine table in the longitudinal and traverse directions simultaneously with the aid of two operators.[1] On the basis of his experience in machining profiled parts, John Parsons, of the Parsons Corporation, proposed that the three-dimensional tool-position data be generated for the profiled part and then be used to control the machine tool motion. On the basis of his proposal and the technology achieved at that time, William T. Webster, a man of great vision, and a few engineers in the Air Material Command concluded that an integration of the digital computer and high-performance servo-mechanisms was required for the new high-precision profile machining technology.[2]

The initial contract to study the feasibility of computer control of a machine tool was awarded to the Parsons Corporation, which subcontracted to the Servomechanisms Laboratory of the Massachusetts Institute of Technology (MIT) in October 1949. The MIT study confirmed the feasibility of a system that could meet the performance required by high-precision profile machining. The first vertical milling machine with three-axis simultaneous tool movement, controlled by a new type of control system, was built by MIT in 1952. It had an analog-digital, hybrid control unit that used a binary perforated tape as the medium storing the machining program (i.e., the sequence of machine instructions in numerical codes) and was called a *numerical(ly) control(led) machine*.

This machine demonstrated that a part could be made at much higher speed and with accuracy and repeatability three to five times greater than could be achieved by any previous machine.[2] In addition, no template or other change in machine element was needed to manufacture a new part; a program stored on the perforated tape was the only thing to be prepared.

During 1952 to 1955, further research was carried out at MIT under contract with the U.S. Air Force to test and evaluate the new NC machine control system and to study its application to other machine tools. Development of NC programming techniques was another important subject of the research because the Air Force foresaw that it would be difficult for industry people to prepare the required NC programs. It was also planned that MIT should demonstrate the system to industry to promote the transfer of this new technology. However, this last objective was not successful. Because of the technical involvement in several new engineering fields, including electronics, digital control, high-precision measurement, and programming, no company was willing to build or purchase the NC system. The Air Force then decided, in 1956, that it would itself sponsor the construction of 100 large NC profilers (milling machines) for manufacture of the aircraft parts for $60 million. Contracts were awarded to four different groups of machine tool–controller builder teams to reduce the risk of acquiring ineffectual systems[2]:

- Kearney and Tracker   Bendix
- Giddings and Lewis   General Electric
- Morey   General Dynamics
- Cincinnati   EMI (British)

The control systems made by EMI were of the analogue type, whereas the others were digital. The analogue design proved to be unsatisfactory and was later replaced by the digital type. The split contract designed to reduce the risk did bring about effective systems; however, it also resulted in diversified designs of the NC controller systems and in the different formats and codes used in the NC program, which became a hindrance to the portability of the NC programs (i.e., use of the NC program on NC machines of different makes to manufacture the same part).

These NC machines were put in operation between 1958 and 1960 at several aircraft companies. They were completely new to the users and required entirely different treatment from that given by machinists and engineers to conventional machine tools. However, the technical personnel in those companies where the NC machines were installed were not aware of this situation. Furthermore, there were very few people with programming experience in the industry at that time. As a result, many of the machines were damaged, and some of them were even torn apart. To add fuel to the flames, troubles often occurred in the NC controllers because their design had not been proved out and the electronic systems were not as reliable as those in use today. Disheartened by such unprecedented problems, many users wanted to discontinue the use of NC machines. It took a great deal of effort on the part of the Air Force to convince those users to continue the project.

The difficulties were gradually being overcome by improving the design of the NC control systems on the machine tool builders' side and by educating programmers, operators, and maintenance technicians on the users' side. The problems were finally brought under control around 1961 to 1962. Convinced of the advantages of the new technology, the aerospace companies started purchasing or building new NC machines at their own expense.

In the meantime, a simpler type of NC machines, namely, the point-to-point type of NC drilling machines, gained wide popularity because of the ease of its programming and the simplicity of its controller.

The first NC machine was developed at MIT by retrofitting a vertical-spindle Cincinnati Hydrotel tracer milling machine and installing in it an NC control unit. The machine was of conventional design, and the NC control unit had vacuum tubes that made it bulky and unreliable. It only demonstrated the feasibility of NC technology. The design of the NC machine-controller system had to be significantly improved to attain the desired machining accuracy for aircraft manufacturing.

Historically, the successful application of NC depended on two major factors: the improvement of the NC machine tool–controller system and the development of software programming aids. The structure, configuration, and design of modern NC metal-cutting machine tools have changed considerably in comparison with those of conventional machine tools. The feed-drive system was the vital part of the NC machine because it determined the positioning and contouring accuracy of the machine. The requirements of the feed-drive system used on NC machine tools were high precision and quick response. Slideways of antifriction design with rolling elements between the slideway and the moving part were used to reduce the friction and the stick-slip effect* occurring with conventional design.[3,4] As the basic driving element, the recirculating ball screw, of antifriction design, replaced the traditional Acme screw. Antibacklash driving mechanisms were developed to minimize the positioning error resulting from the backlash between the driving and the driven ele-

---

Stick-slip effect is a term describing the undesirable small, intermittent motion caused by sticking and jerking during the slow sliding motion of one machine element over the other, a result of the combined effects of static and dynamic frictions.

ments. Separate direct-current motors were used for motions in different axes instead of the central drive design of conventional machine tools.

High-precision position measuring and feedback systems for both angular and linear motions[5,6] were developed and used on the NC machines because they were necessary for realization of closed-loop control.

Additional motions, including linear and rotational, in directions other than the classic X, Y, and Z axes were incorporated into NC machines to cope with complex contour machining. Moreover, automatic tool selection and changing systems were developed to allow the NC machine to cut a part with different cutting tools within one hold. A new kind of machine, called an NC machining center, came on the scene in the late 1950s.[7] It combined the functions of several conventional machines, including milling, turning, boring, and drilling.

In the meantime, the structural configuration of the NC machine tools also went through significant change. Emphasis was given to the stiffness of the machine frame, greater support for the saddle and turret, and easier removal of swarf, as these factors were vital to high-precision machining and to a high rate of material removal. For both versatility and simplicity, the modular design concept* has been the trend in the design of NC machine tools since the late 1960s.[8]

Another important aspect in the development of NC machine tools was related to the NC control system, called the *NC controller*. NC controllers can be divided into two types, namely open- and closed-loop control types. The former does not allow positional errors to be compensated for and is used basically for machines that require positional or point-to-point control of low accuracy. Most modern NC machines are based on closed-loop control, which compensates for the positional error on the basis of feedback from the position measuring unit, thereby ensuring the accuracy and repeatability needed for positional and profile-following control.

The first NC controller used vacuum tubes and electrical relays; its control over the machine was realized through a hydraulic servomechanism. This type of control system was unreliable and inaccurate. With the development of electronic technology, second- and third-generation NC controllers were constructed with digital circuitry using individual transistors and integrated circuit boards. These controllers required that the NC program be written in special codes, stored on a perforated paper tape, and fed into the controller through a tape reader. Furthermore, the basic control functions were realized through the hardware (i.e., electronic circuitry), a fact that resulted in the high cost of NC controllers and prevented them from having more sophisticated functions. The development of computer technology brought about a continual decrease in the cost of computer hardware, and by the end of the 1960s, read-only memory (ROM) technology could be applied to NC controllers. A sequence of operation instructions could be stored in ROM and could be accessed and executed by the machine control unit (MCU). As a result, an NC controller could have more functions without relying on the addition of more hardware; a stored program was the only thing needed. The NC controller became more so-

*According to the modular design concept, a machine is composed of several basic functional units. If the design of these units is standardized, the design and manufacture of a machine can be greatly simplified.

phisticated but less expensive. As the decrease in the size of microprocessors and computers continued, it became possible in the 1970s to incorporate a dedicated computer into an NC controller, a technique called *computer numerical control* (CNC). The tape, either paper or magnetic strip, was no longer the only means to store the NC program for a CNC controller. A program could either be stored in the memory unit of the controller or be received from a separate computer. The program stored in the CNC controller could also be modified, a convenience that an older type of NC controller could never have. The CNC controller also provided on-line diagnosis of machine status and easy communication with various input-output devices and computers.

Two approaches were developed to realize communications between a CNC controller and a computer. With *distributed numerical control* (DNC), a complete part-machining program can be sent from a computer and stored in the CNC controller before it is executed. It can also be sent to the NC controller statement by statement while being executed; the method of sending an NC program in real time while the part is being machined is called *direct numerical control* (also DNC). The difference between these two approaches is that the program size for distributed numerical control is limited by the capacity of the memory unit of the NC controller. With direct numerical control, the operation of the NC machine depends on the signals sent by the computer. Usually a central computer is used for control of several NC machines. Thus the operation of the NC machines relies on the satisfactory operation of the control computer.

The achievement attained by NC technology today would not be possible if the software necessary to facilitate NC programming had not been developed. As can be seen from the above description, precise data regarding consecutive tool positions in a machining operation should be given in the NC program. For parts of complicated shapes, it is not possible to calculate the tool position data manually. Even for those parts whose NC cutter paths can be calculated manually, the time required for such programming is considerable. The difficulty in programming arises also from the need for translating the data into the codes required by different NC controllers.

In 1955 a prototype NC programming system developed by MIT was tested on the Whirlwind computer to demonstrate the feasibility of using computers to assist in NC programming. In 1957 a joint effort was initiated by member organizations of the Aerospace Industries Association to develop a computer program that could assist in preparing NC tapes for all kinds of NC systems. The development of this program was carried out by a group of research mathematicians from those aircraft companies under the coordination of MIT. This computer program, completed in late 1958, was named *APT* (Automatically Programmed Tool). It was designed for use on an IBM computer system. The initial version had many errors and "bugs", and the work of testing, debugging, and improving the APT program was not completed until June 1960. Later it was further refined as the well-known APT III program or processor. As the largest software package used in the industry in the early 1960s, APT was also the most used software on the mainframe computer installed in those aircraft plants (over 30% of the load).[2]

One important decision during the development of APT was that it should be designed for use by all four NC systems supported by the Air Force. Thus the output from the APT processor, which is a description of the tool positions and the desired sequence of operations and is known as the *cutter-location data* (CLDATA), should be in a standard format that is independent of the NC systems. It must further be translated by another computer program, called the *postprocessor,* into the specific NC codes required by the NC machine to be used.

In 1961 a nonprofit group, the APT Long Range Program, was formed to improve and maintain the APT software package. The task was assigned to the Illinois Institute of Technology Research Institute. Later the program became international in scope, with participants from a number of European countries and Japan.[9]

The early version of APT was designed basically for three-dimensional or multiaxis profiling. A part machining program was written in an English-like language, called the *APT language,* to describe the geometry of the part, the cutter paths, and the machining specifications. For many simple machining operations involving point-to-point motion and the motion cycles of two-dimensional machining, the language was not so convenient for the user. Another disadvantage of APT for smaller users was the requirement for a large mainframe computer, which was very expensive in the 1960s, to run APT. Therefore a number of NC programming and processing programs were developed in both the United States and European countries to simplify the APT processor, to lessen the computer requirements, and to include some functions that were not available in APT. Many of these languages were based on APT[10] or on concepts similar to APT. Those include, for example, ADAPT, EXAPT, IFAPT, MINIAPT, NELAPT, and COMPACT II (which is not a derivative of APT). Only APT and COMPACT II have been in general use in this country. Today, APT is the only NC programming language that has gained worldwide acceptance and has been standardized in the United States since 1974.[11] Currently, APT is supported by the IBM corporation.

The successful use of higher-level NC programming languages to generate NC programs depends also on postprocessing of the CLDATA. A postprocessor usually is designed by the user according to the characteristics of the NC machine–controller system and the user's computer system. Because the postprocessor is a complicated program of considerable size, its design is too big a problem for the ordinary users; this has prevented many users from successfully using the automatic programming system.

APT as a higher-level NC programming language did simplify the NC programming work and tape preparation. An NC cutter path can be defined on the basis of the geometric entities specified by the APT geometric definition statements. The rapid development of computer-aided design (CAD) technology in the 1960s made it possible for designers to construct engineering drawings on a cathode ray tube (CRT) screen and to create a geometric model within the computer. A geometric entity defined by a CAD system within a computer is represented by a group of data, the parameters of the mathematical functions or equation(s) describing the geometric entity. In other words, such data are a representation of a part within the computer and are called the *CAD model* of the part. The data of a CAD model can be further used to define an NC cutter path with the aid of certain NC programming software. Thus if a CAD system is provided with the necessary function to define an NC cutter path based on the CAD model, one can also define an NC machining process directly on the drawing displayed on the CRT screen. Such a system is usually called a computer-aided design/computer-aided manufacturing (CAD/CAM) system. CAD/CAM systems were not widely used before 1980 because of their high cost and the unreliability of the CAD/CAM software. Currently a good number of general and NC-oriented CAD/CAM systems are available on the market.

NC technology was initially developed for controlling metal-profiling operations. Today its application has already gone beyond this scope. Besides its applications in metal cutting, which includes turning, milling, drilling, grinding, and electric discharge machining, NC technology is also used in welding, flame cutting, metal forming (which also has a large field of applications, such as sheet-metal working, rolling, and forging[12]), inspection, and measurement processes. NC has also been used in industries other than metalworking, such as woodworking, textiles, plastics, and electronics.[6] Robotics can also be thought as one of the recent important applications of NC since the same principles of programming and control philosophies apply.

1.2 THE CURRENT STATUS OF NC TECHNOLOGY

The history of NC technology indicates that the United States was the nation that contributed most to all aspects of NC. The number of NC machine tools made in the United States far exceeded that in any other country before the middle of the 1970s. Since then, the production of NC machine tools in the United States has gradually fallen behind that of Japan because of the low cost and electronic-reliability advantages of the Japanese product. In 1985 the United States imported 53.3% of its NC machine tool consumption.[13] Another statistic[14] shows that during 1980 to 1985, 55% of the machine tools introduced in Japan were CNC machines, compared with only 18% in the United States. Of the total number of the machine tools in U.S. metalworking industries, NC machine tools currently occupy only a small share*; however, the rate of increase in this share is significant. In 1985 the annual NC machine tool consumption, in dollars, reached as high as 43.8% of the total U.S. machine tool consumption.[13] Since NC machine tools have much higher productivity than conventional ones, their effect on manufacturing is much greater than that reflected by the current machine-unit percentage value.

After more than 30 years' development, the design and performance of NC machine tools have reached a new level. Most NC machines built today are equipped with CNC controllers characterized by a number of new features. All CNC controllers have a CRT character display that makes it possible for the operator to review and modify NC programs more easily. The controller incorporates more and more software for automatic compensation of the cutter radius and for easy manual programming of machining cycles and symmetrical cutter paths (mirror-image function). Simple decision and calculation statements can also be included in an NC program for controllers with variable (or MACRO) programming capability. The CNC controllers of newest design, such as the Mazatrol CAM M-2 CNC controller,[15] also include graphic display and software that greatly simplify the manual programming of die-machining and some three-dimensional contouring processes. The graphic dis-

---

*The figure was 4.7% for 1983.[13]

play can also show the cutter path, thereby permitting a program to be examined before it is actually run on the NC machine. Communication with robots, measuring devices, and computers can also be realized through the CNC controller. A recent trend in NC controller design is the use of adaptive control systems, which sense and evaluate those characteristic parameters, such as cutting force, tool temperature, motor torque, and tool wear, that are not controlled by position and velocity feedback loops, and such adaptive systems also alter the NC commands so that optimal metal removal or safety conditions are maintained.

Modern CNC machines can attain high positioning accuracy in a single axial direction (about ±0.008 to ±0.015 mm/full stroke or ±0.0003 to ±0.0006 in./full stroke) and positional repeatability of ±0.001 to ±0.002 mm (or ±0.00004 to ±0.00008 in.).[16] The accuracy of multiaxis positioning depends also on the form accuracy of the guideway and the angular-position relationship among various axes; it is relatively lower. Maximum spindle speed is about 4000 to 8000 rpm; for some CNC machines, the designed maximum spindle speed reaches as high as 75,000 rpm.[7] Cutting and rapid feedrates can attain 5000 mm/min (or 197 in./min) and 15,000 mm/min (or 590 in./min), respectively.[16] Therefore a high material removal rate can be attained.

The pallet system, a type of automatic part loading and unloading system, has also become a basic feature in machining centers of recent design. Some NC machines even use a dedicated robot or manipulator for loading and unloading parts to increase productivity.

Automatic measuring devices, such as an electronic surface-sensing probe, are used on NC machines to detect the position of the part and to set the origin of the machine coordinate system.

Numerical control as a technology for controlling both the cutter path and machining operations of a machining process shows its new splendor when NC machines are used in combination with material handling systems and measuring devices under computer control, forming a *flexible manufacturing system* (FMS), which is an automatic manufacturing system that adjusts, within the design limit, its production scheme according to the part design and manufacturing requirements under the central control of a computer. Foreseeing that this is a critical new area in modern manufacturing automation, the Japanese have given great emphasis to what they have been calling *mechatronics* and have heavily invested in it during the past ten years. As a result, they have achieved a significant edge over other countries in this new technological frontier.[14]

With respect to NC programming software, more functions are included in the new versions of APT processors. The recently released IBM APT-AC Numerical Control Processor* allows the definition of bounded geometric entities and turning cycles and the definition of a set of elementary geometric entities as a contour. It also permits programming of a machining process through communication with a computer graphics system.

Currently a large number of NC-oriented CAD/CAM systems are available with different levels of capability. Some of them also provide aids for postprocessor design.

*See Section 8.5 of this book.

The output from the NC programming systems (i.e., CLDATA) has been standardized to simplify the design of postprocessors and communication between different systems. The postprocessor continues to present a significant hurdle to the successful use of CAD/CAM systems for NC program generation, and to communication between different NC programming systems. Postprocessor design aids have also become a useful feature of some computer-aided NC programming systems. The recent trend in postprocessor design is to provide a sophisticated program within the NC programming system, allowing users to design a postprocessor by just stating the answers to a series of questions.

Automatic verification of NC programs is still an open problem. According to a conservative estimate made by Welch[17] in 1980, a typical aerospace company spent more than a million dollars annually on NC program verification for every 75 NC machines. In general, an NC program verification system should examine the program syntax, check the cutter positions and the generated profile against the designed part profile, and determine whether there is any possible collision of moving parts of the machine tool, including the cutting tool, with the fixture, the part material, and the machine table. Considerable worldwide effort is in progress to develop systems for verifying part programs.[18] Most verification systems use a computer graphics system to display the cutter path, primarily in two or two-and-one-half dimensions.* There is still no satisfactory system capable of simulating the dynamic performance of the part–machine tool–controller system. Usually a test run should be carried out on the machine before a part is actually cut on it.

The automatic generation of an NC program directly based on a CAD model and through use of a manufacturing data base, including data on tools, machines, fixtures, material, and production schedules, represents a major development trend in NC programming technology.

1.3 THE INFLUENCE OF NC TECHNOLOGY ON DESIGN AND MANUFACTURING

The history of NC technology indicates that numerical control, as a type of automation tool, evolved from conventional methods of manufacturing in response to the demand for diversified products and rapidly changing product design. NC can be seen as a technology that bridges the conventional high-volume hard automation and the computer-controlled flexible one. "In the last two decades," as Milner and Vasiliou indicated, "few other new engineering and manufacturing processes have created as keen an interest, and forced so many changes in so many sectors of industry."[19]

NC is a manufacturing method that makes it possible, for the first time, to fabricate a mechanical part of arbitrary form without relying on the operational skill of a human being. Designers can specify, in their design, the forms, surfaces, or profiles

---

*A two-and-one-half dimensional cutter path is one wherein the cutter moves in two axial directions simultaneously most of the time, with motion in the third direction being used only for changing the depth of machining.

that were previously considered economically unjustified or not manufacturable. Because of the high positioning accuracy, multiaxial motions, and multiple functions of modern NC machines, it is also possible to cut easily the parts that have a combination of geometric elements with strict requirements for their relative positions and orientations.

NC machining and CAD/CAM systems are characterized by the fact that parts and cutter paths are defined on the basis of mathematical formulas and a single-reference coordinate system. This affects the way that dimensions and tolerances are specified on the design drawing. Definition of part features based on a basic coordinate system, description of surfaces and profiles by means of mathematical formulas, and bilateral tolerance are among the recommended guidelines for dimensioning and determining the tolerance of parts to be designed and manufactured with CAD/CAM and NC systems.[20]

The effect of NC on product or part design is only partial, because the design of a part traditionally depends first on its functional requirements and next on the requirement of ease of manufacturing. The influence of NC on manufacturing, however, is manifold. First, it changes the way a machining or manufacturing process is designed and planned. For example, a process plan for traditional machining provides information regarding machines, tools, fixtures, time rates, the sequence of processes involved, and the necessary machining specifications, such as feedrates and cutting speeds. However, the operation of each individual process, such as turning or milling, is specified by the operator, usually on a trial-and-error basis. With NC and modern computer-integrated manufacturing (CIM) systems, which involve minimum intervention by human beings, a process plan and all the processes involved should be planned and defined in every detail so that they can be designed as a routine or program to be executed by a computer and followed exactly by the NC machine. Consequently a process planner should possess much more in-depth knowledge of individual processes, and the NC programmer should, like an operator or machinist, know every detail of the machining operation. Second, NC also changes the way that an operation is performed. Operations are no longer controlled by operators but by the NC program. The skill of an operator is replaced by information processing (i.e., defining the operation steps in the NC program); as a result, the demand for skilled machine operators is decreasing. Much of the preparation work can be removed from the machine as it is done by the programmer now; the adjustment time in an operation is also greatly reduced. Third, NC changes the way that a process or operation is controlled. Manual control over an operation or process is replaced with control by the NC controller or a computer.

The impact of NC on the manufacturing industry is profound. It changes the philosophy of automation, the way that a process is designed, performed, and controlled, and the structure of the labor force in the manufacturing industry. It is now possible to realize unmanned operation and the integration of design and manufacturing; thus flexible automation—that is, the type of automation that can adapt to changes in design and manufacturing easily and rapidly through fast processing of information by computers—can be achieved.

As we review the history of NC and look into the prospect indicated by the numerous research papers on computer-aided manufacturing, we are convinced that

NC, as one approach to automation, did revolutionize manufacturing in the past and will be an important constituent of modern computer-aided manufacturing, which is characterized by rapid information processing and exchange.

REVIEW QUESTIONS

1.1 Give the definition of numerical control in your own terms.

1.2 Compared with the traditional method of automation, numerical control can be considered as an approach to flexible or programmable automation. Compare these two types of automation and explain why the automation through the application of numerical control is flexible.

1.3 What factors caused the appearance and evolution of NC machines and NC technology?

1.4 To obtain an effective NC system, the U.S. Air Force made an important decision to build competitive NC systems of different design. What was the profound effect of this decision on the development of NC and CIM technology?

1.5 List the major improvements on machine tool design that are necessary for the realization of modern NC machines.

1.6 Why is the software that provides NC programming aids so important to the successful application of NC technology and machines?

1.7 What is the basic difference between NC and CNC machines? Why have NC machines been superseded by CNC machines?

1.8 Explain the meaning of "distributed numerical control" and "direct numerical control" and the difference between them.

1.9 Why is a higher-level language, such as APT, needed for NC programming? What kind of aid should it provide to NC programming?

1.10 What are the three approaches to NC programming? Compare them with one another.

1.11 What is the current trend in the development of NC technology?

1.12 What is the role that NC plays in the development of CIM systems and CAM technology?

1.13 What is the problem caused by the presence of the postprocessor in the realization of a CIM system? What are the approaches used to resolve it?

1.14 What is NC program verification? Why is it important to the manufacturing based on NC technology?

1.15 What is the effect of NC on product design and manufacturing?

chapter 2

Introduction to Manual NC Programming of Machining Processes

The operation of an NC machine tool is controlled by a program written in NC code, called an *NC program*, which consists of a series of statements, or blocks, specifying the operations to be executed and the cutter motion to be realized by the NC machine in order to machine a specific part. On classic metal-cutting machine tools, a machining process plan is usually provided and should be followed by the operator in order to make the part to the required dimensions and tolerances. Correspondingly, an NC program is the translation of a machining process plan from English into NC codes that are understandable to the NC machine controller. The program is read and then executed by the NC machine-controller system. Although there are NC code standards, such as the international standard ISO 6983/1, controllers made by different manufacturers use different NC codes. Therefore an NC program is not portable to controllers of different design. Nevertheless, the NC code formats and the programming of modern CNC controllers are very similar. It is thus possible to explain the general principles and the characteristics of manual NC programming by studying the programming of one specific CNC controller system. In this book, we use the NC codes for FANUC 6MB (milling machine) and FANUC 6T (lathe) controller systems, these being among the most widely used NC controller systems in the world, to explain the fundamentals of manual NC programming.

Figure 2-1 shows a simple NC program that is used to cut the periphery of a square part. The machining plan is also included. It can be seen from this example that an NC program consists of the following:

1. A declaration of the program number
2. Statements defining the origin of the coordinate system and the kind of coordi-

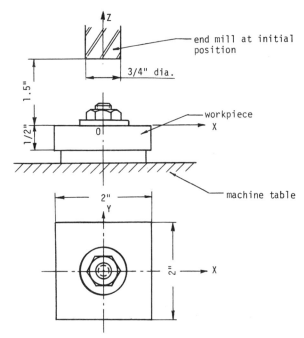

NC PROGRAM

```
00001;
N1G20G90G92X0Y0Z1.5;
N2G00X-1.375Y-1.5M03S20;
N3Z0.05;
N4G01Z-0.55F4.0;
N5Y1.375;
N6X1.375;
N7Y-1.375;
N8X-1.375;
N9G00Z1.5;
N10X0Y0M05;
N11M30;
```

MACHINING PLAN No. 0001

Step 1: Place the tool at the center of the workpiece, 1.5" above the surface. Use absolute coordinates, in inches, to define a tool position (statement N1).

Step 2: Move the tool horizontally to the lower left corner of the part (x = -1.375, y = -1.5) and turn on the spindle. Set the spindle speed at 20% of the maximum speed (statement N2).

Step 3: Move the tool down rapidly to the point (z = 0.05) and then lower it to the desired depth (z = -0.55) at a speed of 4 in./min (statements N3 and N4).

Step 4: Cut the periphery of the part (statements N5 to N8).

Step 5: Move the tool up rapidly (statement N9). Return to the starting position and turn off the spindle (statement N10).

Step 6: End of this machining process plan (statement N11).

Figure 2-1   A simple NC program in word address block format with explanation.

nates (incremental or absolute) used to describe the tool position or path in the program

3. Statements defining operations, other than tool movements, to be performed by the NC machine, such as turning the coolant and spindle on or off, setting the spindle speed and feedrate, and changing or selecting tool(s)

4. Statements defining the movement and position of the tool

5. Statements defining operations to be performed by the controller, such as "Stop reading and execution," "Stop reading and execution and rewind the memory," and "Call a subprogram"

Each statement or command occupies one line and is usually called a *block*. At the end of a statement, there should be an end-of-block code symbol, this being ";" for the FANUC 6MB controller and "*" for the FANUC 6T controller. Four kinds of block formats have been used during the development of NC technology; they will be described later.

2.1 THE COORDINATE SYSTEMS OF NC MACHINES

A machine tool is characterized by the motions it can perform. Such motions as changing the relative position of the tool and workpiece consist of linear translations and rotations about different axes. However, they do not include the rotation of the cutter or workpiece for maintaining cutting action. For example, an engine lathe has only two translation motions along axes parallel and perpendicular to the spindle axis (i.e., the X and Z axes shown in Fig. 2-2[a]), whereas a vertical milling machine has three motion axes (Fig. 2-2[b]). The rotation of the spindle (i.e., the workpiece for a lathe and the cutter for a milling machine) is usually not considered as a characteristic motion.

A right-hand rectangular coordinate system is used to describe the position and motion of the tool or workpiece. The axes of motions have been standardized both

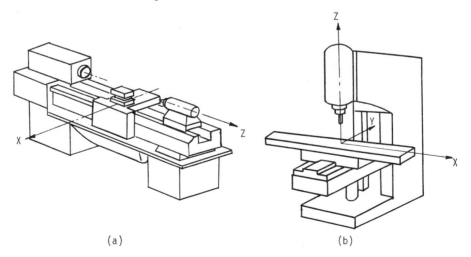

(a)   (b)

Figure 2-2   The axes of motion: (a) lathe; (b) vertical milling machine.

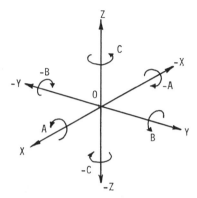

Figure 2-3 The right-hand coordinate system is used to define the cutter motion of NC machines.

by the Electronic Industries Association (EIA) in the United States and by the International Organization for Standardization (ISO).[21,22] Generally, the axis of the spindle that imparts cutting power is identified as the Z axis of motion, with the positive direction of motion increasing the clearance between the workpiece and the tool holder. The X axis, perpendicular to the Z axis, is the principal axis of motion in the positioning plane of the tool or workpiece. Once the Z and X axes have been determined, the Y axis can be located on the basis of the right-hand rule. Some of the NC machines also have rotatory motions other than those required for maintaining cutting action about the X, Y, and Z axes; they are defined as motions in the A, B, and C axes, respectively (Fig. 2-3). For example, the rotation of the rotary table of a vertical milling machine can be defined as motion in axis C if the axis of the rotary table is parallel to the spindle axis. Additional translation and rotation in different axes other than those listed above may also be provided for machining centers, boring machines, gantry profilers, and so forth, with the respective axis designations given in references 21 and 22.

2.2 THE MACHINE CODE

As can be seen from Fig. 2-1, a program consists of a series of sequentially listed statements (or blocks) specifying the cutter motion and operations to be completed. A statement can be further divided into words. Characters, delimiters, and numerical digits are the elements that constitute a word. For example, in the program in Fig. 2-1, the statement

N1G20G90G92X0Y0Z1.5;

consists of the words "N1," "G20," "G90," "G92," "X0," "Y0," "Z1.5"; the elements of the words in this statement are "N," "G," "X," "Y," "Z," "1," "2," "5," "9," "0," ".", and ";".

As is well known, a telegraph message is transmitted and understood by a person through the use of Morse code. Similarly, an NC program is made understandable to an NC machine through the use of the binary-coded decimal code. Two kinds of codes are currently in use on NC machines: the EIA RS-244-B code[23] and the

ASCII code (American Standard Code for Information Interchange). They are shown in Fig. 2-4. The traditional media by which an NC program is transferred to an NC controller is paper tape. With the increasing use of computers in manufacturing, an NC program can also be down-loaded from a computer or an NC program preparation system directly to a CNC (computer numerical control) controller through an RS-232 cable. The ASCII code has also been accepted by the EIA as the RS-358-B code,[24] and by the ISO as well. Both codes are accepted by many of the CNC machines. They are known as the EIA code (which is the RS-244-B code) and ISO code (which originated from the ASCII code).

The detailed description of these two codes can be found in many books (see for example, references 23, 24, and 25). The basic difference in these two codes is that the number of holes across the tape is *odd* for the EIA code and *even* for the ISO code. Hence they have odd or even parity, respectively. Channel 5 in the EIA RS-244-B code and Channel 8 in the ISO code are reserved for parity use. A hole is punched in these channels only when the number of holes representing a character or digit does not satisfy the required odd or even parity. In addition to the difference mentioned above, different combinations of holes are used in these two codes for specifying the same character or digit. An example of a punched paper tape is given in Fig. 2-5.

2.3 THE BLOCK FORMATS OF AN NC PROGRAM

There have been different ways to specify NC words in a statement (or block), depending on the design of the NC controller. One of the earliest NC formats is *fixed sequential*. Here, a specific sequence is required for every word in a block, and all the words that are used by the controller are required in a block whether or not they specify meaningful changes in the NC machine state. Each word consists of a number specifying a state (e.g., position in the X direction, feedrate, or spindle speed) of the NC machine. Therefore a statement is composed of a fixed number of numerals that are listed in a fixed order and that represent different NC words. For example, if the order for specifying NC words is

statement No. — motion type — X coordinate — Y coordinate — Z coordinate — feedrate — spindle speed — miscellaneous function (e.g., spindle on/off),

then a statement in fixed sequential format may appear as

003 01 1.00 2.00 3.00 2.0 500 3

meaning that the statement labeled 003 requires movement of the tool from its current position to a position (1.0,2.0,3.0) linearly with a feedrate of 2.0 in./min and with the spindle rotating clockwise at 500 rpm. Thus a complete NC program consists of a sequence of numbers. This is where the term "numerical control" originated.

The second type of block format used during the development of NC technology is the *block address format*, in which a block code is specified at the beginning of a block to inform the machine as to which words are going to be changed, specified, or both, in this particular command. Table 2-1 gives an example of the block

Char (ISO)	8	7	6	5	4	F	3	2	1	Char (EIA)	8	7	6	5	4	F	3	2	1		Meaning
0			O	O		O				0			O			O					Numeral 0
1	O		O	O		O			O	1						O			O		Numeral 1
2	O		O	O		O		O		2						O		O			Numeral 2
3			O	O		O		O	O	3				O		O		O	O		Numeral 3
4	O		O	O		O	O			4						O	O				Numeral 4
5			O	O		O	O		O	5				O		O	O		O		Numeral 5
6			O	O		O	O	O		6				O		O	O	O			Numeral 6
7	O		O	O		O	O	O	O	7						O	O	O	O		Numeral 7
8	O		O	O	O	O				8					O	O					Numeral 8
9			O	O	O	O			O	9				O	O	O			O		Numeral 9
A		O				O			O	a		O	O			O			O		Address A
B		O				O		O		b		O	O			O		O		?	Address B
C	O	O				O		O	O	c		O	O	O		O		O	O	?	Address C
D		O				O	O			d		O	O			O	O				Address D
E	O	O				O	O		O	e		O	O	O		O	O		O		Address E
F	O	O				O	O	O		f		O	O	O		O	O	O			Address F
G		O				O	O	O	O	g		O	O			O	O	O	O		Address G
H		O			O	O				h		O	O		O	O				?	Address H
I	O	O			O	O			O	i		O	O	O	O	O			O		Address I
J	O	O			O	O		O		j		O		O		O			O	?	Address J
K		O			O	O		O	O	k		O		O		O		O			Address K
L	O	O			O	O	O			l		O				O		O	O		Address L
M		O			O	O	O		O	m		O		O		O	O				Address M
N		O			O	O	O	O		n		O				O	O		O		Address N
O	O	O			O	O	O	O	O	o		O				O	O	O			Address O
P		O		O		O				p		O		O		O	O	O	O		Address P
Q	O	O		O		O			O	q		O		O	O	O					Address Q
R	O	O		O		O		O		r		O			O	O			O		Address R
S		O		O		O		O	O	s			O	O		O		O			Address S
T	O	O		O		O	O			t			O			O		O	O		Address T
U		O		O		O	O		O	u			O	O		O	O				Address U
V		O		O		O	O	O		v			O			O	O		O	?	Address V
W	O	O		O		O	O	O	O	w			O			O	O	O			Address W
X	O	O		O	O	O				x			O	O		O	O	O	O		Address X
Y		O		O	O	O			O	y			O	O	O	O				?	Address Y
Z		O		O	O	O		O		z			O		O	O			O		Address Z
DEL	O	O	O	O	O	O	O	O	O	Del	O	O	O	O	O	O	O	O	O	*	Delete (cancel an error punch).
NUL						O				Blank						O				•	Not punched. Can not be used in significant section in EIA code.
BS	O				O	O				BS			O		O	O		O		*	Back space
HT					O	O			O	Tab			O	O	O	O	O	O	O	*	Tabulator
LF or NL					O	O		O		CR or EOB	O					O					End of block
CR	O				O	O	O		O											*	Carriage return
SP	O		O			O				SP					O	O				*	Space
%	O		O			O	O		O	ER			O	O		O	O	O		*	Absolute rewind stop
(			O		O	O				(2-4-5)				O	O	O		O			Control out (a comment is started)
)	O		O		O	O			O	(2-4-7)		O			O	O		O			Control in (the end of a comment)
+			O		O	O		O	O	+		O	O	O		O				*	Positive sign
−			O		O	O	O		O	−			O			O					Negative sign
:			O	O	O	O		O												*	Colon
/	O		O		O	O	O	O	O	/			O	O		O			O		Optional block skip
.			O		O	O	O	O		.			O	O		O		O	O		Period (A decimal point)
#	O		O			O		O	O											*	Sharpe
$			O			O	O													*	Dollar sign
&	O		O			O	O	O		&					O	O	O	O		*	Ampersand
'			O			O	O	O	O											*	Apostrophe
*	O		O		O	O		O												*	Asterisk
,	O		O		O	O	O						O	O	O	O		O	O	*	Comma
;	O		O	O	O	O		O	O											*	Semicolon
<			O	O	O	O	O													*	Left angle bracket
=	O		O	O	O	O	O		O											*	Equal
>	O		O	O	O	O	O	O												*	Right angle bracket
?			O	O	O	O	O	O	O											*	Question mark
@	O	O				O														*	Commercial at mark
"			O			O		O												*	Quotation
{		O	O	O	O	O		O	O											*	Left brace
}		O	O	O	O	O	O		O											*	Right brace

Figure 2-4   The ISO and EIA codes for the NC perforated tape.[26] (Courtesy of FANUC Ltd.)

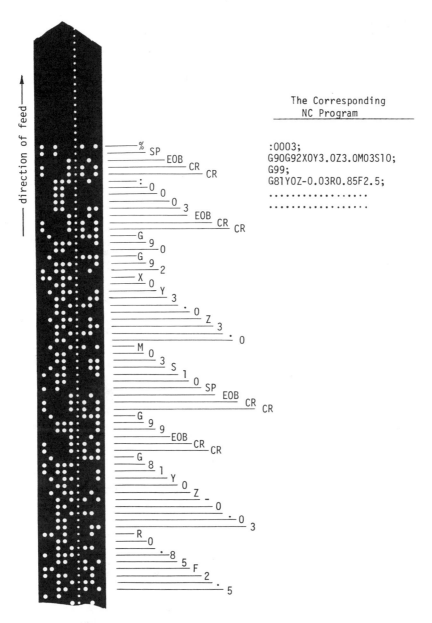

The Corresponding
NC Program

:0003;
G90G92X0Y3.0Z3.0M03S10;
G99;
G81Y0Z-0.03R0.85F2.5;
....
....

Figure 2-5   An NC perforated tape.

codes for different operation requirements. According to this table, a block address code "20" in an NC statement means that solely a change in position in the X and Y directions will be specified in this block. As in the fixed sequential block format, all the words in a block are specified in fixed order. Because the unchanged words are not necessary in a statement, the block address format is more compact than the fixed sequential one.

TABLE 2-1 EXAMPLE OF BLOCK ADDRESS CODES

Change in	Change of Position in Direction(s)				
	X	Y	X, Y	Z	No Change
Position only	00	10	20	30	
Feed	01	11	21	31	41
Speed	02	12	22	32	42
Coolant	03	13	23	33	43
Feed, speed	04	14	24	34	44
Feed, coolant	05	15	25	35	45
Speed, coolant	06	16	26	36	46
Feed, speed, coolant	07	17	27	37	47

Example: If the feedrate and position in the X and Y directions are to be changed in a program, the block address code is 21.

From: C.R. Asfahl, *Robots and Manufacturing Automation*. Copyright © 1985 by John Wiley & Sons, Inc., and reprinted by permission of John Wiley & Sons, Inc.

The third format is *tab sequential*, which is also a variation of the fixed sequential format. In this format, a tab character must be specified before each word. Words must also be presented in a specific order, and all the characters in a word can be omitted when there is no change in the content of the word with respect to the previous statement. However, the tab code before a particular word is still needed, even though that word has been omitted.

The three formats mentioned above are used in old hard-wired numerical control systems. With the advent of computer numerical control systems, they have been gradually superseded by the word address format and may be considered obsolete.

The *word address format* is the one used in modern CNC controller systems. Each word in a statement (block) consists of a character indentifying the meaning or address of the word and a number representing its content. For example, the word specifying a tool position of $X = 1.0$ is

X1.0

Addresses and their meanings in FANUC 6MB and 6T controller systems are listed in Tables 2-2 and 2-3, respectively.[26,27] The number after a character can be an integer or a real number, depending on the type of word (or code), as can be seen in these two tables. The codes X, Y, Z, I, J, K, R, Q, U, and W, indicating coordinate position, can also accept a number without a decimal point. In this case the magnitude of the number is considered as the specified number multiplied by the resolution of the controller (0.001 mm for metric programming, and 0.0001 in. for programming in inches). For example, if programming is done in inches,

X1.	means	$X = 1.0$ in.
X1	means	$X = 0.0001$ in.
X100	means	$X = 0.0100$ in.

TABLE 2-2 **SELECTED WORD ADDRESS CODES AND THEIR FORMATS FOR FANUC 6MB NC CONTROLLER SYSTEM**

Address Code	Meaning	Type of Number after Address Code / Number of Digits
:(ISO)/O(EIA)	Program number	Integer/4 digits
N	Statement label	Integer/from 1 to 4 digits
G	Code defining motion	Integer/2 digits
X, Y, Z*	Coordinates in X, Y, and Z directions	Inch input: real/4 digits to the left and right of decimal point
R*	Radius of arc or circle	Metric input: real/5 and 3 digits to left and right of decimal
I, J, K*	Coordinate values of arc center	point, respectively A number n specified without
Q, R*	Positional parameters in cycle statement	decimal point considered as $n \times 0.0001$ in. (inch input) or $n \times 0.001$ mm (metric input)
F*	Feedrate or thread lead	Inch/min input: real/3 digits before and 2 digits after decimal point, respectively mm/min input: integer/5 digits Inch/rev input: real/2 digits before and 4 digits after decimal point, respectively mm/rev input: real/3 digits before and 2 digits after decimal point, respectively
S	Spindle speed	Integer/2 or 4 digits depending on design of NC machine
M	Operation of machine tool	Integer/2 digits
D	Offset number	Integer/2 digits
P, X*	Dwell time (sec.)	Real/5 digits before and 3 digits after decimal point, respectively
P	Designation of program or statement number	Integer/4 digits
L	Repetition count	Integer/1 to 4 digits
;	End of block (statement)	

Example: A statement can be as follows: N1G92X2.0Y1.3542Z0.5M03S40;

*Trailing zeros after a decimal point and leading zeros before a decimal point can be suppressed.

In metric programming,

X1.	means	$X = 1.0$ mm
X1	means	$X = 0.001$ mm
X100	means	$X = 0.100$ mm

It is suggested that the decimal point be used in manual NC programming to avoid making errors.

TABLE 2-3 SELECTED WORD ADDRESS CODES AND THEIR FORMATS FOR FANUC 6T NC
CONTROLLER SYSTEM

Address Code	Meaning	Type of Number After Address Code / Number of Digits
:(ISO)/O(EIA)	Program number	Integer/4 digits
N	Statement label	Integer/from 1 to 4 digits
G	Code defining cutter motion	Integer/2 digits
X, Z[†]	Coordinates in X and Z directions	Inch input: real/4 digits to the left and right of decimal point
U, W[†]	Incremental coordinates in X and Z axes	Metric input: real/5 and 3 digits to left and right of decimal point, respectively
R[†]	Radius of arc or circle	A number n specified without decimal point is considered
I, K[†]	Coordinate values in x and z direction of arc center, or radial and axial retraction	as $n \times 0.0001$ in. (inch input) or $n \times 0.001$ mm (metric input)
D[†]	Depth of cut	
F[†]	Feedrate or thread lead	Inch/rev input: real/2 digits before and 4 digits after decimal point mm/rev input: real/3 digits before and 2 digits after decimal point
E[†]	Thread lead	Inch/rev input: real/1 digit before and 6 digits after decimal point mm/rev input: real/3 digits and 4 digits after decimal point
S	Spindle speed	Integer/2 or 4 digits
T	Tool and offset number	Integer/4 digits; first two digits are tool number, and last two digits are offset number
M	Operation of machine tool	Integer/2 digits
X, U, P[†]	Dwell time (sec.)	Real/5 digits and 3 digits after decimal point; following inputs are equivalent: $3.0 = 3000 = 3$ seconds
P, Q	Designation of statement or program number	Integer/4 digits
A	Thread angle	Integer (discrete value)
L	Repetition count	Integer/1 to 4 digits
*	End of block	

[†]Trailing zeros after a decimal point and leading zeros before a decimal point can be suppressed.

The F, E, and P codes can also accept real number input. They will be discussed in the following sections. All the other codes can accept only integer input.

One of the important features of the word address block format is that the words in a block need not be specified in fixed order because the meaning of a word is clearly indicated by the first alphabetic character (i.e., the address). In addition, if a code is the same as that specified in the preceding statement, normally it can be omitted.

2.4 THE BASIC CODES USED IN NC PROGRAMMING FOR TURNING AND MILLING OPERATIONS

An NC program usually consists of two kinds of commands. The first kind is used for defining a cutter motion with respect to a workpiece. Cutter motion is defined on the basis of the following parameters:

1. The type (incremental or absolute), unit (inch or metric system), and relative positions of the coordinate systems used in the NC program
2. The type of cutter motion: positioning, linear interpolation, circular interpolation, thread cutting, or others
3. Feedrate, dwell time, and amount of offset to be made for a specific tool to cut a workpiece

From Table 2-2, it is evident that one or several of the following codes are needed to define the cutter motion on a three-axis milling machine with a FANUC 6MB controller system: G, X, Y, Z, R, I, J, K, H, D, P, Q, R, and F.

The second kind of commands is used to control the operation of an NC machine, to select tool and spindle speed, and to indicate the program and statement sequence number. The following codes are used on the FANUC 6MB controller system for these purposes: M, T, L, O, P, and S.

The codes used for a lathe with a FANUC 6T controller system (Table 2-3) are slightly different from those used for a milling machine. A lathe can have motions in only two dimensions (X and Z). Accordingly, codes defining the position of the cutter are needed only in these two directions. The FANUC 6T controller system uses two separate codes, U and W, to denote the incremental coordinates in the X and Z directions, respectively. In addition, the offset required of a lathe cutting tool is specified by a T code. Therefore the codes used to define a cutter motion on a FANUC 6T controller system are G, X, Z, U, W, R, I, K, P, Q, D, and F. The codes used to define the operation of a lathe are basically the same as those used for a milling machine except that the T code is used to define the tool to be used and the offset as well.

Each of the codes other than G and M has a unique function, with the exception of X, U, and P, which have more than one function. The G and M codes have many functions, as can be seen from Tables 2-4, 2-5, and 2-6.

It must be pointed out that there are two kinds of G codes, depending on whether such a code is effective only in the block in which it is specified or is effective until explicitly canceled. Once specified, a code in groups other than group 00 is effective until it is canceled by specifying another code in the same group. They are said to be modal. Any code in group 00 is not modal. It can be seen that each group of the G code defines one type of motion or one mode of operation. When the con-

TABLE 2-4 SELECTED G CODES USED ON FANUC 6MB NC CONTROLLER SYSTEM

G Code	Group No.	Function
G00	01	Rapid positioning
G01		Linear interpolation motion
G02		Circular interpolation motion (clockwise)
G03		Circular interpolation motion (counterclockwise)
G04	00	Dwell
G10		Setting offset value
G17*	02	*X-Y* plane selection
G18		*Z-X* plane selection
G19		*Y-Z* plane selection
G20	06	Inch input
G21		Metric input
G28	00	Return to reference point
G40*	07	Canceling cutter compensation
G41		Cutter-radius compensation left
G42		Cutter-radius compensation right
G80*	09	Canceling cycle motion defined by G80s code
G81		Spot drilling cycle
G83		Peck drilling cycle
G85		Boring cycle
G90	03	Programming in absolute coordinates
G91*		Programming in incremental coordinates
G92	00	Setting origin of machine coordinate system
G94	05	Per minute feed
G95		Per revolution feed
G98	10	Return to initial point in cycle motion
G99		Return to R point in cycle motion

*When power is turned on, the codes marked with an asterisk are effective.

troller power is turned on, the G codes marked with an asterisk are effective. G codes that do not belong to the same group can be specified in a block. If more than one code of the same group are specified in a block, only the last specified code is effective.

For example, the first statement in the NC program shown in Fig. 2-1 consists of three G codes of different groups (group 06, 03, and 00). The statement

G02G00X_Y_I_J_;

has two G codes of group 01. Thus its motion is defined by the last specified G code: G00. Accordingly, the I and J codes are neglected; hence,the motion is defined by

G00X_Y_;

It must be pointed out that, as a rule, specifying more than one code of the same group should be avoided. In addition, there are exceptions to the rule that G

TABLE 2-5 SELECTED G CODES USED ON FANUC 6T NC CONTROLLER SYSTEM

G Code	Group No.	Function
G00	01	Rapid positioning
G01		Linear interpolation motion
G02		Circular interpolation motion (clockwise)
G03		Circular interpolation motion (counterclockwise)
G04	00	Dwell
G10		Set offset value
G20	06	Inch input
G21		Metric input
G28	00	Returning to reference point
G32	01	Thread cutting (single cycle)
G40*	07	Cancel cutter nose radius compensation
G41		Cutter nose radius compensation on the left
G42		Cutter nose radius compensation on the right
G50	00	Set origin of machine coordinate system
G70		Finishing cycle
G71		Stock removal in turning
G72		Stock removal in facing
G73		Repeating pattern
G76		Thread cutting (multiple cycle)
G98	05	Per minute feed
G99		Per revolution feed

*When power is turned on, the code marked with an asterisk is effective.

TABLE 2-6 SELECTED M CODES USED ON FANUC CONTROLLER SYSTEMS

M Code*	Function
M00	Program stop (usually used in middle of program)
M01	Optional stop (This function is effective only when optional stop button on control panel is depressed.)
M02	End of program
M03	Spindle start clockwise (milling machine) Spindle start forward (lathe)
M04	Spindle start counterclockwise (milling machine) Spindle start reverse (lathe)
M05	Spindle stop
M08	Coolant on
M09	Coolant off
M11	Tool change (Cadillac NC-100 lathe)
M30	End of program and memory (or tape) rewind
M41	Lower range of spindle speed
M42	Higher range of spindle speed
M98	Call a subprogram
M99	Return to main program

*M codes might be different for machines of different manufacturers. For example, an M06 code is normally used to specify a tool change on other machines. The reader should consult the programming manual of his particular machine.

codes of different groups can be specified in a block. For example, in the statement

G04G00X_Y_;

there are two G codes of different groups. The G04 code orders the tool to dwell at its current position, whereas the G00 code orders it to move from its current position to the position defined by the X and Y codes. The effects of these two codes are in conflict with each other. It is suggested that two G codes that belong to groups 00 and 01 not be specified in the same statement.

Codes other than the G codes can be specified only once in a block. With the word address format, codes in a block can be specified in any order.

It should be noted that an X code for a lathe has a different meaning from that for a milling machine. Generally, the cross sections of the parts cut on a lathe are circular. On the drawing, the radial dimension of a cylindrical surface is usually defined not by its radius but by its diameter, so that it can be easily measured. For this reason, the X code is defined as the *diametric* value (i.e., twice the amount of the X coordinate), so that a cutter position can be specified more easily on the basis of an engineering drawing. In Fig. 2-6(a), the codes defining the tool position A (2.0, 2.0) are

X4.0Z2.0

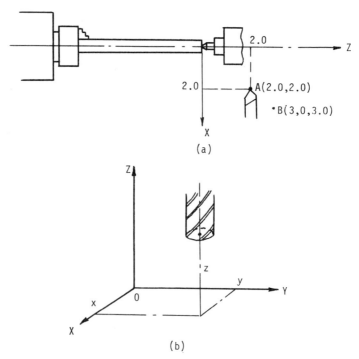

(a)

(b)

Figure 2-6   (a) The coordinate system and X and Z codes for a lathe. The origin of the coordinate system can be set by the statement G50X4.0Z2.0, based on tool position A. (b) The origin of a coordinate system is set by the statement G92XxYyZz for a milling machine.

The U code, which is the incremental coordinate of X, should also be specified in the *diametric* value. Thus, the U and W codes for moving the tool from point A to point B are

U2.0W1.0

2.5 DEFINING THE COORDINATE SYSTEM

A coordinate system should be defined before a motion statement is specified. On an NC machine, the directions of the coordinate axes are fixed. However, the origin of the coordinate system can be set at any desired position. A cutter motion can be correctly defined only after the origin of the coordinate system has been set at the right position. The code G92 is used to define the origin of a coordinate system with the following format (Fig. 2-6[b]):

G92XxYyZz;

where x, y, and z are the coordinates of the current tool position in the coordinate system to be defined. Thus the origin is defined on the basis of the current tool position. In the example given in Fig. 2-1, the initial position of the tool is (0,0,1.5) in the coordinate system to be defined, so that the statement for defining the coordinate system is

G92X0Y0Z1.5;

For a lathe with the FANUC 6T controller system, a G50 code is used, instead of G92, to set the origin of the coordinate system. An example is shown in Fig. 2-6(a).

Another way to define a coordinate system is to set its origin on the basis of a fixed reference point on the NC machine (Fig. 2-7). In an NC program the tool is first moved to the reference point, and then a G50 (for a lathe) or G92 (for a mill) code is used. In this way the origin of the coordinate system can always be defined at the same point, even though the starting positions of the tool differ from one op-

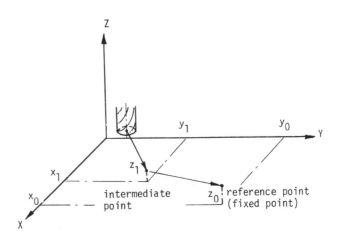

Figure 2-7 Setting up a coordinate system, based on a reference point, through the use of statements $G28Xx_1Yy_1Zz_1$ and $G92Xx_0Yy_0Zz_0$.

eration to another. This method is especially useful for cutting a number of identical parts on the NC machine. The tool can be moved to any desired position before a program is executed. A special code (G28) for moving the tool to the reference point* is provided on the FANUC CNC controller system, its format being

G28XxYyZz   (for a milling machine)
$$G28 \begin{Bmatrix} XxZz \\ UuWw \end{Bmatrix}$$ (for a lathe)

where x, y, and z are the coordinates of an intermediate point. The tool movement defined by this code is a rapid motion from its current position, via the intermediate point, to the reference point. In incremental programming, x, y, z, u, and w should be the incremental coordinates from the current tool position to an intermediate point. If no intermediate point is needed, these coordinates should be zero. The required coordinate system in Fig. 2-7 can be set up by the following two statements:

G90G28Xx_1Yy_1Zz_1;
G92Xx_0Yy_0Zz_0;

The reference point of an NC machine is usually set at one corner of the machine table. It is also a safe retreat position for changing a tool. The specified intermediate point defines a cutter path by which an obstacle (either the part or the fixture) between the starting and the reference points can be circumvented.

For an easy way to define a cutter path, the coordinate system used in an NC program can be changed. Figure 2-8 shows an example in which two sets of holes can be defined more easily in two different coordinate systems, namely, X-Y-Z and X'-Y'-Z'. Therefore, in an NC program, we can first define the machine coordinate system as X-Y-Z. After having defined the cutter motion for drilling the first set of holes, we can specify the statement

G92Xx_2Yy_2Zz_2;

*For many NC machines, the reference point is a position to change the tool.

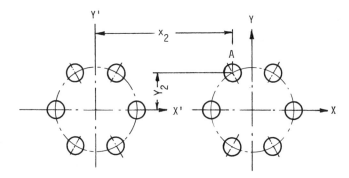

Figure 2-8   Defining a cutter path can be simplified by using different coordinate systems.

to redefine the machine coordinate system as X'-Y'-Z' on the basis of the last tool position A with coordinates x_2, y_2, and z_2 in the X'-Y'-Z' coordinate system. In the example shown in Fig. 2-8, point A is the position of the tool before the statement for defining the new coordinate system X'-Y'-Z' is executed, and the coordinate value x_2, y_2, and z_2 can be determined easily.

For a lathe with the FANUC 6T controller system, a G50 code is used instead of G92, and only the X and Z codes are included in the statement. The end surface of a rotational part machined on a lathe is often used as the reference surface for programming the cutter path. Thus the origin of the machine coordinate system can be set at the center of the end surface after it has been machined.

It must be pointed out that redefining the machine coordinate system on the basis of the current tool position results in an accumulation of tool position errors. The reason is that the actual tool position can never be the programmed one because of positioning error of the machine tool. Therefore this method can be used only when the accumulated error does not exceed the specified tolerance.

A cutter position can be specified in absolute coordinates or in incremental coordinates. On a milling machine with the FANUC 6MB controller system, these two states are defined by codes G90 and G91, respectively. Thus the statement

G90G00X3.0Y4.0;

means that the coordinates specified in and after this block are absolute coordinates, whereas the statement

G91;

means that the coordinates specified in and after this block are incremental.

For the lathe, codes U and W are used to represent the incremental coordinates in the X and Z directions. Accordingly, codes G90 and G91 do not have the same meaning as for the milling machine. There are many cases when incremental programming can be used to simplify and shorten an NC program significantly. Example 2 in Section 2.9 of this chapter represents a typical case.

2.6 DEFINING CUTTER MOTION

2.6.1 The Basic Motion of an NC Machine

The majority of NC machines can perform three kinds of basic motion: rapid positioning motion, linear interpolation motion, and circular interpolation motion.

Rapid Positioning.   This motion is defined by a G00 code with the following format:

G00X*x*Y*y*Z*z*;

where x, y, and z are the coordinates of the destination point. Some NC machines can move the tool simultaneously in only two directions. In that case, only two coordinates can be specified for the G00 code. If one of the codes X, Y, or Z is not specified, it is the same as in the former statement in absolute programming. In incremental programming, an omitted code means that the incremental coordinate is zero.

The purpose of a rapid positioning motion is to move the tool rapidly from its current position to a desired destination position (x, y, z) with no requirement on the cutter path. Only the final position is of importance to the machining operation. A typical application of this code consists in rapidly moving a drill from one position to another in a drilling operation. Under a G00 code, a cutter moves at the same *rapid* speed, which is set by a parameter of the NC controller, in the X, Y, and Z directions until one of the three coordinates of the destination point is reached. Then it moves at the same speed in the other two directions until one of the remaining two coordinates is reached. Finally, it moves unidirectionally to the destination point. Therefore the cutter path is determined by the increments of the coordinates, from the current tool position to the destination position. Figure 2-9 shows the cutter motions for various destination points. Generally, the cutter path is not a straight line except when the incremental coordinates in different directions are the same or when the motion is only in one direction. During rapid motion the velocity of motion (feedrate) is set by the controller, so that a feedrate code F is not needed.

When incremental programming is in effect, the coordinate codes X, Y, and Z should be given in incremental values. Thus the statement for rapidly moving the tool from its current position (1.0,2.0,3.0) to point (2.5,-4.2,3.0) is

G91G00X1.5Y-6.2;

For the lathe, the U and W codes should be used instead of codes X and Z. The motion of the tool from position (1.0,3.0) to point (2.5,-2.5) is defined by the statement

G00U3.0W-5.5*

Linear Interpolation Motion.   This motion is defined by a G01 code with the following format for a milling machine:

G01X*x*Y*y*Z*z*F*f*;

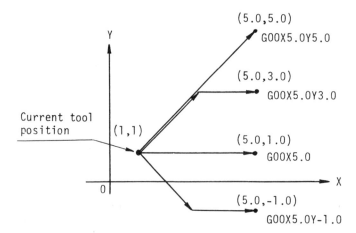

(5.0,5.0)
G00X5.0Y5.0

(5.0,3.0)
G00X5.0Y3.0

Current tool position

(1,1)

(5.0,1.0)
G00X5.0

(5.0,-1.0)
G00X5.0Y-1.0

Figure 2-9   The cutter path specified by the G00 code is determined by the difference in distances from the current tool position to the destination point in the X and Y axes.

where x, y, and z are the absolute or incremental coordinates of the destination point, and f is the feedrate in inches per minute or millimeters per minute. If only one or two of the coordinates are specified, then the motion will be one or two dimensional, respectively. Some controllers can control simultaneous motion in only two axes; in such cases, only two coordinates can be specified in a statement. For the lathe, the following format is used:

G01XxZzFf*   (absolute programming)
G01UuWwFf*   (incremental programming)

When a G01 code is specified, the NC controller generates three sets of pulses for the three motors that drive the tool or table in the X, Y, and Z directions, respectively. The ratio of the numbers of pulses per second for the three motors is such that the resulting tool motion will be on the straight line connecting the starting and end positions with the speed equal to the specified F code. The F code can be omitted if its value is the same as that specified in the previous statement.

The G01 code is a contouring motion code because the cutter path is under continuous control. Normally, we can use it to define a cutter path for cutting a linear profile.

Circular Interpolation Motion.   Most NC milling machines and lathes are capable of moving a tool along a circular path in the X-Y, Y-Z, or X-Z planes, an operation that can be specified by the circular interpolation codes G02 and G03 with the following formats:

- in the X-Y plane:

$$G17G\begin{Bmatrix}02\\03\end{Bmatrix}XxYy\begin{Bmatrix}Rr\\IiJj\end{Bmatrix}Ff;$$

- in the Y-Z plane:

$$G19G\begin{Bmatrix}02\\03\end{Bmatrix}YyZz\begin{Bmatrix}Rr\\JjKk\end{Bmatrix}Ff;$$

- in the Z-X plane:

$$G18G\begin{Bmatrix}02\\03\end{Bmatrix}XxZz\begin{Bmatrix}Rr\\IiKk\end{Bmatrix}Ff;$$

where

G17, G18, and G19 define, respectively, the planes X-Y, Z-X, and Y-Z in which the circular movements are performed.

G02 and G03 define clockwise (CW) and counterclockwise (CCW) circular movements, respectively (Fig. 2-10). Clockwise and counterclockwise senses of rotations are distinguished on the basis of the rule that when one looks from the positive direction of the axis perpendicular to the plane on which the circular motion is performed, the motion is in clockwise and counterclockwise directions, respectively.

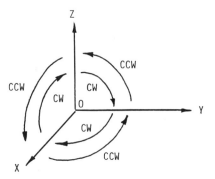

Figure 2-10   The definition of clockwise (CW) and counterclockwise (CCW) directions of circular interpolation motion.

x, y, and z denote the coordinates of the destination point in either absolute (G90) or incremental (G91) values. For a lathe, the X and Z codes should be changed to the U and W codes in incremental programming.

i, j, and k denote the incremental coordinates from the center of the circle to the initial tool position.

r denotes the radius of the circle. If the circular cutter path is greater than 180 degrees, then r should be negative. For a lathe, because of the characteristics of the turning operation, the circular motion can only be less than 180 degrees.

Examples of the use of G02 and G03 codes are shown in Fig. 2-11. In NC programming, mistakes can be found very often in the sign and value of i, j, and k. Since the radius r is normally given on the drawing, it is less easy to make a mistake by using an R code instead of using the I, J, and K codes. When a circular interpolation motion is programmed, the X, Y, or Z code can be omitted if the coordinate of the end point is the same as that of the starting point. Omitting the I, J, or K code means that the incremental coordinate from the circle center to the starting point in the X, Y, or Z axis is zero. Figure 2-12 shows several examples of programming circular paths when one or more of the codes can be omitted. When a complete circle is to be programmed, statements with the following formats can be used:

$$G \begin{Bmatrix} 02 \\ 03 \end{Bmatrix} G \begin{Bmatrix} 17IiJj \\ 18IiKk \\ 19JjKk \end{Bmatrix}$$

However, if an R code is used instead of I, J, and K in the above statements, no motion will take place.

Dwelling at the Current Position.   A G04 code in one of the following formats

$$G04 \begin{Bmatrix} Xn \\ Pn \\ Un \text{ (for lathe only)} \end{Bmatrix}$$

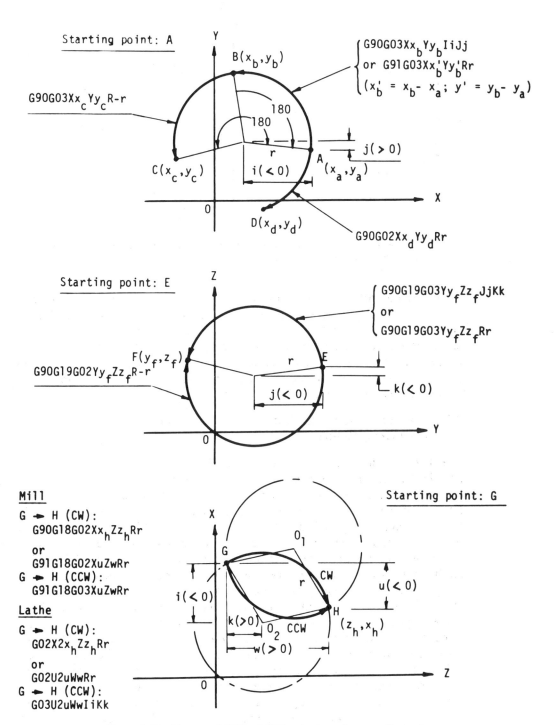

Figure 2-11 The use of G02 and G03 codes in various coordinate planes.

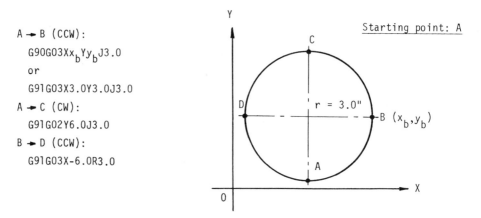

A → B (CCW):

    G90G03Xx$_b$Yy$_b$J3.0

or

    G91G03X3.0Y3.0J3.0

A → C (CW):

    G91G02Y6.0J3.0

B → D (CCW):

    G91G03X-6.0R3.0

Figure 2-12   Examples showing that some of the codes can be omitted in programming circular interpolation motion.

can be used to order the tool to stop at its current position for n seconds. One of the uses of this code is to get a sharp corner on the profile of the workpiece. During the cutter motion, a deceleration at the end of the motion specified by one statement and an acceleration at the start of the motion specified by the next statement are usually applied automatically by the NC controller (Fig. 2-13). Thus a round corner results even though the programmed cutter path is rectangular. A G04 code can be inserted between the two statements to make a sharp corner.

2.6.2 The Compensation for Cutter Radius in Defining a Cutter Contouring Motion

The statements in the former section are used to define the path of the cutter center line. In the example shown in Fig. 2-1, the cutter path in the X-Y plane cannot be programmed on the basis of the coordinates of the part profile: (1.0,1.0), (1.0,-1.0), (-1.0,-1.0), and (-1.0,1.0). Indeed, the tool must be shifted outside by 0.375 in. (i.e., the radius of the end mill) to yield the correct profile. For the turning operation, the positions and cutter path defined by the statements are those of the tool nose center. Therefore a cutter path for contouring motion cannot be defined directly on the basis of the dimensions specified on a part drawing. The coordinates of the end position in each contouring motion statement of an NC program must be calculated. This calculation is time consuming and error prone. On modern CNC machines, special calculation functions or cutter-radius compensation codes are provided to allow a user to utilize part-profile coordinates obtainable from the part drawing to program a contouring motion. These are the G41 and G42 codes for tool-radius compensation on the left- and right-hand sides of a profile, respectively. A left or right compensation is based on the fact that the tool is on the left- or right-hand side when one goes along the part profile in the direction specified by the contouring motion statements in the program. A G40 code is provided to cancel the cutter-radius compensation. Together with the cutter compensation codes, a radius offset code D must be specified. In an NC program, a G41 or G42 code should be

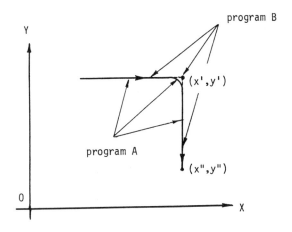

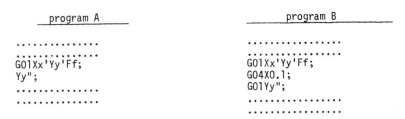

Figure 2-13   Deceleration at the end of the movement specified by one statement and acceleration at the start of the next movement are applied automatically by the NC controller, resulting in a round corner. A G04 code can be inserted between the two blocks to cut a sharp corner.

specified to start a contouring motion (i.e., to position a tool from its current position to a ready position for contouring motion). The G41 and G42 codes also tell the controller that the motion defined in the following statement is in the tool-radius compensation mode until canceled. For example, in Fig. 2-14, which depicts the top view of the part shown in Fig. 2-1, the desired profile is ABCD. The following program defines the required cutter contouring motion:

```
. . . . . . . . . . . . . . . . . .
G90G20. . . . . . . . . . . .
. . . . . . . . . . . . . . . . . .
N0010G17G41G01X-1.0Y-1.0D01F2.0;   (start-up motion from P to A₁)
N0015Y1.0;
N0020X1.0;                         (contouring motion statements in the
N0025Y-1.0;                         cutter-radius compensation mode)
N0030X-1.0;
N0040G40G00X-2.0Y-1.5;             (move the tool to point P and cancel
                                    the cutter-radius compensation)

. . . . . . . . . . . . . . . . . .
. . . . . . . . . . . . . . . . . .
. . . . . . . . . . . . . . . . . .
```

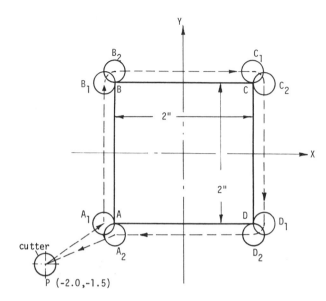

Figure 2-14 The use of the cutter-radius compensation code to program a contouring motion.

The statement labeled N0010 defines a start-up motion for the contouring motion with cutter-radius compensation on the left. A number of offset values can be stored, through manual input, into the memory of the controller system (i.e., in the No. 1, 2, ... offset registers). The code D01 means that the radius offset stored in No. 1 position should be used in the cutter compensation. The other way to input an offset is to use a G10 code with the following statement format:

G10PpRr;

where G10 causes to input an offset value of r into offset register No. p. Evidently, this statement must be specified before a D code. In our case the statement for inputting an offset of 0.375 in. is

G10P01R0.375;

The D01 code in the statement labeled N0010 in the program shown above means that the radius offset stored in the No. 1 position should be used in the cutter-radius compensation. The statement labeled N0010 specifies:

- That the start-up motion should be a linearly interpolated motion with a feedrate of 2.0 in./min
- That the cutter compensation should be on the left-hand side with a cutter-radius compensation given by the value stored in the No. 1 offset register

The direction (left or right) of compensation is determined by two statements: the start-up motion statement (labeled N0010) and the statement (labeled N0015) immediately following it. At the end of the start-up motion, the tool should move to a position where it is tangent to the desired path defined by the next motion statement. The cutter path in the following statements is defined according to the part profile.

The statement labeled N0040 cancels the radius compensation gradually from point A_2 to point P.

Various situations that might be encountered in defining a contouring motion with cutter-radius compensation and might confuse a beginner are explained below:

1. The direction of compensation specified by G41 and G42 codes are determined by the fact that when traveling from one point to another along the profile in the direction specified by the contouring motion statements, one finds that the tool is on the left- and right-hand sides, respectively. In the program shown above, when traveling from point A through B, C, and D to A, the tool should be on the left-hand side.

2. Only the G00 or G01 code can be used in the start-up motion statement.

3. The statement right after the start-up motion statement should define a motion on the plane specified in the start-up motion statement. If a statement other than a motion statement or one specifying motion not on the plane indicated by the start-up motion statement is specified, the controller will not be able to determine the direction of compensation. From the second statement of a contouring motion with cutter-radius compensation, the direction of compensation for each statement in the G41 or G42 mode can be determined on the basis of the next two statements. Therefore a nonmotion statement or one specifying a motion not in the cutter-radius compensation plane can be inserted. For example, the statement

   N0018Z-0.5;

   can be inserted between statements labeled N0015 and N0020 to move the tool down by 0.5 in., and then the cutter will continue the contouring motion in the X-Y plane. The NC controller will fail to determine a compensation direction if two nonmotion statements or two statements defining motion not in the compensation plane are specified in succession.

4. The included angle between the cutter paths of a start-up motion and the next motion should be greater than or equal to 90 degrees. Otherwise, the start-up motion will fail.

5. For the FANUC 6MB controller system, the tool movement at a corner of a profile is defined by the design and the parameter setting of the controller; it can be in the form shown in Fig. 2-15.

6. The G40 code, together with a G00 or G01 code in the last statement (labeled N0040) in the cutter-radius compensation mode, gradually cancels the cutter-radius compensation from the starting point of this statement to the end point. Note that the initial position of the tool for this statement is the end point for the preceding statement and that the tool is tangent to the part profile. As a result, an undesirable undercut will result in some cases (Fig. 2-16). Depending on the plane on which the compensation applies, two codes, either I and J, J and K, or I and K, can be added in the last statement to specify a part profile in which an undercut is to be avoided (Fig. 2-16).

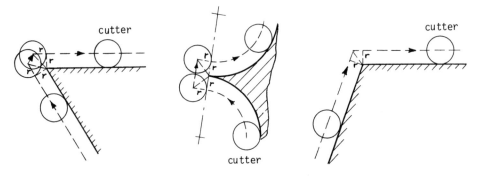

Figure 2-15   The cutter motion at the corner of various workpieces.

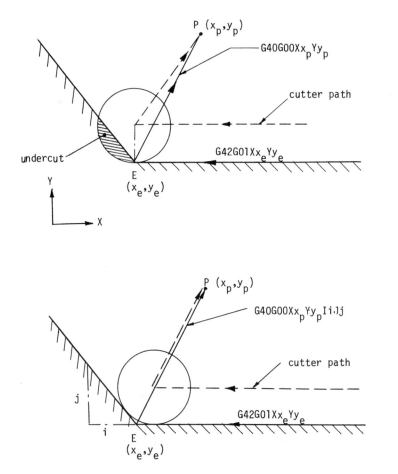

Figure 2-16   An undercut might result in the last cutter compensation motion; I and J codes are added in the last statement to avoid this behavior.

The cutter compensation for a lathe is more complicated than that for a milling machine for the following reason. The nose of a lathe cutter is only a section of a circle and does not rotate, like an end mill, during the cutting process. Therefore different cutter compensation vectors (or directions) must be applied with different types of cutting tools (Fig. 2-17). In addition, the tools installed on the turret have different relative positions with respect to the turret center. To compensate for these differences, one should set the offsets in the X and Z directions for different tools (Fig. 2-18). Thus the X and Z offsets, the tool nose radius r, and the tool vector numbers should be input into a tool offset register before a program is run. For the FANUC 6T controller system, a G10 code can also be used to input the above data into the offset registers. An example of the tool offset display on the FANUC 6T controller is shown in Table 2-7.

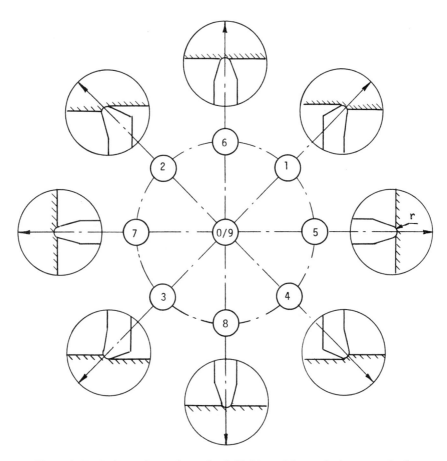

Figure 2-17   Lathe cutting tools can be divided into eight standard types, each of which is assigned a number representing a standard tool nose vector (e.g., **1, 2,** ..., or **8**). The No. **0** or No. **9** vector is used to cancel the tool radius compensation.

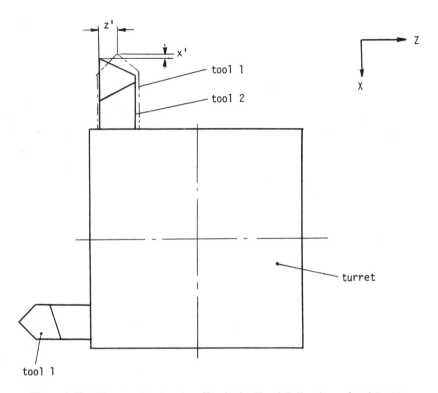

Figure 2-18   Diagram showing the offset in the X and Z directions of tool 1 with respect to tool 2. The offsets in the X and Z directions for this case are determined on the basis of the fact that the coordinate system of the cutting process is set on the basis of tool 2, and that a program is written according to that coordinate system. To have, in addition, the correct cutter paths for the other tools necessitates moving these tools to the exact position of tool 2 (e.g., tool 1 must move by amounts $-z'$ and $+x'$ in the Z and X directions, respectively) when they are in working position. These amounts of movements are the X and Z offsets, respectively.

Programming a turning process in the compensation mode is similar to programming a milling process. The same codes, G40, G41, and G42, are used. However, a T code should be used instead of a D code for the offset. Four digits are specified in the T code, with the first two digits defining the turret position (or tool number) and the last two digits defining the offset number.

TABLE 2-7   TOOL OFFSET DISPLAY

Offset Number	Offset in X Axis	Offset in Z Axis	Tool Nose Radius	Tool Vector Number
1	−0.02	0.1	0.035	4
2	0	0	0.025	2
. . . .	. . . .	. . . .	. . . .	. . . .
. . . .	. . . .	. . . .	. . . .	. . . .

When one is programming in compensation mode, various situations might arise as different cutter paths are programmed. The reader should consult the programming manual of the machine to be used for further information.

2.7 CODES DEFINING MACHINING CYCLES AND REPETITIVE MACHINING CYCLES

Certain machining operations are performed on a lathe or milling machine in a specific order and are cyclic in nature. For example, the following tool movements must be carried out to drill a hole at a given position:

- Rapidly moving the tool to the desired position
- Rapidly lowering the tool to a position a small distance (clearance) above the part surface
- Moving the tool down in programmed feedrate to the desired depth (i.e., drilling the hole)
- Moving the tool up to the desired position

Therefore a subroutine can be written for the drilling operation. When calling the subroutine, a user needs only to assign the proper parameters: the coordinates of the desired position, the depth of drilling, and the amount of clearance.

In the FANUC 6MB and 6T controller systems, a number of subroutines that can be used to perform different machining operations, such as drilling and boring, are stored and ready for execution. A specific combination of codes is needed to call and execute these stored subroutines.

2.7.1 Machining Cycle Operations on Milling Machines

There are a number of machining cycle operations that can be programmed by statements with special G codes. These are drilling, peck drilling, tapping, and boring cycles. The common codes used for programming these cycles are as follows:

G80	Cancel cycle function
G81	Spot drilling cycle
G82	Counter boring cycle
G83	Peck drilling cycle
G84	Tapping cycle
G85	Boring cycle
G86 to G89	Special boring cycle

When the power is turned on, the NC controller is in G80 mode (i.e., a G80 code is in effect). A machining cycle operation can be programmed by using G81, ..., or G89 codes, and can be canceled by a G80 code or by any G code in group 01.

A machining cycle specified by the above codes generally comprises some or all of the six operations shown in Fig. 2-19. The tool first rapidly moves horizontally to a point specified by the X and Y codes, where it rapidly descends to the level defined by the R code. Then the tool moves down in a programmed feedrate to the level specified by the Z code. There are two kinds of retreating motions, which are

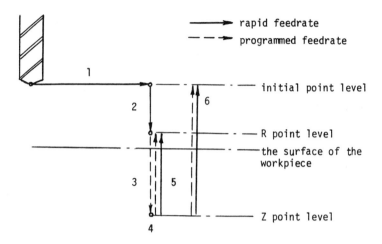

Operation 1: Rapid movement of the tool to the desired position.
2: Rapid movement of the tool down to a point (R) above the part surface with the desired clearance.
3: moving the tool down at a specified feedrate (performing drilling, boring, etc.).
4: operation performed at the bottom (e.g., stop running the spindle, dwelling, etc.).
5: returning rapidly or at programmed feedrate to point R (G99 mode).
6: returning rapidly or at programmed feedrate to initial point level (G98 mode).

Figure 2-19   The operations in a cycle (milling machine).

chosen by specifying either a G98 or G99 code in or before the statement containing a G81, G82, . . . , or G89 code. When a G98 code is in effect, the tool returns to the initial point level; in the G99 mode, the tool returns to the *R* point level.

The machining cycle operation can also be programmed in the G90 or G91 mode. The definitions of R and Z codes in these two modes of programming are different.

G81 Code (Spot Drilling).   The G81 code is used to call a stored spot-drilling-cycle subroutine, wherein the X, Y, Z, R, F, and L codes should be used to assign the necessary parameters. The format of a statement defining a spot-drilling cycle is

G81X*x*Y*y*Z*z*R*r*F*f*L*l*';

where (Fig. 2-20)

x and *y* = coordinates of the hole to be drilled. In G91 mode, *x* and *y* are the incremental coordinates from the current tool position to the hole position.

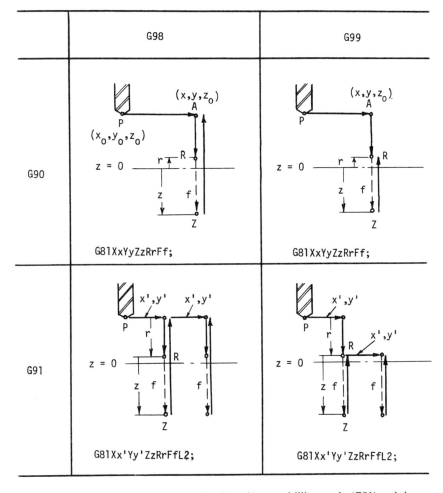

Figure 2-20   The tool motion defined by the spot drilling cycle (G81) and the meaning of the codes used in conjunction with a G81 code.

z = coordinate of the hole bottom. It is the incremental coordinate from the R point when code G91 takes effect.

r = the z coordinate of point R (in G90 mode) or the incremental z coordinates from the initial point to point R (in G91 mode).

f = the feedrate in inches or millimeters per minute.

l' = the number of times this cycle subroutine is to be executed. In G90 mode, the L code has no effect on the drilling operation; the tool will move down and up l' times at the position (x, y). In G91 mode, an L code will generate l' holes with equal spacing between holes. If the L code is omitted in the statement, the default value of one is used instead.

The cutter motions defined by this statement in different modes are shown in Fig. 2-20.

The part shown in Fig. 2-21(a) is used as an example to explain the use of the spot drilling cycle. The program for drilling the 12 holes with the initial tool position at $(-4.0, 0, 3.0)$ can be as follows (Fig. 2-21[b]):

```
O0001;
N1G90G92X-4.Y0Z3.M03S35; (set the origin, turning on the spindle*)
N2G91G99G81X4.Z-0.9R-2.9F2.0L4; (drill the 4 holes on the X axis)
N3Y3.L2; (drill the two holes on the right)
N4X-4.L3; (drill the three holes on the top row in the top view)
N5Y-3.;   (drill one central hole on the left)
N6G00Z2.9; (cancel the cycle function and move the tool up)
N7G81X3.0Z-1.2R-0.9; (drill the left top hole in the front view)
N8X6.0; (drill the right top hole in the front view)
N9G90G00X-4.Y0Z3.M05; (move back to starting position and stop the
                       spindle)
N10M30; (the end of this program)
```

*The M and S codes will be discussed in Section 2.8 on p. 60.

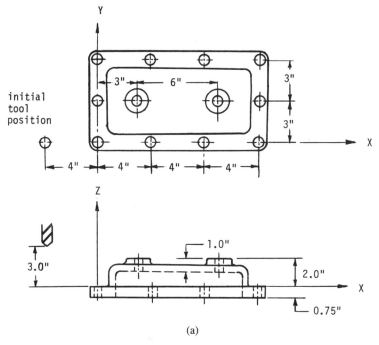

(a)

Figure 2-21 (a) A part with 12 holes to be drilled on an NC milling machine. (b) The tool path defined by the NC program (0001). The starting position of the tool (drill) is at P1 (statement N1). The tool paths defined by different statements are as follows: statement N2: P1-P2-P3-....-P6 (in plane $y = 0$); statement N3: P6-P7-P8 (in plane $x = 12$); statement N4: P8-P9-P10-P11 (in plane $y = 6$); statement N5: P11-P12 (in plane $x = 0$); statement N6: P12-P13 (in plane $x = 0$); statement N7: P13-P14-P15 (in plane $Y = 3$); statement N8: P15-P16 (in plane $y = 3$); statement N9: P16-P1.

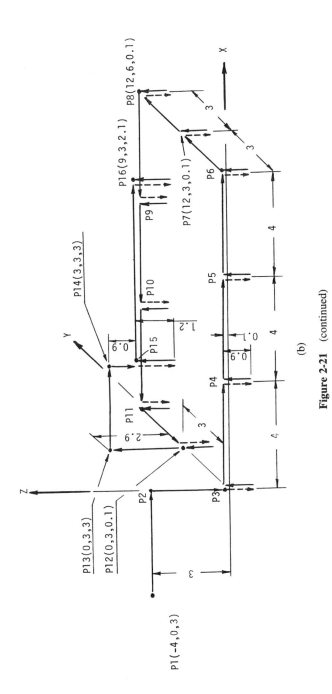

Figure 2-21 (continued)

(b)

47

The first statement defines the coordinate system shown in the figure on the basis of the initial tool position and turns the spindle on. It also sets the spindle speed at 35% of the maximum spindle speed. The second statement changes the programming mode to incremental and uses the G81 code to define the drilling operations for the four holes on the X axis. The next three statements specify the drilling operations for the other six holes on the same level. The statement labeled N6 moves the tool up to the same Z level as the initial position; the next two statements finish drilling the two holes on the top. The ninth statement orders the tool to return to the initial position and turns off the spindle. All the drilling operations are in the G99 mode (i.e., the drill returns to point R at the end of each drilling cycle).

G83 Code (Peck Drilling Cycle).   It is common practice that when one is drilling a deep hole (whose depth/diameter ratio is greater than 3), the drill should be retracted occasionally to avoid congestion of chips between the hole and the drill. The subroutine for this specific operation is called by a G83 code. The format of the calling statement is

G83XxYyZzQqRrFfLl';

where (Fig. 2-22)

> x and y = the x and y coordinates of the hole to be drilled (in G90 mode) *or* the incremental coordinates from the current tool position to the hole position (in G91 mode)
>
> z = the depth of the hole, which is the absolute coordinate in G90 mode and the incremental coordinate from point R in the G91 mode
>
> r = the z coordinates of the R point (in G90 mode) *or* the incremental z coordinates from the initial point to the R point (in G91 mode)
>
> q = the depth of cut for each peck drill — always a positive incremental value
>
> f = the feedrate in inches or millimeters per minute
>
> l' = number of times this cycle is to be repeated

The motion defined by this statement is shown in Fig. 2-22. It should be noted that when moving down, the tool changes its speed from rapid to programmed feedrate at a position that is d inches above the end position of the previous cut. The value of d is determined by the parameter setting of the NC controller. The use of the G83 code is basically the same as that of the G81 code.

G85 Code (Boring Cycle).   The boring operation requires that the tool move at a programmed feedrate when it is between points R and Z. The format of the statement for calling the boring cycle subroutine is

G85XxYyZzRrFfLl';

where the codes x, y, z, r, f, and l' have the same meanings as for the spot drilling cycle (G81). The only difference between codes G81 and G85 is in the retreat motion (Fig. 2-23). The G85 code can be used in the same way as the G81 code.

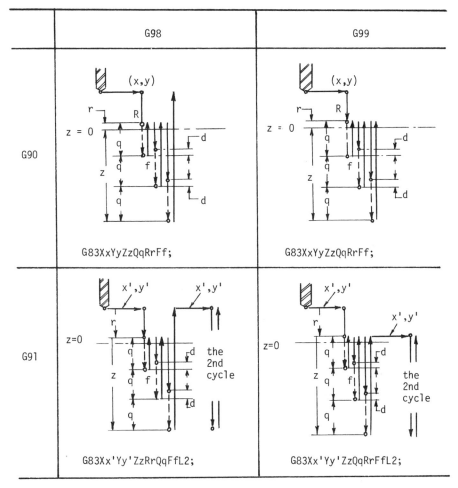

Note: ——► rapid motion; ----► tool motion at programmed feedrate

Figure 2-22 The tool motion of a peck drilling cycle defined by the G83 code.

The function of other G80 codes are similar to those of the three discussed above. The reader should consult the programming manual (e.g., reference 26) of the NC machine in use for further detail.

2.7.2 Machining Cycle Operations on a Lathe

A cutting cycle on a lathe generally requires motions in both the X and Z directions (with the exception of a drilling cycle) and is different from that on a milling machine, which requires cutting motion basically in the Z direction. On an NC lathe with FANUC 6T controller, two kinds of codes can be used, respectively, to pro-

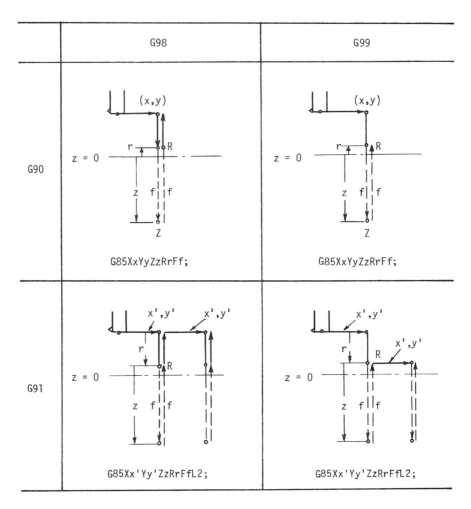

	G98	G99
G90	G85XxYyZzRrFf;	G85XxYyZzRrFf;
G91	G85Xx'Yy'ZzRrFfL2;	G85Xx'Yy'ZzRrFfL2;

Note: ⟶ rapid motion; ------➤ cutter motion at programmed feedrate

Figure 2-23   The tool motion of a boring cycle defined by the G85 code.

gram a single cycle and a repetitive cycle, which is used to remove the excessive material that cannot be cut away with one pass.

G90 Code (Cylindrical or Conical Surface Cutting Cycle).   The general format of the statement defining this cycle is (Fig. 2-24)

$$\text{G90}\begin{Bmatrix} XxZz \\ UuWw \end{Bmatrix}\text{liFf}*$$

where $x/2$ and z = the coordinates of the destination point
$u/2$ and w = the incremental coordinates from the starting point to the destination point B

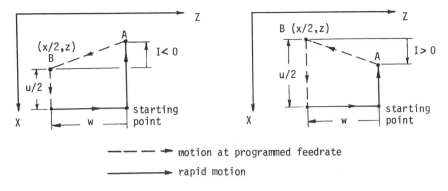

Figure 2-24   The cycle motion defined by the G90 code (lathe).

i = the incremental coordinates measured from the destination
    point B to point A (If i = 0, it can be omitted.)

f = feedrate in inches or millimeters per revolution

* = the end-of-block sign for the FANUC 6T controller

It should be noted that the first and third motions in this cycle are perpendicular to the Z axis at rapid and programmed feedrate, respectively. The last motion is in the direction parallel to the Z axis. An example of the use of the G90 code to define the cutting operation is given in Fig. 2-25.

G94 Code (Facing Cycle).   Facing is an operation used to reduce the length of a turned part as opposed to the turning operation (defined by a G90 code) whose function is to reduce its diameter. The general format of the statement defining a facing cycle is (Fig. 2-26)

$$\text{G94} \begin{Bmatrix} \text{X}x\text{Z}z \\ \text{U}u\text{W}w \end{Bmatrix} \text{K}k\text{F}f*$$

where x, z, u, w, and f have the same meaning as in the G90 code, and k denotes the incremental coordinates from point B to point A. The motion of the facing cycle is in the reverse direction to that of the G90 cycle. The first- and third-step movements are parallel to the Z axis; the last step of the motion is perpendicular to the Z axis. The direction of the motion from A to B can be defined by a code K with $k < 0$, if point A is on the negative side of the Z axis with respect to point B, and $k > 0$ if point A is on the positive side. The K code can be omitted if the motion from A to B is parallel to the X axis.

G92 Code (Thread Cutting Cycle).   A threading cycle is a turning cycle that maintains a precise feed/spindle speed ratio and a precise relative position between the tool and the workpiece. The general format of the statement defining a thread cutting cycle is

$$\text{G92} \begin{Bmatrix} \text{X}x\text{Z}z \\ \text{U}u\text{W}w \end{Bmatrix} l_i \begin{Bmatrix} \text{F}f \\ \text{E}e \end{Bmatrix} *$$

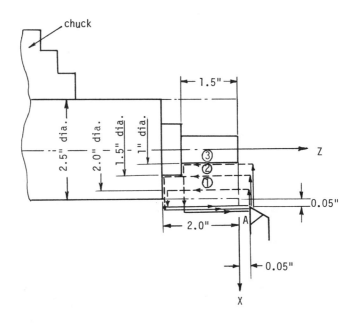

N21............ (the tool is at position A)
N22G90X2.0Z-2.0F0.01* (cycle 1)
N23X1.5* (cycle 2)
N24X1.0Z-1.5* (cycle 3)
. . . .

Programming in incremental coordinates

N21....... (the tool is at position A)
N22G90U-0.6W-2.05F0.01* (cycle 1)
N23U-1.1* (cycle 2)
N24U-1.6W-1.55* (cycle 3)
. . . .

Figure 2-25 An example of lathe programming using the G90 code.

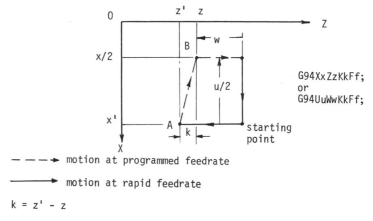

$$G94XxZzKkFf;$$
or
$$G94UuWwKkFf;$$

- - - → motion at programmed feedrate

——→ motion at rapid feedrate

$k = z' - z$

Figure 2-26 The facing cycle defined by the G94 code. Note that k is the incremental coordinate from point B to point A. Here, k is less than zero.

where the parameters x, z, u, w, i, and f are identified in Fig. 2-27. The E code, which accepts an input with six digits after the decimal point, is used when higher accuracy in the thread lead is needed.

A thread can never be cut in a single pass; several passes, and hence several consecutive thread-cutting statements, are required to complete a thread. For example, the following program will cut the thread shown in Fig. 2-28. Seven threading cycles (statements N31 to N37) are needed here to cut the thread (M24×3):

```
. . . . . . . . . . . . .
. . . .G21. . . . . .
. . . . . . . . . . . . .
N31G92X23.0Z-40.0F3.0*
N32X22.3*
N33X21.8*
N34X21.4*
N35X21.1*
N36X20.8*
N37X20.752*
. . . . . . . . . . . . .
. . . . . . . . . . . . .
. . . . . . . . . . . .
```

The controller will synchronize the tool motion with the motion of the spindle to ensure that the tool will not be fed into a wrong position in the second through seventh cycles.

2.7.3 Repetitive Cycle Codes for the Lathe[27]

A turning operation on a lathe normally requires several passes for the rough cut and one pass for the finishing cut. Furthermore, the cutter paths for the rough and finishing cuts can be similar or different. It is possible to have a subroutine that directs the tool to make several cutting passes based on the cutter path specified in the NC pro-

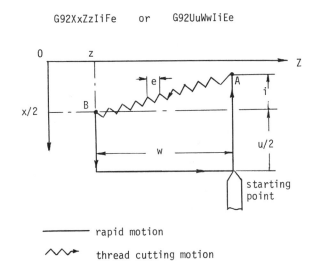

G92XxZzIiFe   or   G92UuWwIiEe

rapid motion

thread cutting motion

Figure 2-27   The thread cutting cycle defined by a G92 code.

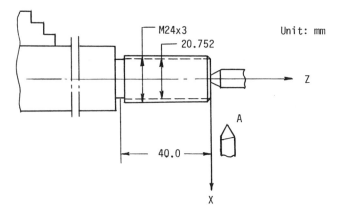

M24x3

20.752

Unit: mm

Z

A

40.0

X

Figure 2-28   A threaded part.

gram. With the FANUC 6T controller system, the codes for calling and executing these subroutines are as follows:

G70	Finishing
G71	Repetitive turning cycle (stock removal)
G72	Repetitive facing cycle (stock removal)
G73	Pattern repeating cycle (stock removal)
G76	Repetitive threading cycle

G71 Code (Repetitive Turning Cycle).   The format and the statements used for defining a repetitive stock removal turning operation are as follows (Fig. 2-29):

```
G71PpQqUuWwDdFfSsTt*
Np.............*
  .............*
  .............*
Nq.............*
G70PpQq*
  .............
```

Where

p = sequence number of the statement defining the cutter motion from its current position P to point A

q = sequence number of the last statement in this cutting cycle (In the case shown in Fig. 2-29, the last statement is the one defining the tool motion from point A_5 to point B.)

$u/2$ = allowance, in diametric value, left for the finishing cut in the X direction

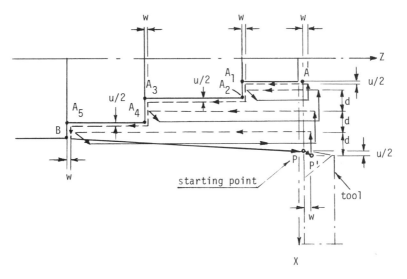

Figure 2-29   The tool motion specified by a G71 code: - - - - - → programmed feed-rate; ⟶ rapid motion.

w = allowance left for the finishing cut in the Z direction

d = the depth of cut in each pass

f = feedrate (This feedrate will override the feedrate(s) specified in the statements labeled p through q.)

s = spindle speed (This spindle speed will override the speed specified in the statements defining the cutter path.)

t = tool code and offset number

The statement defining the finishing cut, namely

G70PpQq*

can be specified after the last statement defining the cycle cutter path. Once it is specified, the tool will make a finishing cut based on the cutter path, spindle speed, and feedrate specified in the statements labeled p through q.

A common mistake in NC programming is that a cutter motion in a direction other than that perpendicular to the Z axis is defined in the first statement (labeled p) of the cutting cycle. The motion should be perpendicular to the Z axis. Furthermore, the end position of the cycle, or more specifically the end position of the statement labeled q, should be defined as B. It is not necessary to have a statement specifying the motion from point B to the starting point P. It is also worth noting that all the cutter paths in the rough cut are parallel to the Z axis.

Example

The cutter path shown in Fig. 2-29 can be programmed as follows (assuming $u = 0.01$ and $d = 0.12$):

...G20.........

....

N101G71P102Q108U0.01W0.005D0.12F0.005*
N102G00Xx_a*
N103G01Zz_{a1}*
N104Xx_{a2}*
N105Zz_{a3}*
N106Xx_{a4}*
N107Zz_{a5}*
N108Xx_b*
N109G70P102Q108* (The cutter path of this statement is $A A_1 A_2 A_3 A_4 A_5 B$)

....

G72 Code (Repetitive Facing Cycle).   The format of the statement with a G72 code is the same as that with a G71 code. The differences are as follows (Fig. 2-30):

1. The cutter motion is in the reverse direction to that defined by the G71 code.
2. The D code defines the depth of cut in the Z direction.
3. The first statement in the cycle should define a motion from point P to point A, which is parallel to the Z axis.

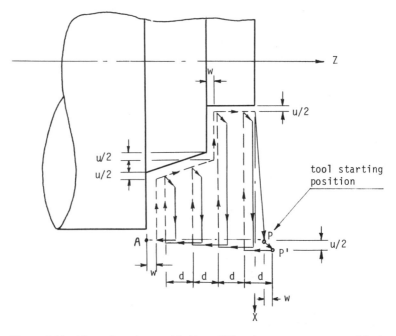

Figure 2-30   The tool motion specified by a G72 code: - - - - → programmed feed-rate; ⟶ rapid motion.

Programming in the G72 code is the same as in the G71 code.

G73 Code (Pattern Repeating Cycle).   As can be seen from Figs. 2-29 and 2-30, the effective cutter paths (the paths along which a cutter moves in programmed feed-rate) in each pass of the multiple cycles are not the same as those defined in the cycle statements labeled p through q except for the last pass. They are either parallel (G71) or perpendicular (G72) to the Z axis. There are two cases when it is not appropriate to use these two codes to define a multiple stock removal operation:

1. The stock material is a drop-forged part or casting (Fig. 2-31) whose profile is an enlarged form of, or is similar to, the finished part. When a G71 or G72 code is used, most of the time the cutter will not be in contact with the workpiece.
2. The part has undercuts on its surface (Fig. 2-32). Since the effective cutter paths in every pass other than the last one is parallel to the Z axis for the G71 code, a heavy cut will result in the last cut.

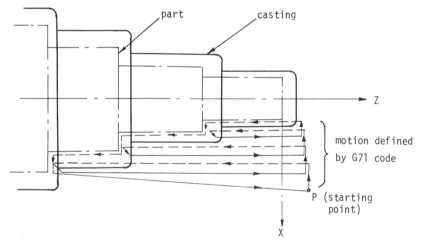

Figure 2-31   A casting or drop-forged stock cannot be cut effectively by a G71 multiple repetitive cycle motion: $----\rightarrow$ programmed feedrate; $\longrightarrow$ rapid motion.

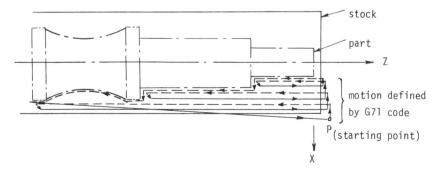

Figure 2-32   A heavy cut results in the final cycle at the undercut when a G71 code is used: $-----\rightarrow$ programmed feedrate; $\longrightarrow$ rapid motion.

To resolve these problems, one can use a G73 code to call a stored subroutine that orders the tool to move in the pattern defined by the cycle statements labeled p through q. Statements in the following format can be used to define a pattern-repeating cycle (Fig. 2-33):

```
G73PpQqIiKkUuWwDdFfSs*
Np..............*
..............*
..............*
Nq..............*
G70PpQq*
```

Where

p and q = the statement labels for the first and last statements defining the cutter path ($P \rightarrow A \rightarrow B \rightarrow C \rightarrow D$ in Fig. 2-33), respectively

i and k = radial and axial retraction for multiple cycle cut

$u/2$ and w = radial and axial allowance for finishing cut

d = number of repetitive cuts

f = feedrate in inches or millimeters per revolution

s = spindle speed

As can be seen from Fig. 2-33, in contrast to codes G71 and G72, there is no limitation on the direction of the movement specified by the statement labeled p. Very often, a beginner will have difficulty in determining the values of i and k. As a matter of fact, i and k are the total feed-in amount in the X and Z directions in the

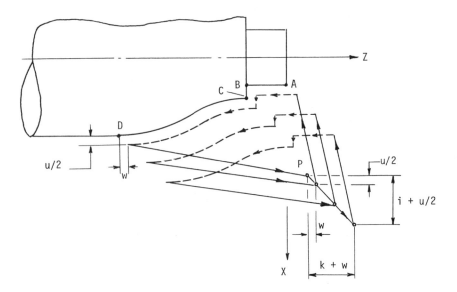

Figure 2-33   The cutter motion in the pattern-repeating cycle defined by a G73 code: - - - - - → programmed feedrate; ⟶ rapid motion; starting point — P.

multiple passes. They can be determined easily by defining in advance the first and last cutter paths in the cycle. When a repeating pattern is basically a turning process, the amount of movement in the Z direction is larger than that in the X direction (i.e. $\Delta Z > \Delta X$ in Fig. 2-34), k should be less than i; otherwise, $k > i$. Although it is not required to specify both the I and K codes in a statement, an NC programmer should choose the right combination of I and K codes to ensure that the amounts of material cut in different passes are as even as possible. If one of the I and K codes is omitted, it is considered as zero by the NC controller.

G76 Code (Repetitive Threading Cycle).   The format of this statement is as follows (Fig. 2-35):

$$G76XxZzIiKkDd\begin{Bmatrix}Ff\\Ee\end{Bmatrix}Aa*$$

Where

> x and z = coordinates of the end point B of the thread
> i = difference in thread radii at both ends
> k = height of thread (radius value)
> d = depth of first cut

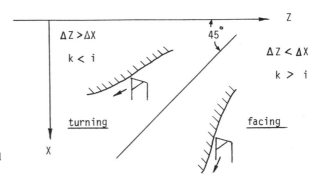

Figure 2-34   The determination of I and K codes for a G73 code.

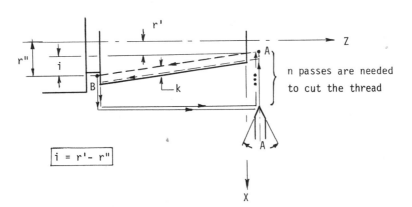

Figure 2-35   Multiple threading cycle defined by a G76 Code.

f and e = lead of thread

a = angle of thread

Once this statement is specified, the NC controller will automatically calculate the depth of cut with respect to the outer diameter of the thread, $(x + 2k)$, for each pass according to the following formula:

$$\text{Depth of cut} = d \cdot (n)^{1/2}$$

where n is the sequential pass number. Thus, for the example in Fig. 2-28, if the first depth of cut is 0.5 mm, then the following depths of cut will be

$$(1/2) \cdot 2^{1/2} = 0.7071 \qquad \text{(second cut)}$$
$$(1/2) \cdot 3^{1/2} = 0.8660 \qquad \text{(third cut)}$$

and so on. The height of the thread is 1.624 mm. The number of passes is determined by the following formula:

$$n = (k/d)^2$$

Substituting k and d, we have

$$n = (1.624/0.5)^2 = 10.549$$

Therefore eleven passes are needed to finish the thread cutting. The above calculation and decision processes are carried out automatically by the NC controller. The program in Section 2.7.2, above, for cutting this thread can be replaced by a single statement:

G76X20.752Z-40.K1.624D0.5F3.A60∗

Subroutines for peck drilling (G74) and peck grooving (G75) are also available for the FANUC 6T controller. Programming of these two cycles on a lathe is similar to that for a milling machine. The reader should consult the programming manual of his particular NC machine for further details.

2.8 CODES SPECIFYING MISCELLANEOUS FUNCTIONS

In addition to defining cutter paths, a number of codes are needed to define the mode of operation, machining specification, and so forth.

M Code (Miscellanous Function Code).   Like the G code, the M code has various uses. Usually the meanings of M codes vary with the NC machines of different manufacturers. The selected M codes shown in Table 2-6 are used for the Cadillac NC-300 milling machine with the FANUC 6MB controller system and for the Cadillac NC-100 lathe with the FANUC 6T controller system. Except for codes M98 and M99, each of the codes listed in the table can be specified without other codes. The M98 code is specified in a main program to call a subprogram. The statement format is

M98P*p*L*l'*;

where p is the program number specified as O code at the beginning of a subprogram; and l' is the number of times the subprogram is to be executed.

The M99 code is specified in the last statement of the subprogram to return control back to the statement in the main program immediately after the statement calling the subroutine (Fig. 2-36). A P code can also be specified after an M99 code to direct control to a statement after the calling one in the main program. Thus, if the statement

M99P0032;

is specified instead of a single M99 code in the statement labeled *12* in the subprogram, control will go to the statement labeled *0032* in the main program.

Subprogramming is often very useful to shorten a program and hence to save memory space in the CNC controller. Typical examples are as follows:

1. A part profile should be cut in three passes on a milling machine. The cutter paths in these three passes are the same in the X-Y plane, but their depths of cut are different. A subprogram can be written for directing the tool in the X-Y plane. The subprogram is called each time the tool has been positioned to the correct Z position.

2. A part profile should be cut in two passes, one rough cut and one finishing cut on a milling machine by using the same end mill. Two offsets, one of which is the sum of the cutter radius and the allowance and the other equal to the cutter radius, are stored in the offset registers. A subprogram can be written by using the cutter-radius compensation code to define the cutter path. The rough and the finishing cuts can be realized by executing the same subprogram twice, using initially the first and then the second offset.

S Code.   The S code controls the spindle speed. Depending on the machine, two digits or four digits can be specified after the address S. The unit of spindle speed varies with the machine tool builder. For example, on the Cadillac NC-100 lathe, two digits are used to represent the spindle speed in percentage value of the maximum speed.

T Code.   Two or four digits can be specified after the address T, depending on the design of the NC machine. On the Cadillac NC-100 lathe with the FANUC 6T controller system, four digits can be specified. The first two digits represent the turret position or tool number, and the last two digits, the offset number.

Main program	Subprogram
....	O0008;
....	....
N0030M98P0008L2;	....
N0031..........;	....
....	N12M99;
....	

Figure 2-36   The use of a subprogram: subprogram No. 0008 is executed twice, and then control goes back to the statement labeled N0031 in the main program.

2.9 EXAMPLES OF NC PROGRAMMING

What is normally called NC programming consists in writing down the machining operations step by step in terms of NC codes. In fact, it is the process of translating a machining process plan from English into NC codes. Depending on the experience and knowledge of NC programmers, programs written by different programmers for the same part might differ. The criteria for evaluating an NC program are correctness, accuracy, and the memory space needed for storing the program in the CNC controller.

An NC program must generate a correct cutter path and meet the dimensional and tolerance requirements specified for the part. Since most CNC controllers have limited memory capacity (80 m or 160 m, converted into tape length), it is important to make the program as concise as possible. Choosing or shifting to the most convenient coordinate system, effectively using subprograms, and omitting the unnecessary codes (e.g., the N code can be omitted if it is not required) and those codes which have been defined in previous statements can significantly reduce the length of a program.

An NC program should be written on the basis of a sound machining process plan that considers, for example, the characteristics and structure of the NC machine, the method of clamping, toolings, the form and material of the workpiece, ease of adjustment and operation, the reduction of machining time, and safety. An NC program can never be satisfactory if the machining process plan is inadequate.

During manual programming of an NC cutter path, the major routine work that an NC programmer should do is to calculate the coordinates of the consecutive cutter positions. It should be noted that when the G40 code is programmed (i.e., when cutter-radius compensation is canceled), the coordinates specified in an NC statement define the position of the cutter end center for the end mill or of the nose radius center for the lathe cutting tool. Programming in the cutter-radius compensation mode allows the use of the part profile to program the cutter path; in other words, the coordinates specified in the NC statements (blocks) are the coordinates of the part profile. Thus the additional work needed to calculate the tool positions on the basis of the part profile can be saved. In either case, an NC programmer must calculate the coordinates of the destination position for each NC statement. This is a heavy job, especially when a part profile is complex and consists of sections whose forms are other than linear and circular. Very often, the calculation work is so complicated that it is unreasonable to program the NC cutter path manually. NC programming in higher-level languages, such as APT or COMPACT II, or through the use of a computer graphics system should then be selected instead.

Before an NC program is written, the usual practice is to list all the necessary information — including the tool motion, the tools used, and the machining specifications — that is needed to define the consecutive machining steps in a tabular form called the operation sheet (Table 2-8). A separate tooling sheet (Table 2-9) is also needed, especially if more than one tool is used in a machining program. The table lists the tool numbers, sizes, offset and radius compensation values, carbide insert types, holders, and the offset register numbers (i.e., the addresses of the registers

TABLE 2-8   A BLANK OPERATION SHEET

Operation No.	Tool No.	Cutting Speed	Feed Rate	Spindle Speed	Operation Description	Work Holding

TABLE 2-9   A BLANK TOOLING SHEET

Tool No.	Tool Offset No.	Nominal Size, Tool Description	Tool Offset			Cutter Diameter or Nose Radius Compensation	Insert Type	Holder Type
			X	Y	Z			

used to store the offset and radius compensation values). Then a manually written NC program can be written in a tabular form, for example, as shown in Table 2-10, according to the information given in the operation and tooling sheets.

The following examples[28] have been selected to explain the details of NC programming. They were assigned as exercises for the NC laboratory course at UCLA (University of California, Los Angeles). In the first example, programming in the cutter-radius compensation mode is explained. The second example shows the combined use of several techniques, including shifting of the programming mode (G90/G91), changing of the coordinate system, and multiple execution of the subprogram and machining cycle statement, to program a drilling process. A very concise program results. NC lathe programming is explained in the third example.

Example 1

The V-block clamp (Fig. 2-37) is cut in two steps. In the first step the stock material is held by clamping units at points A, B, and C, and the inner profile of the part is cut in three passes with different depths of cut, respectively. Thus a main program

TABLE 2-10   AN EXAMPLE OF A MANUAL NC PROGRAMMING SHEET

O	N	G	X	Y	Z	F	S	T	M	P	.....	Comment
0001												Program number
	10	90;92	0	−1.7000	2.0000		25		03			Absolute programming; set origin and spindle speed; turn on spindle
	15	00			0.1000				08			Rapidly move downward: turn on coolant

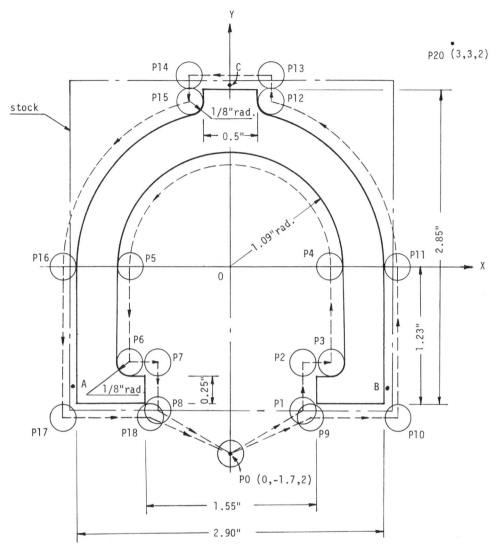

Figure 2-37 A V-block clamp and the cutter paths for programs No. 0002 and No. 0004 in Example 1 (Section 2.9). Thickness of the part: $\frac{1}{2}$ in.; cutter: $\frac{1}{4}$ in. end mill; material: aluminum plate, $3 \times 3 \times \frac{1}{2}$ in.

(No. 0001) and a subprogram (No. 0002), listed below, are used to direct the cutting process:

```
O0001;  (program No. 0001)
N10G90G92X0Y-1.7Z2.0M03S25;  (absolute coordinate programming; tool
                starting position: P0; set origin of machine coordinate
                system at O; set spindle speed as 25% of maximum speed;
                turn on spindle in clockwise direction.)
N15G00Z0.1M08;  (lower tool rapidly to level z = 0.1 in.; turn on
                coolant.)
N20G01Z-0.18F2.0;  (lower tool to level z = -0.18 in. at programmed
                feedrate of 2 in./min.)
N30G10P01R0.125;  (set value of No. 1 offset register as radius of
                cutter, 0.125 in.)
N40M98P0002;  (call subprogram No. 0002, which directs tool moving
                along path P0-P1-P2-...-P8-P0)
N50G01Z-0.36;  (move tool farther down to level z = -0.36 in.)
N60M98P0002;  (call subprogram No. 0002 and repeat cutting process)
N70G01Z-0.52;  (move cutter to required depth of cut, z = -0.52 in.)
N80M98P0002;  (call subprogram No. 0002 and repeat cutting process)
N90G00Z2.0M09;  (move cutter up to initial level; turn off coolant)
N100X3.0Y3.0M05;  (move cutter to point P20)
N110M30;  (end of program and memory rewind)
```

```
O0002;  (program No. 0002)
N10G17G41D01G01X0.775Y-1.3;  (set X-Y plane as cutter-radius
    compensation plane; set cutter-radius compensation on left-hand
    side of the profile; use value stored in offset register No. 1 as
    as offset needed for radius compensation; order tool to move from
    P0 to P1)
N20Y-0.98;       (from P1 to P2)
N30X1.09;        (from P2 to P3)
N40Y0;           (from P3 to P4)
N50G03X-1.09R1.09;  (circular interpolation motion: from P4 to P5)
N60G01Y-0.98;    (from P5 to P6)
N70X-0.775;      (from P6 to P7)
N80Y-1.3;        (from P7 to P8)
N90G40G01X0Y-1.7;  (from P8 to P0; cutter radius compensation is
                canceled at end of this motion)
N100M99;           (return control to statement, in the main program,
                immediately after that calling this subprogram)
```

Because the slug left in the middle of the part will hinder the tool movement from point $P8$ to point $P0$ in the last pass, it is better to add a rough cut to remove the center part of the material.

The outer profile is also cut in three passes when the part is clamped at the inner profile and the same starting point and coordinate system, as in program No. 0001, are selected. The cutter path, $P0$-$P9$-$P10$-....-$P18$-$P0$, defined by program No. 0004, is also shown in Fig. 2-37. The program No. 0003 is similar to program No. 0001 given above. These programs are as follows:

```
O0003;
N10G90G92X0Y-1.7Z2.0M03S25;
N15G00Z0.1;
N20G01Z-0.18F2.0;
N30G10P01R0.125;
N40M98P0004;
N50G01Z-0.36;
N60M98P0004;
N70G01Z-0.52;
N80M98P0004;
N90G00Z2.0;
N100Y2.0M05;
N110M30;
```

```
O0004;
N10G17G41D01X0.75Y-1.23;
N20X1.45;
N30Y0;
N40G03X0.25Y1.4283R1.45;
N50G01Y1.62;
N60X-0.25;
N70Y1.4283;
N80G03X-1.45Y0R1.45;
N90G01Y-1.23;
N100X-0.75;
N110G40G01X0Y-1.7;
N120M99;
```

Example 2

The part shown in Fig. 2-38(a) is designed for a special nozzle. It has 500 holes with a diameter of 0.01 in. These holes are divided into five groups of the same pattern. The cutter motion defined by subprogram (No. 0002) is shown in Fig. 2-38(b). By repeatedly executing the program 10 times, the complete 100 holes in a group can be drilled. Note that the end position of the tool after the subprogram is executed 10 times is at point $P7$. For the drilling of each group of holes, the machine coordinate system is set at the center of that group. Thus redefinition of the coordinate system is required in this program. The NC programs for drilling these 500 holes are as follows:

```
O0001; (main program)
G90G92X0Y0Z0.5M03S50;   (programming in absolute coordinates; starting
      point is (0,0,0.5); set origin of machine coordinate system at O;
      set spindle speed as 50% of maximum speed; turn on spindle in
      clockwise direction)
G00X0.0354Y-0.3536Z0.05M08;   (move tool to point [0.0354,-0.3536,0.05];
      turn on coolant)
M98P0002L10;   (execute subprogram 10 times and drill central pattern)
G90G92X0.3889Y-1.0Z0.05; (programming in absolute coordinates; reset
      origin of machine coordinate system at P12)
G00X0.0354Y-0.3536;   (move tool to point [0.0354,-0.3536,0.05])
M98P0002L10;   (execute subprogram 10 times and drill top pattern)
```

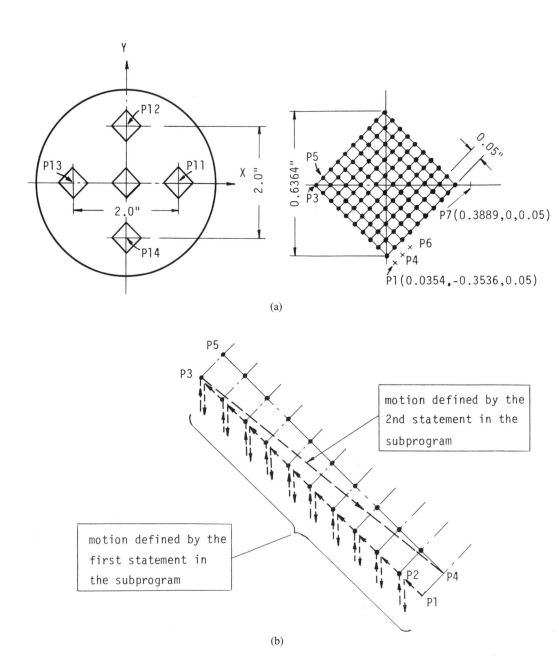

(a)

(b)

Figure 2-38 (a) A nozzle plate with 500 holes divided into five groups of the same pattern. Thickness: 0.15 in.; $z_{surface} = 0$. (b) The cutter motion defined by the subprogram. Note that points $P1$, $P4$, $P6$, and $P7$ are not the hole positions.

G90G92X0.3889Y2.0Z0.05; (programming in absolute coordinates; reset
    origin of machine coordinate system at P14)
G00X0.0354Y-0.3536;   (move tool to point [0.0354,-0.3536,0.05])
M98P0002L10;   (execute subprogram 10 times and drill bottom pattern)
G90G92X1.3889Y-1.0Z0.05; (programming in absolute coordinates; reset
    origin of machine coordinate system at P13)
G00X0.0354Y-0.3536;   (move tool to point [0.0354,-0.3536,0.05])
M98P0002L10;   (execute subprogram 10 times and drill left pattern)
G90G92X-1.6111Y0Z0.05;   (programming in absolute coordinates; reset
    origin of machine coordinate system at P11)
G00X0.0354Y-0.3536;   (move tool to point [0.0354,-0.3536,0.05])
M98P0002L10;   (execute subprogram 10 times and drill right pattern)
G90G92X1.3889Y0Z0.05M09;   (programming in absolute coordinates; reset
    origin of machine coordinate system at point O; turn off the
    coolant)
G00X0Z0.5M05;   (returning to starting position; turn off spindle)
M30;   (end of program and memory rewind)

O0002; (subprogram)
G91G99G83X-0.0354Y0.0354Z-0.21Q0.02R0F1.0L10; (programming in
    incremental coordinates; peck-drilling ten holes on line with
    slope of -1, e.g., line P2-P3; see Fig. 2-38[b])
G00X0.3889Y-0.3182;   (return to position ready to start another
    drilling cycle)
M99; (return control to statement, in calling program, immediately
    after that calling this subprogram)

Example 3

The part shown in Fig. 2-39(a) is to be machined on the NC-100 lathe with the FANUC
6T controller. Two programs are needed to cut this part: program O0001 for cutting the
left end (Fig. 2-39[b]) and program O0002 for cutting the right end (Fig. 2-39[c]). A
rough cutting tool and a finishing cutting tool, installed at turret positions Nos. 1 and 3,
respectively, are used in this machining operation. At the beginning of each program,
the origin of the coordinates system is set at the center of the end surface while the
rough cutting tool is in working position. In these two programs, the tool nose radii of
the two tools are assumed to be zero; thus no tool radius compensation is needed. The
finishing cutting tool has a tool number of 03; and an offset number of 01 is specified
to compensate for the differences in X and Z coordinates between the two tools. In the
first statement of these two programs, the tool is ordered to return to the reference
point. Then the coordinate system is set by the second statement in each program. For
an illustration of the use of various codes, several codes are used in this program to define the rough and the finishing cutter paths. The programs, with explanation, are as
follows:

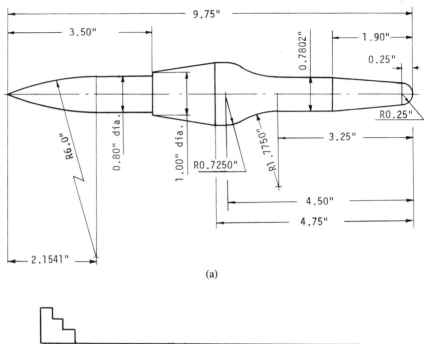

(a)

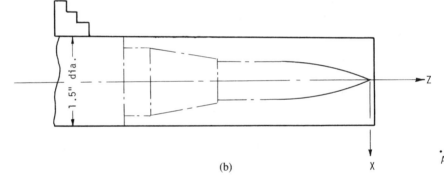

(b)

chuck jaw

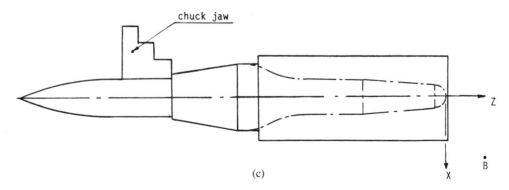

(c)

Figure 2-39 (a) A formed shaft. (b) The setup diagram for cutting the left end of the part shown in Figure 2-39(a). Point A: (15.8328, 12.6055). (c) The setup diagram for cutting the right end of the part shown in (a). Point B: (15.8328, 12.3641).

O0001* (program for cutting left end, Fig. 2-39[b])
N1G20G28U0W0M42S40* (programming in inch unit; return to reference
                      point point A; set spindle speed at 40% of
                      maximum speed)
N2G50X15.8328Z12.6055* (set origin of coordinate system at center point
                        of end surface)
N3G00X1.6Z0.1M03* (move tool to position ready to start cutting; start
                   running spindle)
N4G94X0Z0F0.003*   (face end surface)
N5G90X1.6Z-3.0I-0.1*   (stock removal by using G90 code begins)
N6Z-3.5I-0.2*
N7I-0.3*
N8Z-3.75I-0.4*
N9Z-4.0I-0.5*   (stock removal ends)
N10G00Z1.0*   (cancel G90 code and move tool to the position
                defined by x = 0.8, z = 1.0)
N11G73P12Q18I0.20K0U0.02W0.01D5F0.003S40* (start repeating pattern;
                radial retraction -- 0.2 in.; axial retraction -- 0;
                radial allowance -- 0.01 in.; axial allowance -- 0.01;
                the cutter path is defined in statements labeled N12
                through N18)
N12G00X0Z0.05S50*
N13G01Z0*
N14G03X0.8Z-2.1541I-5.6K-2.1541*
N15G01Z-3.5*
N16X1.0*
N17X1.45Z-5.0*
N18Z-5.5*   (end of repeating pattern)
N19M11*   (turn turret by 90 degrees)
N20M11T0301*   (Turn turret by another 90 degrees; finishing tool
                enters working position; No. 1 offset is used)
N21G70P12Q18F0.002* (finishing cut)
N22G00X3.0Z3.0M05*   (retract tool to safe position; stop spindle)
N23M11*
N24M11*   (rotate turret back to starting position)
N25M30*   (program stop)

O0002*   (program for cutting right end, Fig. 2-39[c])
N1G20G28U0W0*   (return to the reference point B)
N2G50X15.8328Z12.3641M42S40* (set origin and spindle speed)
N3G00X1.6Z0.1M03*
N4G94X0Z0F0.003*   (facing cycle)
N5G71P6Q9U0D700F0.015*   (start multiple repetitive turning cycle)
N6G00X0.5*
N7G01X0.9Z-1.9*
N8Z-3.25*
N9X1.4Z-4.1375*   (end of multiple repetitive turning cycle)
N10G73P11Q17I0.075K0U0.02W0.01D2F0.003S40* (start repeating pattern)
N11G00X0Z0.1S50*
N12G01Z0*
N13G03X0.5Z-0.25I0K-0.25*
N14G01X0.7802Z-1.9S45*
N15Z-3.25*
N16G02X1.2557Z-4.1375I1.7750*

N17G03X1.45Z-4.5I-0.6279K-0.3625S40*   (end of repeating pattern)
N18M11*
N19M11T0301*
N20G70P11Q17F0.002*   (finishing cut)
N21G00X3.0Z3.0M05*
N22M11*
N23M11*
N24M30*

PROBLEMS

2.1 An extended part of a machine element with a hexagonal outer profile and a square inner profile is shown in Fig. P2-1. Suppose that the top surface of this part has been machined to the required dimensions. The following parameters are suggested:

Cutting tool	$\frac{1}{2}$ in. end mill
Spindle speed	1000 rpm (or NC code S34)
Feedrate	4 in./min (rough cut)
	2.5 in./min (finishing cut)
	2.0 in./min (plunging)
Maximum depth of cut	0.4 in.

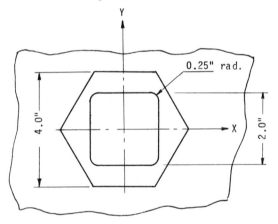

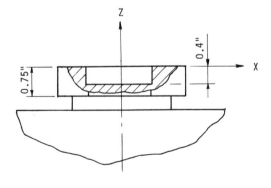

Figure P2-1

The machine does not have an automatic tool changer. Write an NC program for machining the outer profile in three passes (two passes for rough cut and one for finishing cut). The starting position of the tool is (0,0,2.0).

2.2 A flat surface (Fig. P2-2) is to be machined by a facing mill $2\frac{1}{2}$ in. in diameter. Write as concise an NC program as possible to machine this surface, using the following given conditions:

Spindle speed	900 rpm (or NC code S30)
Feedrate	4 in./min
Starting position	(-5.0,0,3.0)
Depth of cut	0.1 in.

The surface is to be finished within one pass.

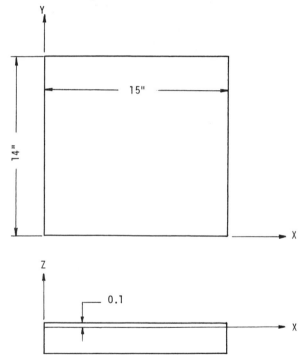

Figure P2-2

2.3 The profile of a Wankel (rotary) engine rotor is shown in Fig. P2-3. Two NC programs are needed to cut the part: one for cutting the top surfaces and the hole, the other for cutting the outer profile. The following machining conditions are assumed:

a. The starting and end points are at (0,0,50.0).

b. An end mill $\frac{3}{4}$ in. in diameter is used to cut the top surfaces of the part with dimensions of 20.0 mm and 18.0 mm from the bottom surface and to cut the hole 50.05 mm in diameter. The hole is finished by two rough and two finishing cuts. The surfaces are finished in one pass only.

c. The same cutter is also used to cut the outer profile in four passes (two for rough cut and two for finishing cut). The depth of cut for each pass is 10 mm. The part is held through the center hole. The up-milling and climb-milling methods are used for the rough and finishing cuts, respectively.

d. The feedrates are 100 and 70 mm/min for rough and finishing cuts, respectively; the feedrate for plunging is 40 mm/min. The spindle speed is 600 rpm (or NC code S20).

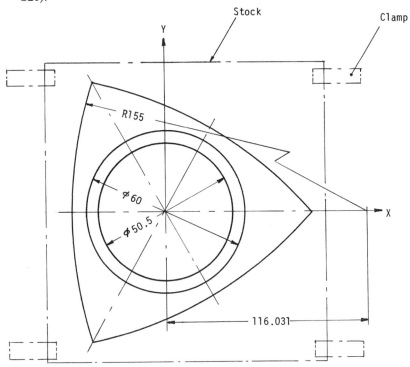

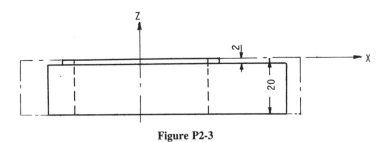

Figure P2-3

2.4 A gear with a simplified tooth profile is shown in Fig. P2-4. The gear profile is to be cut in eleven passes (seven for rough cut with the depth of cut equal to 1.5 mm, four for finishing cut with the depth of cut equal to 3.0 mm) by using an end mill $\frac{1}{8}$ in. in diameter. Write an NC program with the use of the following parameters:

Spindle speed	1500 rpm (or NC code S50)
Feedrate	40 mm/min
Allowance for finishing cut	0.062 mm
Starting point	(0,0,50.0)
Z coordinate of the top surface	0

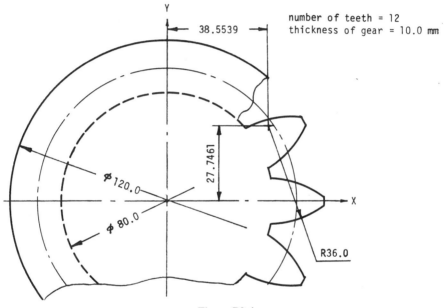

number of teeth = 12
thickness of gear = 10.0 mm

Figure P2-4

2.5 A matrix (Fig. P2-5) is to be cut on a three-axis NC milling machine with no automatic tool changer. The surface, with a distance of 0.2 in. from the top surface and the circular profile (3 in. radius) is to be cut by an end mill $\frac{3}{4}$ in. in diameter in one pass. An end mill $\frac{1}{4}$ in. in diameter is used to cut the rest of the surfaces in three passes. Suppose that the stock material has been machined to the required dimension 4.25 × 4.25 × 0.75 in.

and is held by a machine vise. The starting point is at (0,0,1.0). Write the NC programs to cut the matrix, using the following parameters:

CUTTER DIAMETER	SPINDLE SPEED	FEEDRATE
$\frac{1}{4}''$	900 rpm (or S30)	2.0 in./min
$\frac{3}{4}''$	300 rpm (or S10)	2.0 in./min

R1 = 0.125" rad.

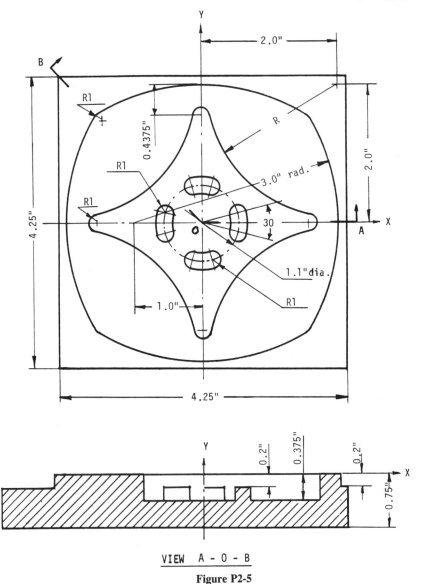

VIEW   A - O - B

Figure P2-5

2.6   A bell-shaped part (Fig. P2-6) is to be cut on a three-axis NC milling machine with a ball-end mill $\frac{1}{2}$ in. in diameter. For simplification of the problem, the part profile is approximated by moving the cutter around the horizontal cross sections (circles), shown in the figure, with incremental distance between two cross sections equal to 0.2 in. Write an NC program to direct the finishing cut, using the following parameters:

Spindle speed   1000 rpm (or NC code S34)
Feedrate   2 in./min

Assume that the starting point is at (0,0,5.0).

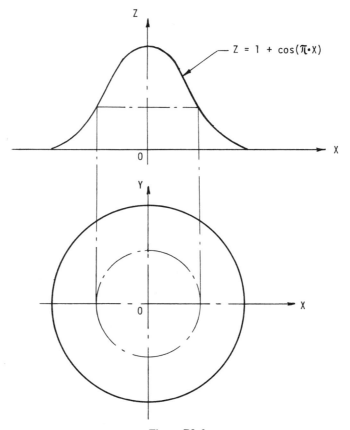

Figure P2-6

2.7 The driven wheel of a Malta (intermittent) mechanism has the profile shown in Fig. P2-7. The part is held by the fixture through the center hole, which has been machined to the specified dimension in a previous operation. Write an NC program to direct the machining operation. The following conditions are given:

Cutter (T01)	$\frac{3}{4}$ in. flat-end mill (for cutting holes and outer profile)
Cutter (T02)	$\frac{1}{4}$ in. flat-end mill (for cutting slots)
Feedrate	100 mm/min (for rough cut with tool T01)
	70 mm/min (for finishing cut with tool T01)
	50 mm/min (for tool T02)
Finishing allowance on outer profile	0.2 mm
Starting and end points	(150.0,150.0,30.0)

Tool T02 is longer than tool T01, and the difference is 30.988 mm. The origin of the coordinate system is at the center of the top surface.

Note: On the NC milling machine with a FANUC 6MB controller, the compensation for the difference (offset) of the tool lengths of two tools, Δz, can be programmed by using codes

G43Hnn

where the number nn in the H code represents the number nn register for storing the tool length offset Δz. The effect of this statement is to add the value Δz to the z coordinates of the tool positions defined by the following statements. A G44 code can also be used together with an Hnn code to compensate for the tool length. However, its effect is to subtract the offset Δz from the z coordinates defined by the following NC statements. A G49 or H00 code can be used to cancel the tool length compensation.

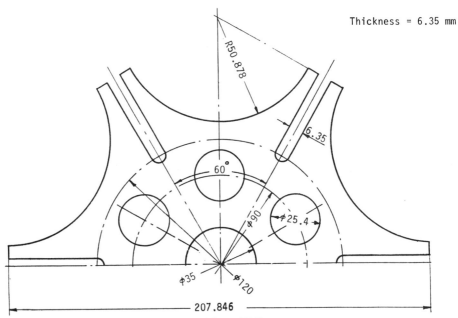

Figure P2-7

2.8 The shaft shown in Fig. P2-8 is to be cut on an NC lathe and is supported between centers. Four tools T01, T02, T03, and T04 are used for rough turning, finishing, grooving, and threading, respectively. The coordinate system is set on the basis of the roughing tool; the X and Z offsets, tool radii, and tool vector numbers for the four tools are listed as follows:

REGISTER NO.	X OFFSET (mm)	Z OFFSET (mm)	RADIUS (mm)	TOOL VECTOR NO.
1	0	0	0.25	2
2	+3.304	+3.995	0.13	2
3	−5.110	+13.573	0	6
4	−4.912	+12.517	0	6

unspecified chamfer 1.5x45
width of grooves 2.5
R 2.0

1x45°

MT5x1

φ21.0

M32x1.5

φ27.0

φ29.5

φ36.0

φ46.5

φ51.0

φ25.0

φ22.0

2x45°

φ24.90

18

21

63.75

111.75

264.0

39.0

8.25

7.5

7.5

7.5 7.5

7.5 7.5

7.5 7.5

7.5

R

R

44.0

23

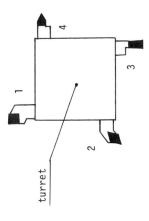

1

4

3

2

turret

Figure P2-8

79

The stock material has a diameter of 55 mm. The part is to be cut with the use of two NC programs. With the first program, the part is positioned as shown in the figure and the part profile, from $z = 0$ to $z = -220$ mm, is cut. Then the shaft is turned around with the left end of the shaft shown in the figure supported by the tail stock. A second program is used to cut the part profile from $z = 0$ (the end surface) to $z = -45.5$ mm. The origin of the coordinate systems for the two programs are set at the center of the end surface on the right side. Assume that the two center holes have been drilled and the following parameters are used:

Feedrate	0.13 mm/rev (rough cut)
	0.06 mm/rev (finishing cut)
	0.02 mm/rev (grooving)
Maximum depth of cut	3 mm (rough cut)
Spindle speed	800 rpm or S40 (rough cut)
	1200 rpm or S60 (finishing)
	400 rpm or S20 (grooving and threading)
Finishing allowance	0.2 mm
Coordinates of the reference point	(201.076,191.8)

2.9 The part given in Fig. P2-9 has both an internal and an external profile. An NC turning center with FANUC 6T controller is used to produce this part. The tools and their offset

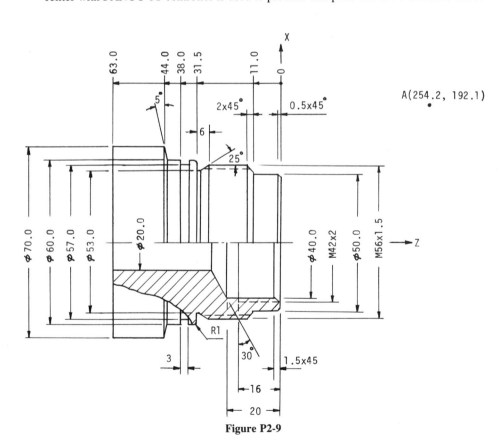

Figure P2-9

registers used in the machining operation are as follows:

T0101   External rough turning and facing
T0202   Drilling (20.0 mm diameter)
T0303   Internal rough boring
T0404   Internal finishing boring
T0505   Internal threading
T0606   External finishing and cutting groove with tapered surface of 25 degrees
T0707   External threading
T0808   External grooving and cutoff tool (width = 3 mm)

The stock material has a diameter of 75 mm and is held in the chuck. The part is cut off from the stock after it has been finished.

Write an NC program for machining this part, using the following parameters:

OPERATION	SPINDLE SPEED (IN NC CODE)	FEEDRATE (mm/rev)
Rough turning		
External	S700	0.13
Internal	S900	0.10
Finishing	S1000	0.05
Grooving and cutting off	S300	0.04
Threading	S300	
Drilling	S300	0.08

A tool change should be programmed in two stops:
1. Returning to reference point A (G28U_W_)
2. Changing the tool (M06T____)

2.10 The cavity of the matrix shown in Fig. P2-10 is to be cut on a three-axis NC milling machine. Write an NC program for directing the machine process. The following parameters are given:

Cutting tool $\frac{1}{2}$ in. flat-end mill with 0.1 in. corner radius
Feedrate   2.5 in./min
Spindle speed   S30
Starting point   (-5.0,5.0,3.0)

The allowable scallop height on the surface is 0.005 in.

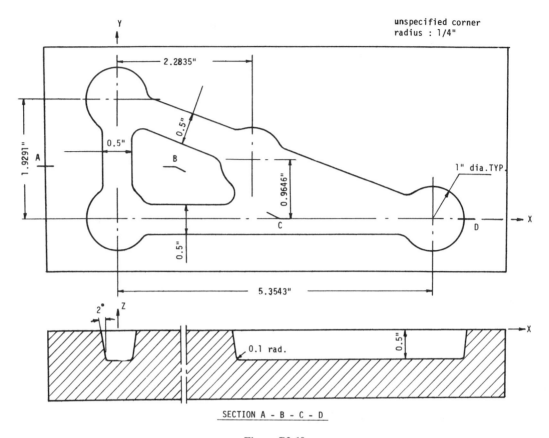

SECTION A – B – C – D

Figure P2-10

REFERENCES FOR PART I

1. *Modern Machine Shop 1987 NC/CIM Guidebook*. Cincinnati, Ohio: Gardner, 1987.
2. Rosenberg, J., *A History of Numerical Control 1949–1973*, DoD DAHC-15-72-C-0308, 1973.
3. Ogden, H., "Mechanical Design Aspects of Electronically Controlled Machine Tools," *Proc. 4th Intl. MTDR Conf., 1963*, p. 37. Oxford: Pergamon Press, 1963.
4. Leathan-Jones, B., *Introduction to Computer Numerical Control*. London: Pitman, 1986.
5. Koren, Y., *Computer Control of Manufacturing Systems*. New York: McGraw-Hill, 1983.
6. Pressman, R. and Williams, J., *Numerical Control and Computer Aided Manufacturing*. New York: Wiley, 1977.
7. Wick, C., "Advances in Machining Centers," *Manufacturing Engineering*, Oct. 1987, p. 24.

8. Koenigsberger, F., "Trends in the Design of Metal Cutting Machine Tools," *CIRP Annals*, Vol. 23, No. 2, 1974, p. 255.

9. Drayton, D. E. et al., "Automatically Programmed Tools," *Numerical Control Programming Languages, Proc. 1st Intl. IFIP/IFAC PROLAMAT Conf.*, 1969. Amsterdam: North-Holland, 1970.

10. Shah, R., *NC Guide—Numerical Control Handbook*, 2nd ed. Zurich, Switzerland: NCA Verlag, 1979.

11. ANSI Standard ANSI X3.37-1980: *Programming Language APT*. New York: American National Standards Institute, 1980.

12. Lange, K. et al., "Contribution of Numerical Control to the Development of Metal Forming Processes," *CIRP Annals*, vol. 31, No. 2, 1982, p. 511.

13. *1986–1987 Economic Handbook of the Machine Tool Industry*. McLean, Va.: National Machine Tool Builders' Association, 1987.

14. Jaikumar, R., "Postindustrial Manufacturing," *Harvard Business Review*, Nov.–Dec., 1986, No. 6, pp. 69–76.

15. *Mazak CAM-CNC system: Mazatrol CAM M-2* [manufacturer's catalogue]. Oguchi-cho, Japan: Yamazaki Machinery Works, 1983.

16. *MC-DC Series Vertical Machining Center* [manufacturer's catalog No. M0984-DC]. Fukui, Japan: Matsuura Machinery Corp.

17. Welch, A., "NC Verification Using Computer Simulation," CASA/SME paper MS80-233, Computer and Automated Systems Association/Society of Manufacturing Engineers, 1980.

18. Shpitalni, M. and Lemaistre, C., "The Problem of NC Program Verification: Analysis and Approaches," *Computer Integrated Manufacturing*, M. Martinez, et al., ed. New York: American Society of Mechanical Engineers, 1983.

19. Milner, D. A. and Vasiliou, V. C., *Computer-Aided Engineering for Manufacture*. New York: McGraw-Hill, 1987.

20. ANSI Standard (ANSI Y14.5M-1982): *Dimensioning and Tolerancing*. New York: American Society of Mechanical Engineers, 1983.

21. EIA Standard RS-267-B: *Axis and Motion Nomenclature for Numerically Controlled Machines (ANSI/EIA RS-267-B-83)*. Washington, D.C.: Electronic Industries Association, June 1983.

22. ISO Standard 841-1974: *Axis and Motion Nomenclature for Numerical Controlled Machines*. Geneva, Switzerland: International Organization for Standardization, 1974.

23. EIA Standard RS-244-B: *Character Code for Numerical Machine Control Perforated Tape*. Washington, D.C.: Electronic Industries Association, Oct. 1976.

24. EIA Standard RS-358-B: *Subset of American National Standard Code for Information Interchange for Numerical Machine Control Perforated Tape (ANSI/EIA RS-358-B-80)*. Washington, D.C.: Electronic Industries Association, Feb. 1979.

25. Childs, J. J., *Principle of Numerical Control*, 3rd ed. New York: Industrial Press, 1982.

26. *FANUC System 6M—Model B Operator's Manual*. Elk Grove Village, Ill.: General Numeric Corp.

27. *FANUC System 6T Operator's Manual*, Elk Grove Village, Ill.: General Numeric Corp.

28. Chang, C.-H., Lecture Notes for Course MANE 195A: "Numerical Control Manufacturing Machinery Laboratory," Manufacturing Engineering Program, University of California, Los Angeles, 1985.

Note: Works cited in references 21, 23, and 24 can be purchased from the Electronic Industries Association, 2001 Eye St., N.W., Washington, D.C. 20006, works cited in references 11 and 20 from the American National Standards Institute, Inc., 1430 Broadway, New York, NY 10018.

part II

Computer-Aided NC Programming in the APT Language

The language through which human beings communicate with CNC machines consists of NC codes in word address format. From the discussion in Chapter 2, it is clear that precise coordinates of tool positions are needed to write an NC program. There are two disadvantages of programming in NC codes. First, tool position coordinates are not easy to calculate for many parts. Indeed, manual calculation is time consuming and error prone. Second, an NC program written manually in NC codes has to be input manually into the NC machine or a tape-preparing device, or it has to be manually keyed into a computer and thence sent to the NC controller. Since NC programs of hundreds or even thousands of statements are common, manual input takes excessive time and generates typing errors that are difficult to detect. Although efforts have been made to incorporate more functions into NC codes, such as the multiple repetitive cycle commands, mirror commands (symmetrical reflection of the cutter path about one axis), and some simple calculation and decision statements (for NC controllers with variable programming capability), the situation has essentially not changed. Therefore a higher-level NC language is needed that permits a cutter path to be defined on the basis of the part geometry (the profile of the workpiece) and allows calculation routines and decision statements to be included in the NC program. A debugging facility is also needed to help programmers quickly locate NC programming errors.

One might suggest using an existing high-level language, such as FORTRAN or PASCAL, to write NC programs, because they are well designed and have all the above-mentioned capabilities. The basic problems that arise when NC programs are written in any high-level language are as follows:

1. Defining the geometric entities that are used to specify a cutter path
2. Calculating the cutter positions (coordinates) or path
3. Defining the machining specifications needed for completing the machining process
4. Outputing the cutter path and machining specifications in machine-understandable language (i.e., NC codes)

NC programming in high-level languages of general use is possible but cumbersome and inconvenient for the following reasons. The cutter path is determined on the basis of the profile of the workpiece, which is composed of a number of geometric entities. Calculation of cutter positions (coordinates) consists of finding the intersections of one geometric entity with another or others and computing the compensation for the cutter radius and allowance. These calculations are sometimes very complicated and require a solid knowledge of mathematics. They should be simplified so that they can be carried out easily by NC programmers without extensive mathematical training. On the other hand, writing an NC program requires a great deal of work to define the geometric entities and to translate and output the results in a format (i.e., NC codes) that is acceptable by the NC machine to be used. Therefore it is necessary to have a special high-level NC language that can simplify the definition of geometric entities and save NC programmers from having to design complex calculation routines and from having to write translation and output routines.

APT (Automatically Programmed Tool) is a high-level NC language that was designed to solve the above-mentioned problems. English-like words are used to make the language more user-friendly. A program (e.g., IBM APT-AC [automatically programmed tool-advanced contouring] numerical control processor) is needed to process part machining programs in APT. The processor creates the necessary geometric environment according to the APT geometric definition statements and calculates the cutter positions and path according to APT cutter motion statements. The calculated results and the statements regarding machining specifications are output, in the order defined by the APT program, as CLDATA (cutter center line data or cutter location data), which is then translated, by a postprocessor specific to the NC machine to be used, into machine-understandable NC codes.

APT is a language designed for machining processes. Therefore sound knowledge of machining processes is a prerequisite to writing correct APT programs. In addition to a thorough understanding of the APT programming language itself, some knowledge of the postprocessors to be used is also required.

Although APT has been standardized, there are always some differences in various computer systems. The following discussion applies to the IBM computer system with the assumption that the APT program will be processed by version 1, modification level 3, of the IBM APT-AC numerical control processor.*

One might suggest using an existing high-level language, such as FORTRAN or PASCAL, to write NC programs, because they are well designed and have all the above-mentioned capabilities. The basic problems that arise when NC programs are written in any high-level language are as follows:

is certainly much easier for the guide (NC programmer) to point out (define) the route (cutter path) in the former way.

*In the following discussion, the IBM APT-AC numerical control processor is referred to as the *APT-AC NC processor*, or simply the *NC processor*.

chapter 3

The Elements of the APT Language

Like any computer program, an APT part machining program is a set of orderly listed instructions, called statements, that direct a computer to carry out a desired sequence of operations. A few examples of APT statements are as follows:

```
MACHIN/GN5CC,9,OPTION,2,0
PT0=POINT/0,0,1.5
A3)SPINDL/500,CLW
```

As can be seen from these examples, each statement consists of one or more of the following elements:

- Vocabulary words (e.g., MACHIN, OPTION, POINT)
- Symbols (e.g., GN5CC and PT0)
- Numbers (e.g., 9, 2, 1.5)
- Punctuation and delimiters (e.g., slash, comma)
- Statement label (e.g., the characters A3 preceding the closing parenthesis)

Symbols, vocabulary words, and statement labels are made up of English letters (A, B, ..., Z in upper case) or English letters and numerical digits (0, 1, 2, ..., 9). Numbers are defined as numerical decimal digits. The characters used in

TABLE 3-1   THE CHARACTERS USED IN THE APT LANGUAGE

Character	Meaning and Usage
	Blank or null character, ignored by the NC Processor except in literal alphanumeric text
$	Dollar sign — statement continued in next line
$$	Double dollar sign — end of statement
;	Semicolon, dividing two statements in same line
: or)	Colon or closing parenthesis, indicating that preceding alphanumeric characters constitute a statement label
( )	Opening and closing parentheses — nested or parenthetical expression
' '	Pair of apostrophes — alphanumeric literal delimiter
*	Asterisk — multiplication
**	Double asterisk — exponentiation
+	Plus — positive sign, or addition
−	Minus — negative sign, or subtraction
.	Period — decimal point
/	Slash — division or major-minor word separator
=	Equals — assigning operator
,	Comma — word or number separator
" "	Pair of double quotation marks, delimiting a complex MACRO variable specification
| |	Double bar or stroke — extraction of canonical parameter
A, B, . . . , Z	English alphabet (upper case), used to construct vocabulary words, symbols, variables, and statement labels
E or D	English alphabet (upper case) — used to define a number in exponential format
PI	The symbol for π (3.14159265359879310)
0, 1, . . . , 9	Numeric characters — used to construct a number, symbol, or statement label

APT as punctuation symbols and delimiters are shown in Table 3-1. In the following discussion, an alphanumeric character is defined as either a digit or an English letter; an alphanumeric string consists of one or more alphanumeric characters.

3.1 APT VOCABULARY WORDS

APT vocabulary words are strings of one to six alphanumeric characters. They can be divided into five categories (Table 3-2):

- Those used to define geometric entities
- Those defining mathematical operations
- Those used to define cutter motion
- Those defining modes of computer operation, such as CLPRNT (ordering the NC processor to print the CLDATA) or NOPOST (suppressing execution of the postprocessor)
- Postprocessor words, which are processed directly by the postprocessors, most of them being related to machining specifications, remarks or comments, and so forth

TABLE 3-2 SELECTED APT VOCABULARY WORDS

Geometric Entity Definition	Mathematical Operation	Cutter Motion	Mode of Computer Operation	Postprocessor Word
ABSLTE	ABSF	AUTOPS	CALL	ARCSLP
ARC	ACOSF	CUT	CANON	AUXFUN
AT	ANGLF	DNTCUT	CLPRNT	BRKCHP
ATANGL	ASINF	FROM	CONTIN	CCLW
ATTACH	ATANF	GO	COPY	CLEARP
AVOID	ATAN2F	GOBACK	DO	CLRSRF
BISECT	CBRTF	GODLTA	FINI	CLW
CENTER	COSF	GODOWN	IF	COOLNT
CCLW	COTANF	GOFWD	INSERT	CUTCOM
CIRCLE	DISTF	GOLFT	JUMPTO	CUTTER
CLW	DOTF	GORGT	LOOPST	CYCLE
CONE	EXPF	GOTO	LOOPND	DEEP
CONST	INTGF	GOUP	MACHIN	DELAY
CROSS	LNTHF	INDIRP	MACRO	DRILL
CYLNDR	LOGF	INDIRV	NOPOST	DWL
DECR	LOG10F	NOPS	PARTNO	END
DELTA	MAX1F	OFFSET	PPRINT	FEDRAT
ELLIPS	MIN1F	ON	PRINT	FLOOD
FUNOFY	MODF	PAST	REMARK	GOCLER
GCONIC	NUMF	POCKET	RESERV	GOHOME
HYPERB	SIGNF	PSIS	RESET	HIGH
IN	SINF	REFSYS	TERMAC	INCHES
INCR	SQRTF	TANTO	TITLE	INTOL
INTERC	TANF	THICK		IPM
INTOF	TYPEF	TLLFT		IPR
INVX	'EQ'	TLON		LEADER
INVY	'GE'	TLRGT		LOADTL
LARGE	'GT'	TRACUT		LOW
LCONIC	'LE'			MCHTOL
LEFT	'LT'			MIST
LINE	'NE'			MM
LINEAR				OFF
MATRIX				ON
MINUS				OPSKIP
NEGX				OPSTOP
NEGY				OPTION
NEGZ				ORIGIN
NORMAL				OUTTOL
NOX				PLABEL
NOY				PREFUN
NOZ				RAPID
NUMPTS				RETRCT
OBTAIN				ROTABL
OMIT				ROTREF
OUT				SELCTL
PARLEL				SEQNO
PATERN				SETOOL
PERPTO				SPINDL
PLANE				STOP

Note: For a complete list of APT vocabulary words, see Appendix A.

TABLE 3-2 CONTINUED

Geometric Entity Definition	Mathematical Operation	Cutter Motion	Mode of Computer Operation	Postprocessor Word
POINT				TAPKUL
POLCON				TOLER
POSX				UNITS
POSY				
POSZ				
PLUS				
PTNORM				
PTSLOP				
QADRIC				
RADIUS				
RETAIN				
RIGHT				
RLDSUF				
RTHETA				
SLOPE				
SMALL				
SPHERE				
SPLINE				
STEP				
TABCYL				
TANTO				
THETAR				
THRU				
TIMES				
TORUS				
TRANPT				
TRANSL				
TRFORM				
TWOPT				
UNIT				
VECTOR				
XAXIS				
XCOORD				
XLARGE				
XSMALL				
XYPLAN				
XYZ				
YAXIS				
YCOORD				
YLARGE				
YSMALL				
YZPLAN				
ZAXIS				
ZCOORD				
ZLARGE				
ZSMALL				
ZSURF				
ZXPLAN				
3PT2SL				
4PT1SL				

A few vocabulary words must be specified in columns 1 to 6 of an input line and are named *fixed-field words*. They are PARTNO, REMARK, PPRINT, INSERT, and TITLE.

There are a great number of vocabulary words (about 600) accepted by the IBM APT-AC NC processor. In this book, we shall mention only about two hundred of them, which are in common use (Table 3-2). Appendix A gives a list of vocabulary words accepted by the IBM APT-AC NC processor for reference purposes only.

Vocabulary words, other than those defining mathematical operations, can be divided into two groups, namely, *major words* and *minor words*. Major words are those placed before a slash, specifying the type of geometric entity, the required computer operation, the machining specification, or the type, direction, and cutter-profile relationship of the cutter motion. Minor words are mostly modifiers and are used to further define geometric entities, cutter motion, or machining specifications within the scope specified by the major words. The meaning and usage of the vocabulary words are discussed in detail in the following chapters.

3.2 SYMBOLS

A symbol is composed of one to six alphanumeric characters, at least one of which must be nonnumerical. It is assigned to a geometric entity, mathematical expression, subprogram, or number that can then be referenced easily in the succeeding part of a program. For example, in the statement

L1=LINE/0,0,0,1,1,1

the line passing through two points (0,0,0) and (1,1,1) is defined and assigned the symbol L1. In the succeeding statement(s), the line can be referred to by simply specifying the symbol L1.

The use of any of the vocabulary words defined in APT as symbols is not allowed. Since there are several hundred vocabulary words in APT, and very few of them contain numerical characters, it is recommended that at least one numerical character be included in a symbol. Blanks and punctuation marks cannot be constituents of a symbol.

3.3 NUMBERS AND SCALARS

All numbers in an APT program are treated by the APT-AC NC processor as real numbers. Thus the number 5 can be expressed in any one of the following forms:

$$5$$
$$+5.$$
$$5.0$$
$$0.5E1 \qquad (i.e., \ 0.5 \cdot 10^1)$$
$$.5E+1 \qquad (i.e., \ 0.5 \cdot 10^1)$$

50.E-1   (i.e., $50 \cdot 10^{-1}$)

.05E2

0.5D1   (i.e., $0.5 \cdot 10^{1}$)

0.05D+2

etc.

where D and E are symbols used to specify a number in the exponential form and have the same meaning and usage. Note that the number preceding the symbols E and D must have a decimal point; otherwise, the whole set of characters defining the number will be treated as a variable symbol.

The maximum number of digits allowed in a number is 75. The range of a number that can be specified in an APT program runs from 10^{-75} to 10^{75} ; however, only sixteen decimal digits after the decimal point of a number are considered significant during processing by the NC processor.

An integer is not treated as a separate class of number. If a number is used as an index or a subscript (in such a case, an integer is required), only its integer part is used. A positive number can be specified with or without a plus sign; a negative one must be preceded by a minus sign.

A number specifying the magnitude of an angle should be in degrees, decimal fractions of a degree, or both. An angle is positive when measured in counterclockwise direction, and negative, in clockwise direction (see Fig. 2-3).

In the following, the term "scalar" is used to denote a single-valued entity. It may be an expression, a symbol, or a number that, when reduced, is single valued.

3.4 PUNCTUATION AND DELIMITERS

3.4.1 Blank or Null Character

A blank or null character is ignored by the APT-AC NC processor and can be specified at any place in a statement except for the following cases:

- A blank in a literal text that is delimited by a pair of apostrophes will be considered as a part of the literal text.
- A blank cannot be placed before and within a fixed-field vocabulary word (e.g., PARTNO, INSERT, REMARK, PPRINT, or TITLE).
- A blank placed after a $ or $$ sign, or after a fixed-field vocabulary word, will be treated as part of a commentary text.

Normally, a blank or blanks are specified to improve the readability of an NC program. It is suggested that a blank not be used in a statement label because it will cause confusion.

Example 1

Blanks are used to improve the readability of an APT program:

```
. . . . . . . . . . . . . . . . . . . .
. . . . . . . . . . . . . . . . . . . .
L1   = LINE/PT1,PT2
C(1) = CIRCLE/CENTER,PT3,RADIUS,1.0
PL(1)= PLANE/0,0,1,3
. . . . . . . . . . . . . . . . . . .
. . . . . . . . . . . . . . . . . . .
TLLFT,GOLFT/L1,TO,C(1)
        GOLFT/C(1),TO,L2
. . . . . . . . . . . . . . . . . . .
. . . . . . . . . . . . . . . . . . .
```

Example 2
A blank or blanks will be considered as part of a comment or literal text in the following statements:

```
PARTNO A 3-D PROFILE CUTTING PROGRAM
REMARK TOOL DIAMETER = 0.75 IN.
REMARK/'TOOL DIAMETER = 0.75 IN.'
```

3.4.2 $ (Dollar Sign)

This symbol means that the statement preceding it is continued on the next line. The characters that follow this symbol on the same line will be ignored by the NC processor and can be used as comments. However, after a fixed-field vocabulary word, this symbol is considered as part of the commentary text. The dollar sign is used when a statement consists of more than 72 characters (the maximum number of characters that can be input in a line).

Example 1
The statement

```
L1=LINE/P1,LEFT,$
TANTO,C1
```

is the same as the following:

```
L1=LINE/P1,LEFT,TANTO,C1
```

Example 2
The characters coming after the dollar sign in the following statement constitute a comment:

```
C1=CIRCLE/XLARGE,L1,$ THIS CIRCLE IS TANGENT TO THREE LINES L1, L2, L3
            YLARGE,L2,XSMALL,L3
```

3.4.3 $$ (Double-Dollar Sign)

The double-dollar sign represents the end of a statement (unless it is placed after a fixed-field vocabulary word). The information after this sign on the same line is treated as commentary. As in some other high-level programming languages, a statement in APT is considered to occupy one line unless a statement-continuation sign ($) is specified. Therefore the use of the end-of-a-statement sign, $$, is optional; it is needed only when a comment is to be specified on the same line.

Example

The following statement defines a point, P1, with a comment explaining that it is the center of circle C1:

P1=POINT/1.0,2.0,3.0$$ THIS IS THE CENTER OF CIRCLE C1

3.4.4 ; (Semicolon)

This sign is used to divide two statements on a single line. However, it does not have the same effect if specified after a single dollar sign or a fixed-field word on the same line; in that case, it is treated as part of the commentary text. If a semicolon is specified within an alphanumeric string delimited by a pair of apostrophes or double apostrophes, it does not become a part of the alphanumeric string, as it is intended to be, but acts as an end-of-statement sign, dividing the statement into two parts.

Example 1
The input line

INDIRV/V1;GO/S1,S2

is equivalent to two statements specified on two consecutive lines:

INDIRV/V1
GO/S1,S2

Example 2
The input line

INDIRV/V1$$V1 IS SHOWN ON THE PART DRAWING;GO/S1,S2

is divided by the semicolon into two statements, one of which has a commentary text:

INDIRV/V1$$V1 IS SHOWN ON THE PART DRAWING
GO/S1,S2

Example 3
It is intended that the following statements be used to define three circles C1, C2, and C3:

C1=CIRCLE/YSMALL,L1,YLARGE,L2,RADIUS,R$FIRST CIRCLE;C2=CIRCLE/1.,2.,1.5
C3=CIRCLE/1.,2.,1.5

However, the input after the single-dollar sign in the first line is ignored. The NC processor reads the two lines as one statement

C1=CIRCLE/YSMALL,L1,YLARGE,L2,RADIUS,RC3=CIRCLE/1.,2.,1.5

which is syntactically incorrect.

Example 4
The semicolon after the fixed-field word REMARK in the following statement is part of the commentary text:

REMARK TOOL DIAMETER 1/2 IN FOR OUTER PROFILE; 1/4 IN FOR INNER PROFILE

3.4.5 : (Colon) or) (Closing Parenthesis)

The colon is used to indicate that the preceding alphanumeric string is a statement label. When not used after an opening parenthesis, a closing parenthesis has the same meaning as a colon. A statement label should be specified at the beginning of a statement.

Example
A1 and 1B are the statement labels of the following two statements:

A1)CALL/C1,P=1.0,Q=2.0

1B:JUMPTO/3C

3.4.6 () (Opening and Closing Parentheses)

Opening and closing parentheses can be used for the following purposes:

- To enclose the arguments of a mathematical function; for example,

K=TANF(B)

where B is the argument of the tangent function.

- To enclose the subscript or index of a variable or symbol; for example,

M(1)=MATRIX/XYROT,15

where the character 1 is the subscript of the symbol M.

- To enclose the argument of a conditional branching statement IF. The argument may be a variable, a mathematical expression, or a relational expression.
  Example 1
  The enclosed argument A is a variable:

IF(A)1A,2A,3A

  Example 2
  The enclosed argument A-B is a mathematical expression:

IF(A-B)C1,C2,C3

  Example 3
  The enclosed argument A'EQ'B is a relational expression:

IF(A'EQ'B),JUMPTO/A1

- To enclose a list of statement labels for the conditional branching statement JUMPTO:

JUMPTO/(C1,C2,C3,C4),N

where C1, . . . , C4 are the statement labels.
- To enclose a nested definition.

Example

It is intended to define line L1 with two defined points, P1 and P2, as follows:

L1=LINE/P1,P2

However, it may happen that P1 has not yet been defined in the program. To resolve this problem, one can use the nested definition as follows:

L1=LINE/(P1=POINT/1.0,0,2.0),P2

- To enclose something that should be considered as a whole or to enclose a part of the mathematical expression that should be given higher precedence in calculation than the rest.

  Examples

  A=B*(−C)**(−2.5)
  A=(B*(C/D)+F)**G

3.4.7 / (Slash)

The slash sign is used to separate the major elements in a statement or is used as a symbol of arithmetic division.

Example 1

The slash in the following statement is used to separate the vocabulary word defining the type of the geometrical entity and its parameters:

POINT/1,0,2

Example 2

The slash in the following statement is used as an arithmetic operator (division):

K=(A+B)/(C+D)

3.4.8 = (Equal Sign)

The equal sign is used to assign a name to a geometric entity (see the example in Section 3.2) or a subprogram (called MACRO in APT); or it can be used to assign a value to a variable.

Example

The following statements assign K1 as the name of the defined subprogram (MACRO) and the value 2.75 to variable X1, respectively:

K1=MACRO/A,B,C
X1=2.75

3.4.9 + (Plus Sign), − (Minus Sign), * (Asterisk), and ** (Double Asterisk)

The plus and minus signs and the single and double asterisks are the arithmetic operators representing addition, subtraction, multiplication, and exponentiation, respectively. Together with the slash sign, they constitute the whole set of simple arithmetic operators. The signs + and - are also used to specify a positive and a negative number, respectively. A number without a sign is considered as positive.

Examples

A = B + C − M/N
F = (G**(−H/2))*I

The NC processor evaluates a parenthesis-free expression, that is, an expression without parentheses, using the following precedence table:

First precedence: Exponentiation (**)
Second precedence: Multiplication (*) and division (/)
Third precedence: Addition (+) and subtraction (−)

The computer scans a parenthesis-free arithmetic expression from left to right three times. First, it looks for exponentiations and calculates the result(s); then it looks for multiplications and divisions; finally, it carries out the additions and subtractions. The NC processor also permits the use of parentheses in the conventional way; that is, parentheses have precedence over all arithmetic operations, and the innermost pair of parentheses is treated first.

Example 1
The statement

A=B/C+D*F**G

is equivalent to the following conventional mathematical expression

$$A = \frac{B}{C} + D \cdot F^G$$

Example 2
If the mathematical expression

$$A = \frac{B}{C} + (D \cdot F)^G$$

is to be specified in a program, the statement should be written as

A=B/C+(D*F)**G

Example 3
Mistakes and confusion may be avoided by separating two consecutive operators having the same precedence with a pair of parentheses. The following table lists some examples of APT statements and their meaning:

APT STATEMENT	MEANING
$A = B * C/D$	$A = \dfrac{B \cdot C}{D}$
$A = B*(C/D)$	$A = B \cdot \dfrac{C}{D}$
$A = B/C/D$	$A = \dfrac{\frac{B}{C}}{D}$
$A = B/(C/D)$	$A = \dfrac{B}{\frac{C}{D}}$

3.4.10 . (Period or Decimal Point)

A period is used to separate the integer from the fractional part of a number (e.g., 3.52, 0.002).

3.4.11 , (Comma)

In addition to the slash sign and parentheses, commas can also be used to separate elements of a statement.

Examples

```
MACHIN/LATH1,6,OPTION,2,0
TLRGT,GORGT/(L1=LINE/P1,P2),ON,L2
```

3.4.12 ‖ (Double Bar)

The double-bar sign is used to extract an element of a canonical form for a geometric entity. For example, the canonical form for a circle C1 is x, y, z, 0, 0, 1, r (see Appendix B), where x, y, and z are the coordinates of the center; 0, 0, 1 are the directional coefficients of the normal to the plane containing the circle; and r is the radius. The statement defining the center of the circle C1 as point P1 is as follows:

```
P1=POINT/C1|1|,C1|2|,C1|3|
```

where the numbers 1, 2, and 3 enclosed in double-bar signs indicate that the first, second, and third elements, respectively, of the canonical form should be used.

3.4.13 " (Pair of Apostrophes)

A pair of apostrophes is used to enclose a literal character string (see example 2 in Section 3.4.1).

3.4.14 "" (Pair of Double Quotation Marks)

A pair of double quotation marks is used to enclose a complex specification for a MACRO (subprogram) variable.

Example

```
A1=MACRO/C1,C2,C3="TLLFT,GOLFT"
```

3.5 STATEMENT LABELS

If a statement is to be referenced in other statements, it should be assigned a label. A *statement label* consists of one to six alphanumeric characters, all of which may be numerical or alphabetical. It should be placed at the beginning of a statement and is separated from the statement by a closing parenthesis or a colon.

Examples

```
1A)CUTTER/0.5,0
15:FEDRAT/2.0
A31)GOTO/P1
AN1CY2)GODLTA/−0.2
```

It is not necessary to assign a label to a statement if it is not referred to in other statements.

3.6 STATEMENT SIZE LIMITATION

A statement consists of one or several elements, as described above. The maximum number of elements that can be specified in an APT statement is 4092. For example, the statement

A1)GOTO/(POINT/0,0,0)

consists of 11 elements, that is, two vocabulary words: GOTO and POINT; six punctuation marks: two slashes, two commas, and opening and closing parentheses; and three numerical characters. Note that the statement label and its closing parenthesis (or colon) are not considered as constituents of a statement but as an attached part. Therefore they are not counted as elements of a statement.

3.7 THE NOTATION FOR THE APT STATEMENT FORMAT

The following symbols and word conventions are used in this book to describe the APT statement format:

1. All the APT vocabulary words are printed in capital letters.
2. A word that represents a numerical value or a scalar that should be provided in programming is shown by a lower-case English character or characters with or without a digit or digits (e.g., r, r1, rr12).
3. Information, other than a numerical value or scalar, supplied by a programmer is represented by a symbol that consists of both upper-case English character(s) and digit(s).
4. The possible choices for one element of an APT statement are enclosed in braces, and one of them must be chosen during programming. However, braces themselves should not be specified in a statement.
5. The optional element(s) (i.e., elements that can be specified if necessary or neglected) is (are) enclosed in brackets. However, brackets themselves should not be specified in a statement.
6. All the punctuations and delimiters in a statement are required, and their positions in the statement should not be changed.

The examples in Fig. 3-1 show the convention described above.

Example 1:

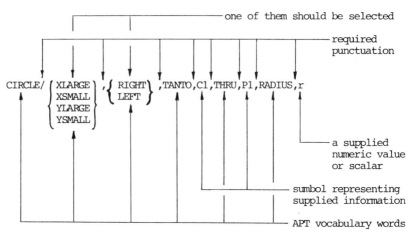

Example 2:

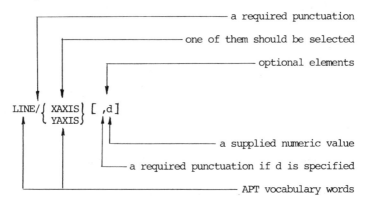

Figure 3-1   Two examples showing the symbols and word conventions.

chapter 4

The Definition of Geometric Entities

From a kinematics point of view, programming a machining process basically consists of directing a tool to move along the profile of a part while maintaining deviation within a specified tolerance. Thus the profile of a part is the guideline for the cutter profiling motion. Usually the outline of a part is made up of various geometric entities, namely, straight lines, circular arcs, and curves in two-dimensional machining, and planes and profiled surfaces in three-dimensional machining. These elements must be defined before a cutter motion is specified or programmed.

In the APT computer language, a geometric entity is defined by a major vocabulary word, indicating the type of the entity, followed by one or more minor words and/or parameters specifying the position, dimension, and/or limiting condition of the entity. For example, the statement defining a circle whose center is at the point (1.0,2.0,2.5), with a radius equal to 3.0, is as follows:

C1=CIRCLE/CENTER,1.0,2.0,2.5,RADIUS,3.0

where C1 is the symbol assigned to the defined circle; CIRCLE is a major vocabulary word; and CENTER and RADIUS are minor words. The APT-AC NC processor allows definition of a number of geometric entities, including point, line, vector, plane, circle, cylinder, cone, ellipse, hyperbola, sphere, ruled surface, and torus. For most of these entities, a number of statement formats with different parameters are provided, thus permitting definition of a geometric entity based on various known conditions. For example, a point can be defined by its coordinates, by its position relative to a given point, or by the intersection of two or more given geometric entities. The NC processor uses the given condition to calculate a solution (i.e., the desired point) and then stores it in a standard format known as *canonical form* (see Section 4.11 and Appendix B). There may be more than one solution for a given set of conditions. In such a case, additional modifier(s) should be specified to select the correct solution among all the possible ones. These modifiers include the following:

- Directional modifiers: XLARGE, XSMALL, YLARGE, YSMALL, ZLARGE, ZSMALL, POSX, POSY, POSZ, NEGX, NEGY, NEGZ; LEFT and RIGHT; CLW, CCLW;

- Dimensional modifiers: LARGE and SMALL

- Relational modifiers: IN and OUT

For example, a point in the X-Y plane can be defined as the intersection of two given circles, C1 and C2, by the following statement (see Fig. 4-1 and the No. 4 statement format in Table 4-3):

$$\text{POINT/} \begin{Bmatrix} \text{XLARGE} \\ \text{XSMALL} \\ \text{YLARGE} \\ \text{YSMALL} \end{Bmatrix} \text{,INTOF,C1,C2}$$

Generally, there exist two solutions (points) for the given condition. A directional modifier is then selected to indicate the desired one. The meanings of the directional modifiers used in APT are as follows (Fig. 4-2):

DIRECTIONAL MODIFIER	MEANING
POSX, XLARGE POSY, YLARGE POSZ, Z LARGE	The desired solution is the one with the larger x, y, or z coordinate, or it lies in the X-, Y-, or Z-increasing direction, in comparison with the other possible solution.
NEGX, XSMALL NEGY, YSMALL NEGZ, ZSMALL	The desired solution is the one with the smaller x, y, or z coordinate, or it lies in the X-, Y-, or Z-decreasing direction, in comparison with the other possible solution.

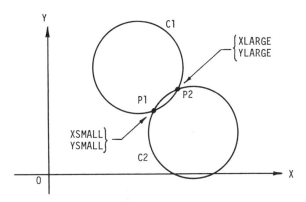

Figure 4-1 Use of directional modifiers to indicate the desired solution among the two possible ones. A directional modifier can be selected by comparing the coordinates of the desired solution with those of the other solution. Thus, for point P1, modifier XSMALL or YSMALL can be selected; for point P2, XLARGE or YLARGE applies.

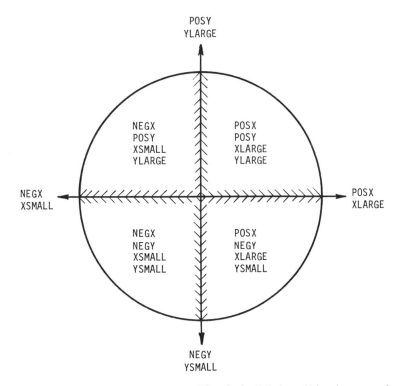

Figure 4-2 Selection of directional modifiers in the *X-Y* plane. Using the center of the circle as a starting point or reference position, draw a vector toward the end position or defined geometric entity. The possible choices of modifiers are given in the resulting quadrant.

Therefore either XSMALL or YSMALL can be specified for point P1 in Fig. 4-1. Sometimes a directional modifier preceding a given geometric entity is also used to indicate on which side the geometric entity to be defined is located with respect to the given one. For example, the statement defining a circle tangent to three given lines reads (Fig. 4-3)

CIRCLE/ $\begin{bmatrix} \text{XLARGE} \\ \text{XSMALL} \\ \text{YLARGE} \\ \text{YSMALL} \end{bmatrix}$,L1, $\begin{bmatrix} \text{XLARGE} \\ \text{XSMALL} \\ \text{YLARGE} \\ \text{YSMALL} \end{bmatrix}$,L2, $\begin{bmatrix} \text{XLARGE} \\ \text{XSMALL} \\ \text{YLARGE} \\ \text{YSMALL} \end{bmatrix}$,L3

The directional modifiers preceding L1, L2, and L3 can be selected to indicate the position of the defined circle with respect to lines L1, L2, and L3, respectively. As a rule, the selected directional modifier should be able to show a significant difference between two possible solutions. For example, in Fig 4-3, the use of a modifier is suggested for the Y direction (YLARGE or YSMALL) preceding line L3, because the difference in X coordinates of the solutions on both sides is not significant.

The directional modifiers RIGHT and LEFT are used to indicate that a tangent line is, respectively, on the right- or left-hand side of a given circle when one is looking from a reference point (or circle) toward the center of the given circle (Fig. 4-4).

The directional modifiers CLW (clockwise) and CCLW (counterclockwise) indicate the direction in which an arc distance or angle is measured from a reference position.

The meaning of the dimensional modifiers is explicit. LARGE or SMALL means that the desired solution is the one with a larger or smaller dimension, in comparison with the other possible solution (Fig. 4-5).

The relational modifier IN or OUT indicates whether or not the geometric entity defined by the statement shares a common space, or area, with the geometric entity specified after the modifier (Fig. 4-6).

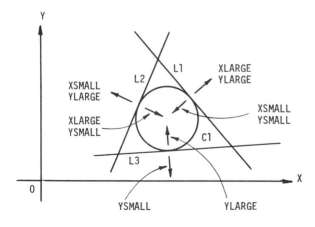

Figure 4-3 Use of directional modifiers to indicate on which side of a given line the defined geometric entity (circle C1) is located. The circle C1 is on the XSMALL or YSMALL side of L1. The modifiers of X direction, XLARGE and XSMALL, are not recommended for line L3 because they do not show a significant difference in the coordinates of the solutions on the two sides.

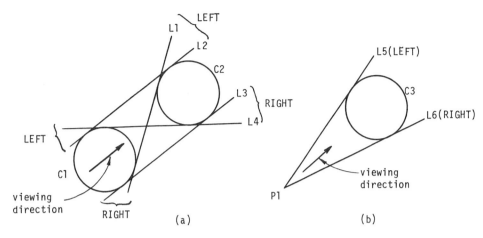

Figure 4-4 Use of directional modifiers LEFT and RIGHT to indicate on which side of the given circle the defined line is located. (a) As one looks from the center of circle C1 toward that of circle C2, line L1 is on the right-hand side of C1 and left-hand side of C2. (b) As one looks from point P1 toward circle C3, lines L5 and L6 are, respectively, on the left- and right-hand sides of circle C3.

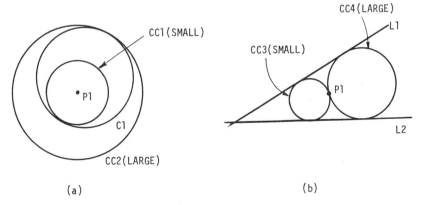

Figure 4-5 Dimensional modifiers LARGE and SMALL indicate the relative size of a desired solution with respect to that of the other. (a) There are two solutions (circles CC1 and CC2) for a given center, P1, and a given tangent circle, C1. A dimensional modifier can be selected according to the relative size of the desired circle. (b) Circles CC3 and CC4 are the two solutions for two given tangent lines, L1 and L2, and a given point, P1, on the circumference of the desired circle.

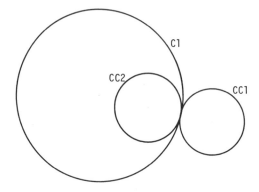

Figure 4-6 Relational modifiers IN and OUT. Circle CC2 is IN circle C1, or circle C1 is IN circle CC2, because they share a common area. Circle CC1 is OUT of circle C1 and CC2.

It should be noted that a modifier affects the geometric entity that follows it; the following diagram shows the relationships between modifiers and their related geometric entities:

This statement defines a circle, C1, passing through a given point, P1. This circle is on the XSMALL side of the given tangent line L2 and on the YLARGE side of the tangent line L1. Furthermore, this circle is the larger one of the two possible tangent circles.

Sometimes, a modifier is followed by more than one geometric entity (see statement format No. 4 in Table 4-3). In such a case the modifier is used only to indicate the correct solution among the possible ones resulting from those geometric entities.

As will be seen later, many of the geometric definition statements in APT are used to define geometric entities in the X-Y plane. The reason is that for NC milling machines, the Z axis is always the spindle (or cutter) axis, and in most cases the motion of the cutter is in the X-Y plane.

A geometric entity can be assigned a name (symbol) so that reference can be made to it in other statements. Normally, once having been defined, a symbol cannot be used as the name of another geometric entity. For example, if P1 is defined as a point by the statement

P1=POINT/1,2,3

it is not permissible to use P1 as a symbol for any geometric entity other than point (1,2,3). If a symbol has to be redefined because of the programming requirement, a statement that instructs the NC processor to allow redefinition of symbols of geometric entities should be specified (see Section 4.11).

In the following, we discuss the formats available for various geometric definition statements; the reader should consult the IBM APT-AC NC processor manuals[1,2] for additional formats that are not discussed here.

4.1 STATEMENTS DEFINING A POINT

In the APT language, there are 26 statement formats that can be used to define a *point* in space. They can be classified into three types: those which define a point by its coordinates; those which define a point by its position on, or relative to, a given geometric entity; and those which define a point as the intersection of two or more geometric entities. These statements are given in Tables 4-1, 4-2, and 4-3, respectively.

TABLE 4-1   STATEMENTS DEFINING A POINT BY ITS COORDINATES

No.	Known Condition	Statement Format
1	The Cartesian coordinates (x, y, z) of the point are given.	$\text{POINT}/\begin{Bmatrix} x,y,z \\ x,y \end{Bmatrix}$

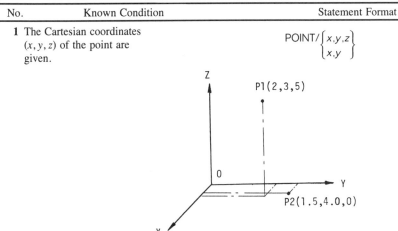

Comment   If the second option is chosen (i.e., only the x and y coordinates are specified), the z coordinate is determined by the statement ZSURF (see Section 4.10) preceding this statement, or is zero if no ZSURF statement is specified.

Examples
P1=POINT/2,3,5
P2=POINT/1.5,4.0,0 (or P2=POINT/1.5,4.0)
where P1 and P2 are symbols.

| 2 | The point is in the X-Y, Y-Z, or Z-X plane and is described by its polar coordinates (i.e., polar distance r and angle a). | $\text{POINT/RTHETA,} \begin{bmatrix} \text{XYPLAN} \\ \text{XZPLAN} \\ \text{ZXPLAN} \end{bmatrix} ,r,a$ $\text{POINT/THETAR,} \begin{bmatrix} \text{XYPLAN} \\ \text{YZPLAN} \\ \text{ZXPLAN} \end{bmatrix} ,a,r$ |

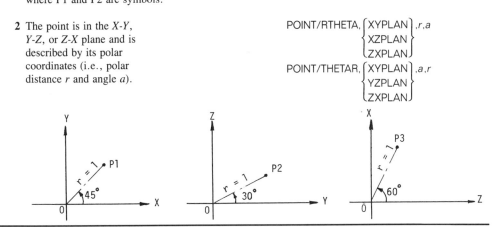

(continued)

TABLE 4-1   STATEMENTS DEFINING A POINT BY ITS COORDINATES   (continued)

No.	Known Condition	Statement Format

Comment   Vocabulary words XYPLAN, YZPLAN, and ZXPLAN indicate that the point is on the X-Y, Y-Z, and Z-X planes, respectively. The words RTHETA and THETAR indicate the order in which the parameters are specified.

Examples
P1=POINT/RTHETA,XYPLAN,1,45
P2=POINT/THETAR,YZPLAN,30,1
P3=POINT/RTHETA,ZXPLAN,1,60

3 The point is in the *X-Y*
plane and is defined by its
polar coordinates (r, a) with
respect to a given point,
P1 (x, y).

$$\text{POINT}/\left\{\begin{matrix} x,y \\ \text{P1} \end{matrix}\right\},\text{RADIUS},r,a$$

Examples
P2=POINT/P1,RADIUS,2.0,30
or
P2=POINT/1.0,1.0,RADIUS,2.0,30

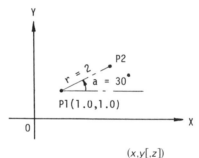

4 The point is defined by
nested Cartesian coordinates
(x, y, z).

$(x,y[,z])$

Comment   This format is used in a geometric definition statement to replace a point symbol that has not been defined in the previous statements. The *z* coordinate is zero if omitted.

Example   In statement format No. 3 we can use (1.0,1.0,0) to replace the symbol P1:

P2=POINT/(1.0,1.0,0),RADIUS,2.0,30

**TABLE 4-2   STATEMENTS DEFINING A POINT BY ITS POSITION ON, OR RELATIVE TO,
A GEOMETRIC ENTITY**

No.	Known Condition	Statement Format
1	The point is on a given circle, C1, in the X-Y plane, and the radius passing through it has an angle, a, with the X axis.	POINT/C1,ATANGL,a

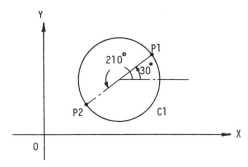

Comment   ATANGL is a vocabulary word meaning "at an angle of."

Examples
P1=POINT/C1,ATANGL,30
P2=POINT/C1,ATANGL,210

No.	Known Condition	Statement Format
2	The point is on a given circle, C1, in the X-Y plane and at a circular arc distance, s, from the reference position on the circle, determined by a given point, P1.	POINT/P1,DELTA,$\begin{Bmatrix} \text{CLW} \\ \text{CCLW} \end{Bmatrix}$,ON,C1,ARC,$s$

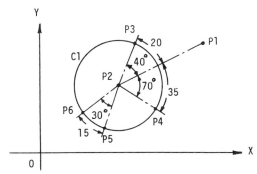

Comment   Vocabulary words DELTA and ARC indicate that the length, s, is an incremental arc distance measured from the reference point. The given point P1 can be located anywhere on the X-Y plane. The reference point is the intersection of the line connecting P1 and the center of the given circle C1. If point P1 is the center of the given circle C1, then the reference point is the intersection of the given circle and the horizontal line passing through the center and extending in the $+X$ direction.

(continued)

TABLE 4-2 **STATEMENTS DEFINING A POINT BY ITS POSITION ON, OR RELATIVE TO,**
              A GEOMETRIC ENTITY (continued)

No.	Known Condition	Statement Format

Example

P3=POINT/P1,DELTA,CCLW,ON,C1,ARC,20
P4=POINT/P1,DELTA,CLW,ON,C1,ARC,35
P6=POINT/P5,DELTA, CLW,ON,C1,ARC,15

3 The point is on a given   POINT/P1,DELTA, $\begin{Bmatrix} \text{CLW} \\ \text{CCLW} \end{Bmatrix}$,ON,C1,ATANGL,a
circle, C1, and at an angular
distance, a, from a given
point, P1.

Comment This statement is similar to the previous one except that an angle is used instead of an arc
length. The discussion concerning the previous statement format applies.

Example (see figure in statement format No. 2)

P3=POINT/P1,DELTA,CCLW,ON.C1,ATANGL,40
P6=POINT/P5,DELTA,CLW,ON,C1,ATANGL,30

4 The point is the center of   POINT/CENTER,C1
a given circle, sphere,
torus, cone, or cylinder C1.

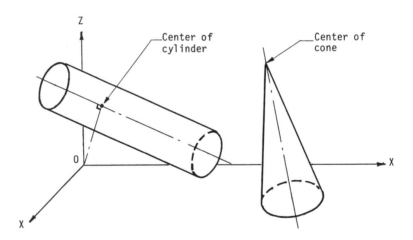

Comment The center of a cone is its vertex. In APT, the center of a cylinder is the point that has
been defined as the center of the cylinder in a previous CYLNDR statement. If the center has not been
defined, the intersection of the cylinder axis and its normal passing through the origin of the coordinate
system is the center of the given cylinder.

5 The point is defined by   POINT/P1,DELTA,n,TIMES,[UNIT,]V1
vectorial displacement, $n \cdot V1$,
from a given point, P1.

TABLE 4-2 **STATEMENTS DEFINING A POINT BY ITS POSITION ON, OR RELATIVE TO, A GEOMETRIC ENTITY   (continued)**

No.	Known Condition	Statement Format

Comment   If the optional word UNIT is specified, the vector V1 is considered as a unit vector, that is, $|V1| = 1$.

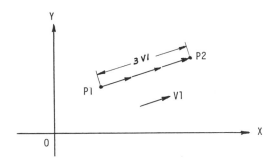

Example
P2=POINT/P1,DELTA,3,TIMES,V1

6 The point, in the X-Y plane, is on a line, L1, and at a distance, d, from a given point, P1.

POINT/$\begin{Bmatrix} \text{XLARGE} \\ \text{XSMALL} \\ \text{YLARGE} \\ \text{YSMALL} \end{Bmatrix}$,ON,L1,DELTA,$d$,P1

Comment   The given point P1 can be located anywhere in the X-Y plane. If it is not on the line L1, then a projection line normal to the given line L1 is drawn to determine the reference point from which the distance, d, is measured.

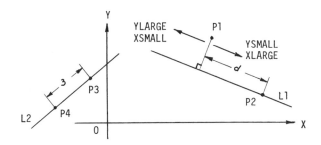

Example
P2=POINT/YSMALL,ON,L1,DELTA,d,P1
P4=POINT/XSMALL,ON,L2,DELTA,3,P3

7 The point is the nth point of a pattern, PT1.*

POINT/PT1,n

*See Chapter 5 for the pattern definition.

(continued)

TABLE 4-2 STATEMENTS DEFINING A POINT BY ITS POSITION ON, OR RELATIVE TO,
A GEOMETRIC ENTITY (continued)

No.	Known Condition	Statement Format

Example
P1=POINT/PT1,4 $$PT1 IS A PATTERN CONSISTING OF SIX POINTS.

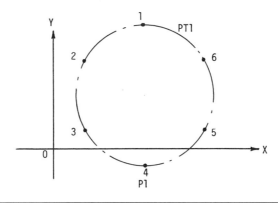

TABLE 4-3 STATEMENTS DEFINING A POINT AS THE INTERSECTION OF TWO
OR MORE GEOMETRIC ENTITIES

No.	Known Condition	Statement Format
1	The point is the intersection of two given lines, L1 and L2.	POINT/INTOF,L1,L2

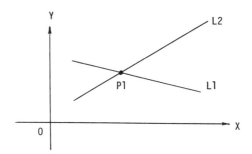

Example
P1=POINT/INTOF,L1,L2

2	The point is defined as the intersection of a plane, PL1, and a vector, V1, passing through a given point, P1.	POINT/INTOF,PL1,V1,P1

**TABLE 4-3   STATEMENTS DEFINING A POINT AS THE INTERSECTION OF TWO
OR MORE GEOMETRIC ENTITIES   (continued)**

No.	Known Condition	Statement Format

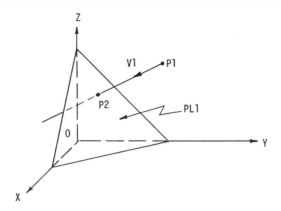

Example
P2=POINT/INTOF,PL1,V1,P1

3 The point, in the *X-Y* plane,
is the intersection of a
given line, L1, and a given
circle, C1.

POINT/⎡XLARGE⎤,INTOF,L1,C1
⎢XSMALL⎢
⎢YLARGE⎢
⎣YSMALL⎦

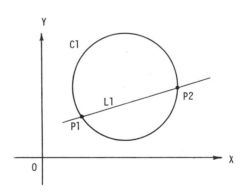

Comment   Generally, there are two solutions. A directional modifier is selected to indicate the
desired solution.

Example
P1=POINT/XSMALL,INTOF,L1,C1
P2=POINT/YLARGE,INTOF,L1,C1

TABLE 4-3 STATEMENTS DEFINING A POINT AS THE INTERSECTION OF TWO
                OR MORE GEOMETRIC ENTITIES (continued)

No.	Known Condition	Statement Format
4	The point is the intersection of two given circles, C1 and C2.	POINT/ $\begin{Bmatrix} \text{XLARGE} \\ \text{XSMALL} \\ \text{YLARGE} \\ \text{YSMALL} \end{Bmatrix}$,INTOF,C1,C2

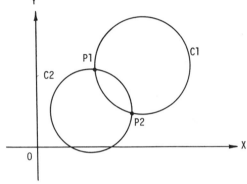

Example
P1=POINT/XSMALL,INTOF,C1,C2
P2=POINT/YSMALL,INTOF,C1,C2

5 The point is the nth
intersection of a given
point vector, or space line,
V1, with an APT surface, S1.

POINT/n,INTOF,V1,S1

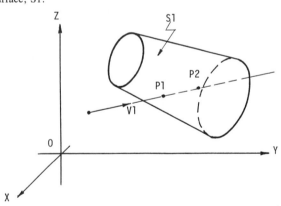

Comment   A point vector, or space line, is a vector passing through a fixed point in space (see Section 4.4). S1 can be one of the following surfaces: circle, cone, cylinder, ellipse, hyperbola, GCONIC, LCONIC (see Section 4.5), plane, quadric surface, or sphere.

Example
P2=POINT/2,INTOF,V1,S1

**TABLE 4-3   STATEMENTS DEFINING A POINT AS THE INTERSECTION OF TWO
OR MORE GEOMETRIC ENTITIES (continued)**

No.	Known Condition	Statement Format
6	The point is on a curve (surface) S1 in the X-Y plane where the tangent is at a specified angle, a, with the X axis.	POINT/ $\begin{Bmatrix} \text{XLARGE} \\ \text{XSMALL} \\ \text{YLARGE} \\ \text{YSMALL} \end{Bmatrix}$,ON,S1,ATANGL,a

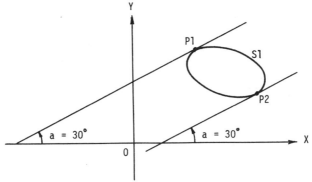

Comment   The surface S1 can be one of the following surfaces: circle, ellipse, hyperbola, GCONIC, and LCONIC.

Examples
P1=POINT/YLARGE,ON,S1,ATANGL,30
P2=POINT/XLARGE,ON,S1,ATANGL,30

4.2 STATEMENTS DEFINING A LINE

A *line* can be defined by the statement

LINE/......

There are 27 formats for the LINE statement in APT. They can be divided into three groups:

- A line defined by two given points or by its relationship with a given point, a given line, or both
- A line defined by using a circle or circles as the references
- A line defined by a point on it and its relationship with an APT geometric entity, such as ellipse, conic, and TABCYL (tabulated cylinder)*

The three groups of statements are listed in Tables 4-4, 4-5, and 4-6, respectively.
    According to the APT programming manuals,[1,2] "a line [defined by a LINE statement] is the intersection of an X-Y plane with a plane perpendicular to it." This

---

*See Sections 4.5 and 4.7 for the definitions of a conic and a tabulated cylinder.

TABLE 4-4 THE STATEMENTS DEFINING A LINE BY TWO GIVEN POINTS, OR BY A GIVEN POINT,
A GIVEN LINE OR LINES, OR BOTH

No.	Known Condition	Statement Format
1	The line passes through two given points, P1 $(x1, y1, z1)$ and P2 $(x2, y2, z2)$.	LINE/ $\begin{cases} x1,y1,z1,x2,y2,z2 \\ x1,y1,x2,y2 \\ P1,P2 \end{cases}$

Comment Within the NC processor, the lines defined by the first and second options of the formats listed above are the same. When the third option is selected, symbols P1 and P2 should have been defined previously.

| 2 | The line passes through two points, in the *X-Y* plane, which are defined by their polar coordinates $(r1, a1)$ and $(r2, a2)$, respectively. | LINE/ $\begin{cases} \text{RTHETA},r1,a1,r2,a2 \\ \text{THETAR},a1,r1,a2,r2 \end{cases}$ |

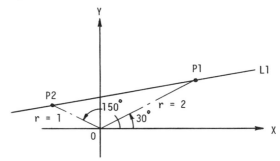

Examples
L1=LINE/RTHETA,2,30,1,150
L1=LINE/THETAR,30,2,150,1

| 3 | The line is parallel to and at a distance, *d*, from the *X* or *Y* axis. | LINE/ $\begin{cases} \text{XAXIS} \\ \text{YAXIS} \end{cases}$ [,*d*] |

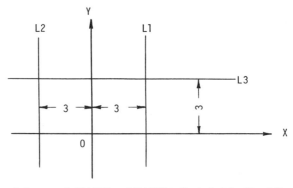

Comment The vocabulary words XAXIS and YAXIS indicate that the *X* and *Y* axes are used as reference lines, respectively. The sign of the distance, *d*, indicates the side on which the defined line is located with respect to the reference axis. The distance is zero if it is omitted in the statement.

TABLE 4-4 THE STATEMENTS DEFINING A LINE BY TWO GIVEN POINTS, OR BY A GIVEN POINT, A GIVEN LINE OR LINES, OR BOTH   (continued)

No.	Known Condition	Statement Format

Examples
L1=LINE/YAXIS,3.0
L2=LINE/YAXIS,−3.0
L3=LINE/XAXIS,3.0

4 The line is parallel to and at a distance, *d*, from a given line, L1.

$$LINE/PARLEL,L1,\begin{Bmatrix} XLARGE \\ XSMALL \\ YLARGE \\ YSMALL \end{Bmatrix},d$$

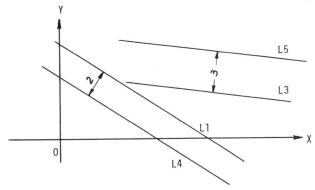

Comment   The vocabulary word PARLEL means "parallel to."

Examples
L4=LINE/PARLEL,L1,XSMALL,2
L5=LINE/PARLEL,L3,YLARGE,3

5 The line has a given slope, *s*,* with respect to the +*X* axis and an intercept value, *d*, on the *X* or *Y* axis.

$$LINE/SLOPE,s,INTERC,\begin{Bmatrix} XAXIS, \\ [YAXIS,] \end{Bmatrix}d$$

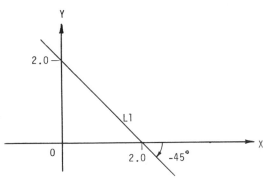

Comment   The vocabulary word INTERC means "intercept."

*In references 1 and 2 an angular value is also used for *s* in this statement. This usage is incorrect.

(continued)

TABLE 4-4 **THE STATEMENTS DEFINING A LINE BY TWO GIVEN POINTS, OR BY A GIVEN POINT,
A GIVEN LINE OR LINES, OR BOTH** (continued)

No.	Known Condition	Statement Format

Examples

L1=LINE/SLOPE,−1,INTERC,2

L1=LINE/SLOPE,−1,INTERC,XAXIS,2

6 The line has an angle, a,
with the $+X$ axis and an
intercept value, d, on the
X or Y axis.

Examples (see the figure in statement format No. 5.)

L1=LINE/ATANGL,−45,INTERC,2

L1=LINE/ATANGL,135,INTERC,2

L1=LINE/ATANGL,−45,INTERC,XAXIS,2

L1=LINE/ATANGL,135,INTERC,XAXIS,2

7 The line passes through a
given point P1 (x, y) and is
at an angle, a, with the
$+X$ or $+Y$ axis.

$$\text{LINE/}\begin{Bmatrix} x,y \\ \text{P1} \end{Bmatrix}\text{,ATANGL,}a\begin{Bmatrix} [,\text{XAXIS}] \\ ,\text{YAXIS} \end{Bmatrix}$$

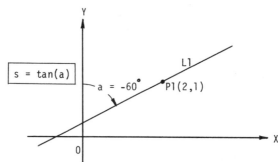

Examples

L1=LINE/2,1,ATANGL,−60,YAXIS

L1=LINE/2,1,ATANGL,30

L1=LINE/P1,ATANGL,120,YAXIS

8 The line passes through a
point, P1, and has a slope, s,
with respect to the $+X$ or
$+Y$ axis.

$$\text{LINE/P1,SLOPE,}s\begin{Bmatrix} [,\text{XAXIS}] \\ ,\text{YAXIS} \end{Bmatrix}$$

Example (see the figure in statement 7)

L1=LINE/P1,SLOPE,−1.7321,YAXIS

or

L1=LINE/P1,SLOPE,0.5774

9 The line passes through a
given point, P1 (x, y), and
is at an angle, a, with
a given line, L1.

$$\text{LINE/}\begin{Bmatrix} x,y \\ \text{P1} \end{Bmatrix}\text{,ATANGL,}a\text{,L1}$$

TABLE 4-4 THE STATEMENTS DEFINING A LINE BY TWO GIVEN POINTS, OR BY A GIVEN POINT,
             A GIVEN LINE OR LINES, OR BOTH   (continued)

No.	Known Condition	Statement Format

Comment   The angle is measured from the given line, with the positive value in the counterclockwise direction.

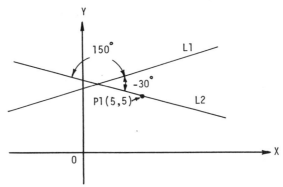

Example
L2=LINE/P1,ATANGL,−30,L1
or
L2=LINE/5,5,ATANGL,150,L1

10 The line passes through a
given point, P1, and has a
given slope, s, with respect
to a given line, L1.

LINE/P1,SLOPE,s,L1

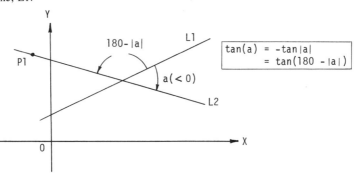

$$\tan(a) = -\tan|a|$$
$$= \tan(180 - |a|)$$

Example
If a = −60 degrees, we have
L2=LINE/P1,SLOPE,−1.7321,L1

11 The line passes through a
given point, P1 (x, y), and is
parallel to a given line, L1.

LINE/$\begin{Bmatrix} x,y \\ P1 \end{Bmatrix}$,PARLEL,L1

12 The line passes through a
given point, P1 (x, y), and is
perpendicular (PERPTO) to
a given line, L1.

LINE/$\begin{Bmatrix} x,y \\ P1 \end{Bmatrix}$,PERPTO,L1

(continued)

TABLE 4-4 THE STATEMENTS DEFINING A LINE BY TWO GIVEN POINTS, OR BY A GIVEN POINT, A GIVEN LINE OR LINES, OR BOTH (continued)

No.	Known Condition	Statement Format
13	The line is perpendicular to a given line, L1, and at a distance, d, from a given point, P1.	LINE/ $\left\{\begin{array}{l}\text{XLARGE}\\\text{XSMALL}\\\text{YLARGE}\\\text{YSMALL}\end{array}\right\}$,PERPTO,L1,DELTA,$d$,P1

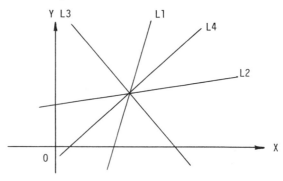

Comment   The directional modifiers are used to indicate on which side of the given point P1 the desired line is located.

Examples
L2=LINE/XSMALL,PERPTO,L1,DELTA,3,P1
L2=LINE/YLARGE,PERPTO,L1,DELTA,3,P1
L3=LINE/XLARGE,PERPTO,L1,DELTA,5,P1
L3=LINE/YSMALL,PERPTO,L1,DELTA,5,P1

No.	Known Condition	Statement Format
14	The line is the bisector of the angle subtended by two given lines, L1 and L2.	LINE/ $\left\{\begin{array}{l}\text{XLARGE}\\\text{XSMALL}\\\text{YLARGE}\\\text{YSMALL}\end{array}\right\}$,BISECT,L1,L2

Comment   Generally, there are two solutions when the given lines L1 and L2 are not parallel to each other. A directional modifier can be selected to indicate that the desired solution is the one with a larger or smaller intercept on the coordinate axis X (for options XLARGE and XSMALL) or Y (for options YLARGE and YSMALL).

TABLE 4-4 THE STATEMENTS DEFINING A LINE BY TWO GIVEN POINTS, OR BY A GIVEN POINT, A GIVEN LINE OR LINES, OR BOTH (continued)

No.	Known Condition	Statement Format

Examples

L3=LINE/XLARGE,BISECT,L1,L2

L3=LINE/YLARGE,BISECT,L1,L2

L4=LINE/XSMALL,BISECT,L1,L2

L4=LINE/YSMALL,BISECT,L1,L2

TABLE 4-5 THE STATEMENTS USING A CIRCLE OR CIRCLES AS REFERENCES TO DEFINE A LINE

No.	Known Condition	Statement Format
1	The line, passing through a given point, P1 (x, y), is tangent to a given circle, C1.	LINE/$\begin{Bmatrix} x,y \\ P1 \end{Bmatrix}$, $\begin{Bmatrix} \text{RIGHT} \\ \text{LEFT} \end{Bmatrix}$,TANTO,C1

Comment   The directional modifier LEFT or RIGHT indicates that the line is on the left or right of the circle, C1, when one looks from point P1 to the circle.

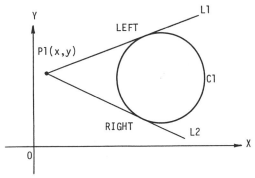

Examples

L1=LINE/1.2,2.5,LEFT,TANTO,C1 $$ X=1.2, Y=2.5

L2=LINE/P1,RIGHT,TANTO,C1

2	The line passes through a given point, P1, and the center of a given circle, C1. Thus the line is the normal to given circle C1.	LINE/P1,PERPTO,C1

(continued)

TABLE 4-5 THE STATEMENTS USING A CIRCLE OR CIRCLES AS REFERENCES TO DEFINE A LINE
(continued)

No.	Known Condition	Statement Format

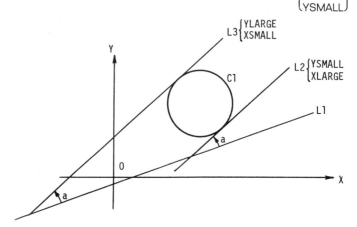

3 The line is tangent to a
given circle, C1, and makes an
angle, *a*, with a given line, L1.

LINE/ATANGL,*a*,L1,TANTO,C1, $\begin{Bmatrix} \text{XLARGE} \\ \text{XSMALL} \\ \text{YLARGE} \\ \text{YSMALL} \end{Bmatrix}$

Comment Angle *a* is measured from L1 to the line to be defined. The directional modifier can be determined by comparing the coordinates of the two tangent points made by the two lines with the circle.

Examples
L2=LINE/ATANGL,*a*,L1,TANTO,C1,XLARGE
L3=LINE/ATANGL,*a*,L1,TANTO,C1,XSMALL

4 The line is tangent to a
given circle, C1, and has a
slope, *s*, with respect to
the *X* or *Y* axis.

LINE/ $\begin{Bmatrix} \text{XLARGE} \\ \text{XSMALL} \\ \text{YLARGE} \\ \text{YSMALL} \end{Bmatrix}$,TANTO,C1,SLOPE,$

s $\begin{Bmatrix} \text{[,XAXIS]} \\ \text{,YAXIS} \end{Bmatrix}$

TABLE 4-5 THE STATEMENTS USING A CIRCLE OR CIRCLES AS REFERENCES TO DEFINE A LINE
              (continued)

No.	Known Condition	Statement Format

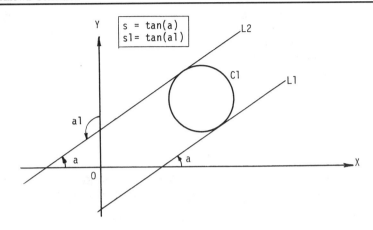

$$s = \tan(a)$$
$$s1 = \tan(a1)$$

Examples
L1=LINE/XLARGE,TANTO,C1,SLOPE,s
L2=LINE/YLARGE,TANTO,C1,SLOPE,s1,YAXIS

5 The line is tangent to a
given circle, C1, and makes
an angle, *a*, with the *X*
axis.

LINE/ $\begin{Bmatrix} \text{XLARGE} \\ \text{XSMALL} \\ \text{YLARGE} \\ \text{YSMALL} \end{Bmatrix}$,TANTO,C1,ATANGL,*a*

Examples   (see the figure in statement format No. 4)
L1=LINE/YSMALL,TANTO,C1,ATANGL,*a*
L2=LINE/XSMALL,TANTO,C1,ATANGL,*a*

6 The line is a common
tangent to two circles,
C1 and C2.

LINE/ $\begin{Bmatrix} \text{LEFT} \\ \text{RIGHT} \end{Bmatrix}$,TANTO,C1, $\begin{Bmatrix} \text{LEFT} \\ \text{RIGHT} \end{Bmatrix}$,TANTO,C2

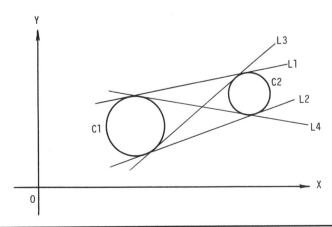

(continued)

TABLE 4-5   THE STATEMENTS USING A CIRCLE OR CIRCLES AS REFERENCES TO DEFINE A LINE (continued)

No.	Known Condition	Statement Format

Comment   The reference point for selecting a directional modifier, LEFT or RIGHT, is the center of the first circle C1 specified in the statement. Thus, for example, when one looks from the reference point toward circle C2, line L1 will be on the left-hand side of circles C1 and C2; line L4 will be on the left-hand side of circle C1 and on the right-hand side of circle C2.

Examples

L2=LINE/RIGHT,TANTO,C1,RIGHT,TANTO,C2
L3=LINE/RIGHT,TANTO,C1,LEFT,TANTO,C2

TABLE 4-6   SELECTED STATEMENTS DEFINING A LINE PASSING THROUGH A GIVEN POINT AND HAVING A SPECIFIED RELATIONSHIP WITH A SPECIAL APT SURFACE

No.	Known Condition	Statement Format
1	The line passes through a given point, P1, and is tangent to a given curve, C1, in the X-Y plane. The plane curve C1 can be an ellipse, hyperbola, LCONIC, or GCONIC.	LINE/P1, $\begin{Bmatrix} \text{LEFT} \\ \text{RIGHT} \end{Bmatrix}$,TANTO,C1

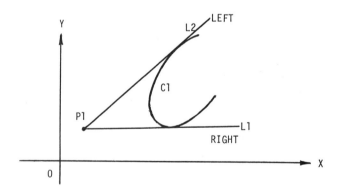

Comment   Point P1 should not be on the given curve C1.

Examples

L1=LINE/P1,RIGHT,TANTO,C1
L2=LINE/P1,LEFT,TANTO,C1

No.	Known Condition	Statement Format
2	The line is tangent to a given curve, C1, at point P1. Curve C1 can be an ellipse, hyperbola, LCONIC, or GCONIC.	LINE/P1,TANTO,C1

TABLE 4-6 SELECTED STATEMENTS DEFINING A LINE PASSING THROUGH A GIVEN POINT AND
HAVING A SPECIFIED RELATIONSHIP WITH A SPECIAL APT SURFACE   (continued)

No.	Known Condition	Statement Format

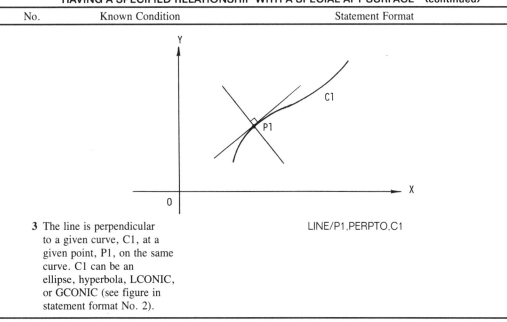

3	The line is perpendicular to a given curve, C1, at a given point, P1, on the same curve. C1 can be an ellipse, hyperbola, LCONIC, or GCONIC (see figure in statement format No. 2).	LINE/P1,PERPTO,C1

seems to suggest that no matter how a line is defined, it is always in the X-Y plane. This is true when one tries to locate a point on the defined line by using the NC processor; the located point is always on the X-Y plane. For example, the intersection point of two lines found by the NC processor is always on the X-Y plane, even though these two lines are defined (by statement format No. 1 in Table 4-4, with the use of two sets of point coordinates) out of the X-Y plane or do not intersect. However, in general, a line is considered by the NC processor as a plane containing the line and being perpendicular to the X-Y plane. This is useful for defining cutter contouring motion.

4.3 STATEMENTS DEFINING A CIRCLE

A *circle* is the locus of a point that moves in a plane and maintains a constant distance from a fixed point. In APT, a circle is defined by the statement

CIRCLE/......

with vocabulary word(s) and parameter(s) after the slash. Within the NC processor, a circle is defined as a cylinder of infinite length, whose axis is perpendicular to the X-Y plane. The z coordinate of the circle center is assumed to be zero unless it is specified in the statement defining the circle.

The CIRCLE statement in the NC processor has 27 formats, and they can be divided into four groups:

- Those using a point or points as the reference or references to define a circle
- Those using points and lines as references to define a circle
- Those using lines as references to define a circle
- Those using a circle or circles as the reference or references to define a circle

Tables 4-7 to 4-10 give the statement formats and the examples for each group of statements.

TABLE 4-7   THE STATEMENTS USING A POINT OR POINTS AS REFERENCES TO DEFINE A CIRCLE

No.	Known Condition	Statement Format
1	The circle has its center at P1 (x, y, z), with a given radius, r.	CIRCLE/CENTER, $\begin{Bmatrix} x,y,z \\ x,y \\ P1 \end{Bmatrix}$,RADIUS,r

Comment   If the z coordinate of the center is not specified, it is taken as zero.

Example
C1=CIRCLE/CENTER,P1,RADIUS,1.5
or
C1=CIRCLE/CENTER,1,2,RADIUS,1.5

2	The circle has its center at point P1 and passes through a given point, P2.	CIRCLE/CENTER,P1,P2

Comment   The center point, P1, should be specified immediately after the vocabulary word CENTER.

Example   (see the figure in statement format No. 1)
C1=CIRCLE/CENTER,P1,P2

TABLE 4-7 THE STATEMENTS USING A POINT OR POINTS AS REFERENCES TO DEFINE A CIRCLE
              (continued)

No.	Known Condition	Statement Format
3	The circle passes through two given points, P1 and P2, with a given radius, r.	CIRCLE/ $\begin{Bmatrix} \text{XLARGE} \\ \text{XSMALL} \\ \text{YLARGE} \\ \text{YSMALL} \end{Bmatrix}$,P1,P2,RADIUS,r

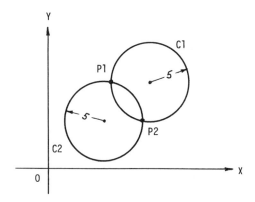

Comment A directional modifier can be selected by comparing the coordinates of the centers of the two possible solutions.

Example
C1=CIRCLE/XLARGE,P1,P2,RADIUS,5.0
or
C1=CIRCLE/YLARGE,P1,P2,RADIUS,5.0

No.	Known Condition	Statement Format
4	The circle passes through three given points, P1, P2, and P3.	CIRCLE/P1,P2,P3

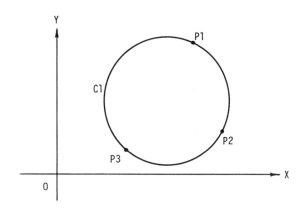

TABLE 4-8 THE STATEMENTS USING POINTS AND LINES AS REFERENCES TO DEFINE A CIRCLE

No.	Known Condition	Statement Format
1	The circle is tangent to a given line, L1, with its center at a given point, P1 (x,y).	CIRCLE/CENTER, $\begin{cases} x,y \\ P1 \end{cases}$,TANTO,L1

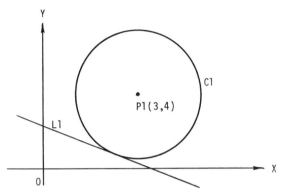

Example

C1=CIRCLE/CENTER,3,4,TANTO,L1

or

C1=CIRCLE/CENTER,P1,TANTO,L1

No.	Known Condition	Statement Format
2	The center of the circle is on a given line, L1, and at a distance, d, measured in the direction of line L1, from a given point, P1. The radius of the circle is r.	CIRCLE/ $\begin{cases} \text{XLARGE} \\ \text{XSMALL} \\ \text{YLARGE} \\ \text{YSMALL} \end{cases}$,ON,L1,DELTA,d,P1,RADIUS,r

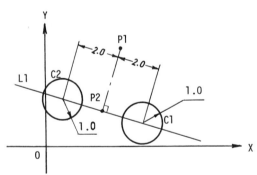

Examples

C1=CIRCLE/XLARGE,ON,L1,DELTA,2.0,P1,RADIUS,1.0

C2=CIRCLE/YLARGE,ON,L1,DELTA,2.0,P2,RADIUS,1.0

No.	Known Condition	Statement Format
3	The circle is tangent to a given line, L1, and passes through a given point, P1, with a given radius, r.	CIRCLE/TANTO,L1, $\begin{cases} \text{XLARGE} \\ \text{XSMALL} \\ \text{YLARGE} \\ \text{YSMALL} \end{cases}$,P1,RADIUS,r

TABLE 4-8 THE STATEMENTS USING POINTS AND LINES AS REFERENCES TO DEFINE A CIRCLE
(continued)

No.	Known Condition	Statement Format

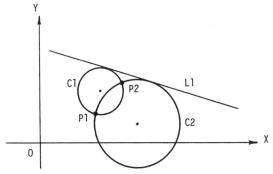

CIRCLE/ {XLARGE / XSMALL / YLARGE / YSMALL} ,P1,TANTO,L1,RADIUS,r

Comment The directional modifier indicates the center position of the defined circle with respect to that of the other possible solution.

Examples
C1=CIRCLE/TANTO,L1,XLARGE,P1,RADIUS,3.0
C3=CIRCLE/YLARGE,P2,TANTO,L1,RADIUS,2.5

4 The circle passes through
two given points, P1 and
P2, and is tangent to a
given line, L1.

CIRCLE/ {XLARGE / XSMALL / YLARGE / YSMALL} ,TANTO,L1,THRU,P1,P2

Comment Points P1 and P2 cannot both be on given line L1. In general, there are two possible solutions.

Examples
C1=CIRCLE/YLARGE,TANTO,L1,THRU,P1,P2
C2=CIRCLE/XLARGE,TANTO,L1,THRU,P1,P2

(continued)

No.	Known Condition	Statement Format

5 The circle passes through a given point, P1, and is tangent to given line L1, with its center on another given line, L2.

CIRCLE/ $\begin{Bmatrix} \text{XLARGE} \\ \text{XSMALL} \\ \text{YLARGE} \\ \text{YSMALL} \end{Bmatrix}$,TANTO,L1,THRU,P1,ON,L2

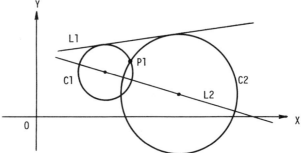

Comment Generally, there are two possible solutions for the given condition.

Examples
C1=CIRCLE/YLARGE,TANTO,L1,THRU,P1,ON,L2
C2=CIRCLE/XLARGE, TANTO,L1,THRU,P1,ON,L2

6 The circle passes through a given point, P1, and is tangent to two given lines, L1 and L2.

CIRCLE/ $\begin{Bmatrix} \text{LARGE} \\ \text{SMALL} \end{Bmatrix}$, $\begin{Bmatrix} \text{XLARGE} \\ \text{XSMALL} \\ \text{YLARGE} \\ \text{YSMALL} \end{Bmatrix}$,TANTO,$

L1, $\begin{Bmatrix} \text{XLARGE} \\ \text{XSMALL} \\ \text{YLARGE} \\ \text{YSMALL} \end{Bmatrix}$,TANTO,L2,THRU,P1

Comment The two given lines divide the X-Y plane into four sections. The two directional modifiers indicate the part of the X-Y plane where the defined circle is located. The dimensional modifiers can be selected to determine the desired solution from the two possible ones. If P1 is on one of the given lines, L1 and L2, then there will be only one solution. In such a case, either one of the dimensional modifiers, LARGE or SMALL, yields the same result.

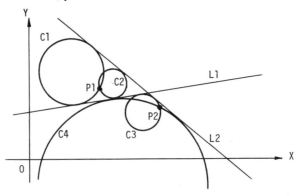

Examples
C1=CIRCLE/LARGE,YLARGE,TANTO,L1,YSMALL,TANTO,L2,THRU,P1
C3=CIRCLE/SMALL,YSMALL,TANTO,L1,XSMALL,TANTO,L2,THRU,P2

TABLE 4-9   THE STATEMENTS USING LINES AS REFERENCES TO DEFINE A CIRCLE

No.	Known Condition	Statement Format
1	The circle is tangent to three given lines, L1, L2, and L3.	CIRCLE/ $\begin{Bmatrix} \text{XLARGE} \\ \text{XSMALL} \\ \text{YLARGE} \\ \text{YSMALL} \end{Bmatrix}$,L1, $\begin{Bmatrix} \text{XLARGE} \\ \text{XSMALL} \\ \text{YLARGE} \\ \text{YSMALL} \end{Bmatrix}$,\$ L2, $\begin{Bmatrix} \text{XLARGE} \\ \text{XSMALL} \\ \text{YLARGE} \\ \text{YSMALL} \end{Bmatrix}$,L3

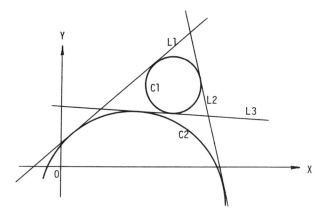

Examples

C1=CIRCLE/YSMALL,L1,XSMALL,L2,YLARGE,L3
C2=CIRCLE/YSMALL,L1,XSMALL,L2,YSMALL,L3

2	The circle is tangent to two given lines, L1 and L2, with its center on a third given line, L3.	CIRCLE/ $\begin{Bmatrix} \text{XLARGE} \\ \text{XSMALL} \\ \text{YLARGE} \\ \text{YSMALL} \end{Bmatrix}$,TANTO,L1,TANTO,\$ L2,ON,L3

Comment   In general, there are two solutions for the given condition (Fig. [a]). A directional modifier can be selected by comparing the center coordinates of the two possible solutions. If the third line, L3, is the bisector of the angle formed by lines L1 and L2, or if the three given lines are parallel to each other and line L3 is midway between the other two, then there will be an infinite number of solutions. In this case, other statement formats should be used to define the circle. If the two tangent lines are parallel and the third line is not parallel to the other two (Fig. [b]), then only one solution exists. In such a case, all the modifiers listed above yield the same result.

(continued)

TABLE 4-9 **THE STATEMENTS USING LINES AS REFERENCES TO DEFINE A CIRCLE**   **(continued)**

No.	Known Condition	Statement Format

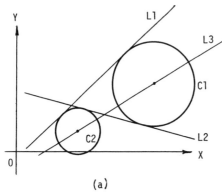
(a)

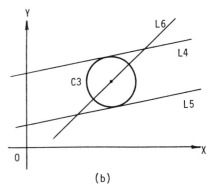
(b)

Examples

C1=CIRCLE/YLARGE,TANTO,L1,TANTO,L2,ON,L3
C1=CIRCLE/XLARGE,TANTO,L1,TANTO,L2,ON,L3
C2=CIRCLE/YSMALL,TANTO,L1,TANTO,L2,ON,L3
C2=CIRCLE/XSMALL,TANTO,L1,TANTO,L2,ON,L3
C3=CIRCLE/XLARGE,TANTO,L4,TANTO,L5,ON,L6
C3=CIRCLE/XSMALL,TANTO,L4,TANTO,L5,ON,L6

3 The circle is tangent to two intersecting lines, L1 and L2, with a given radius, *r*.

CIRCLE/ $\left\{\begin{matrix} \text{XLARGE} \\ \text{XSMALL} \\ \text{YLARGE} \\ \text{YSMALL} \end{matrix}\right\}$,L1, $\left\{\begin{matrix} \text{XLARGE} \\ \text{XSMALL} \\ \text{YLARGE} \\ \text{YSMALL} \end{matrix}\right\}$,$

L2,RADIUS,*r*

Comment In general, there are four possible solutions for the given condition. Therefore two directional modifiers are needed in this statement.

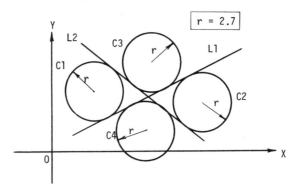

Examples

C1=CIRCLE/YLARGE,L1,YSMALL,L2,RADIUS,2.7
C4=CIRCLE/XLARGE,L1,XSMALL,L2,RADIUS,2.7

TABLE 4-10   STATEMENTS USING A CIRCLE OR CIRCLES AS THE REFERENCE OR REFERENCES TO DEFINE A CIRCLE

No.	Known Condition	Statement Format
1	The circle is tangent to a given circle, C1, with its center at point P1.	CIRCLE/CENTER,P1,$\begin{Bmatrix} \text{LARGE} \\ \text{SMALL} \end{Bmatrix}$,TANTO,C1

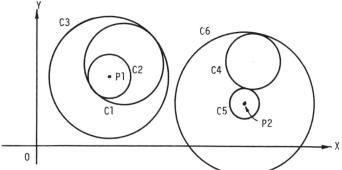

Comment   In general, there are two possible solutions. A dimensional modifier must be specified to indicate the desired solution.

Examples
C2=CIRCLE/CENTER,P1,SMALL,TANTO,C1
C6=CIRCLE/CENTER,P2,LARGE,TANTO,C4

No.	Known Condition	Statement Format
2	The circle passes through a given point, P1, and is tangent to a given circle, C1, with a given radius, r.	CIRCLE/$\begin{Bmatrix} \text{XLARGE} \\ \text{XSMALL} \\ \text{YLARGE} \\ \text{YSMALL} \end{Bmatrix}$,$\begin{Bmatrix} \text{RIGHT} \\ \text{LEFT} \end{Bmatrix}$,\$
		TANTO,C1,THRU,P1,RADIUS,r

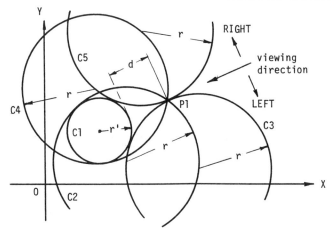

(continued)

TABLE 4-10 STATEMENTS USING A CIRCLE OR CIRCLES AS THE REFERENCE OR REFERENCES TO DEFINE A CIRCLE (continued)

No.	Known Condition	Statement Format

Comment Generally there are four possible solutions when point P1 is outside the given circle C1. Modifiers RIGHT and LEFT mean that in looking from P1 toward the center of circle C1, one finds that the center of the defined circle is on the right- or left-hand side, respectively. Directional modifiers XLARGE, XSMALL, YLARGE, and YSMALL are used to indicate the desired solution from the two possible ones selected by modifiers RIGHT and LEFT. The number of solutions under various conditions are as follows:

r	No. of solutions
$<d/2$	None
$=d/2$	1
$d/2 < r < (r' + d/2)$	2
$= r' + (d/2)$	3
$>r' + (d/2)$	4

An error message will be issued by the NC processor if r is less than $d/2$. See Section 4.11 for further discussion.

Examples
C2=CIRCLE/XSMALL,LEFT,TANTO,C1,THRU,P1,RADIUS,r
C5=CIRCLE/XLARGE,RIGHT,TANTO,C1,THRU,P1,RADIUS,r

3 The circle is tangent to a given circle, C1, and a given line, L1, with radius r.

CIRCLE/ ⎡XLARGE⎤,L1, ⎡XLARGE⎤,$
⎢XSMALL⎥ ⎢XSMALL⎥
⎢YLARGE⎥ ⎢YLARGE⎥
⎣YSMALL⎦ ⎣YSMALL⎦
⎡IN ⎤,C1,RADIUS,r
⎣OUT⎦

Comment See discussion in Section 4.11.

Examples
C2=CIRCLE/YSMALL,L1,XSMALL,IN,C1,RADIUS,1.0
C8=CIRCLE/YLARGE,L1,XLARGE,OUT,C1,RADIUS,1.0
C9=CIRCLE/YLARGE,L1,XSMALL,OUT,C1,RADIUS,1.0

**TABLE 4-10   STATEMENTS USING A CIRCLE OR CIRCLES AS THE REFERENCE OR REFERENCES TO
DEFINE A CIRCLE   (continued)**

No.	Known Condition	Statement Format
4	The circle passes through a given point, P1, and is tangent to a given circle, C1, at a given point, P2.	CIRCLE/P1,TANTO,C1,P2

Example
C2=CIRCLE/P1,TANTO,C1,P2

5 The circle passes through a given point, P1, and is tangent to a given line, L1, and a given circle, C1. P1 and C1 are located on the same side of line L1.

$$\text{CIRCLE/}\begin{Bmatrix}\text{LARGE}\\\text{SMALL}\end{Bmatrix},\begin{Bmatrix}\text{XLARGE}\\\text{XSMALL}\\\text{YLARGE}\\\text{YSMALL}\end{Bmatrix},\text{TANTO,\$}$$

$$\text{L1,}\begin{Bmatrix}\text{IN}\\\text{OUT}\end{Bmatrix},\text{TANTO,C1,THRU,P1}$$

Comment   Generally, there are four possible solutions if P1 is not on C1 or L1, and L1 does not share a common point with C1. The modifier IN or OUT is used to indicate that the desired circle does or does not share a common area with the given circle C1; the dimensional modifier LARGE or SMALL means that the desired solution is the circle with a larger or smaller radius, respectively. The function of the directional modifiers XLARGE, . . . , and YSMALL is not well defined, and they have no effect on the solution if the other two modifiers in the statement have been correctly specified. See Section 4.11 for further discussion on the usage of this statement.

(continued)

TABLE 4-10 STATEMENTS USING A CIRCLE OR CIRCLES AS THE REFERENCE OR REFERENCES TO DEFINE A CIRCLE (continued)

No.	Known Condition	Statement Format

Examples

C2=CIRCLE/SMALL,YLARGE,TANTO,L1,OUT,TANTO,C1,THRU,P1
C3=CIRCLE/LARGE,YLARGE,TANTO,L1,OUT,TANTO,C1,THRU,P1
C4=CIRCLE/SMALL,XSMALL,TANTO,L1,IN,TANTO,C1,THRU,P1
C5=CIRCLE/LARGE,XSMALL,TANTO,L1,IN,TANTO,C1,THRU,P1

6 The circle is tangent to two given lines, L1 and L2, and a given circle, C1.

$$CIRCLE/\begin{Bmatrix} XLARGE \\ XSMALL \\ YLARGE \\ YSMALL \end{Bmatrix},TANTO,L1,\$$$

$$\begin{Bmatrix} XLARGE \\ XSMALL \\ YLARGE \\ YSMALL \end{Bmatrix},TANTO,L2,\begin{Bmatrix} XLARGE \\ XSMALL \\ YLARGE \\ YSMALL \end{Bmatrix},\$$$

$$\begin{Bmatrix} IN \\ OUT \end{Bmatrix},TANTO,C1$$

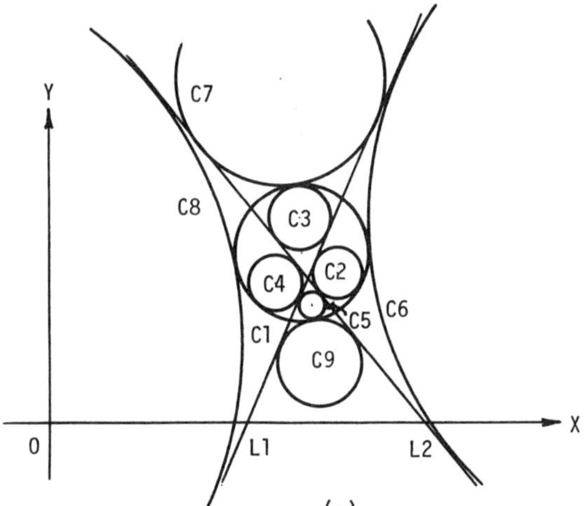

(a)

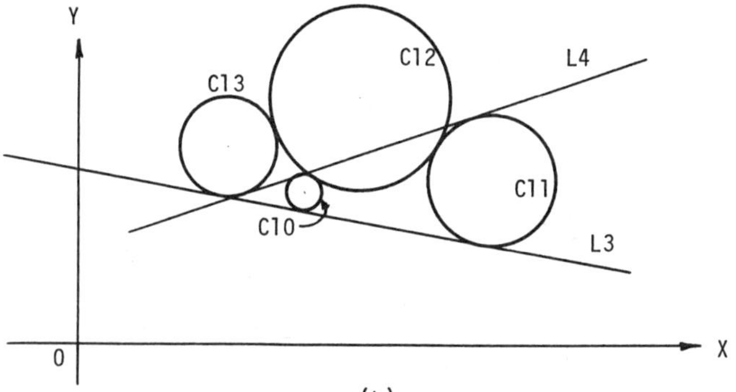

(b)

TABLE 4-10   STATEMENTS USING A CIRCLE OR CIRCLES AS THE REFERENCE OR REFERENCES TO DEFINE A CIRCLE   (continued)

No.	Known Condition	Statement Format

Comment In general, there are eight possible solutions. The number of solutions varies with the relative positions of the given lines and circle. It should be noticed that in some cases (e.g., for circle C2 in Fig. (a), the directional modifier preceding given circle C1 in this statement does not have any effect on the definition of the desired circle because circle C2 has been clearly defined by the other three modifiers. In this case, any one of the modifiers, XLARGE, . . . , YSMALL, can be specified. However, in Fig. (b), the directional modifier preceding C12 acts as a selector of the right solution from the two possible ones.

Examples

C2=CIRCLE/XLARGE,TANTO,L1,YLARGE,TANTO,L2,XLARGE,IN,TANTO,C1

C10=CIRCLE/YLARGE,TANTO,L3,YSMALL,TANTO,L4,XSMALL,OUT,TANTO,C12

C11=CIRCLE/YLARGE,TANTO,L3,YSMALL,TANTO,L4,XLARGE,OUT,TANTO,C12

7 The circle is tangent to two given circles, C1 and C2, with a given radius, r.

$$\text{CIRCLE}/\begin{Bmatrix} \text{XLARGE} \\ \text{XSMALL} \\ \text{YLARGE} \\ \text{YSMALL} \end{Bmatrix}, \begin{Bmatrix} \text{IN} \\ \text{OUT} \end{Bmatrix}, \text{C1},\$$$
$$\begin{Bmatrix} \text{IN} \\ \text{OUT} \end{Bmatrix}, \text{C2}, \text{RADIUS}, r$$

Comment Depending on the given condition, the number of possible solutions varies from 0 to 8.

Examples

CC1=CIRCLE/YLARGE,OUT,C1,IN,C2,RADIUS,r

CC2=CIRCLE/YSMALL,OUT,C1,IN,C2,RADIUS,r

CC5=CIRCLE/YSMALL,OUT,C3,OUT,C4,RADIUS,r

CC6=CIRCLE/YLARGE,OUT,C3,OUT,C4,RADIUS,r

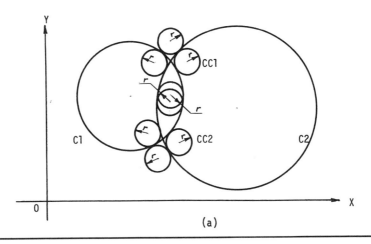

(a)

(continued)

TABLE 4-10 STATEMENTS USING A CIRCLE OR CIRCLES AS THE REFERENCE OR REFERENCES TO DEFINE A CIRCLE **(continued)**

No.	Known Condition	Statement Format

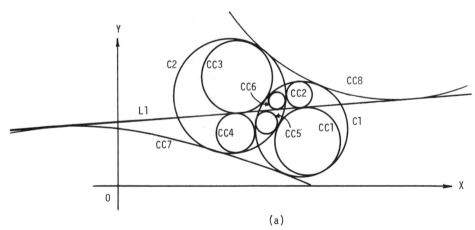

(b)

8 The circle is tangent to two given circles, C1 and C2, and a given line, L1.

$$\text{CIRCLE/} \left\{ \begin{matrix} \text{LARGE} \\ \text{SMALL} \end{matrix} \right\}, \left\{ \begin{matrix} \text{XLARGE} \\ \text{XSMALL} \\ \text{YLARGE} \\ \text{YSMALL} \end{matrix} \right\}, \text{TANTO,\$}$$

$$\text{L1,} \left\{ \begin{matrix} \text{IN} \\ \text{OUT} \end{matrix} \right\}, \text{TANTO,C1,} \left\{ \begin{matrix} \text{IN} \\ \text{OUT} \end{matrix} \right\}, \text{\$}$$

$$\text{TANTO,C2}$$

Comment   A general case has as many as eight solutions (CC1 to CC8 in Fig. [a]). Circle CC5 can be defined by the statement

CC5=CIRCLE/LARGE,YSMALL,TANTO,L1,IN,TANTO,C1,IN,TANTO,C2

or

(a)

TABLE 4-10   STATEMENTS USING A CIRCLE OR CIRCLES AS THE REFERENCE OR REFERENCES TO
DEFINE A CIRCLE   (continued)

No.	Known Condition	Statement Format

(b)   (c)

CC5=CIRCLE/SMALL,YSMALL,TANTO,L1,IN,TANTO,C1,IN,TANTO,C2

Therefore the dimensional modifier has no effect in this case. The following examples show the
selection role of a dimensional modifier when it is needed (Fig. [b]):

C11=CIRCLE/SMALL,YSMALL,TANTO,L2,IN,TANTO,C4,OUT,TANTO,C3
C10=CIRCLE/LARGE,YSMALL,TANTO,L2,IN,TANTO,C4,OUT,TANTO,C3

If C10 and C11 have the same diameter (Fig. [c]), this statement cannot determine which is the desired
solution. As a result, an error message will be issued.

4.4 STATEMENTS DEFINING A VECTOR

A *vector* is a directed line segment that has both magnitude and direction and is defined by the APT statement

VECTOR/......

Mathematically, any two vectors that have the same magnitude and direction are
considered to be the same. In other words, a vector is not located at a specific point
in space. In the NC processor, a vector can be used to indicate a direction or an incremental displacement from a given point, or both. There are 11 statement formats
in APT to define a vector. They are listed in Table 4-11.
   On some occasions it is necessary to define a vector that has a fixed starting
position in space. Such a vector is called a *point vector* in APT (Fig. 4-7) and is defined by the statement

$$\text{PNTVCT/} \begin{Bmatrix} P1 \\ x,y,z \end{Bmatrix}, \begin{Bmatrix} V1 \\ i,j,k \end{Bmatrix}$$

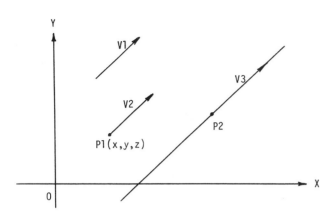

Figure 4-7 Defining vectors that have a fixed position in space, with finite (V2) or infinite (V3) length:
V2=PNTVCT/P1,V1
V3=SPALIN/P2,V1

where the given point P1 (x, y, z) is the starting point of the vector to be defined; V1 is a given vector, with i, j, and k as its components in the X, Y, and Z directions, respectively.

A vector passing through a given point in space and having infinite length is called a *space line* in APT (Fig. 4-7) and is defined by the statement

$$SPALIN/P1, \begin{Bmatrix} P2 \\ V1 \end{Bmatrix}$$

where V1 is a vector through which the direction of the desired vector (space line) is defined. If two given points, P1 and P2, are specified in this statement, then the direction of the defined vector is from point P1 to point P2.

TABLE 4-11 STATEMENTS DEFINING A VECTOR

No.	Known Condition	Statement Format
1	The X, Y, and Z components of the vector are x, y, and z, respectively.	VECTOR/x,y,z

TABLE 4-11 STATEMENTS DEFINING A VECTOR (continued)

No.	Known Condition	Statement Format

Example   The components of vector V1 are
$$x = 1 - 5 = -4$$
$$y = 5 - 1 = 4$$
and $$z = 5 - 1 = 4$$
respectively. Hence the statement defining V1 is
V1=VECTOR/−4,4,4

2 The starting and end points   VECTOR/$\begin{Bmatrix} x1,y1,z1,x2,y2,z2 \\ P1,P2 \end{Bmatrix}$
are P1 $(x1, y1, z1)$ and
P2 $(x2, y2, z2)$, respectively.

Example   (see the figure in statement format No. 1)
V1=VECTOR/P1,P2
V1=VECTOR/5,1,1,1,5,5

3 The vector is in the X-Y VECTOR/ATANGL,a,L1, $\begin{Bmatrix} \text{POSX} \\ \text{NEGX} \\ \text{XLARGE} \\ \text{XSMALL} \\ \text{POSY} \\ \text{NEGY} \\ \text{YLARGE} \\ \text{YSMALL} \end{Bmatrix}$
plane and at an angle, a,
with a given line, L1.

Comment   The angle that the defined vector makes with the given line can be one of two values,
namely a and $(a + 180°)$. Either one can be selected. A directional modifier is specified to indicate
the desired direction for the defined vector.

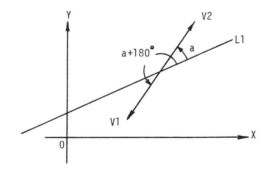

Examples   Assume $a = 30$ degrees in the figure, we have
V1=VECTOR/ATANGL,30,L1,NEGX
V1=VECTOR/ATANGL,30,L1,YSMALL
V1=VECTOR/ATANGL,210,L1,NEGX
V2=VECTOR/ATANGL,30,L1,POSX
V2=VECTOR/ATANGL,30,L1,POSY
V2=VECTOR/ATANGL,210,L1,POSX

(continued)

TABLE 4-11   STATEMENTS DEFINING A VECTOR   (continued)

No.	Known Condition	Statement Format
4	The vector is perpendicular to a given plane, PL1.	VECTOR/PERPTO,PL1, $\begin{Bmatrix} \text{POSX} \\ \text{POSY} \\ \text{POSZ} \\ \text{NEGX} \\ \text{NEGY} \\ \text{NEGZ} \end{Bmatrix}$

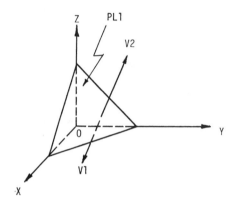

Examples

V1=VECTOR/PERPTO,PL1,NEGX
V2=VECTOR/PERPTO,PL1,POSZ

5 The vector is parallel to the X-Y, Y-Z, or Z-X plane and at an angle, a, with the +X (for X-Y plane), +Y (for Y-Z plane) or +Z (for Z-X plane) axis, with magnitude k.

VECTOR/LENGTH,k,ATANGL,a, $\begin{Bmatrix} \text{XYPLAN} \\ \text{YZPLAN} \\ \text{ZXPLAN} \end{Bmatrix}$

Examples

V1=VECTOR/LENGTH,1.5,ATANGL,60,XYPLAN
V2=VECTOR/LENGTH,1.5,ATANGL,240,XYPLAN
V2=VECTOR/LENGTH,1.5,ATANGL,−120,XYPLAN

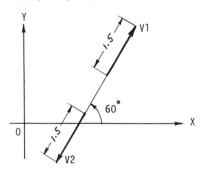

TABLE 4-11   STATEMENTS DEFINING A VECTOR   (continued)

No.	Known Condition	Statement Format
6	The vector is parallel to the intersection of two given planes, PL1 and PL2.	VECTOR/PARLEL,INTOF,PL1,PL2, $\begin{Bmatrix} \text{POSX} \\ \text{POSY} \\ \text{POSZ} \\ \text{NEGX} \\ \text{NEGY} \\ \text{NEGZ} \end{Bmatrix}$

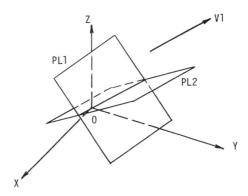

Example
V1=VECTOR/PARLEL,INTOF,PL1,PL2,NEGX

No.	Known Condition	Statement Format
7	The vector is a unit vector that passes through a given point, P1, and is perpendicular to a quadric surface,* Q1.	VECTOR/ $\begin{Bmatrix} \text{POSX} \\ \text{POSY} \\ \text{POSZ} \\ \text{NEGX} \\ \text{NEGY} \\ \text{NEGZ} \end{Bmatrix}$,PERPTO,Q1,P1

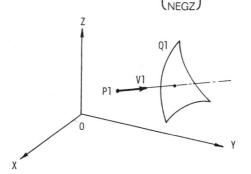

Example
V1=VECTOR/POSY,PERPTO,Q1,P1

No.	Known Condition	Statement Format
8	The vector is a given vector, V1, multiplied by a scalar, n.	VECTOR/n,TIMES, $\begin{Bmatrix} \text{P1} \\ \text{V1} \end{Bmatrix}$

Comment   If a point P1, is given instead of a given vector, then the reference vector stretches from the origin of the coordinate system to point P1.

---

*See Section 4.7 for the definition of quadric surfaces.

                                                                                         (continued)

TABLE 4-11   STATEMENTS DEFINING A VECTOR   (continued)

No.	Known Condition	Statement Format

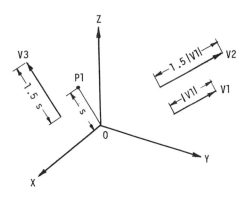

Examples
V2=VECTOR/1.5,TIMES,V1
V3=VECTOR/1.5,TIMES,P1

9 The vector is a unit vector, whose direction is defined by a given vector, V1, or a point P1 (x, y, z).

$$\text{VECTOR/UNIT,} \begin{Bmatrix} x,y,z \\ P1 \\ V1 \end{Bmatrix}$$

Comment   See the comment for statement format No. 8.

10 The vector is the vector sum or difference of two given vectors, V1 and V2.

$$\text{VECTOR/V1,} \begin{Bmatrix} \text{PLUS} \\ \text{MINUS} \end{Bmatrix} ,\text{V2}$$

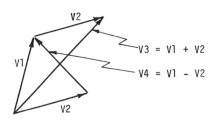

Examples
V3=VECTOR/V1,PLUS,V2
V4=VECTOR/V1,MINUS,V2

11 The vector is the cross product of two given vectors, V1 and V2.

$$\text{VECTOR/} \begin{Bmatrix} \text{V1} \\ \text{P1} \end{Bmatrix} ,\text{CROSS,} \begin{Bmatrix} \text{V2} \\ \text{P2} \end{Bmatrix}$$

TABLE 4-11   STATEMENTS DEFINING A VECTOR   (continued)

No.	Known Condition	Statement Format

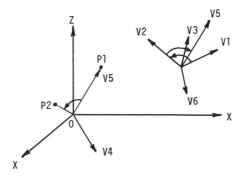

Comment   If two points, P1 and P2, are given instead of two vectors, then the two vectors are those with the origin of the coordinate system as their starting point and the given points as their end points, respectively.

Examples
V3=VECTOR/V1,CROSS,V2
V4=VECTOR/P1,CROSS,P2
V6=VECTOR/V2,CROSS,P1

4.5 STATEMENTS DEFINING PLANE CURVES (ELLIPSE, HYPERBOLA, GCONIC, AND LCONIC)

The ellipse and the hyperbola are two common plane curves used in engineering design. In general, an *ellipse* is the intersection of a plane and a cone or a circular cylinder, or the projection of a circle on a plane that is not parallel to that containing the circle. An ellipse can also be defined as a plane curve that is the locus of a moving point, P1, which maintains a constant sum of the distances from that point to two fixed points, P2 and P3, in the plane (see S1 in Fig. 4-8).

The standard form of the mathematical expression for an ellipse (S1 in Fig. 4-8) in the *X-Y* plane is

$$\frac{x^2}{a^2} + \frac{y^2}{b^2} = 1$$

where *a* and *b* are the semimajor and semiminor axes of the ellipse. Any ellipse in the *X-Y* plane, other than that defined by the expression presented above (e.g., S2 in Fig. 4-8), can also be defined by this expression through transformation of the coordinate system.

An ellipse is defined by the statement

ELLIPS/......

In APT, two statement formats define an ellipse; they are listed in Table 4-12.

A *hyperbola* is the locus of a moving point, P1, that maintains a constant difference of the distances from that point to two given points (focuses), P2 and P3, in

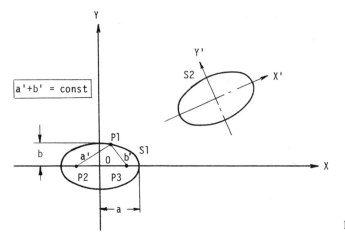

Figure 4-8   The definition of an ellipse.

TABLE 4-12   STATEMENTS DEFINING AN ELLIPSE

No.	Known Condition	Statement Format
1	The ellipse has a semimajor axis, $r1$, a semiminor axis, $r2$, and its center at a given point, P1, with the major axis being at an angle, a, with the $+X$ axis.	ELLIPS/CENTER,P1,$r1$,$r2$,a

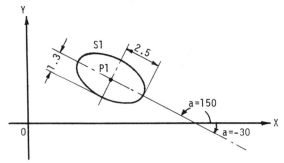

Examples
S1=ELLIPS/CENTER,P1,2.5,1.3,150
S1=ELLIPS/CENTER,P1,2.5,1.3,−30

2	The ellipse is the intersection of a given plane, PL1, and a given cylinder or cone, S1.	ELLIPS/INTOF,PL1,S1

Comment   If S1 is a cylinder, plane P11 should not be parallel to its axis; if S1 is a cone, the angle between plane PL1 and the axis of the cone should be greater than the half vertex angle of the cone.

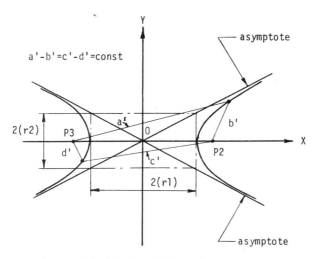

Figure 4-9   The hyperbola and its parameters.

the plane (Fig. 4-9). The standard form of the mathematical expression for a hyperbola in the X-Y plane, with the X axis as its transverse axis, is

$$\frac{x^2}{(r1)^2} - \frac{y^2}{(r2)^2} = 1$$

where $r1$ and $r2$ are the transverse and conjugate axes, respectively. By transformation of the coordinate system, a hyperbola at any position and in any direction can be defined by the expression shown above. Thus, in APT, the given condition for defining a hyperbola consists of the center point, P1, which can be other than the origin of the coordinate system, the half-transverse and half-conjugate axes, represented by $r1$ and $r2$, and the angle, a, that the transverse axis makes with the X axis. The statement format is as follows:

HYPERB/CENTER,P1,*r*1,*r*2,*a*

There are cases when a curve in the X-Y plane is defined by a second-order equation. In APT, such a curve is called a *general conic* and includes circles, ellipses, hyperbolas, and parabolas; it can be defined by the statement

GCONIC/......

Three statement formats are provided in APT to define general conics. They are listed in Table 4-13.

Sometimes, curves other than those described above — for example, curves defined by higher-order equations — are used in design; as an alternative, several discrete points on a curve and the slopes of the curve at some of these points may be given. In other cases, a curve might be drawn by means of plastic curves. It is possible that the curve cannot be described by a simple mathematical expression. However, we know the coordinates of several discrete points and, perhaps, the slopes at some points. Therefore we need to fit a smooth curve passing through the given or calculated points with given slopes at specific points. Mathematically, this consists

TABLE 4-13 STATEMENTS DEFINING A GENERAL CONIC

No.	Known Condition	Statement Format
1	The curve is defined by the equation	GCONIC/a,b,c,d,e,f

$$ax^2 + bxy + cy^2 + dx + ey + f = 0$$

Example   If a curve, C1, is defined by the equation

$$3x^2 + 2xy + 4y^2 + 1.5x - 5 = 0$$

the APT statement defining this curve is

C1=GCONIC/3,2,4,1.5,0,−5

| 2 | The curve is defined by the equation | GCONIC/p,q,r,s,t |

$$y = px + q \pm (rx^2 + sx + t)^{\frac{1}{2}}$$

| 3 | The curve is defined by the equation | GCONIC/p,q,r,s,t,FUNOFY |

$$x = py + q \pm (ry^2 + sy + t)^{\frac{1}{2}}$$

in creating a mathematical formulation of a corresponding curve, with the use of suitable interpolation functions, to match the given constraint (points and slopes).[3] In APT, such curves are called *loft conics;* the statement defining them follows:

LCONIC/......

Three statement formats are provided (Table 4-14). When a loft conic is tangent to a given curve or line at one end, it is better to use the No. 2 or No. 3 format to attain slope continuity (Fig. 4-10).

4.6 STATEMENTS DEFINING A PLANE

Planes are among the most frequently used geometric entities for defining the cutter path. There are 13 statement formats (Table 4-15) in APT, allowing a user to define a *plane* on the basis of various given conditions.

TABLE 4-14   STATEMENTS DEFINING LOFT CONICS IN THE *X-Y* PLANE

No.	Known Condition	Statement Format
1	The curve passes through five given points, P1 $(x1, y1)$, P2 $(x2, y2)$, P3 $(x3, y3)$, P4 $(x4, y4)$, and P5 $(x5, y5)$, in the specified order. P1 and P5 are the end points of the curve.	LCONIC/5PT$\left\{\begin{array}{l} x1,y1,x2,y2,x3,y3,x4,y4,x5,y5 \\ P1,P2,P3,P4,P5 \end{array}\right\}$

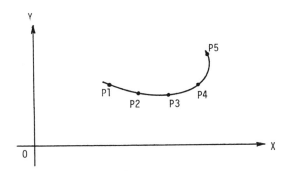

2	The curve passes through four given points, P1 $(x1, y1)$, P2 $(x2, y2)$, P3 $(x3, y3)$, P4 $(x4, y4)$, in the specified order, with slope, s, at the first given point, P1. P1 and P4 are the end points.	LCONIC/4PT1SL,$\left\{\begin{array}{l} x1,y1,s,x2,y2,x3,y3,x4,y4 \\ P1,s,P2,P3,P4 \end{array}\right\}$

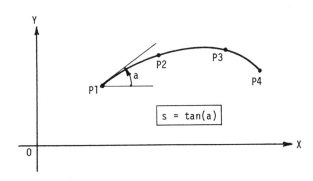

(continued)

TABLE 4-14 STATEMENTS DEFINING LOFT CONICS IN THE *X-Y* PLANE (continued)

No.	Known Condition	Statement Format
3	The curve passes through three given points, P1 $(x1, y1)$, P2 $(x2, y2)$, and P3 $(x3, y3)$, in the specified order, with slopes at the end points P1 and P3 equal to $s1$ and $s3$, respectively.	LCONIC/3PT2SL, $\left\{ \begin{array}{l} x1,y1,s1,x3,y3,s3,x2,y2 \\ P1,s1,P3,s3,P2 \end{array} \right\}$

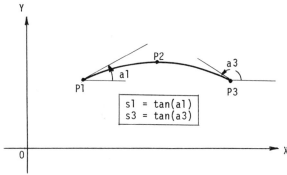

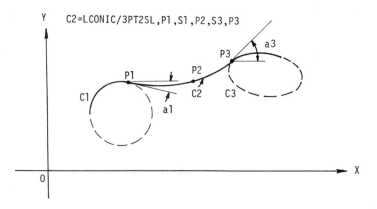

Figure 4-10 Defining a loft conic by three points and two slopes. P1, P2, and P3 are the three given points. P1 and P3 are also the tangent points of the loft conic C2 to the given curves C1 and C3, respectively. S1 = tan(*a*1); S3 = tan(*a*3)

TABLE 4-15 STATEMENTS DEFINING A PLANE

No.	Known Condition	Statement Format
1	The plane is defined by the equation $ax + by + cz = d$	PLANE/a,b,c,d

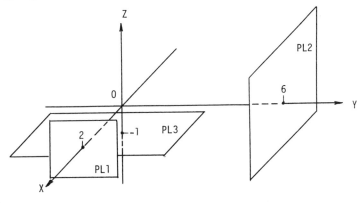

Examples
PL1=PLANE/1,0,0,2
PL2=PLANE/0,1,0,6
PL3=PLANE/0,0,1,−1

2 The plane passes through three given points, P1, P2, and P3.   PLANE/P1,P2,P3

Comment   P1, P2, and P3 are symbols of the three points defined previously.

3 The plane passes through a given point, P1, and is perpendicular to a given vector, V1.   PLANE/P1,PERPTO,V1

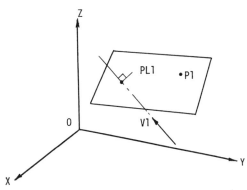

Example
PL1=PLANE/P1,PERPTO,V1

(continued)

TABLE 4-15 **STATEMENTS DEFINING A PLANE** (continued)

No.	Known Condition	Statement Format
4	The plane passes through a given point, P1, and is parallel to a given plane, PL1.	PLANE/P1,PARLEL,PL1

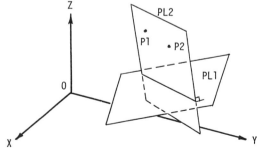

Example
PL2=PLANE/P1,PARLEL,PL1

No.	Known Condition	Statement Format
5	The plane passes through a given point, P1, and is perpendicular to two given planes, PL1 and PL2, which are not parallel to each other.	PLANE/P1,PERPTO,PL1,PL2

No.	Known Condition	Statement Format
6	The plane passes through two given points, P1 and P2, and is perpendicular to a given plane, PL1.	PLANE/ { PERPTO,PL1,P1,P2 / P1,P2,PERPTO,PL1 }

TABLE 4-15 STATEMENTS DEFINING A PLANE (continued)

No.	Known Condition	Statement Format

Comment If the line passing through points P1 and P2 is perpendicular to the given plane, then there will be an infinite number of solutions. In such a case, other statement formats should be selected.

7 The plane passes through   PLANE/P1,P2,ATANGL,*a*,PL1
two given points, P1, and P2,
in a given plane, PL1, and
makes an angle, *a*, with
the plane.

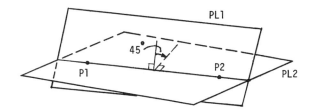

Comment Angle *a* is measured from the given plane PL1 to the defined plane PL2 in a plane perpendicular to the line connecting points P1 and P2. A positive angle is measured in the counterclockwise direction when one looks from point P1 to point P2.

Examples
PL2=PLANE/P1,P2,ATANGL,45,PL1
PL2=PLANE/P2,P1,ATANGL,−45,PL1

8 The plane is parallel to a   PLANE/PARLEL,PL1, ⎧XLARGE⎫ ,*d*
given plane, PL1, and at a   ⎪XSMALL⎪
given distance, *d*, from it.   ⎨YLARGE⎬
                                                            ⎪YSMALL⎪
                                                            ⎪ZLARGE⎪
                                                            ⎩ZSMALL⎭

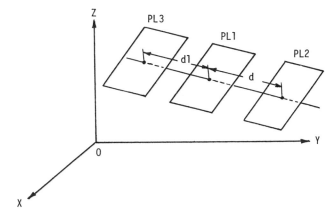

Comment The directional modifier in this statement indicates the motion direction from the given plane to the defined plane along the normal to the planes.

(continued)

TABLE 4-15   STATEMENTS DEFINING A PLANE (continued)

No.	Known Condition	Statement Format

Example

PL2=PLANE/PARLEL,PL1,YLARGE,*d*
PL3=PLANE/PARLEL,PL1,YSMALL,*d*1

9 The plane is parallel to
and at a distance, *d*, from
coordinate plane *X-Y*,
Y-Z, or *Z-X*.

$$\text{PLANE/} \begin{Bmatrix} \text{XYPLAN} \\ \text{YZPLAN} \\ \text{ZXPLAN} \end{Bmatrix} [,d]$$

Comment   If *d* is omitted in this statement, its value is assumed to be zero. Then the defined plane is
the specified coordinate plane. A positive *d* means that the defined plane is on the positive side of the
given coordinate plane; a negative *d* is on the negative side.

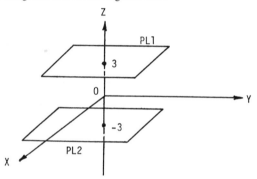

Examples

PL1=PLANE/XYPLAN,3
PL2=PLANE/XYPLAN,−3

10 The plane passes through a
given point, P1, and is
tangent to a given cylinder,
C1.

$$\text{PLANE/P1,} \begin{Bmatrix} \text{XLARGE} \\ \text{XSMALL} \\ \text{YLARGE} \\ \text{YSMALL} \\ \text{ZLARGE} \\ \text{ZSMALL} \end{Bmatrix} ,\text{TANTO,C1}$$

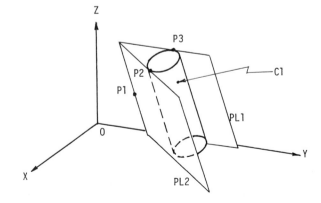

TABLE 4-15 STATEMENTS DEFINING A PLANE (continued)

No.	Known Condition	Statement Format
	Comment The directional modifier can be selected in the same way as for the circle definition statement by comparing the coordinates of the two tangent points, P2 and P3, in the plane perpendicular to the cylinder axis. The reference point is P1, and the viewing direction is from point P1 to cylinder C1.	

Examples
PL1=PLANE/P1,XSMALL,TANTO,C1
PL2=PLANE/P1,XLARGE,TANTO,C1

No.	Known Condition	Statement Format
11	The plane passes through a given point, P1, and is perpendicular to a given cylinder, C1 (i.e., the plane is perpendicular to the axis of the given cylinder).	PLANE/P1,PERPTO,C1
12	The plane is tangent to a given sphere, C1, at a given point, P1, on the sphere.	PLANE/P1,TANTO,C1
13	The plane passes through two given points, P1 and P2, and is tangent to a given sphere, S1.	PLANE/ { XLARGE / XSMALL / YLARGE / YSMALL / ZLARGE / ZSMALL } ,TANTO,S1,THRU,P1,P2

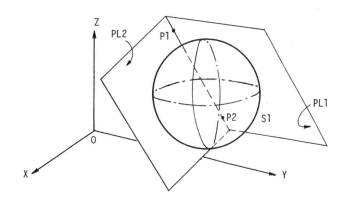

Comment The positional modifier indicates the position of the tangent point with respect to the center of the sphere.

Examples
PL1=PLANE/XSMALL,TANTO,S1,THRU,P1,P2
PL2=PLANE/YSMALL,TANTO,S1,THRU,P1,P2

4.7 STATEMENTS DEFINING QUADRIC SURFACES, CYLINDERS, AND SURFACES OF ROTATION

Although a part or machine element should be designed in as simple a form as possible, there are many cases when a designer has to use curved surfaces as segments of the profile. Mathematically, a surface is the locus of a point (x, y, z) that moves in space while maintaining the relationship (Fig. 4-11)

$$f(x, y, z) = 0$$

A group of surfaces in common use in engineering design are those defined by the general second-degree equation

$$ax^2 + by^2 + cz^2 + fyz + gzx + hxy + px + qy + rz + d = 0 \qquad (4.1)$$

which are known as *quadric surfaces*. The APT statement defining a quadric surface is

QADRIC/a,b,c,f,g,h,p,q,r,d

where a, b, ..., d are the coefficients in equation 4.1.

A quadric surface defined in APT must be a real surface (i.e., the coordinates x, y, and z satisfying equation 4.1 must be real numbers). A general quadric equation can be rewritten in a standard form known as canonical form, as follows:

$$ax^2 + by^2 + cz^2 + 2f'yz + 2g'zx + 2h'xy + 2p'x + 2q'y + 2r'z + d = 0$$

$$(4.2)$$

where

$$f' = f/2;$$
$$g' = g/2;$$
$$h' = h/2;$$
$$p' = p/2;$$
$$q' = q/2;$$
$$r' = r/2.$$

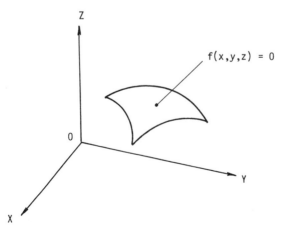

Figure 4-11   The definition of a surface.

As is well known, the four quantities

$$I_1 = a + b + c$$

$$I_2 = \begin{vmatrix} a & h' \\ h' & b \end{vmatrix} + \begin{vmatrix} b & f' \\ f' & c \end{vmatrix} + \begin{vmatrix} c & g' \\ g' & c \end{vmatrix}$$

$$I_3 = \begin{vmatrix} a & h' & g' \\ h' & b & f' \\ g' & f' & c \end{vmatrix}$$

$$I_4 = \begin{vmatrix} a & h' & g' & p' \\ h' & b & f' & q' \\ g' & f' & c & r' \\ p' & q' & r' & d \end{vmatrix}$$

are invariants with respect to transformations of the coordinate system, and I_4 is called the *discriminant* of equation 4.2.[5] The type of surfaces described by quadric equation 4.2 can be determined by the values of I_1, I_2, I_3 and I_4, as shown in Table 4-16.

In general, a quadric equation that defines a surface other than an imaginary one can be simplified, by means of transformation of the coordinate system, into a standard form:

$$a' x^2 + b' y^2 + c' z^2 + d' = 0 \tag{4.3}$$

which defines, generally, an *ellipsoid* or a *hyperboloid*, or one of the following forms

$$\left. \begin{aligned} a'' x^2 + b'' y^2 + r'' z = 0 \\ b'' y^2 + c'' z^2 + s'' x = 0 \\ c'' z^2 + a'' x^2 + t'' y = 0 \end{aligned} \right\} \tag{4.4}$$

which defines generally an *elliptic* or *hyperbolic paraboloid*.[5] The quadric surfaces that can be processed by the NC processor are listed in Fig. 4-12(a) and (b). As can be seen from the figures, they are the same surfaces as those defined by equations 4.3 and 4.4.

A QADRIC statement should not be used to define some of the degenerate quadrics, such as point, line, and plane; the respective statements defining points, lines, and planes, described in the previous sections, should be selected.

Two types of surfaces that are of particular interest in engineering design are cylindrical surfaces and surfaces of revolution.

TABLE 4-16   THE CLASSIFICATION OF QUADRIC SURFACES

	$I_3 \neq 0$			$I_3 = 0$	
I_4	$I_1 \cdot I_3 > 0, I_2 > 0$	$I_1 \cdot I_3 \leq 0, I_2 \leq 0$	I_4	I_2	Type of surface
< 0	Ellipsoid	Hyperboloid of two sheets	< 0	> 0	Elliptic paraboloid
> 0	Imaginary ellipsoid	Hyperboloid of one sheet	> 0	< 0	Hyperbolic paraboloid
$= 0$	Imaginary cone	Cone	$= 0$	Any	Cylinder* or pairs of planes

*The cylinder listed here is a cylindric surface, which can be an elliptic, hyperbolic, parabolic, or circular cylinder.

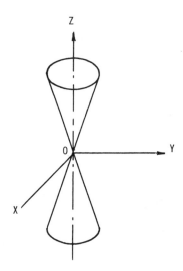

Elliptic cone:

$$\frac{x^2}{a^2} + \frac{y^2}{b^2} - \frac{z^2}{c^2} = 0$$

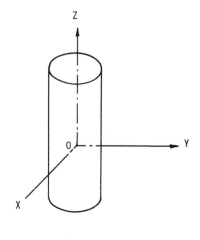

Elliptic cylinder:

$$\frac{x^2}{a^2} + \frac{y^2}{b^2} = 1$$

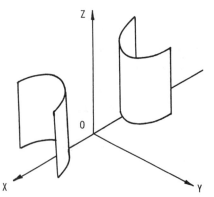

Hyperbolic cylinder:

$$\frac{x^2}{a^2} - \frac{y^2}{b^2} = 1$$

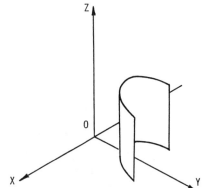

Parabolic cylinder:

$$x^2 = py$$

(a)

Figure 4-12 (a) The degenerated quadrics allowed in APT. (b) The proper quardrics defined in APT.

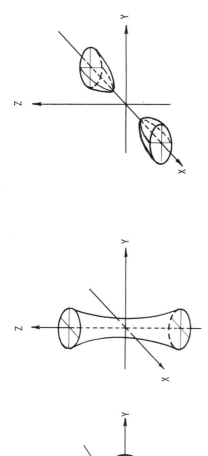

Hyperboloid of two sheets:
$(x/a)^2 - (y/b)^2 - (z/c)^2 = 1$

Hyperboloid of one sheet:
$(x/a)^2 + (y/b)^2 - (z/c)^2 = 1$

Ellipsoid: $(x/a)^2 + (y/b)^2 + (z/c)^2 = 1$

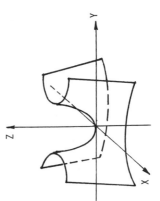

Hyperbolic paraboloid:
$(x/a)^2 - (y/b)^2 = 2cz$

(b)

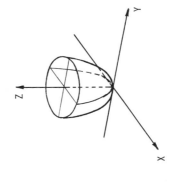

Elliptic paraboloid:
$(x/a)^2 + (y/b)^2 = 2cz$

Figure 4-12 (continued)

161

A cylindrical surface is generated by the translation of a straight line along a given plane curve, called a *directrix*, wherein the line is not in the plane containing the curve. If the coordinate system is rotated to set one of its axes, X, Y, or Z, parallel to the generating line, the equation of a cylindrical surface can be simplified so as to contain only two variables, namely

$$f(x, y) = 0$$
$$f(x, z) = 0$$

or

$$f(y, z) = 0$$

Besides the cylinders defined by the QADRIC statement, APT also provides the statement

CYLNDR/......

to define a circular cylinder (Table 4-17) and the statement

TABCYL/......

to define a *tabulated cylinder* (i.e., a cylinder with its directrix defined by a set of given points and slopes [Fig. 4-13]). The TABCYL statement allows us to define a cylindrical surface that cannot be defined by the statements mentioned previously. An explanation of the formats and usage of the TABCYL statement is given in Tables 4-18 and 4-19.

TABLE 4-17   STATEMENTS DEFINING CIRCULAR CYLINDERS

No.	Known Condition	Statement Format
1	The cylinder has a radius, r, with its axis passing through a given point, P1 (x, y, z), and parallel to a unit vector, V1 (a, b, c).	$\begin{Bmatrix} P1 \\ x,y,z \end{Bmatrix}, \begin{Bmatrix} V1 \\ a,b,c \end{Bmatrix}, r$

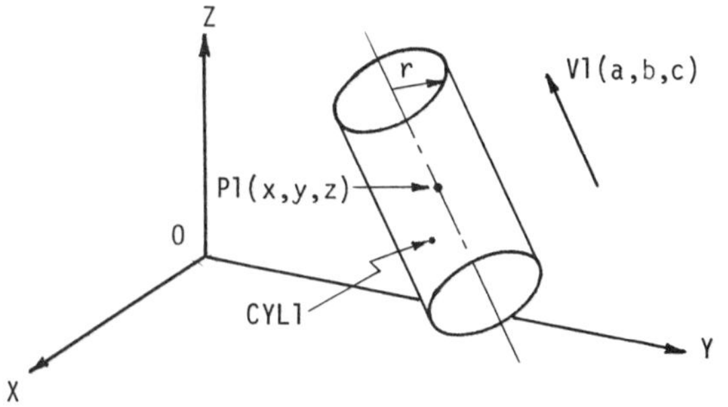

Examples
CYL1=CYLNDR/P1,V1,r
CYL1=CYLNDR/x,y,z,a,b,c,r
CYL1=CYLNDR/P1,a,b,c,r
CYL1=CYLNDR/x,y,z,V1,r

TABLE 4-17   STATEMENTS DEFINING CIRCULAR CYLINDERS (continued)

No.	Known Condition	Statement Format
2	The cylinder is tangent to two given planes, PL1 and PL2, with a radius, r.	CYLNDR/ $\begin{Bmatrix} \text{XLARGE} \\ \text{XSMALL} \\ \text{YLARGE} \\ \text{YSMALL} \\ \text{ZLARGE} \\ \text{ZSMALL} \end{Bmatrix}$,TANTO,PL1,\$ $\begin{Bmatrix} \text{XLARGE} \\ \text{XSMALL} \\ \text{YLARGE} \\ \text{YSMALL} \\ \text{ZLARGE} \\ \text{ZSMALL} \end{Bmatrix}$,TANTO,PL2,RADIUS,r

Comment   The directional modifier before the symbol of a plane indicates on which side of the plane the defined cylinder is located.

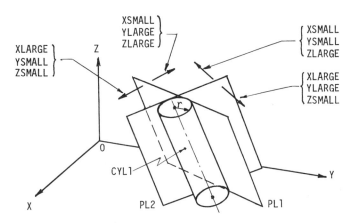

Examples
CYL1=CYLNDR/XLARGE,TANTO,PL1,YLARGE,TANTO,PL2,RADIUS,r
CYL1=CYLNDR/ZSMALL,TANTO,PL2,ZSMALL,TANTO,PL1,RADIUS,r

No.	Known Condition	Statement Format
3	The cylinder is tangent to a given plane, PL1, at the line defined by the two given points P1 and P2, with a radius, r.	CYLNDR/ $\begin{Bmatrix} \text{XLARGE} \\ \text{XSMALL} \\ \text{YLARGE} \\ \text{YSMALL} \\ \text{ZLARGE} \\ \text{ZSMALL} \end{Bmatrix}$,TANTO,PL1,THRU,\$ P1,P2,RADIUS,r

Example
CYL1=CYLNDR/ZLARGE,TANTO,PL1,THRU,P1,P2,RADIUS,r

(continued)

TABLE 4-17 STATEMENTS DEFINING CIRCULAR CYLINDERS (continued)

No.	Known Condition	Statement Format

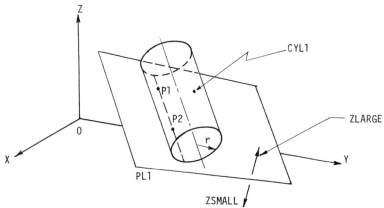

4 The cylinder is tangent to a given cylinder, CYL1, and a given plane, PL1, which is parallel to the axis of the given cylinder, with a radius, r.

CYLNDR/ $\begin{Bmatrix} \text{XLARGE} \\ \text{XSMALL} \\ \text{YLARGE} \\ \text{YSMALL} \\ \text{ZLARGE} \\ \text{ZSMALL} \end{Bmatrix}$,PL1, $\begin{Bmatrix} \text{XLARGE} \\ \text{XSMALL} \\ \text{YLARGE} \\ \text{YSMALL} \\ \text{ZLARGE} \\ \text{ZSMALL} \end{Bmatrix}$,$

$\begin{Bmatrix} \text{IN} \\ \text{OUT} \end{Bmatrix}$,CYL1,RADIUS,r

Comment This statement format is similar to circle statement format No. 3 in Table 4-10.

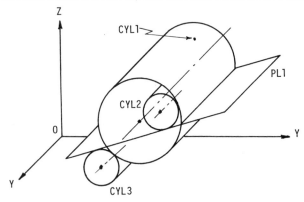

Examples

CYL2=CYLNDR/ZLARGE,PL1,ZLARGE,IN,CYL1,RADIUS,r

CYL3=CYLNDR/ZSMALL,PL1,ZSMALL,OUT,CYL1,RADIUS,r

5 The cylinder is tangent to a given cylinder, CYL1, with its axis passing through a given point, P1.

CYLNDR/CENTER,P1, $\begin{Bmatrix} \text{LARGE} \\ \text{SMALL} \end{Bmatrix}$,TANTO,CYL1

TABLE 4-17   STATEMENTS DEFINING CIRCULAR CYLINDERS   (continued)

No.	Known Condition	Statement Format

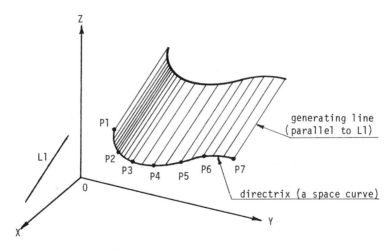

Examples

CYL2=CYLNDR/CENTER,P1,SMALL,TANTO,CYL1
CYL3=CYLNDR/CENTER,P1,LARGE,TANTO,CYL1

Figure 4-13   A tabulated cylinder.

TABLE 4-18   STATEMENTS DEFINING A TABULATED CYLINDER

No.	Known Condition	Statement Format
1	The cylinder is parallel to the X, Y, or Z axis, and its directrix is defined on the Y-Z, Z-X, or X-Y plane by a set of given points (yi, zi), (xi, zi) or (xi, yi), $i = 1, \ldots, n$, and the respective slopes si, $i = 1, \ldots, n$, or normal angles ni, $i = 1, \ldots, n$, at selected given points. The angle of a tangent or normal to the directrix is measured from the $+X$ (on X-Y plane), $+Y$ (on Y-Z plane), or $+Z$ (on Z-X plane) axis in counterclockwise direction.	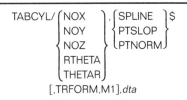

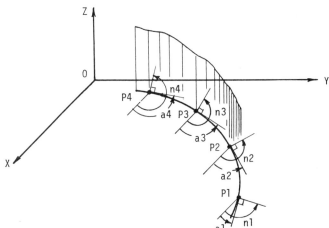

Comment The data regarding the given points, slopes, and normal angles are specified in the last element of this statement, *dta* (see Table 4-19). The first vocabulary word specifies both the direction of the cylinder and the plane containing the data regarding the given set of points:

> NOZ, RTHETA, and THETAR: The cylinder is parallel to the Z axis. The given points listed in *dta* are in the X-Y plane and are specified by the Cartesian coordinates (xi, yi), $i = 1, \ldots, n$, or cylindrical coordinates (ri, ai) and (ai, ri).

> NOX and NOY: The given points listed in *dta* are on the Y-Z and X-Z planes, respectively, they are specified by Cartesian coordinates (yi, zi) and (zi, xi), respectively.

The second vocabulary word specifies the routine to be used by the NC processor to calculate and generate the directrix:

> SPLINE: The directrix is generated by interpolating the given points (xi, yi, zi), $i = 1, \ldots, n$, with or without the slope or normal angle at selected given points. Each point (xi, yi, zi), slope si, or

TABLE 4-18   STATEMENTS DEFINING A TABULATED CYLINDER   (continued)

No.	Known Condition	Statement Format

normal angle ni is counted as a data element. The minimum number of given data elements is three (e.g., three given points, or two given points with a given slope or normal angle). This option produces the smoothest curve (directrix) for the given condition.

PTSLOP: The directrix is generated by interpolating the given points, with its slope at each given point, Pi, equal to the given value, si. At least two points and two slopes should be specified in *dta*.

PTNORM: The directrix is generated by interpolating the given points, with its normal angle at each given point, Pi, equal to the given value ni. At least two points and two normal angles should be specified in *dta*.

TRFORM: This vocabulary word specifies the transformation (translation, rotation, or both), represented by a defined matrix, M1, needed to convert the tabulated cylinder into the desired one.

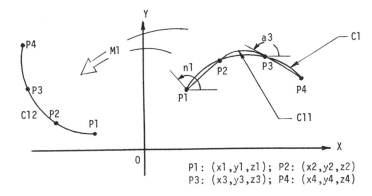

P1: $(x1,y1,z1)$; P2: $(x2,y2,z2)$
P3: $(x3,y3,z3)$; P4: $(x4,y4,z4)$

Examples
C1=TABCYL/NOZ,SPLINE,$x1,y1,x2,y2,x3,y3,x4,y4$
C11=TABCYL/NOZ,SPLINE,$x1,y1$,NORMAL,$n1,x2,y2,x3,y3$,SLOPE,$s3,x4,y4$
C12=TABCYL/NOZ,SPLINE,TRFORM,M1,$x1,y1,x2,y2,x3,y3,x4,y4$

In the first two examples, the same set of given points is used to define cylinders C1 and C11. However, the results are not the same, because there are additional constraints (slope $s3$ and normal angle $n1$) on the directrix of the cylinder C11. In the third example, M1 is the rotation matrix needed to transform cylinder C1 into cylinder C12.

No.	Known Condition	Statement Format
2	The cylinder is parallel to a given vector, V1, with its directrix defined by a set of given points (xi, yi, zi), $i = 1, \ldots, n$, and the respective slopes si or normal angles ni $(i = 1, \ldots, n)$, at selected given points.	TABCYL/XYZ,SPLINE,[TRFORM,M1,]V1,*dta*

(continued)

TABLE 4-18   STATEMENTS DEFINING A TABULATED CYLINDER   (continued)

No.	Known Condition	Statement Format

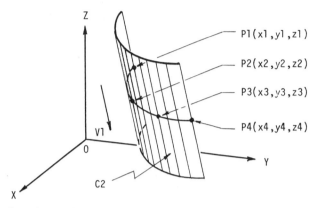

Comment   The discussion on the vocabulary word SPLINE in statement format No. 1 applies. The content of *dta* is

$$x1,y1,z1, \begin{bmatrix} \begin{Bmatrix} \text{SLOPE,}s1, \\ \text{NORMAL,}n1, \end{Bmatrix} \end{bmatrix} x2,y2,z2, \begin{bmatrix} \begin{Bmatrix} \text{SLOPE,}s2, \\ \text{NORMAL,}n2, \end{Bmatrix} \end{bmatrix} \ldots,$$

$$xn,yn,zn, \begin{bmatrix} \begin{Bmatrix} \text{SLOPE,}sn \\ \text{NORMAL,}nn \end{Bmatrix} \end{bmatrix}$$

Example
C2=TABCYL/XYZ,SPLINE,V1,x1,y1,z1,x2,y2,z2,x3,y3,z3,x4,y4,z4

A tabulated cylinder defined by the TABCYL statement, with input points specified in the coordinate planes, has the following characteristics:

1. The generated directrix consists of a series of cubic polynomial curves connecting two adjacent points specified in the statement. The calculation (or curve-fitting) routine in the NC processor ensures that the slopes of the two adjacent curves at a specified point are the same, so that a smooth curve passing through all the specified points is generated. If the slopes or normal angles are specified in the statement, they will be used by the curve-fitting routine to calculate the cubic polynomial curves.

2. If the slopes at the first and last specified points have not been specified in the statement, they are determined in such a manner that a smooth curve passing through the specified points results. Otherwise, the specified slopes or normal angles are used.

3. The generated directrix does not terminate at the first and last points specified in the TABCYL statement but is extended beyond them as straight lines by 10 units. The slopes of these extended lines are, respectively, the same as those of the directrix at the two end points.

TABLE 4-19 THE CONTENTS OF _dta_ IN THE TABCYL STATEMENT

Vocabulary Word	Given Points on Plane	SPLINE	PTSLOP	PTNORM
NOZ		$x1, y1, x2, y2, \ldots, xn, yn$	$x1, y1, s1, x2, y2, s2, \ldots,$ xn, yn, sn	$x1, y1, n1, x2, y2, n2,$ $\ldots, xn, yn, nn$
RTHETA	X-Y	$r1, a1, r2, a2, \ldots, rn, an$	$r1, a1, s1, r2, a2, s2, \ldots,$ rn, an, sn	$r1, a1, n1, r2, a2, n2,$ $\ldots, rn, an, nn$
THETAR		$a1, r1, a2, r2, \ldots, an, rn$	$a1, r1, s1, a2, r2, s2, \ldots,$ an, rn, sn	$a1, r1, n1, a2, r2, n2,$ $\ldots, an, rn, nn$
NOX	Y-Z	$y1, z1, y2, z2, \ldots, yn, zn$	$y1, z1, s1, y2, z2, s2, \ldots,$ yn, zn, sn	$y1, z1, n1, y2, z2, n2,$ $\ldots, yn, zn, nn$
NOY	Z-X	$z1, x1, z2, x2, \ldots, zn, xn$	$z1, x1, s1, z2, x2, s2, \ldots,$ zn, xn, sn	$z1, x1, n1, z2, x2, n2,$ $\ldots, zn, xn, nn$
Remark		1. If needed, a slope, si, or normal angle, ni, at point i can be specified after the coordinates of point i, with a vocabulary word, SLOPE or NORMAL, preceding it, e.g., $x1, y1, x2, y2,$ SLOPE, $s2, \ldots,$ xn, yn or $x1, y1,$ NORMAL, $n1, x2, y2,$ SLOPE, $s2, \ldots, xn, yn$ 2. The minimum number of points is 3. 3. This option generates the smoothest curve for the given set of points.	The minimum number of given points is 2.	The minimum number of given points is 2.

After processing a TABCYL statement, the NC processor prints the information needed to verify, correct, or modify the directrix. A description of the contents
and format of the printout is given in Appendix C.

The cylindrical surfaces defined by statements CYLNDR and TABCYL are of
infinite length.

A surface of rotation is the sweep of a plane curve revolving about a fixed line
or axis that is in the plane containing the curve. Examples are circular cylinders,
cones, spheres, and toruses. The statements defining *cones* and *spheres* are

CONE/......

and

SPHERE/......

respectively. Details concerning statement formats and usage are given in Tables
4-20 and 4-21. A *torus* (Fig. 4-14) can be defined by the statement

TORUS/P1,V1,$r1$,$r2$

where P1 is the center of the torus, V1 is a vector parallel to the axis of the torus, $r1$
is the distance from its center to the center of its generating circle, and $r2$ is the radius of the generating circle.

TABLE 4-20 STATEMENTS DEFINING A CONE

No.	Known Condition	Statement Format
1	The cone has its vertex at a given point, P1 (x, y, z), and a half vertex angle of a, with its axis parallel to a given vector, V1 (i, j, k).	CONE/$\left\{\begin{array}{l} \text{CANON},x,y,z,i,j,k,\cos(a) \\ \text{P1,V1},a \end{array}\right\}$

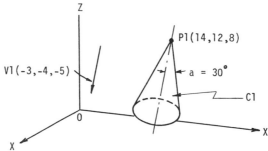

Comment   The first option is the standard (canonical) form in which the NC processor keeps the data
of a defined cone. The values i, j, and k should be, respectively, the X, Y, and Z components of the
unit vector of V1. In the second option, V1 can be a vector other than a unit one.

Examples
C1=CONE/CANON,14,12,8,−0.4243,−0.5657,−0.7071,0.8660
C2=CONE/P1,V1,30

TABLE 4-20   STATEMENTS DEFINING A CONE   (continued)

No.	Known Condition	Statement Format
2	The cone is coaxial with and at a distance, d, from a given cone, C1.	CONE/PARLEL,C1,STEP,$\begin{Bmatrix} \text{IN} \\ \text{OUT} \end{Bmatrix}$,$d$

Comment   The modifiers IN and OUT indicate that the defined cone is inside or outside of given cone C1. The value d is the distance between two surfaces.

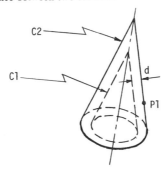

Examples
C2=CONE/PARLEL,C1,STEP,OUT,d
C1=CONE/PARLEL,C2,STEP,IN,d

No.	Known Condition	Statement Format
3	The cone is coaxial with a given cone, C1, and passes through a given point, P1.	CONE/PARLEL,C1,THRU,P1

Example   In the figure for statement format No. 2 we have

C2=CONE/PARLEL,C1,THRU,P1

TABLE 4-21   STATEMENTS DEFINING SPHERES

No.	Known Condition	Statement Format
1	The sphere has its center at a given point, P1 (x, y, z), with a radius, r.	SPHERE/$\begin{Bmatrix} x,y,z,r \\ \text{P1},r \\ \text{CENTER,P1,RADIUS},r \end{Bmatrix}$
2	The sphere has its center at a given point, P1, and passes through another given point, P2.	SPHERE/CENTER,P1,P2

Comment   The symbol P1 immediately after the vocabulary word CENTER should be the center of the sphere.

No.	Known Condition	Statement Format
3	The sphere has its center at a given point, P1, and is tangent to a given plane, PL1.	SPHERE/CENTER,P1,TANTO,PL1

(continued)

TABLE 4-21 STATEMENTS DEFINING SPHERES (continued)

No.	Known Condition	Statement Format
4	The sphere passes through four given points, P1, P2, P3, and P4.	SPHERE/P1,P2,P3,P4
5	The sphere is tangent to three given planes, PL1, PL2, and PL3, with a given radius, *r*.	SPHERE/ $\begin{Bmatrix} \text{XLARGE} \\ \text{XSMALL} \\ \text{YLARGE} \\ \text{YSMALL} \\ \text{ZLARGE} \\ \text{ZSMALL} \end{Bmatrix}$,PL1, $\begin{Bmatrix} \text{XLARGE} \\ \text{XSMALL} \\ \text{YLARGE} \\ \text{YSMALL} \\ \text{ZLARGE} \\ \text{ZSMALL} \end{Bmatrix}$,$ PL2, $\begin{Bmatrix} \text{XLARGE} \\ \text{XSMALL} \\ \text{YLARGE} \\ \text{YSMALL} \\ \text{ZLARGE} \\ \text{ZSMALL} \end{Bmatrix}$,PL3,RADIUS,*r*

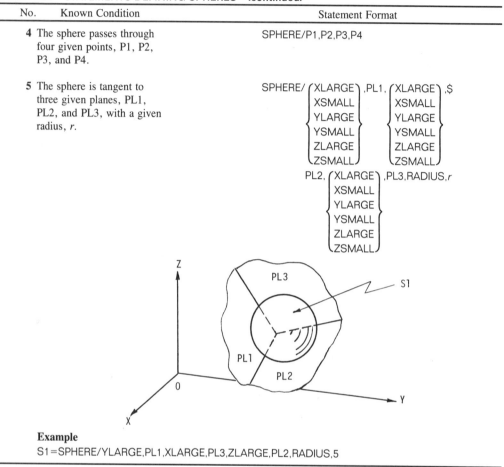

Example

S1=SPHERE/YLARGE,PL1,XLARGE,PL3,ZLARGE,PL2,RADIUS,5

4.8 STATEMENTS DEFINING RULED SURFACES

In engineering design, we often need to have a surface that makes a smooth transition from one cross section to another. The forms of the two cross sections might be completely different, as can be seen in Fig. 4-15. One of the methods by which we design this smooth transition surface is to join the corresponding points on the two planar curves that are the profiles of the two cross sections by straight lines, a method called *ruling* (Fig. 4-16). A pair of corresponding points on two planar curves can be found as follows:

1. Draw the baseline, or chord, connecting the starting and end points on each curve.

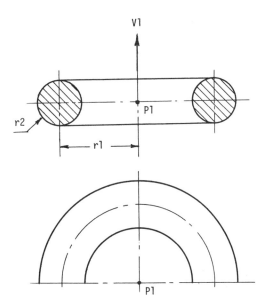

Figure 4-14   A torus and its parameters.

2. Locate the points *n* and *m* on the two baselines, which have the same percentage distance from the respective starting points.
3. Draw the lines perpendicular to the baselines at points *n* and *m*. The intersections of the two lines with the two curves are the corresponding points *p* and *q*.

A surface formed by this method is called a *ruled surface* in APT. It is evident from the description presented above that there are restrictions on the form of the two planar curves. A planar curve, with a certain part of it lying outside the range defined by the starting and end points, or with more than one intersection with the line perpendicular to the baseline, should be avoided (see area D in Fig. 4-17). If it is necessary to form a ruled surface based on such a curve, then it should be divided into two or more sections (see curves C1 and C2 in Fig. 4-18), each of which has only one intersection with the line perpendicular to its baseline. The other planar curve should also be divided into the same number of sections (C3 and C4) according to the design requirement.

The two planar curves used to define a ruled surface are usually the cross sections of two surfaces, that is, the intersections of the two surfaces with two planes (Fig. 4-15[a]). Therefore the elements needed for defining ruled surfaces in APT are the two surfaces, the two planes that intersect the two surfaces, and the starting and end points of the two planar curves. The statement defining a ruled surface is

RLDSRF/......

and its formats and usage are listed in Table 4-22.

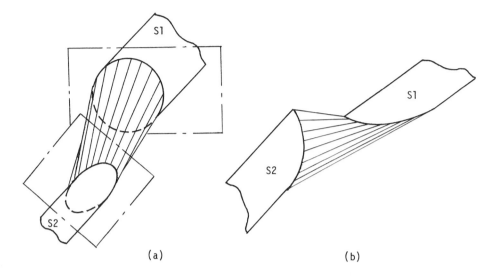

(a)   (b)

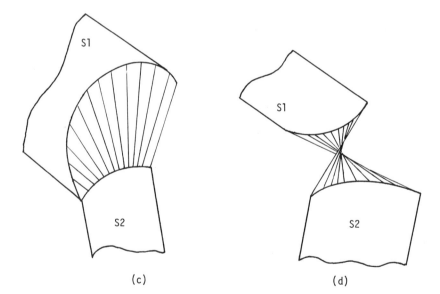

(c)   (d)

Figure 4-15 Transition surfaces.

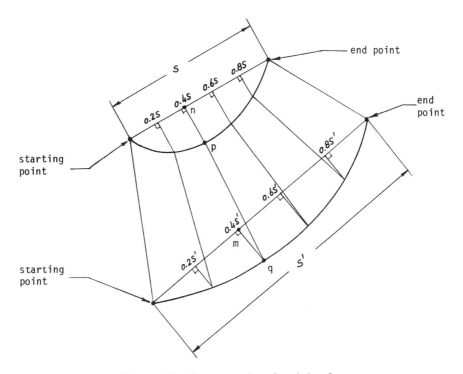

Figure 4-16   The construction of a ruled surface.

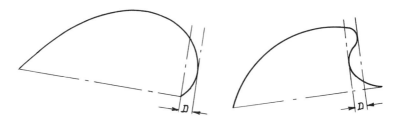

Figure 4-17   Planar curves not to be used to form a ruled surface.

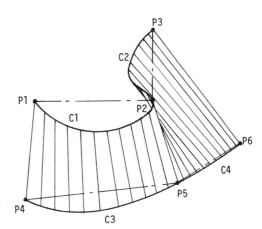

Figure 4-18   The upper planar curve has two intersections with the line perpendicular to the baseline. The ruled surface is formed by dividing each of the planar curves into two sections.

TABLE 4-22 STATEMENTS DEFINING RULED SURFACES

No.	Known Condition	Statement Format
1	The two planar curves are the intersections of two given surfaces, S1 and S2, with two planes, PL1 and PL2, respectively. Plane PL1 is defined by the given starting and end points, P1 and P2, of the planar curve, and another given point, P3, or a given vector, V1; plane PL2 is defined by the given starting and end points, P4 and P5, of the planar curve and another given point, P6, or a vector, V2. Point P3 can be any point other than those on the baseline on plane PL1; point P6 can be any point other than those on the baseline on plane PL2.	RLDSRF/S1,P1,P2, $\begin{Bmatrix} P3 \\ V1 \end{Bmatrix}$,S2,P4,P5, $\begin{Bmatrix} P6 \\ V2 \end{Bmatrix}$

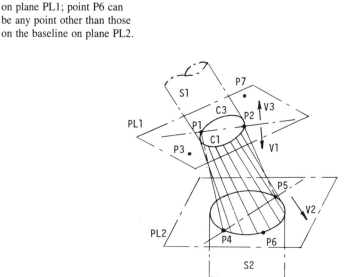

Comment The elements following surface S1 (i.e., points P1, P2, and P3 or V1) should be on the same plane PL1; those following S2 should be on plane PL2. The baseline connecting the starting and end points of a curve divides the curve into two sections. The third element following the surface indicates the section to be chosen as the boundary of the ruled surface. For example, if point P3 or vector V1 is specified, then the section on the same side of the baseline (i.e., the curve C1) is chosen. Conversely, if point P7 or vector V3 is specified, then curve C3 is chosen. Therefore points P3 and P6 in the statement should not be on the baselines. For the same reason, vectors V1 and V2 should not be parallel to the baselines, P1P2 and P4P5, respectively, nor in a direction at a small angle (e.g., less than 5 degrees) with the baseline. A planar curve should have no intersection other than the starting and end points with the baseline; otherwise, it should be divided into sections.

TABLE 4-22   STATEMENTS DEFINING RULED SURFACES   (continued)

No.	Known Condition	Statement Format

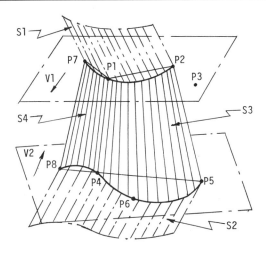

Examples
S3=RLDSRF/S1,P1,P2,P3,S2,P4,P5,P6
S4=RLDSRF/S1,P1,P7,V1,S2,P4,P8,V2

2 One of the surfaces in the
above format is a given
point, P7.

RLDSRF/S1,P1,P2,$\begin{Bmatrix} P3 \\ V1 \end{Bmatrix}$,P7

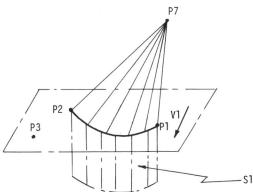

4.9 THE STATEMENT DEFINING A POLYCONIC SURFACE

In engineering design, the profile of a part of a complicated shape can usually be
defined on the basis of a series of cross-sectional curves; the profile along the
longitudinal direction can be determined by connecting the corresponding points
on the cross-sectional curves with straight lines or curves described by polynomial
functions.

The *polyconic surface* in APT is a three-dimensional surface whose cross sections, being perpendicular to the longitudinal axis of the surface, are approximated as conics and whose profile along the longitudinal axis is described by polynomial functions. For example, the winglike surface shown in Fig. 4-19 is to be approximated as a polyconic surface. The longitudinal axis of this part is parallel to the X axis. To describe an arbitrary cross section of the part, for example, at $x = x'$, as a conic curve, one would construct a parallelogram, P1-P4-P2-P3, with its two sides, respectively, being the tangent lines at the two end points, P1 and P2, of the cross-sectional curve. Seven parameters are needed to describe this cross-sectional curve: the coordinates L and H of point P1, the incremental coordinates from point P1 to point P2 and from point P3 to point P2 (i.e., C, B, A and D), and the ratio $K = L_2/L_1$, which determines the intersection of the cross-sectional curve with the diagonal P3-P4.*

The form of the cross section varies with the X coordinate; hence these seven parameters are functions of x. These variations, $A(x)$, $B(x)$, $C(x)$, $D(x)$, $L(x)$, $H(x)$, and $K(x)$, can be described by polynomial functions

$$\left. \begin{aligned} A(x) &= P_{A0}{}' + P_{Aq}{}' x^{\frac{1}{2}} + P_{A1}{}' x + P_{A2}{}' x^2 + P_{A3}{}' x^3 + P_{A4}{}' x^4 + P_{A5}{}' x^5 \\ &\quad + P_{A6}{}' x^6 + P_{A7}{}' x^7 \\ &\quad \cdots\cdots\cdots\cdots\cdots\cdots\cdots\cdots\cdots\cdots\cdots\cdots\cdots\cdots\cdots\cdots\cdots\cdots \\ K(x) &= P_{K0}{}' + P_{Kq}{}' x^{\frac{1}{2}} + P_{K1}{}' x + P_{K2}{}' x^2 + P_{K3}{}' x^3 + P_{K4}{}' x^4 + P_{K5}{}' x^5 \\ &\quad + P_{K6}{}' x^6 + P_{K7}{}' x^7 \end{aligned} \right\}$$

where x varies between $x1$ and $x2$ (i.e., $x \in [x1,x2]$). For simplification of the mathematical representation, it is more convenient to normalize the variable x by substituting it with the function

$$x = t \cdot (x2 - x1) + x1$$

where $t \in [0,1]$. Thus the polynomial functions listed above can be simplified as

$$\left. \begin{aligned} A(t) &= P_{A0} + P_{Aq} t^{\frac{1}{2}} + P_{A1} t + P_{A2} t^2 + P_{A3} t^3 + P_{A4} t^4 + P_{A5} t^5 \\ &\quad + P_{A6} t^6 + P_{A7} t^7 \\ &\quad \cdots\cdots\cdots\cdots\cdots\cdots\cdots\cdots\cdots\cdots\cdots\cdots\cdots\cdots\cdots\cdots\cdots\cdots\cdots \\ K(t) &= P_{K0} + P_{Kq} t^{\frac{1}{2}} + P_{K1} t + P_{K2} t^2 + P_{K3} t^3 + P_{K4} t^4 + P_{K5} t^5 \\ &\quad + P_{K6} t^6 + P_{K7} t^7 \end{aligned} \right\} \quad (4.5)$$

where $t \in [0,1]$.

In APT a polyconic surface is defined by the statement (Fig. 4-19)

```
PLC1=POLCON/CANON,j,m1,m2,m3,m4,m5,m6,m7,m8,m9,m10,m11,m12,t1,x1,x3,$
        PA0, PA1, PA2, PA3, PA4,PA5, PA6,PA7,$
        PB0, PB1, PB2, PB3, PB4,PB5, PB6,PB7,$
        PC0, PC1, PC2, PC3, PC4,PC5, PC6,PC7,$
        PD0, PD1, PD2, PD3, PD4,PD5, PD6,PD7,$
        PL0, PL1, PL2, PL3, PL4, PL5, PL6, PL7, $
        PH0, PH1, PH2, PH3, PH4,PH5, PH6,PH7,$
        PK0, PK1, PK2, PK3, PK4,PK5, PK6,PK7,$
        PAQ,PBQ,PCQ,PDQ,PLQ,PHQ,PKQ
```

*The actual intersection point of the cross-sectional curve of the defined polyconic with the diagonal may not be exactly the same as specified by the K value. The parameter K is used to define approximately the form of the cross-sectional curve. See Problem 8.10 for further discussion.

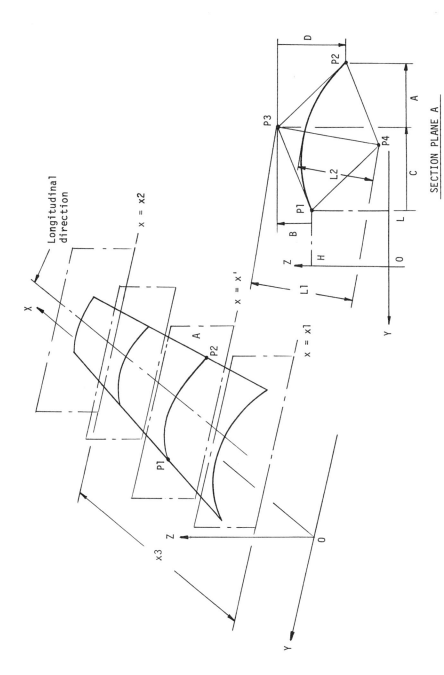

Figure 4-19 A winglike surface with its longitudinal direction parallel to the X axis.

where

$$\text{PLC1} = \text{symbol of the defined surface;}$$
$$j = 1 \text{ if PAQ} = \text{PBQ} = \ldots = \text{PKQ} = 0,$$
$$= 2 \text{ otherwise;}$$
$$m1, \ldots, m12 = \text{the 12 coefficients of the transformation matrix*}$$

$$\begin{bmatrix} m1 & m2 & m3 & m4 \\ m5 & m6 & m7 & m8 \\ m9 & m10 & m11 & m12 \\ 0 & 0 & 0 & 1 \end{bmatrix}$$

which transforms the polyconic surface from its actual position and orientation to the position and orientation appropriate for defining the surface, as illustrated in Fig. 4-19. No transformation is needed if the part longitudinal axis is parallel to the X axis. In such a case, $m1$, $m6$, and $m11$ have the value of 1, and the others are 0; $t1 =$ the machining allowance to be left on the surface; $x1 =$ the x coordinate of the first cross section; $x3 =$ the distance in X direction between the first and last cross sections; and PA0,PA1,...,PKQ = the coefficients in equation 4.5. They are determined on condition that its longitudinal direction is parallel to the X axis. If the part is not in an appropriate orientation, as described above, these coefficients are determined as if the part had been transformed to the required position and orientation by the transformation matrix defined by coefficients $m1$ through $m12$.

Example
The surface PLC1, shown in Fig. 4-20, is to be defined as a polyconic surface. It is first transformed to a normalized orientation (i.e., its longitudinal axis being parallel to the X axis), with the transformation matrix M1. To determine the coefficients of the seven polynomial functions in equation 4.5, we need to project lines C1, C2, and C3 onto the planes t-Z and t-Y. The resulting curves are $C1_y$, $C1_z$, $C2_y$, $C2_z$, $C3_y$, and $C3_z$, respectively. Suppose that these curves are as shown in the figure. Then we have (see also Fig. 4-19)

$$L(t) = C1_y(t) = 1 - 0.2t$$
$$H(t) = C1_z(t) = 1.7 - 0.2t + 0.1t$$
$$A(t) = C2_y(t) - C3_y(t) = -0.4 + 0.3t - (0.5 + 0.1t)$$
$$= -0.9 + 0.2t$$
$$B(t) = C3_z(t) - C1_z(t) = 3.0 - 0.1t^2 - (1.7 - 0.2t + 0.1t^2)$$
$$= 1.3 + 0.2t$$
$$C(t) = C3_y(t) - C1_y(t) = 0.5 + 0.1t - (1 - 0.2t) = -0.5 + 0.3t$$
$$D(t) = C2_z(t) - C3_z(t) = 1.7 + 0.5t - 0.6t^2 - (3.0 - 0.1t^2)$$
$$= -1.3 + 0.5t - 0.5t^2$$
$$K(t) = 0.7 - 0.05t - 0.15t^2$$

Since the square root terms are all zero in these expressions, the parameter j should have the value of 1. In addition, from Fig. 4-20, we have $x1 = 3.2$ and $x3 = 10$. Thus

*See Section 7.6.

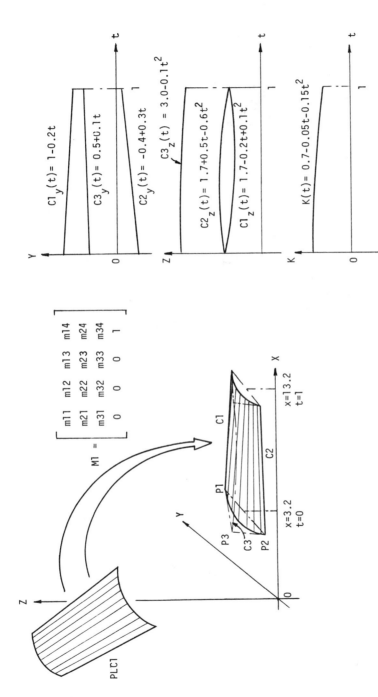

$C1_y(t) = 1 - 0.2t$

$C3_y(t) = 0.5 + 0.1t$

$C2_y(t) = -0.4 + 0.3t$

$C3_z(t) = 3.0 - 0.1t^2$

$C2_z(t) = 1.7 + 0.5t - 0.6t^2$

$C1_z(t) = 1.7 - 0.2t + 0.1t^2$

$K(t) = 0.7 - 0.05t - 0.15t^2$

$$M1 = \begin{bmatrix} m11 & m12 & m13 & m14 \\ m21 & m22 & m23 & m24 \\ m31 & m32 & m33 & m34 \\ 0 & 0 & 0 & 1 \end{bmatrix}$$

PLC1

$x = 3.2$
$t = 0$

$x = 13.2$
$t = 1$

Figure 4-20 A surface to be defined by POLCON statement. The actual orientation and position of the part are at PLC1. The part is transformed to an appropriate position and orientation with the aid of transformation matrix M1. The lines C1, C2, and C3 are then projected onto the t-Y and t-Z planes; the resulting curves are $C1_y$, $C1_z$, $C2_y$, $C2_z$, $C3_y$ and $C3_z$, respectively.

181

the statement for defining this polyconic surface, with zero allowance left on it, is as follows:

```
PLC1=POLCON/CANON,1,m11,m12,m13,m14,m21,m22,m23,m24,m31,m32,m33,m34,$
          0,3.2,10,$
          -0.9,0.2,0,0,0,0,0,0,$
          1.3,0.2,0,0,0,0,0,0,$
          -0.5,0.3,0,0,0,0,0,0,$
          -1.3,0.5,-0.5,0,0,0,0,0,$
          1,-0.2,0,0,0,0,0,0,$
          1.7,-0.2,0.1,0,0,0,0,0,$
          0.7,-0.05,-0.15,0,0,0,0,0,$
          0,0,0,0,0,0,0
```

4.10 DEFINING POINTS WITH STATEMENT ZSURF

It is often necessary to define a number of points located on a plane not parallel to the X-Y plane (Fig. 4-21). On the engineering drawing, the dimensions of these points are normally specified in the X-Y plane; their z coordinates are not explicitly indicated and therefore must be calculated. In such a case, we can make use of the ZSURF statement—together with the POINT statements, which contain the X and Y

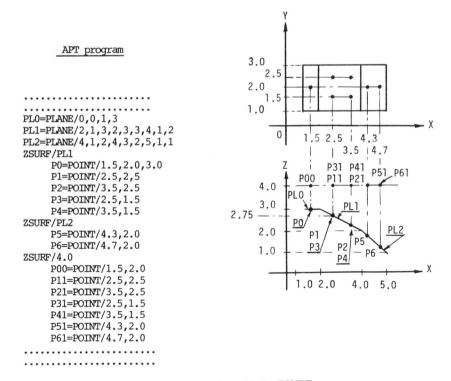

Figure 4-21 Defining points with the ZSURF statement.

coordinates of these points only — to eliminate tedious calculations. The format of the ZSURF statement is

$$\text{ZSURF/}\begin{Bmatrix} \text{PL1} \\ a,b,c,d \\ e \end{Bmatrix}$$

where PL1 is the plane on which the points to be defined are located; a, b, c, and d are the parameters of the equation defining the plane

$$ax + by + cz = d$$

e is the z coordinate of the plane if it is parallel to the X-Y plane.

Once the ZSURF statement is specified, the points located on the plane defined by the ZSURF statement can be defined by their x and y coordinates (i.e., their projections on the X-Y plane) only. The NC processor calculates the z coordinates by projecting lines from the points defined on the X-Y plane to the plane defined by the ZSURF statement. If no ZSURF statement is specified, the statement

ZSURF/0

applies (i.e., the plane is the X-Y plane). The ZSURF statement remains in effect until canceled by another ZSURF statement.

Fig. 4-21 presents an example that explains the meaning and usage of the ZSURF statement. As can be seen from the listed APT program, points P1 through P6 and P00 through P61 are defined by the x and y coordinates; the corresponding ZSURF statements will provide references for calculating the z coordinates of the points defined in the following POINT statements. When a point, such as P1, is referenced in a succeeding statement, the z coordinate, 2.75, is provided according to the calculation.

The ZSURF statement affects the point definition only and does not affect a POINT statement containing a z coordinate.

4.11 SOME COMMENTS ON APT GEOMETRIC DEFINITION STATEMENTS

Within the NC processor, the function of an APT geometric definition statement is to provide the conditions to calculate and define a geometric entity. The conditions vary with the statement formats used. As can be seen from the various formats listed in the previous sections for defining each type of geometric entity, however, the way that the NC processor defines a specific type of geometric entities is unique. This means that regardless of the given conditions or the statement formats used, geometric entities of the same type are always defined by a specific set of parameters, which are the coefficients of a standard equation, called the canonical form (see Appendix B). The calculation of these parameters on the basis of different given conditions is performed by the NC processor. For example, the canonical form of a circle is

$$(x, y, z, a, b, c, r)$$

where x, y, and z are the coordinates of the center point and a, b and c are the X, Y, and Z components of the unit vector, being perpendicular to the plane containing the circle. Since a circle in the NC processor is a cylinder parallel to the Z axis, then a, b, and c are 0, 0, and 1, respectively. The value r in the canonical form is the radius of the circle. A circle may be defined by three given points on the circle. Then the coordinates of the center and radius should be calculated by means of the coordinates of the three given points. Sometimes, a circle is defined by three given tangent lines. Calculation should accordingly be made with the use of a different routine to obtain the parameters of the canonical form. This clearly indicates that the geometric definition statements are powerful means by which we are able to define a geometric entity even without knowledge of the needed mathematical calculations.

For many of the geometric definition statements, there is only one solution to the given conditions. However, there are a number of statements that define geometric entities based on conditions that yield more than one solution. An additional modifier or modifiers are then needed in the statement to select the desired solution among the possible ones. Unfortunately, the use of modifiers is not clearly explained in the APT programming manuals.[1,2] This leads to confusion, in some cases, in selecting the correct modifier(s). The following are some typical examples.

Example 1
For statement format No. 3 in Table 4-7 generally there are two solutions to the given condition: two points on the circle and a radius. Accordingly, a directional modifier is needed, and this statement format is designed to allow for two possible solutions. There is one special case in which the given radius is equal to the half distance between the given points; the number of solutions is then reduced to one. Then how do we select the right modifier?

Example 2
Statement format No. 2 in Table 4-10 defines a circle on the basis of a given point, a given tangent circle, and a given radius. This set of given conditions poses a complicated situation: the number of solutions can range from none to four, depending on the relationship between r and d (see the accompanying figure in Table 4-10). Now the problem is, how do we specify the modifiers in this circle statement when the number of possible solutions is less than four?

There are a few geometric definition statements whose meaning has not been clearly explained in the APT programming manual. A typical example is the point that is defined as the center of a cylinder (statement format No. 4 in Table 4-2). What is the geometic center of a cylinder? Within the NC processor, a cylinder defined by a CYLNDR statement containing the vocabulary word CENTER does have a specified center. However, the center of a cylinder defined by a CYLNDR statement without the word CENTER was not discussed or explained in the APT programming manuals.[1,2] The situation wherein the geometric entity defined by a LINE statement can be a line on the X-Y plane in some cases or a plane perpendicular to the X-Y plane, as described in Section 4.2, is another typical example.

Because of the large number of geometric entities and statement formats, it is not possible to list and explain all of them in detail within the scope of this book. Nevertheless, the rules and method stated below should be useful to a programmer who is writing out the details of a statement.

When choosing a modifier, a programmer must have a clear understanding of the number and locations of the possible solutions. As a rule, each modifier in a statement should discriminate between two possible cases, such as on the left and right, in and out, in X-increasing and X-decreasing directions, and so on. For the case in which the two possible solutions are reduced to one, the modifier does not have any effect on the selection of a specific solution. As a result, any of the modifiers listed in the statement format can be selected. For example, a circle is defined by a tangent line, L1, a tangent circle, C1, and a given radius, r (see statement format No. 3 in Table 4-10). The directional modifier XLARGE, . . . , or YSMALL immediately after L1 in this statement is used to select the desired solution from the two possible ones, C2 and C3, which have been selected by the other two modifiers, that is, YSMALL for L1 and IN for C1 (see the figure accompanying statement format No. 3 in Table 4-10). If the radius, r, is equal to $d_1/2$, then there is only one tangent circle (solution) on the lower side of line L1 and inside circle C1. The two possible solutions, C2 and C3, are reduced to one. In this case, the directional modifier does not have any effect on selecting the result. Hence any of them can be specified in the statement. The example in Fig. 4-22 also shows that the dimensional modifiers SMALL and LARGE do not have an effect when only one solution remains after the other modifiers have been specified.

A useful method for ascertaining that a geometric entity has been correctly defined is to use the statement

PRINT/3,C1[,C2[,C3,[...]]]

to print out the canonical form of the defined geometric entities C1, C2, and so forth. For example, in statement format No. 5 in Table 4-10, the function of the directional modifiers XLARGE, . . . , and YSMALL, specified before line L1, is not well defined nor clearly explained in the manual. It seems that they are used to indicate the location of the defined circle with respect to given line L1. To understand clearly their function and to define the desired circle correctly, we can write a very

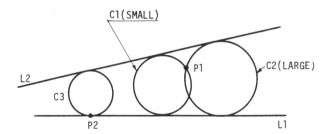

Figure 4-22   A circle defined by two tangent lines, L1 and L2, and a point, P1, on the circle. Generally, there are two solutions, C1 and C2; a dimensional modifier is then needed. If the given point is on one of the lines (e.g., point P2), then there is only one solution. In such a case, either of the dimensional modifiers, LARGE and SMALL, yields the same result, C3.

simple program, including only the needed geometric definition statements, as follows:

```
......................
......................
L1=LINE/X3,Y3,Z3,X4,Y4,Z4
C1=CIRCLE/CENTER,X2,Y2,Z2,RADIUS,R1
P1=POINT/X1,Y1,Z1
C2=CIRCLE/LARGE,M1,TANTO,L1,IN,TANTO,C1,THRU,P1
PRINT/3,C2
......................
......................
```

where M1 can be the modifier XLARGE, ... , or YSMALL. After having processed the program, the NC processor will print the canonical form in the program listing file. By changing solely the modifier M1 in this program, we obtain four canonical forms. Comparing these results, we are able to determine the function of these modifiers. For example, if we have

$$X1 = 0; Y1 = 1; Z1 = 0;$$
$$X2 = 1; Y2 = 3; Z2 = 0;$$
$$X3 = Y3 = Z3 = 0;$$
$$X4 = Y4 = 1; Z4 = 0;$$
$$R1 = 1$$

in the above program, then the canonical forms of circle C2 with different modifiers are as follows (only the parameters X, Y, and R are listed):

	C2		
M1	X	Y	R
XLARGE	−32.3839	45.8525	55.3216
XSMALL	−32.3839	45.8525	55.3216
YLARGE	−32.3839	45.8525	55.3216
YSMALL	−32.3839	45.8525	55.3216

The test results clearly indicate that the directional modifier M1 does not have any effect on defining circle C2. Actually, the direction modifier is used to indicate the relative position of the defined circle with respect to the other solution. It has an effect only when two solutions have the same diameter (see Problem 4.9). The method described here was also used to determine the center of a cylinder; the conclusion in statement format 4 in Table 4-2 was reached on the basis of the test results.

    A geometric entity that is used to define another geometric entity must have been defined previously. A nested geometric definition statement can be used, instead, in a statement to replace the symbol of a geometric entity that has not been defined. For example, in the program listed above, if the statement

L1=LINE/X3,Y3,Z3,X4,Y4,Z4

is missing, then we can write the statement defining circle C2 as follows:

C2=CIRCLE/LARGE,XLARGE,TANTO,(LINE/X3,Y3,Z3,X4,Y4,Z4),$
                IN,TANTO,C1,THRU,P2

if the line is not to be referenced in the following statements, or

C2=CIRCLE/LARGE,XLARGE,TANTO,(L1=LINE/X3,Y3,Z3,X4,Y4,Z4),$
                IN,TANTO,C1,THRU,P2

if the line, L1, is to be referenced in the following statements.

Besides errors in syntax and numbers, the most frequent errors in geometric definition statements are as follows:

1. The defined geometric entity is irrational. For example, a point is defined as the intersection of a line and a circle that do not intersect each other.
2. A geometric entity is defined on the basis of other geometric entities, one or several of which have not been defined.
3. A geometric entity has been redefined. The NC processor does not allow the symbol of a geometric entity to be redefined. For example, if P1 has been defined as the point (3,2,4), then it cannot be redefined or used as a symbol for other points or other types of geometric entities. If the programming requirement necessitates redefinition of a symbol, the statement

   CANON/ON

   should be specified before the symbol is redefined. This statement changes the mode of processing of the NC processor. As a result, the symbols assigned to geometric entities can be redefined in subsequent statements. The statement

   CANON/OFF

   has the opposite effect of CANON/ON. Once specified, it will return the NC processor to the default mode, whereby a symbol for a geometric entity is not allowed to be redefined.

4.12 THE MATHEMATICAL CALCULATION FUNCTIONS

In this section, we introduce the mathematical calculation functions that have been included in the APT-AC NC processor. As can be seen from the previous discussion, a geometric entity is eventually defined by one or several geometric parameters. In many cases, manual calculation of these parameters is complicated and time-consuming. Therefore the mathematical calculation functions incorporated in the NC processor can be used.

Table 4-23 gives a list of mathematical functions that can be performed by the NC processor, together with the equivalent mathematical expression for each statement. An argument that should be operated by a mathematical function should be included in the pair of parentheses after the vocabulary word defining the function. An argument can be a number, a symbol that represents a variable, or a mathematical expression. The following are some examples:

SINF(30)
SINF(A)   (*A* is a variable whose value is assigned by a previous statement.)
SINF(A-B) (The values of *A* and *B* are assigned by previous statements.)

Some of the statements can accept only the symbols of geometric entities as their arguments; they are DOTF, LNTHF, ANGLF, and DISTF.

Examples of the use of mathematical calculation statements in defining geometric entities are given in Section 4.13.

The use of the mathematical functions is not limited to the definition of geometric entities. They can also be used for any calculation needed in programming.

TABLE 4-23 THE STATEMENTS DEFINING MATHEMATICAL OPERATION

Statement	Equivalent Mathematical Expression and Meaning
Trigonometric functions	
SINF (A)	$\sin A$
COSF (A)	$\cos A$
TANF (A)	$\tan A$
COTANF (A)	$\cot A$
ASINF (A)	$\arcsin A$ $(-90° \leq \text{ASINF (A)} \leq 90°)$
ACOSF (A)	$\arccos A$ $(0° \leq \text{ACOSF (A)} \leq 180°)$
ATANF (A)	$\arctan A$ $(-90° \leq \text{ATANF (A)} \leq 90°)$
ATAN2F (A, B)	$\arctan (A/B)$ $(-90° \leq \text{ATAN2F (A, B)} \leq 90°)$
Other functions	
ABSF (A)	Absolute value of A
CBRTF (A)	$A^{\frac{1}{3}}$
EXPF (A)	e^A
INTGF (A)	The input argument A is truncated to an integer.
LOGF (A)	$\ln A$
LOG10F (A)	$\log_{10} A$
MAX1F (A1, A2, ..., An)	The largest of $A1, A2, ..., An$
MIN1F (A1, A2, ..., An)	The smallest of $A1, A2, ..., An$.
MODF (A1, A2)	The remainder in the division $A1/A2$.
SIGNF (A1, A2)	Assigning the sign of the second argument to the absolute value of the first argument, that is, $(\text{sign of A2}) \cdot |A1|$.
SQRTF (A)	$A^{\frac{1}{2}}$
DOTF (V1, V2)	Scalar product of two defined vectors V1 and V2. If $V1 = a_1 \cdot \mathbf{i} + b_1 \cdot \mathbf{j} + c_1 \cdot \mathbf{k}$ and $V2 = a_2 \cdot \mathbf{i} + b_2 \cdot \mathbf{j} + c_2 \cdot \mathbf{k}$, DOTF (V1, V2) is equivalent to $a_1 a_2 + b_1 b_2 + c_1 c_2$.
LNTHF (V1)	The magnitude of vector V1: $(a_1^2 + b_1^2 + c_1^2)^{\frac{1}{2}}$.
ANGLF	The angular relationship among several geometric entities can be obtained by means of this function.

TABLE 4-23 THE STATEMENTS DEFINING MATHEMATICAL OPERATION (continued)

Statement	Equivalent Mathematical Expression and Meaning
1. ANGLF (C1, PT1)	C1 — a defined circle; PT1 — a defined point. The angle defined by this statement is W.

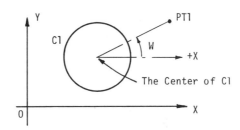

2. ANGLF (PT1, PT2, PT3)	PT1, PT2, PT3 — defined points. This statement defines the smaller included angle (absolute value) between lines PT1PT2 and PT2PT3.
3. ANGLF (L1)	L1 — a defined line. This statement defines the angle (absolute value) that the line L1 makes with the positive X axis in the upper half of the X-Y plane.

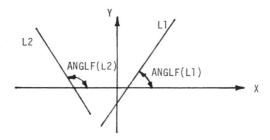

4. ANGLF (PL1, PL2)	The smaller included angle (absolute value) between two defined planes, PL1 and PL2.
5. ANGLF (L1, L2)	The smaller included angle (absolute value) between two defined lines, L1 and L2.
6 ANGLF (d, m, s)	Defines an angle in degrees (d), minutes (m), and seconds (s).
DISTF (G1, G2)	The distance (absolute value) between two parallel lines, two planes, or a line and a plane.
	G1 and G2 — symbols for defined lines or planes.
DISTF (P1, G1)	The distance (absolute value) between a point, P1, and a geometric entity, G1, which can be a point, line, plane, circle, cylinder, or sphere.
TYPEF (A)	A — a number, vocabulary word, or alphanumeric literal string.

TYPEF (A) = 1   If $A \leqslant 1.0 \cdot 10^{-14}$
TYPEF (A) = 2   If $A > 1.0 \cdot 10^{-14}$
TYPEF (A) = 3   If A is a vocabulary word
TYPEF (A) = 4   If A is an alphanumeric literal string

(continued)

TABLE 4-23   THE STATEMENTS DEFINING MATHEMATICAL OPERATION   (continued)

Statement	Equivalent Mathematical Expression and Meaning
Relational operator	
'EQ'	Equal to
'GE'	Greater than or equal to
'LE'	Less than or equal to
'NE'	Not equal to
'GT'	Greater than
'LT'	Less than

Example   The statement

C=A'EQ'B

means that if $A = B$ is true, then $C = 1$; if $A = B$ is false, then $C = 0$.

Figure 4-23 shows an example. The section of the part profile is composed of three lines, L1, L2, and L3. To define the cutter path correctly, one must know which line, L2 or L3, the cutter of radius R will be first in contact with when it moves along line L1 toward the left. The mathematical condition for three different cases is as follows:

1. The cutter will be first in contact with line L2 if $R < R1$
2. The cutter will be first in contact with both line L2 and line L3 if $R = R1$
3. The cutter will be first in contact with L3 if $R > R1$

where $R1$ is the radius of the common tangent circle to the three lines L1, L2, and L3. Thus one of the problems in defining the cutter path is to calculate radius $R1$ on the basis of the three defined lines, L1, L2, and L3.

From Fig. 4-23, we have

$$R1 = S1 \cdot \sin A = S2 \cdot \sin B$$

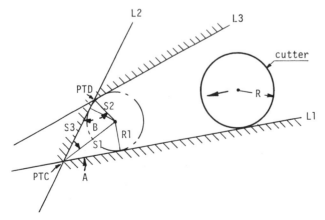

Figure 4-23   Use of mathematical calculation statements to determine radius R1 of the circle that is tangent to the three lines L1, L2, and L3.

and

$$S3 = S1 \cdot \cos A + S2 \cdot \cos B$$

where

$S1 = $ distance between point PTC and the center of the common tangent circle
$S2 = $ distance between point PTD and the center of the common tangent circle
$S3 = $ distance between points PTC and PTD

The solution to the equations listed above is

$$R1 = S3/(\cot A + \cot B)$$

Therefore the section of the APT program for determining radius $R1$ on the basis of the defined lines L1, L2, and L3 can be as follows:

```
. . . . . . . . . . . . . .
. . . . . . . . . . . . . .$$LINES L1, L2 AND L3 ARE DEFINED IN THIS SECTION
. . . . . . . . . . . . . .
A    = (ANGLF(L1,L2))/2
B    = (180 - ANGLF(L2,L3))/2
PTC  = POINT/INTOF,L1,L2
PTD  = POINT/INTOF,L2,L3
V1   = VECTOR/PTC,PTD
S3   = LNTHF(V1)
R1   = S3/(COTANF(A)+COTANF(B))
. . . . . . . . . . . . . .
. . . . . . . . . . . . . .
. . . . . . . . . . . . . .
```

4.13 APPLICATIONS OF GEOMETRIC DEFINITION STATEMENTS

Tables 4-1 through 4-22 list most of the statements and formats needed to define the geometric entities allowed in APT. These various statement formats provide a considerable degree of freedom for a programmer to define geometric entities. A geometric entity can be defined on the basis of information that we already know or, further, on the basis of geometric entities defined in previous statements. The usage of the latter can greatly reduce the number of calculations required. For example, suppose that we need to move an end mill to the innermost position of a corner formed by two straight lines (Fig. 4-24) but do not know the exact position of that point. In APT programming, we can determine the position and order the tool to go to that point without need for calculation. The procedure is as follows:

1. Define the two lines L1 and L2 on the basis of the given condition, using the LINE statement.
2. Define circle C1, which is tangent to these two lines and on the desired side of each line, with its radius, r, equal to that of the end mill.
3. Define point P1 as the center of circle C1.
4. Write a point-to-point motion statement (GOTO)* to order the tool to go to point P1.

*The motion statements are discussed in Chapter 5.

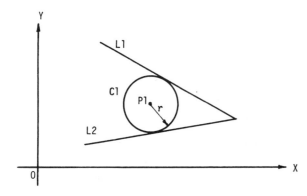

Figure 4-24 A circle tangent to two lines, L1 and L2.

The corresponding program segment may appear as follows:

```
.....................
.....................
L1=LINE/......
L2=LINE/......
C1=CIRCLE/YSMALL,L1,YLARGE,L2,RADIUS,r
P1=POINT/CENTER,C1
.....................
.....................
GOTO/P1
.....................
.....................
```

Such motion can also be realized more easily by using the start-up motion statement described in Chapter 5.

We can also obtain the parameters of the canonical form of a geometric entity through the use of the statement OBTAIN, whose format is

```
OBTAIN, ⎧ POINT  ⎫  /C1,[A1],[A2],....,An
        ⎪ LINE   ⎪
        ⎨ CIRCLE ⎬
        ⎪ ...... ⎪
        ⎩ ...... ⎭
```

The second element in this statement is the vocabulary word indicating the type of geometric entity represented by the symbol C1. The specified geometric entity C1, from which the canonical parameters are to be obtained, should have been defined in a previous statement. Elements A1, ... , An are the symbols to be assigned to the first, second, ... , and last canonical parameters of the geometric entity C1, respectively. If only some of the canonical parameters are required, then only the symbols corresponding to those parameters are specified (the commas and the last symbol, An, cannot be omitted). Thus, in the previous example, the x and y coordinates of the circle center point, P1, can be obtained with the use of the statement

```
OBTAIN,CIRCLE/C1,X1,Y1,,,,,R1
```

which assigns X1 and Y1 as the symbols (scalar variable) of the x and y coordinates of the center point, P1. In subsequent statements, we can reference the x or y coordinates of point P1 by simply specifying X1 or Y1.

We can also make use of the mathematical calculation functions provided in APT to compute the data needed for defining a geometric entity. Fig. 4-25 gives an example of a geometric entity definition using APT mathematical calculation statements. The part is a Wankel (rotary) engine chamber, with an epitrochoidal profile, which can be described by the parametric equation

$$x = (3/2) \cdot R \cdot \cos(a) + (R/4) \cdot \cos(3\,a)$$
$$y = (3/2) \cdot R \cdot \sin(a) + (R/4) \cdot \sin(3\,a)$$

where R is the radius of the stationary circle. Its slope is defined by the expression

$$dy/dx = -[\cos(a) + (1/2) \cos(3\,a)]/[\sin(a) + (1/2) \sin(3\,a)]$$

Because there are no statements in APT for defining an epitrochoid, we calculate the x and y coordinates of points P1 through P36. Then we define the profile using the statement LCONIC. Assume that the incremental parametric angle from one point to another is 10 degrees and R is 40 mm. The program can be as follows:

```
. . . . . . . . . . . . . . . . . . . .
. . . . . . . . . . . . . . . . . . .
X1=60*COSF(10)+10*COSF(30)
Y1=60*SINF(10)+10*SINF(30)
X2=60*COSF(20)+10*COSF(60)
Y2=60*SINF(20)+10*COSF(60)
. . . . . . . . . . . . . . . . . . . .
. . . . . . . . . . . . . . . . . . .
X36=60*COSF(360)+10*COSF(1080)
Y36=60*SINF(360)+10*SINF(1080)
P1=POINT/X1,Y1
P2=POINT/X2,Y2
. . . . . . . . . . . . . . . . . . . .
P36=POINT/X36,Y36
E1=LCONIC/5PT,P1,P2,P3,P4,P5
E2=LCONIC/5PT,P5,P6,P7,P8,P9
. . . . . . . . . . . . . . . . . . . .
E9=LCONIC/5PT,P33,P34,P35,P36,P1
. . . . . . . . . . . . . . . . . . . .
. . . . . . . . . . . . . . . . . . . .
```

In this program the profile of the rotary engine chamber is divided into nine sections (i.e., the loft conics E1 through E9). Each section is defined by an LCONIC/5PT statement. To guarantee the slope continuity at the joining point of two sections, one should use statement format LCONIC/3PT2SL, instead. Further discussion on the definition of the Wankel engine profile is given in Section 8.4.

It is also possible to define the epitrochoidal profile with the geometric definition statement TABCYL.

To conclude this chapter, we use the part given in Fig. 4-26 as an example to explain the procedures for defining a geometric entity. The inner and outer profiles of the part are to be cut on a milling machine. The outer profile comprises lines and circular arcs of different radii; the inner profile comprises circular arcs, elliptical

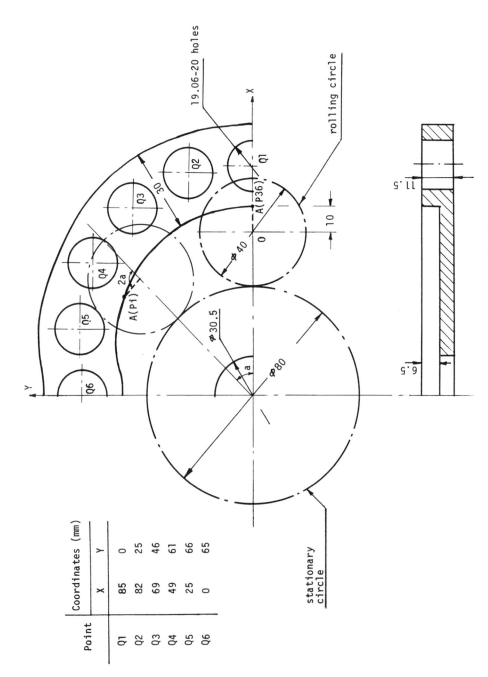

Point	Coordinates (mm)	
	X	Y
Q1	85	0
Q2	82	25
Q3	69	46
Q4	49	61
Q5	25	66
Q6	0	65

19.06-20 holes

rolling circle

stationary circle

Figure 4-25 A Wankel (rotary) engine chamber. The inner profile of the chamber is an epitrochoid, which is generated by the moving point A on the rolling circle.

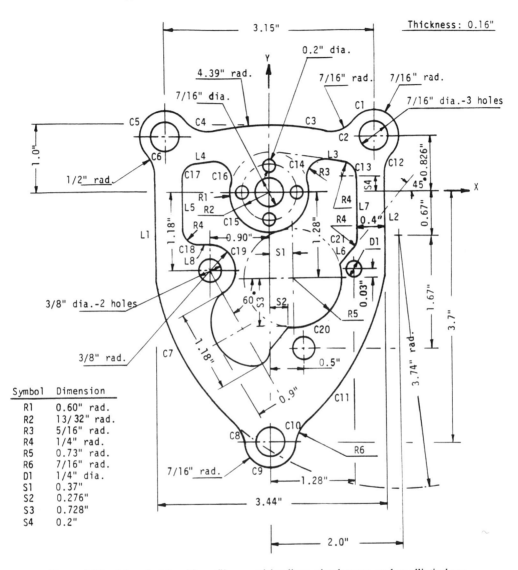

Figure 4-26   A based plate with profile comprising lines, circular arcs, and an elliptical arc.

arcs, and lines. These geometric entities guide the end mill during the machining process. Therefore they should be defined before the cutter path is specified.

First, the programmer must select the most convenient coordinate system for defining the geometric entities or elements of the part. Thus the coordinate system *X-Y* shown in the figure is selected with the origin on the top surface of the part. Then the elements of the profiles can be defined by using the pertinent geometric definition statements and mathematical calculation statements. The nested expression can also be used to replace an undefined element and then to simplify the definition. The segment of the APT program that defines the elements required for specifying the cutter path can be as follows:

```
.....................
.....................
C1=CIRCLE/(3.15/2),0.826,(7/16)
C3=CIRCLE/0,-3.39,4.39
C5=CIRCLE/(-3.15/2),0.826,(7/16)
C2=CIRCLE/YLARGE,OUT,C1,OUT,C3,RADIUS,(7/16)
C4=CIRCLE/YLARGE,OUT,C3,OUT,C5,RADIUS,(7/16)
L1=LINE/YAXIS,-1.72
L2=LINE/YAXIS,1.72
C6=CIRCLE/XSMALL,L1,YSMALL,OUT,C5,RADIUS,0.5
C12=CIRCLE/XLARGE,L2,YSMALL,OUT,C1,RADIUS,0.5
C7=CIRCLE/2.0,-0.67,3.74
C11=CIRCLE/-2.0,-0.67,3.74
C9=CIRCLE/0,-3.7,(7/16)
C8=CIRCLE/XSMALL,OUT,C9,OUT,C7,RADIUS,(7/16)
C10=CIRCLE/XLARGE,OUT,C9,OUT,C11,RADIUS,(7/16)
C13=CIRCLE/(1.72-0.4-0.25),0.2,0.25
C17=CIRCLE/-(1.72-0.4-0.25),0.2,0.25
C15=CIRCLE/0,0,0.6
L3=LINE/LEFT,TANTO,C15,LEFT,TANTO,C13
L4=LINE/RIGHT,TANTO,C15,RIGHT,TANTO,C17
C14=CIRCLE/YSMALL,L3,XLARGE,OUT,C15,RADIUS,(5/16)
C16=CIRLCE/YSMALL,L4,XSMALL,OUT,C15,RADIUS,(5/16)
L7=LINE/YAXIS,1.32
L5=LINE/YAXIS,-1.32
X1=-0.9+(3/8+0.59)*COSF(60)
Y1=-1.18-(3/8+0.59)*SINF(60)
E1=ELLIPS/CENTER,(POINT/X1,Y1),0.59,0.45,-60
C19=CIRCLE/-0.9,-1.18,(3/8)
L8=LINE/YLARGE,TANTO,C19,ATANGL,0
C18=CIRCLE/YLARGE,L8,XLARGE,L5,RADIUS,0.25
C20=CIRCLE/0.37,-1.28,0.73
L6=LINE/0.276,-(1.28+0.728),ATANGL,45,XAXIS
C21=CIRCLE/XSMALL,L7,YLARGE,L6,RADIUS,0.25
.....................
.....................
```

PROBLEMS

4.1 The part shown in Fig. P4-1 has a corrugated surface. Define the geometric entities that constitute this surface.

4.2 The profile of the template shown in Fig. P4-2 is composed of a series of circular arcs. Define these circular elements.

4.3 The tooth form of a sprocket for a roller chain is shown in Fig. P4-3. According to ANSI Standard B29.1-1975, each side of a tooth consists of three circular arcs and one

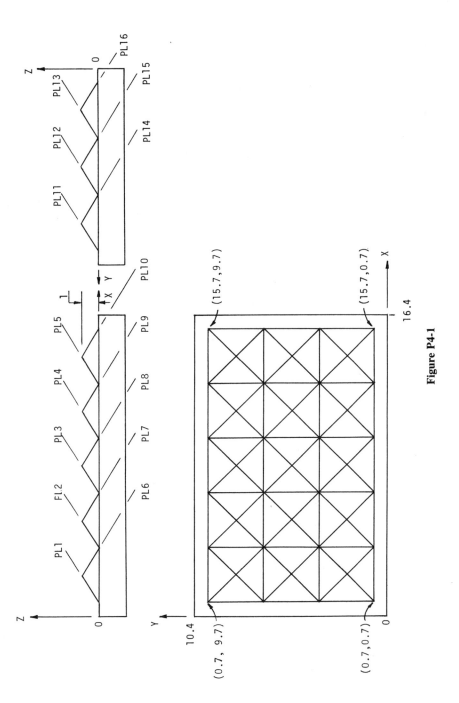

Figure P4-1

R1 = 0.5413"
R2 = 0.1535"
R3 = 1.0630"
R5 = 2.3622"
R6 = 3.5709"
R7 = 1.9567"
R8 = 0.4724"
R10= 1.2795"
R11= 1.0630"
x' = 1.4370"
y' = -0.5787"
x" = 1.2795"
y" = -2.3622"

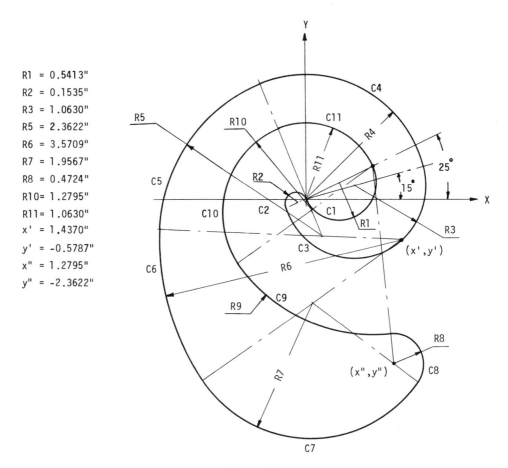

Figure P4-2

straight line (e.g., C3, C4, C5, and L2). Define the tooth profile C1-L1-C2-C3-C4-L2-C5 according to the following given conditions:

$A = 39°$

$B = 14.27°$

Number of teeth $N = 15$

Pitch $\qquad P = \frac{1}{2}$ in.

Norminal roller diameter $D_r = 0.306$ in.

$R1 = (1.005 \cdot D_r + 0.003)/2$ in.

$R2 = D_r [0.8 \cdot \cos(18° - 56°/N) + 1.4 \cdot \cos(17° - 64°/N) - 1.3025] - 0.0015$ in.

$H = [(R2)^2 - (1.4 \cdot D_r - 0.5 \cdot P)^2]^{\frac{1}{2}}$

$D1 = P \cdot \cot(180°/N) + (2 \cdot R1 - D_r) \cos(180°/N) + 2 \cdot H$

$C = 0.8 \cdot D_r$

4.4 An artist's freehand design, shown in Fig. P4-4, is to be cut on an NC milling machine. Suppose that the profile is required to be divided into nine sections by points P1 through

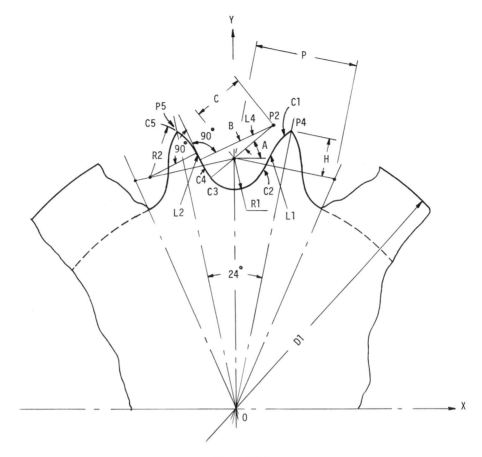

Figure P4-3

P9. Assume that section P1-P2 can be approximated by a line, and section P6-P7, by a circle. The rest can be approximated by LCONIC or TABCYL statements. Estimate visually the coordinates of the points within a section if they are required to define it. The tangents to the profile at points P4 and P8 are parallel to the Y axis. The point of inflection, P9, on the profile is chosen as the dividing point so that an irrational solution can be avoided when the statement LCONIC is used to define the adjacent sections. Assume, also, that the profile is in the plane $z = 0$. Use the appropriate APT statements to define the nine sections of this profile, with a smooth transition from one to another.

4.5 The cavity of a matrix (Fig. P4-5) is composed of two ruled surfaces, S1 and S3, two circular cylinders, S2 and S4, and one sphere, S5. Define the profile of this cavity.

4.6 A part profile (Fig. P4-6) consists of an ellipse (E1), two sheets of a hyperbola (H1), with its center at (7.5,7.5) and its half-transverse and half-conjugate axes equal to 13 and 4.5, respectively, and the transition curves LC1, LC2, LC3, and LC4. It is required that these transition curves be defined by the LCONIC/3PT2SL statement to guarantee their tangency to the adjacent curves. The tangent points are P1 through P8. Define this profile with the use of the appropriate APT statements.

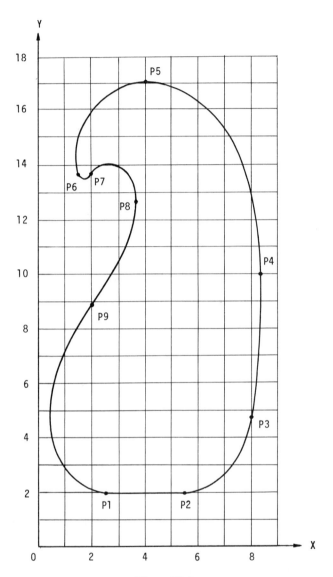

Figure P4-4

4.7 A tabulated cylinder is parallel to the Z axis and is defined according to the points given in Fig. P4-7. The cylinder should also be tangent to the circle C1 at point P1 and to the line L1 at point P9. Define this tabulated cylinder and verify the printed output.

4.8 The cavity of a matrix shown in Fig. P4-8 can be divided into two parts, S1 and S2. Define the surfaces of these two parts, using the POLCON statement. The following conditions are given:

$$C3_z = \begin{cases} -1 & \text{(for surface S1)} \\ -1 + 0.6\,t & \text{(for surface S2)} \end{cases}$$

$$K = \begin{cases} 0.6668 & \text{(for surface S1)} \\ 0.6668 - 0.41\,t & \text{(for surface S2)} \end{cases}$$

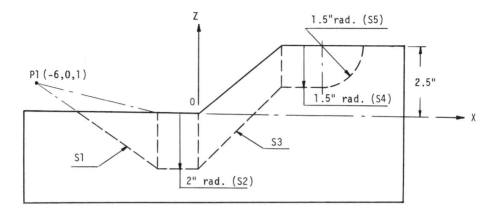

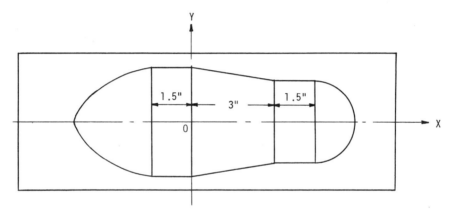

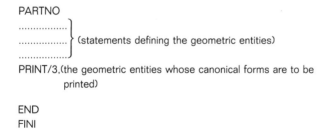

Figure P4-5

4.9 Discuss, in general, the usage of the modifiers in statement format No. 5 in Table 4-10. One or several simple APT programs and their output are needed to support the discussion. The structure of such an APT program may appear as follows:

PARTNO

....
.... } (statements defining the geometric entities)
....

PRINT/3,(the geometric entities whose canonical forms are to be
           printed)

END
FINI

Use the JCL (job control language) statements or EXEC (job execution program — see Chapter 8) used on your own computer system to run the program.

4.10 Define the geometric entities needed for specifying the cutter path for turning the part given in Problem 2.9.

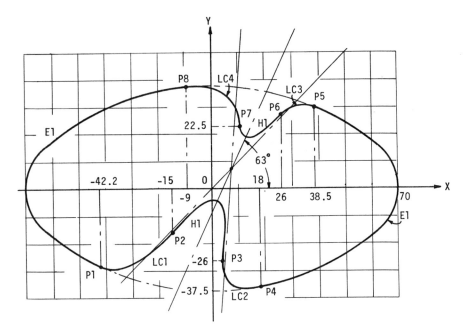

Figure P4-6

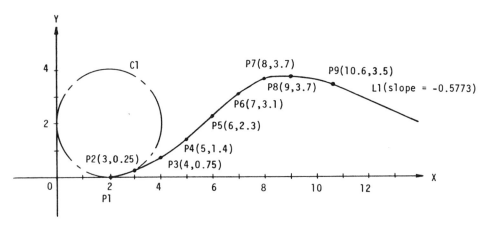

Figure P4-7

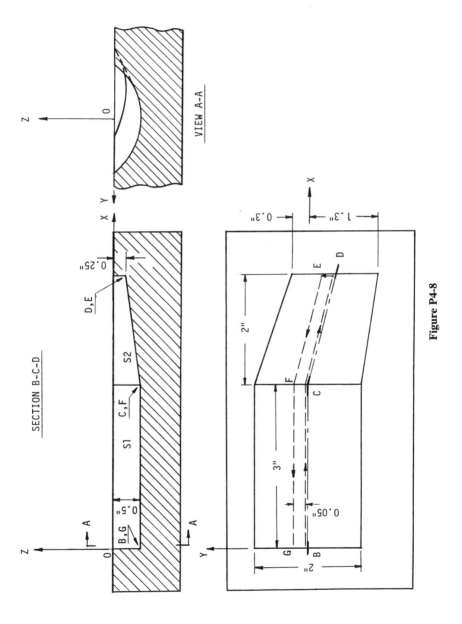

SECTION B-C-D

VIEW A-A

Figure P4-8

chapter 5

The Definition of Cutter Motion

There are two kinds of cutter motions, namely, *point-to-point motion* and *contouring motion*. In point-to-point motion, the path along which the tool travels from the starting point to the destination can be defined as a straight line; otherwise, it may be necessary to control the destination position only, with no particular requirement on the cutter path. Thus, for example, positioning a drill from one hole position to another as part of a sequence of drilling operations is a typical point-to-point motion imposing no special requirement on the tool path, whereas drilling a hole from the starting position to the required depth is a linear point-to-point motion. Contouring motion is the movement of a tool along a designated path while the deviation of the tool from the path is maintained within the specified tolerance. This yields the desired profile.

The APT language allows us to define a cutter path on the basis of geometric entities (i.e., the part profile) if these can be defined either in APT or on the basis of the position coordinates calculated by the NC processor with the use of the defined geometric entities.

Generally, it is easier to define a geometric entity that is part of a profile than to calculate or define the desired tool positions or path. The reason is that a programmer usually cannot find the information needed for defining a cutter path directly from the drawing; he must calculate both the intersection of one geometric entity with another and the offsets needed, in various directions, to compensate for the cutter radius and to ascertain that the deviation of the cutter from the profile is within the specified tolerance. Still, a correct design drawing always provides all the dimensions and parameters needed to define various sections of a part profile. Thus the definition of the cutter path on the basis of the part profile implies that the NC processor incorporates routines that automatically calculate every parameter, such as position coordinates, offsets, and compensation, needed for defining the cutter path.

5.1 THE DEFINITION OF POINT-TO-POINT MOTION

Point-to-point motion, as suggested by its name, consists in positioning a tool from one point to another. In certain types of point-to-point motion, no cutting operation is intended during the motion. The movements needed to cause a tool to retreat from an arbitrary position to a safe one and to position a tool rapidly from one position to another, ready for starting a new cutting operation, are typical examples. No cutting action takes place during the motion. Thus there is no need to be concerned with the motion path, except perhaps for possible collisions with the machine, fixtures, or workpiece. On the other hand, there are also cases in which a tool is required to move in a straight line from the starting point to the destination point.

In APT, a point-to-point motion is defined by either the statement GOTO or the statement GODLTA. The format of the GOTO statement is

$$\text{GOTO/} \begin{Bmatrix} x,y,z[,f] \\ x,y \\ P1[,f] \end{Bmatrix}$$

where

x, y, z = the coordinates of the destination point

$P1$ = the symbol for the destination point if previously defined

f = an optional element specifying the required feedrate (in inches or millimeters per minute, depending on the unit used in the program) for this motion

When a GOTO statement is specified, the tool moves from the position defined by the preceding statement to the specified position; the tool axis is assumed to be parallel to the Z axis. If the z coordinate is omitted in the statement, it is assumed to be zero.

The format of the GODLTA statement is

$$\text{GODLTA/} \begin{Bmatrix} x,y,z \\ z \\ V1 \end{Bmatrix} [,f]$$

where

x, y, z = the incremental distances, in the X, Y, and Z directions, respectively, from the starting point defined by the preceding statement to the destination point (If only the z coordinate is specified, then x and y are assumed to be zero, and the tool moves in the Z direction only.)

V1 = a defined vector specifying the direction and magnitude of the point-to-point motion

f = the required feedrate for this motion

These two statements do not define the actual path of a point-to-point motion. Yet we do need two types of motion, as pointed out above. When the path of a point-to-point motion is required to be a straight line, then a cutting operation is usually involved and the tool should move at a specified feedrate. On the other hand, when the path of a point-to-point motion is immaterial to the cutting operation, then a rapid rate of motion is normally indicated. These two motions are, respectively, specified by G00 and G01 codes in an NC program. Thus the motion designated by a GOTO or GODLTA statement can be controlled by the specified feedrate. When a point-to-point motion statement is processed, the NC processor outputs the specified feedrate and the destination point in the CLDATA file. Then they are transformed by the postprocessor as a G00 code if a rapid motion rate is specified, or as a G01 code if a feedrate other than rapid is specified. If a point-to-point motion statement specifies change of the tool position in only one coordinate axis, the motion is always linear.

A rapid motion rate is defined by the statement

RAPID

which should be specified immediately preceding the point-to-point motion statement GOTO or GODLTA without the feedrate option. A feedrate other than rapid can be specified in the GOTO or GODLTA statement (in inches or millimeters per minute) by specifying the optional element, f, or it can be defined in a preceding statement, such as GOTO, GODLTA, GO, or FROM, or in a separate feedrate statement

FEDRAT/f

where f is the required feedrate. It should be noted that a RAPID statement is effective only up to the next motion statement. However, a specified feedrate in an APT program is effective until it is changed by a statement specifying a new feedrate other than the rapid one.*

At the beginning of a program, when the cutter position has not yet been defined, a FROM statement should be specified to define the initial tool position. The format of that statement is

$$\text{FROM/} \begin{Bmatrix} \text{P1}[,f] \\ x,y,z[,f] \\ x,y \end{Bmatrix}$$

where P1 is a symbol for a defined point, and x, y, and z are its coordinates. The z coordinate is taken as zero if the third option is chosen.

---

*The situation whereby a feedrate applies to one or all of the following motion statements depends on the design of the postprocessor to be used. What we have described here corresponds to a typical design. Please see Part IV of this book for further details.

The drilling operation for the part shown in Fig. 2-21 provides a typical application of the APT point-to-point statement. The following is a segment of the program used for drilling the 12 holes on the part:

```
. . . . . . . . . . . . . . . . . . . . . .
. . . . . . . . . . . . . . . . . . . .
FROM/-4,7,3.0              $$THE STARTING POINT IS (-4,7,3)
RAPID;GOTO/0,0,3           $$RAPIDLY MOVE THE TOOL TO POINT (0,0,3)
RAPID;GODLTA/-2.9          $$RAPIDLY MOVE DOWN TO POINT (0,0,0.1)
FEDRAT/3.0;GODLTA/-1.0     $$DRILL THE HOLE
RAPID;GODLTA/1.0           $$RAPIDLY MOVE UP
RAPID;GODLTA/4,0,0         $$RAPIDLY MOVE TO POINT (4,0,0.1)
GODLTA/-1.0                $$DRILL THE HOLE. THE FEEDRATE, 3.0, IS TAKEN
$$                             FROM THE PREVIOUS STATEMENT
RAPID;GODLTA/1.0           $$RAPIDLY MOVE UP
RAPID;GODLTA/4,0,0         $$RAPIDLY MOVE TO NEXT HOLE POSITION
. . . . . . . . . . . . . . . . . . . . . .
. . . . . . . . . . . . . . . . . . . .
```

5.2 DEFINING POINT-TO-POINT MOTION FOR A SERIES OF POINTS: THE PATERN STATEMENT

In engineering design, holes are among the basic elements that constitute a part. Holes can be distributed in a specific pattern, either symmetrically or asymmetrically. As can be seen from the example presented above, a large number of holes requires a multitude of point-to-point statements. For each point, three statements are needed: one to move the tool to a specific point, and the other two to move the tool downward and then upward. The APT language allows us to define a set of points as a pattern. Then a single GOTO statement is used to define the cutter motion from the initial position to the various points in the pattern.

A set of points either randomly distributed or having some geometric relationship with each other can be defined by using the pattern definition statement

PATERN/......

There are four different kinds of formats for the PATERN statement:

1. That defining an orderly list of previously defined points as a pattern (statement format No. 1 in Table 5-1)
2. Those defining an orderly list of points specified by their coordinates as a pattern (statement format No. 2 in Table 5-1)
3. Those defining an orderly list of points on a circle or circles in the X-Y plane as a pattern (statement format No. 3 in Table 5-1)
4. Those defining an orderly list of points on a line or lines as a pattern (statement format No. 4 in Table 5-1).

Corresponding examples are also given in Table 5-1.

A pattern can be defined by choosing a coordinate system that is most convenient for specifying the points in the pattern. Once a pattern has been defined, one or

TABLE 5-1 STATEMENTS DEFINING A SET OF POINTS AS A PATTERN

No.	Known Condition	Statement Format
1	The pattern consists of given points P1, P2, . . . , Pn in listed order.	PATERN/P1,P2,. . . ,Pn

Comment Any of the symbols P1 through Pn can be replaced by a nested point definition statement. The order of the points in this statement determines the tool path and the sequence of the tool motion from one point to another.

Example The statement
PATERN/P1,(P2=POINT/0,2,4),(POINT/3,7,5),(1,0,0)

defines four points as a pattern. The first point, P1, was defined in a previous statement.

2 a. The points listed in this pattern are defined by a series of polar coordinates (ri, ai), $i = 1,2,\ldots,n$, with the polar center at point P1 (x,y,z). The symbols for polar distances are ri, $i = 1,2,\ldots,n$, and those for angles, ai, $i = 1,2, \ldots,n$. If the z coordinate of the polar center is not specified, all the points are defined in the X-Y plane.

$$\text{PATERN}/\left[\text{AT},\left[\text{P1} \; ; \; \begin{bmatrix} x,y,z \\ x,y \end{bmatrix}\right]\right] \left\{ \begin{array}{l} \text{THETAR,} \left\{ \begin{array}{l} a1,r1,\ldots,an,m \\ \text{CONST},a1,r1,r2,\ldots,rm[\text{CONST},a2,r(m{+}1),\ldots,rm] \\ a1,r1,a2,r2,\text{CONST},a3,r3,r4,\ldots,m \end{array} \right. \\ \text{RTHETA,} \left\{ \begin{array}{l} r1,a1,\ldots,m,an \\ \text{CONST},r1,a1,a2,\ldots,am[\text{CONST},r2,a(m{+}1),\ldots,an] \\ r1,a1,r2,a2,\text{CONST},r3,a3,a4,\ldots,an \end{array} \right. \end{array} \right\}$$

Comment
(1) The polar center point, P1, if specified, is not included in the defined pattern.
(2) If the two elements immediately after the slash defining the polar center are omitted, then the origin of the coordinate system is considered as the polar center. If only coordinates x and y of the polar center are specified, then its z coordinate is assumed to be zero.

(3) The z coordinate for each point is determined by that of the polar center. All the points in the defined pattern are in the plane parallel to the X-Y plane.

(4) The word CONST indicates that the parameter immediately after it applies to all the following points in the pattern until it is canceled by repeating the word CONST.

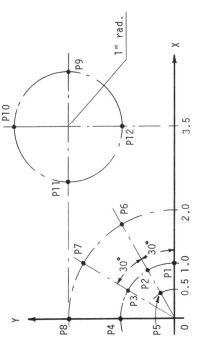

Example

(1) The statement

PT1=PATERN/RTHETA,CONST,1,0,30,60,90

defines a pattern consisting of points P1, P2, P3, and P4, all of which are in the X-Y plane.

(2) The statement

PT2=PATERN/AT,0,0,1,RTHETA,CONST,1,0,30,60,90

defines a pattern consisting of points P1, P2, P3, and P4, all of them being in the plane z = 1.

(3) The statement

PT3=PATERN/RTHETA,0,5,30,CONST,1,0,30,60,90,CONST,2,30,60,90

defines a pattern consisting of points P5, P1, P2, P3, P4, P6, P7, and P8, all of which are in the X-Y plane.

(4) The statement

PT4=PATERN/AT,3.5,2,0,3,THETAR,0,1,180,1,90,1,270,1

defines a pattern consisting of points P9, P11, P10, and P12, in specified order, all of which are in the plane z = 3.

(continued)

TABLE 5-1 STATEMENTS DEFINING A SET OF POINTS AS A PATTERN (continued)

No.	Known Condition	Statement Format
b.	If the option DELTA is chosen, the points listed in the pattern are defined by incremental Cartesian coordinates from one point to the next: $x1, y1, z1, \ldots, xn, yn, zn$. The first point in this pattern is defined with respect to given reference point P1, whose coordinates are (x, y, z) or (x, y) if z is zero. The option ABSLTE indicates that $x1, y1, z1, \ldots, xn, yn, zn$ are the absolute coordinates of the points in the pattern. For options other than XYZ, the omitted coordinates of each point are the same as those of the preceding point in the pattern or as those of point P1 if the word is specified before the first group of points.	$$\text{PATTERN}/\left[\text{AT,}\left[\begin{array}{l}\text{P1}\\x,y,z\\x,y\end{array}\right],\right]\left[\begin{array}{l}\text{DELTA,}\\ [\text{ABSLTE,}]\end{array}\right]\left\{\begin{array}{l}\text{XYZ,}x1,y1,z1,\ldots,xn,yn,zn\\ \text{NOX,}y1,z1,\ldots,yn,zn\\ \text{NOY,}x1,z1,\ldots,xn,zn\\ [\text{NOZ,}]x1,y1,\ldots,xn,yn\\ \text{XCOORD,}x1,x2,\ldots,xn\\ \text{YCOORD,}y1,y2,\ldots,yn\\ \text{ZCOORD,}z1,z2,\ldots,zn\end{array}\right.$$ $$\left[,\left\{\begin{array}{l}\text{XYZ,}x1,y1,z1,\ldots,xm,ym,zm\\ \text{NOX,}y1,z1,\ldots,ym,zm\\ \text{NOY,}x1,z1,\ldots,xm,zm\\ \text{NOZ,}x1,y1,\ldots,xm,ym\\ \text{XCOORD,}x1,x2,\ldots,xm\\ \text{YCOORD,}y1,y2,\ldots,ym\\ \text{ZCOORD,}z1,z2,\ldots,zm\end{array}\right\},\ldots\ldots\right]\ \$$$

Comment

(1) Reference point P1, if specified, is not included in this pattern.

(2) If the reference point is not specified, it is the origin of the coordinate system.

(3) The vocabulary words DELTA and ABSLTE indicate that the supplied coordinates are incremental and absolute, respectively.

(4) The vocabulary word XYZ, NOX, ..., or ZCOORD indicates in which axes the following coordinates are supplied for the points in the pattern. The word XYZ indicates that x, y, and z coordinates are specified for each point. The coordinates for each point after the word NOX, NOY, or NOZ are y and z, x and z, and x and y, respectively. The word XCOORD, YCOORD, or ZCOORD indicates that only the x, y, or z coordinate is given for the points after it.

(5) ABSLTE is the default option; it can be omitted. NOZ, being also a default option, can also be omitted when specified before the first group of coordinates.

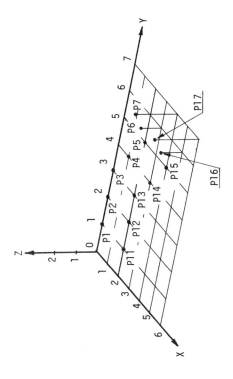

Example

(1) The statement

PT1=PATERN/DELTA,YCOORD,1,1,1,NOZ,1,1,1,1,XYZ,1,1,1,1,1,1,1

defines a point pattern that includes points P1 through P7.

(2) The statement

PT2=PATERN/AT,2,0,0,DELTA,YCOORD,1,1,1,NOZ,1,1,1,1,XYZ,1,1,1,1,1,1,1

defines a point pattern that includes point P11 through P17. Note that in these two examples, the vocabulary word NOZ cannot be omitted.

(3) The statement

PATERN/AT,0,1,0,ABSLTE,YCOORD,1,2,3,NOZ,1,4,2,5,XYZ,3,6,1,4,7,2

defines the same pattern as in Example 1, above.

3 The set of points in the defined pattern are on a circle, C1, in the X-Y plane, whose center is at point (x, y), with a radius, r. The first point is at angular position a and the last point at angular position $a1$. The number of points in this pattern is n, if the word OMIT or RETAIN is not specified.

$$\text{PATERN/ARC,} \begin{Bmatrix} \text{C1} \\ x,y,r \end{Bmatrix} \text{,ATANGL,} a, \begin{Bmatrix} \text{CLW} \\ \text{CCLW} \end{Bmatrix}, \begin{Bmatrix} a1,\text{NUMPTS,}n \\ \text{INCR,}i,\text{NUMPTS,}n \\ \text{INCR,}i1,i2,\ldots,i(n-1) \end{Bmatrix},\$$

$$\left[\begin{Bmatrix} \text{OMIT} \\ \text{RETAIN} \end{Bmatrix}, n1,n2,\ldots, \right]$$

(continued)

211

TABLE 5-1 STATEMENTS DEFINING A SET OF POINTS AS A PATTERN (continued)

No.	Known Condition	Statement Format

Comment

(1) Angles a and $a1$ are always measured from the positive X direction. They have no relationship to the vocabulary words CLW and CCLW, which indicate the direction in which the points are located from the first point to the last one.

(2) Option "$a1$,NUMPTS,n" indicates that there are n points in this pattern and that they are equally spaced between angular positions a and $a1$.

(3) Option "INCR,i,NUMPTS,n" indicates that there are n points in this pattern, with the same incremental angle, i, from one point to another. The angular position for the first point is a.

(4) Option "INCR,$i1,i2,\ldots,i(n-1)$" indicates that there are n points in this pattern, with an angular increment of im from the (m) th point to $(m+1)$ th point.

(5) The optional vocabulary word OMIT inidcates that the $(n1)$ th, $(n2)$ th, $\ldots$ points should be excluded from the defined pattern, whereas the vocabulary word RETAIN indicates that only the $(n1)$ th, $(n2)$ th, $\ldots$ points should be included in the pattern.

Examples

(1) The order of points in the pattern defined by the statement
PT1=PATERN/ARC,0,0,1,ATANGL,0,CCLW,90,NUMPTS,4
is P1, P2, P3, P4.

(2) The order of points in the pattern defined by the statement
PT2=PATERN/ARC,2,1,1,ATANGL,−45,CLW,45,NUMPTS,4
is P5, P6, P7, P8.

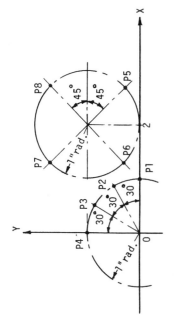

(3) The order of points in the pattern defined by the statement

PT3=PATERN/ARC,0,0,1,ATANGL,0,CCLW,90,NUMPTS,4,RETAIN,1,3,4,ARC,$
2,1,1,ATANGL,−45,CLW,45,NUMPTS,4,OMIT,2

is P1, P3, P4, P5, P7, P8. This example also shows that the whole set of elements after the slash can be concatenated to another set of elements.

4 a. The set of points P1, . . . , P*n*, defined in this pattern, is on a straight line defined by the first and last points, P1 and P*n*, with equal spacing between points.

PATERN/LINEAR,P1,P*n*,NUMPTS,*n*

PATERN/LINEAR,P1,P2,NUMPTS,n

(a)

(*continued*)

TABLE 5-1 STATEMENTS DEFINING A SET OF POINTS AS A PATTERN (continued)

No.	Known Condition	Statement Format

b. The set of points $P1, \ldots, Pn$, defined in this pattern, is on a straight line defined by the first point, $P1$, and a slope angle, a, or a vector, $V1$. Points $P2, \ldots, Pn$ can be located in the increasing (INCR) or decreasing (DECR) direction, with equal spacing, d (for option NUMPTS), or specified spacings, $d1, d2, \ldots, d(n-1)$. If the option

$$\left\{\begin{array}{c}\text{INCR}\\\text{DECR}\end{array}\right\}, d$$

is omitted and vector $V1$ is specified, the vector $V1$ is used as the incremental vector from one point to the next.

$$\text{PATERN/LINEAR}, P1, \left\{\begin{array}{l} \text{ATANGL}, a, \left\{\begin{array}{c}\text{INCR}\\\text{DECR}\end{array}\right\}, \left\{\begin{array}{l} d, \text{NUMPTS}, n \\ d1, d2, \ldots, d(n-1) \end{array}\right\} \\[3ex] V1, \left[\left[\left\{\begin{array}{c}\text{INCR}\\\text{DECR}\end{array}\right\}, d, \right] \text{NUMPTS}, (n-1)\right] \\[3ex] \left\{\begin{array}{c}\text{INCR}\\\text{DECR}\end{array}\right\}, d1, d2, \ldots, d(n-1) \end{array}\right\} \left[, \left\{\begin{array}{c}\text{OMIT}\\\text{RETAIN}\end{array}\right\}, n1, n2, \ldots\right]$$

... LINEAR, P1, V1, $
INCR, NUMPTS, n

... V1, DECR, d1, d2, ...

... ATANGL, a, INCR, d1, d2, ...

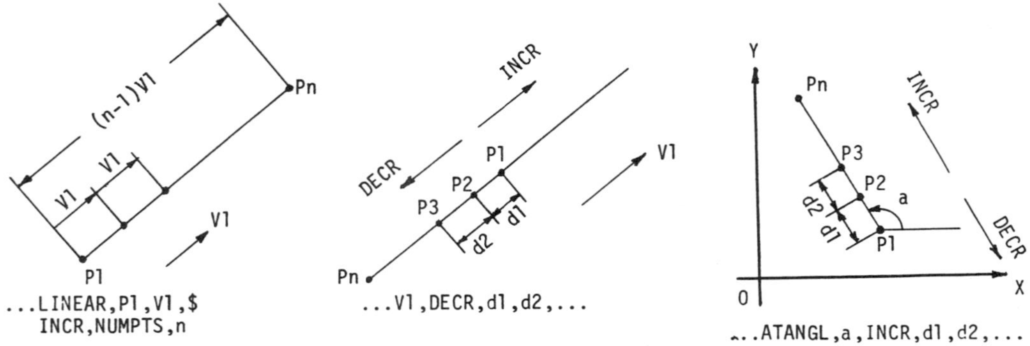

(b)

214

$$\text{PATERN/LINEAR,P1,DELTA,}x1,y1,[z11],\text{NUMPTS,}n-1,\left[\left\{{\text{OMIT} \atop \text{RETAIN}}\right\},n1,n2,\ldots\right]$$

c. The n points, P1, . . . , Pn, defined in this pattern are on a straight line passing through the first point, P1, and in the direction specified by the incremental coordinates, $x1$, $y1$, and $z1$, from one point to the next.

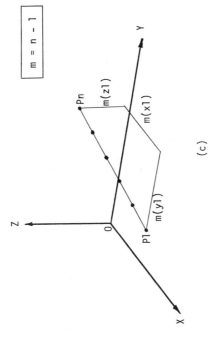

$$\boxed{m = n - 1}$$

(c)

Comment The optional elements $\left\{{\text{OMIT} \atop \text{RETAIN}}\right\},n1,n2,\ldots$ have the same meaning as described previously.

Example If points P1, P2, and P3 are given, then we have

PT1=PATERN/LINEAR,P1,P2,NUMPTS,4
PT2=PATERN/LINEAR,P2,ATANGL,0,INCR,0.5,NUMPTS,4,OMIT,1
PT3=PATERN/LINEAR,P3,ATANGL,150,DECR,0.3,0.35,OMIT,1
PT4=PATERN/LINEAR,(POINT/PT3,2),DELTA,0,−0.2,NUMPTS,3,OMIT,1

The pattern, including the 12 points from P1 to P5 in the order shown in the figure above, can be defined by concatenating the elements in the definition statements shown above:

(continued)

TABLE 5-1 STATEMENTS DEFINING A SET OF POINTS AS A PATTERN (continued)

No.	Known Condition	Statement Format

PT5=PATERN/LINEAR,P1,P2,NUMPTS,4,P3,NUMPTS,4,ATANGL,150,DECR,$
0.3,0.35,DELTA,0,−0.2,NUMPTS,3

216

TABLE 5-2 THE MODIFIERS USED TO MANIPULATE A PATTERN

Modifier	Statement Format and Explanation

TRANSL   PATERN/PT1,TRANSL, $\begin{Bmatrix} V1 \\ x,y,[,z] \end{Bmatrix}$

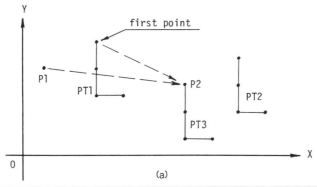

Comment   The pattern is defined by moving the given pattern, PT1, by a vector, V1, or by the incremental coordinates x, y, and z. The z coordinate is zero if it is omitted.

Examples
PT2=PATERN/PT1,TRANSL,V1
PT2=PATERN/PT1,TRANSL,2,1.5,0
PT2=PATERN/PT1,TRANSL,2,1.5

ATTACH   PATERN/PT1,[ATTACH,P1,] $\begin{Bmatrix} \text{TRANPT,P2} \\ \text{XYROT,}a \\ \text{INVX} \\ \text{INVY} \end{Bmatrix}$

Comment
1. The pattern is defined by transforming the given pattern, PT1, as specified. The vocabulary word ATTACH indicates that an arbitrary point, P1, is attached to pattern PT1, and they are considered as a rigid body during the transformation. A transformation specified by the following elements is carried out with respect to this reference point. If "ATTACH,P1," is omitted, the reference point is the first point in pattern PT1.

2. The element "TRANPT,P2" specifies a translation of pattern PT1 obtained by moving point P1 to point P2.

first point

P1

PT1

P2

PT3

PT2

(a)

(continued)

TABLE 5-2 THE MODIFIERS USED TO MANIPULATE A PATTERN (continued)

Modifier	Statement Format and Explanation

Examples

PT2=PATERN/PT1,ATTACH,P1,TRANPT,P2

PT3=PATERN/PT1,TRANPT,P2

3. The element "XYROT,a" specifies a rotation of pattern PT1 about point P1 by an angle, a.

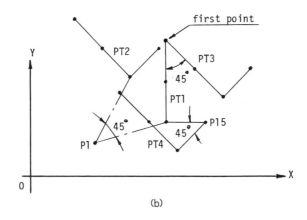

(b)

Examples

PT2=PATERN/PT1,ATTACH,P1,XYROT,45

PT3=PATERN/PT1,XYROT,45

PT4=PATERN/PT1,ATTACH,P15,XYROT,45

4. The words INVX and INVY specify inversions of the given pattern, PT1, about a line passing through point P1, if specified, or the first point in the pattern, if the option "ATTACH,P1" is omitted, and parallel to the Y (INVX) and X (INVY) axes, respectively.

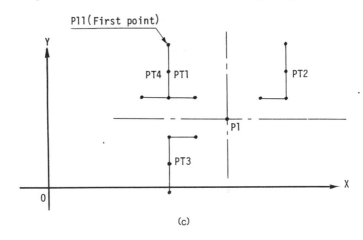

(c)

Example   If PT1 is a defined pattern, with P11 as the first point, we have

PT2=PATERN/PT1,ATTACH,P1,INVX

PT3=PATERN/PT1,ATTACH,P1,INVY

PT4=PATERN/PT1,INVX

TABLE 5-2   THE MODIFIERS USED TO MANIPULATE A PATTERN   (continued)

Modifier	Statement Format and Explanation

MODIFY   PATERN/PT1,MODIFY,M1

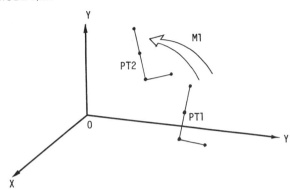

Comment   The defined pattern is the one transformed from the given pattern, PT1, according to a given matrix, M1.*

Example   If the matrix needed for transformation of pattern PT1 into PT2 is M1, then the statement defining PT2 is

PT2=PATERN/PT1,MODIFY,M1

INVERS   PATERN/PT1,INVERS

Comment   The points defined in this pattern are the same as in pattern PT1, but in reverse order.

Example   If pattern PT1 consists of points P1, P2, and P3 in listed order, then the pattern defined by this statement contains the same points in pattern PT1, which are in the following order: P3, P2, P1.

OMIT   PATERN/PT1, $\left\{ \begin{matrix} \text{OMIT} \\ \text{RETAIN} \end{matrix} \right\}$,n1,n2,...
RETAIN

Comment   These two modifiers have the same function as that described in statement formats No. 3 and No. 4 in Table 5-1. For modifier OMIT, the defined pattern has the same set of points as in pattern PT1, except the $(n1)$th, (n)th, ... points; for modifier RETAIN, the defined pattern contains only the $(n1)$th, $(n2)$th, ... points in pattern PT1. The order of points in the defined pattern is the same as in pattern PT1.

Examples   Three point patterns, PT2, PT3, and PT4, are defined on the basis of pattern PT1 by the statements

. . . .
PT1=PATERN/(0,0,0),(0,1,0),(0,2,0),(0,3,0),(0,4,0)
PT2=PATERN/PT1,OMIT,3
PT3=PATERN/PT1,RETAIN,2,4,5
PT4=PATERN/PT1,RETAIN,5,2,4

. . . .
. . . .

Pattern PT2 has four points: (0,0,0), (0,1,0), (0,3,0), (0,4,0) in listed order. Patterns PT3 and PT4 are identical, having the same set of points and order: (0,1,0), (0,3,0), (0,4,0).

*See Chapter 7 for definition of the MATRIX statement.

several of the following modifiers, together with some parameters, may be used to manipulate it further:

TRANSL moves a defined pattern from one place to another without rotating it

ATTACH indicates a point about which a manipulation, such as a translation (TRANPT), rotation (XYROT), or inversion (INVX or INVY), is made

MODIFY indicates that the manipulation is performed according to a defined matrix

INVERS inverts the order of the points in a pattern

OMIT removes some points in a pattern

RETAIN retains some points in a pattern and discard the others

The formats and usage of these modifiers are listed in Table 5-2. One or more modifiers listed in Table 5-2 can be used in a PATERN statement. Thus pattern PT1 can be manipulated by a series of transformations with a single statement:

PATERN/PT1,m1,m2,m3,...

where PT1 must be the symbol for a defined pattern, and $m1$, $m2$, ... can be any of the manipulations listed in Table 5-2. The order of the manipulations is that in which they appear in the statement. There is no restriction on the number of manipulations. For instance (Fig. 5-1), if $m1$, $m2$, and $m3$ are "ATTACH,(0,0,0),INVX," "XYROT,−45," and "ATTACH,P1,TRANPT,P2," respectively, then the statement shown above defines a pattern by first making an inversion of pattern PT1 about the Y axis, then rotating it by −45 degrees about the first point in the pattern, and finally moving point P1 to point P2. The result is pattern PT2. Figure 5-2 shows an application of these modifiers. The part has seven sets of points, each consisting of three points. We can define, first, the general pattern by putting the pattern in the most convenient position and direction; we can then define PT1 through PT7 by means of appropriate transformations. The program may appear as follows:

```
....................
....................
PT0=PATERN/(0,1,0),(P0=POINT/0,0,0),(1,0,0)
PT1=PATERN/PT0,ATTACH,P0,XYROT,35,ATTACH,P0,TRANPT,(d1,d2,0)
PT2=PATERN/PT1,ATTACH,P0,XYROT,60
PT3=PATERN/PT1,ATTACH,P0,XYROT,120
....................
....................
PT7=PATERN/PT0,ATTACH,P0,XYROT,55,ATTACH,P0,TRANPT,P1
....................
....................
```

The NC processor also allows the combining of several previously defined patterns to form a new one. For example, if PT1, PT2, and PT3 are three defined patterns, they can be combined to become a new pattern by using the statement

PT4=PATERN/PT1,PT3,PT2

If the orders of points defined in patterns PT1, PT2, and PT3 are

P1, P2, P3, P4

P5, P6, P7

P8, P9, P10, P11

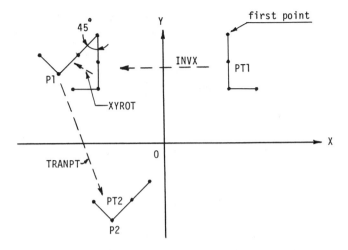

Figure 5-1   A pattern formed by a multiple transformation:
PT2=PATERN/PT1,ATTACH,(0,0,0),INVX,XYROT,−45,ATTACH,P1,TRANPT,P2

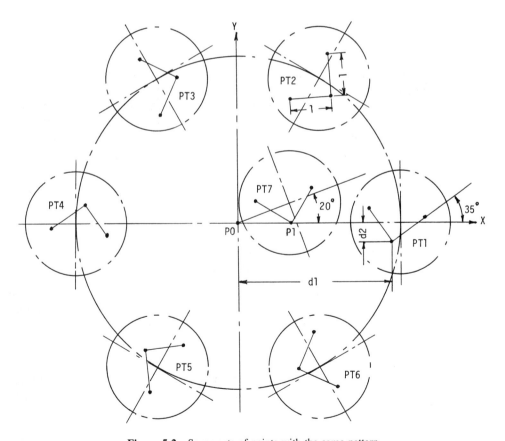

Figure 5-2   Seven sets of points with the same pattern.

respectively, then the order of points in the new pattern PT4 is

P1, P2, P3, P4, P8, P9, P10, P11, P5, P6, P7

A modifier can also be added after a pattern in the statement presented above, to manipulate an individual pattern. For example, the statement

PT5=PATERN/PT1,PT3,TRANPT,(1,0,0)

combines pattern PT1 and the one transformed from pattern PT3 by moving it in the $+X$ direction by one unit, into one pattern (PT5). The statement

PT6=PATERN/PT1,XYROT,90,PT3,TRANSL,1,0

combines the pattern transformed from pattern PT1 by rotating it about its first point by 90 degrees, and the one transformed from pattern PT3 by moving it in the $+X$ direction by one unit, into one pattern (PT6).

A GOTO statement is used to order the tool to go to the successive points in a pattern, its format being

GOTO/PT1

where PT1 is a pattern defined previously. For example, if the pattern PT1 consists of four points, P1, P2, P3, and P4, in listed order, then the statement directs the tool to move from its initial position, which is defined by the previous statement, to points P1, P2, P3, and P4 successively. Although this single statement can define a sequential point-to-point motion, it can hardly be used to program a real operation, such as drilling, punching, or tapping. One reason is that at every point in the pattern, we need to move the tool up and down to realize the cutting operation. In addition, the points in a pattern may not be located on a plane, and there might be obstacles (either a fixture or the part itself) between one point and another, so that the tool cannot be allowed to go directly from one point to the next. To define a real operation, we need additional modifiers.

5.3 THE MODIFIERS DEFINING ADDITIONAL TOOL MOTION TO AVOID OR CIRCUMVENT AN OBSTACLE

Two major modifiers can be used in a PATERN statement to include additional tool movement in a pattern point-to-point movement: AVOID (for raising a tool) and GOTO (for moving a tool around an obstacle).

5.3.1 Modifier AVOID

The modifier AVOID, together with certain parameters, defines the distance by which the tool should be raised at certain specific points of a defined pattern. The statement format is as follows:

PATERN/PT1,AVOID,$z1$, $\begin{Bmatrix} n1,n2,... \\ n1,n2,....,ni,THRU,nj,... \end{Bmatrix}$

where PT1 is a given pattern, and the parameter $z1$ specifies the height by which the tool should be raised at the $(n1)$th, $(n2)$th, ... points of pattern PT1 (Fig. 5-3[b]). The vocabulary word THRU indicates that the additional tool movement should be included at the (ni)th through (nj)th points (Fig. 5-3[c]). The difference in the tool motions for the patterns with and without the AVOID modifier can be seen clearly in the figures. The AVOID modifier generates not only the additional tool motion but a new point as well for each specified point; the X and Y coordinates of the added point is the same as for the specified point, and its z coordinate is the sum of that of the specified point and $z1$. Hence the pattern PT2 in Fig. 5-3(b) contains $n+2$ point,

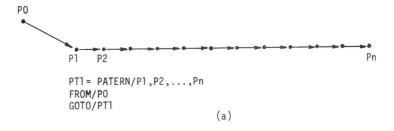

PT1 = PATERN/P1,P2,...,Pn
FROM/P0
GOTO/PT1

(a)

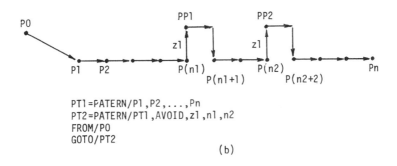

PT1 =PATERN/P1,P2,...,Pn
PT2=PATERN/PT1,AVOID,z1,n1,n2
FROM/P0
GOTO/PT2

(b)

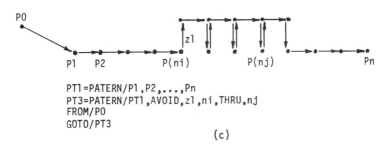

PT1 =PATERN/P1,P2,...,Pn
PT3=PATERN/PT1,AVOID,z1,ni,THRU,nj
FROM/P0
GOTO/PT3

(c)

Figure 5-3   The meaning of the modifier AVOID. The figures show the tool motions without the modifier AVOID (a), with the modifier AVOID (b), and with the modifier AVOID and the word THRU (c).

namely, P1, P2, . . . , P(n1), PP1, P(n1+1), . . . , P(n2), PP2, P(n2+1), . . . , Pn, although the NC processor does not treat the additional points PP1 and PP2 in the same way as it treats points P1, P2, . . . , Pn. This situation should be considered when more than one AVOID modifier is used in a statement.

The format of a PATERN statement containing more than one AVOID modifier is as follows:

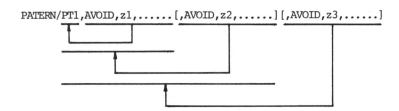

It should be noted that the first modifier applies to the given pattern PT1, and the second modifier, to the pattern modified by the first modifier, and so on. For example, assume that the tool motion shown in Fig. 5-4(c) is required (i.e., the tool should be raised by 1 in. and 1.1 in. at the first and fourth points in pattern PT1, respectively). However, the statement

GOTO/(PT4=PATERN/PT1,AVOID,1,1,AVOID,1.1,4)

generates a wrong motion, as shown in Fig. 5-4(d), because the fourth point in pattern PT1 is the fifth point in the pattern modified by the first AVOID. The correct statement should be (Fig. 5-4[c])

GOTO/(PT3=PATERN/PT1,AVOID,1,1,AVOID,1.1,5)

If the z coordinates of various points in a pattern are not the same, then the tool motion generated by the AVOID modifier will be changed accordingly (Fig. 5-5).

5.3.2 Modifier GOTO

GOTO is a minor word that should be specified after the slash in a PATERN statement and is different from the aforementioned major word GOTO, which defines a point-to-point motion. The modifier GOTO in a PATERN statement specifies an additional tool motion to a safe retreat position so that an obstacle between any two points can be circumvented. Hence it should be specified between the symbols for the two points. For example, pattern PT1 consists of five points, P1 through P5 (Fig. 5-6). To circumvent the obstacle located between points P3 and P4, the following statement can be used:

PT1=PATERN/P1,P2,P3,GOTO,PP1,P4,P5

where PP1 is the symbol for a safe retreat point, which is not treated by the NC processor in the same manner as the points in a pattern (see the example in Section 5.3.3).

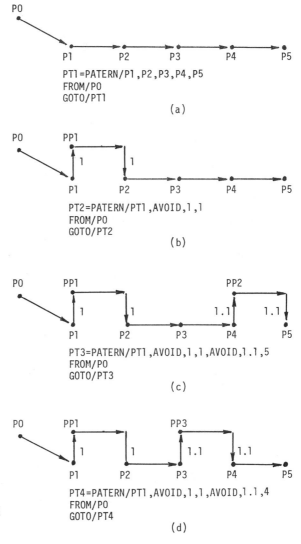

```
PT1=PATERN/P1,P2,P3,P4,P5
FROM/P0
GOTO/PT1
            (a)
```

```
PT2=PATERN/PT1,AVOID,1,1
FROM/P0
GOTO/PT2
            (b)
```

```
PT3=PATERN/PT1,AVOID,1,1,AVOID,1.1,5
FROM/P0
GOTO/PT3
            (c)
```

```
PT4=PATERN/PT1,AVOID,1,1,AVOID,1.1,4
FROM/P0
GOTO/PT4
            (d)
```

Figure 5-4 The use of concatenated AVOID modifiers.

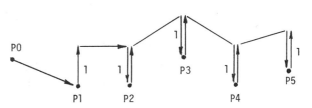

Figure 5-5 The tool motion for the set of points in a pattern defined through the use of the modifier AVOID.

```
PT1=PATERN/P1,P2,P3,P4,P5
FROM/P0
GOTO/(PATERN/PT1,AVOID,1,1,THRU,5)
```

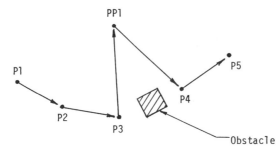

Figure 5-6 The presence of an obstacle requires an additional tool movement.

5.3.3 Examples of Programming a Pattern Point-to-Point Motion

The part shown in Fig. 2-21(a) is a typical example of the use of the pattern statement with the modifier AVOID. The APT program for drilling the 12 holes may be as follows:

```
. . . . . . . . . . . . . . . . . . . . .
. . . . . . . . . . . . . . . . . . . . .
P0=POINT/(-4,0,3)
PP0=POINT/(0,0,0.1)
PT1=PATERN/(0,0,-0.9),(4,0,-0.9),(8,0,-0.9),(12,0,-0.9),(12,3,-0.9),$
          (12,6,-0.9),(8,6,-0.9),(4,6,-0.9),(0,6,-0.9),(0,3,-0.9),$
          (3,3,0.85),(9,3,0.85)
PT2=PATERN/PT1,AVOID,1.0,1,THRU,9,AVOID,3.0,19,AVOID,1.25,21,22
FROM/P0
GOTO/PP0
GOTO/PT2
. . . . . . . . . . . . . . . . . . . . .
. . . . . . . . . . . . . . . . . . . .
```

The complete program and the processing result are shown in Fig. 5-7, and the tool movement defined by this program is given in Fig. 5-8. The program also shows the method for checking and verifying a tool motion defined by the GOTO/(pattern) statement. As can be seen from the listing file, the statement

PRINT/3,......

which was used before, does not print out each individual point defined in a pattern. To verify the tool motion, one can, instead, use the statement

CLPRNT

which requests the printing of the cutter-location data (CLDATA), so that the coordinates of the successive tool positions can be printed in the listing file.

If there is an obstacle between the second and third points of pattern PT1, we can change the statement in the program (Fig. 5-7) as follows:

PT1=PATERN/(0,0,−0.9),(4,0,−0.9),GOTO,(6,−3,−0.9),(8,0,−0.9),......

and run the program again. The tool motion between the second and third points is depicted by the dotted line in Fig. 5-8, which shows that the drilling operation is not performed at the safe retreat point and that the *z* coordinate specified for the modifier AVOID has been added to the *z* coordinates of the safe retreat point.

```
1
  IBM S/370 APT-AC N/C PROGRAM VERSION=1.3.0000                              DATE=07/28/86 TIME=12·43:58 0
                                     ...     BEGIN      TRANSLATION    PHASE...      (SECTION    1)    ...
0
  ISN 00001 PARTNO
  ISN 00002 MACHIN/GN5CC,9,OPTION,2,0
  ISN 00003 CLPRNT
  ISN 00004 FEDRAT/2.0
  ISN 00005 CUTTER/1.0
  ISN 00006 PT1=PATERN/(0,0,-0.9),(4,0,-0.9),(8,0,-0.9),(12,0,-0.9),(12,3,-0.9),$
                       (12,6,-0.9),(8,6,-0.9),(4,6,-0.9),(0,6,-0.9),(0,3,-0.9),$
                       (3,3,0.85),(9,3,0.85)
  ISN 00007 PT2=PATERN/PT1,AVOID,1,1,THRU,9,AVOID,3,19,AVOID,1.25,21,22
  ISN 00008 PRINT/3,PT2

      PT2              PATERN            24 ITEMS      UNITS= INCHES

  ISN 00009 FROM/-4,0,3
  ISN 00010 GOTO/(0,0,0.1)
  ISN 00011 GOTO/PT2
  ISN 00012 END
  ISN 00013 FINI


      NO DIAGNOSTICS ELICITED DURING TRANSLATION PHASE
      11 N/C SOURCE RECORDS (SYSIN)



                                      SECTION 1 ELAPSED CPU TIME IN MIN/SEC IS 0000/00.1433
                                      SECTION  2   ELAPSED   CPU   TIME   IN   MIN/SEC   IS   0000/00.0733
1
  ....SECTION                                                                              3....
0
  ISN
  0001 PARTNO/
  0002      MACHIN/            GN5CC           9.00000000           OPTION         2.00000000
                     0.0
  0004 FEDRAT/     2.00000000
  0005 CUTTER/     1.00000000
  0009   FROM/
                     0.0            0.0            0.0
  0010   GOTO/       0.0            0.0            0.1
  0011   GOTO/
                     0.0            0.0         -0.90000000
  0011 CYCLE /            NOMORE
  0011   GOTO/
                     0.0            0.0          0.10000000
  0011   GOTO/
                  4.00000000        0.0          0.10000000
  0011 CYCLE /              ON
  0011   GOTO/
                  4.00000000        0.0         -0.90000000
  0011 CYCLE /            NOMORE
  0011   GOTO/
                  4.00000000        0.0          0.10000000
  0011   GOTO/
                  8.00000000        0.0          0.10000000
  0011 CYCLE /              ON
  0011   GOTO/
                  8.00000000        0.0         -0.90000000
  0011 CYCLE /            NOMORE
  0011   GOTO/
                  8.00000000        0.0          0.10000000
  0011   GOTO/
                 12.00000000        0.0          0.10000000
  0011 CYCLE /              ON
  0011   GOTO/
                 12.00000000        0.0         -0.90000000
  0011 CYCLE /            NOMORE
  0011   GOTO/
```

Figure 5-7   An APT program and its printed output. The program consists of a pattern statement with concatenated modifiers.   (continued)

```
                          12.00000000      0.0              0.10000000
0011   GOTO/
                          12.00000000      3.00000000       0.10000000
0011 CYCLE /                   ON
0011   GOTO/
                          12.00000000      3.00000000      -0.90000000
0011 CYCLE /                 NOMORE
0011   GOTO/
                          12.00000000      3.00000000       0.10000000
0011   GOTO/
                          12.00000000      6.00000000       0.10000000
0011 CYCLE /                   ON
0011   GOTO/
                          12.00000000      6.00000000      -0.90000000
0011 CYCLE /                 NOMORE
0011   GOTO/
                          12.00000000      6.00000000       0.10000000
0011   GOTO/
                           8.00000000      6.00000000       0.10000000
0011 CYCLE /                   ON
0011   GOTO/
                           8.00000000      6.00000000      -0.90000000
0011 CYCLE /                 NOMORE
0011   GOTO/
                           8.00000000      6.00000000       0.10000000
0011   GOTO/
                           4.00000000      6.00000000       0.10000000
0011 CYCLE /                   ON
0011   GOTO/
                           4.00000000      6.00000800      -0.90000000
0011 CYCLE /                 NOMORE
0011   GOTO/
                           4.00000000      6.00000000       0.10000000
0011   GOTO/
                           0.0             6.00000000       0.10000000
0011 CYCLE /                   ON
0011   GOTO/
                           0.0             6.00000000      -0.90000000
0011 CYCLE /                 NOMORE
0011   GOTO/
                           0.0             6.00000000       0.10000000
0011   GOTO/
                           0.0             3.00000000       0.10000000
0011 CYCLE /                   ON
0011   GOTO/
                           0.0             3.00000000      -0.90000000
0011 CYCLE /                 NOMORE
0011   GOTO/
                           0.0             3.00000000       2.10000000
0011   GOTO/
                           3.00000000      3.00000000       3.85000000
0011 CYCLE /                   ON
0011   GOTO/
                           3.00000000      3.00000000       0.85000000
0011 CYCLE /                 NOMORE
0011   GOTO/
                           3.00000000      3.00000000       2.10000000
0011   GOTO/
                           9.00000000      3.00000000       2.10000000
0011 CYCLE /                   ON
0011   GOTO/
                           9.00000000      3.00000000       0.85000000
0011 CYCLE /                 NOMORE
0011   GOTO/
                           9.00000000      3.00000000       2.10000000
0011   GOTO/
                           9.00000000      3.00000000       2.10000000
0012 END
0013 ***** FINI *****
....END OF SECTION 3....
                                 SECTION 3 ELAPSED CPU TIME IN MIN/SEC IS 0000/00.1299
```

Figure 5-7   (continued)

5.4 DEFINING CUTTER CONTOURING MOTION

A cutter contouring motion is the movement of a cutting tool along the part profile whereby the deviation of the tool position from the designated profile is always maintained within a specified tolerance. As a result, a part of a certain specified form can be produced.

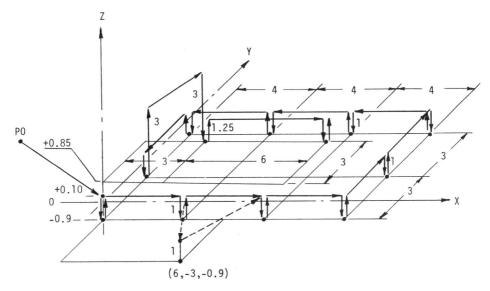

Figure 5-8   The tool motion defined by the program given in Fig. 5-7. The dotted lines with arrows represent the additional tool movement defined by modifier GOTO.

One of the distinguishing features of APT programming is the relative ease of defining cutter contouring motion. During manual NC programming of contouring motion, the coordinates of the starting and end points should be calculated and specified, and the tool motion can be defined only as linear or circular. If the profile of a part consists of sections that are other than linear or circular, these sections must be divided into many smaller ones that are then approximated by straight lines. The calculation of the coordinates should be performed for every small section, which is a lot of work even for a relatively simple workpiece, such as the one shown in Fig. 4-25.

In APT, a contouring motion is defined not on the basis of the end point coordinates but by the given part profile. In addition, the profile of a part can be any one of the geometric entities definable in APT. Therefore the definition of contouring motion is remarkably simplified.

5.4.1 The Three Controlling Surfaces of Cutter Contouring Motion: Drive Surface, Part Surface, and Check Surface

From a geometrical point of view, the starting and end points of a tool motion are the intersections of the current tool path with the preceding and following ones, respectively. They can easily be calculated by the NC processor because each tool path can be defined as a geometric entity. Within one step of a cutter contouring motion, the movement of a cutter in space is characterized by two types of motion: the motion along the tool axis and the motion in the plane perpendicular to its axis. For a three-axis milling machine, for instance, these are the tool motions along the Z axis and in the plane parallel to the X-Y plane. For each step of the cutter motion, a starting point is known: it is the end point of the immediately preceding step of

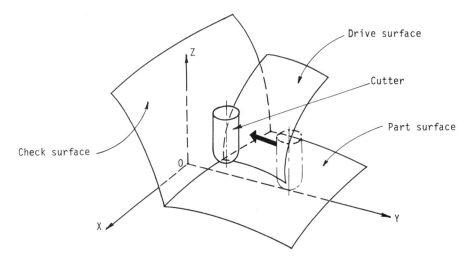

Figure 5-9   The three controlling surfaces of cutter contouring motion.

movement and is defined by the preceding statement. Therefore a cutter motion can be defined by a surface that limits the tool motion along the tool axis, a surface that restrains the tool motion in the plane perpendicular to the tool axis, and a surface that defines the end point of the motion. These three surfaces, in APT called *part surface, drive surface,* and *check surface,* respectively, are defined as follows (Fig. 5-9):

- The *drive surface* is the surface that is in continual contact with the cutter during the cutting operation and guides the cutter motion in the plane perpendicular to the cutter axis.
- The *part surface* is the surface that is in continual contact with the cutter during the cutting operation and controls the cutter motion along the cutter axis or the depth of cut.
- The *check surface* is the surface that determines the end point of the specified cutter contouring motion or the starting point of the next motion.

For each contouring motion statement, these three surfaces must be correctly defined. Differentiating these three surfaces is therefore of primary importance for defining a contouring motion in APT. The following examples illustrate the concept of the part, drive, and check surfaces.

In Fig. 5-10(a), all the limiting surfaces are planes. The tool motion required to cut the box-shaped part is indicated by arrows. The three surfaces for each step of the motion are as follows:

MOTION	PART SURFACE	DRIVE SURFACE	CHECK SURFACE
A to *B*	PL1	PL2	PL3
B to *C*	PL1	PL3	PL4
C to *D*	PL4	PL5	PL6
D to *E*	PL4	PL6	PL7 (not shown)

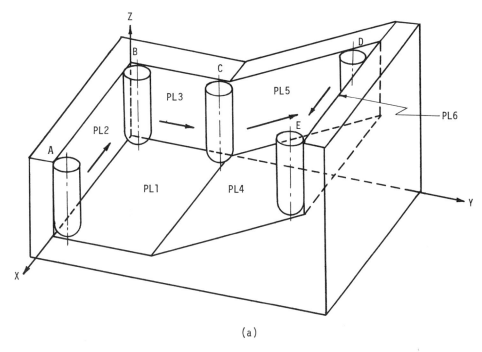

(a)

Figure 5-10(a)   The cutter motion for a pocketing operation.

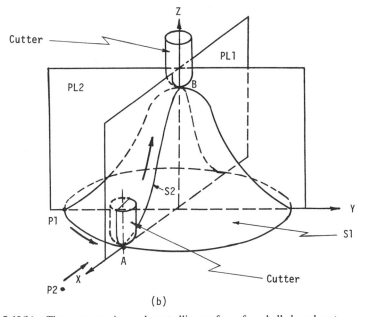

(b)

Figure 5-10(b)   The cutter motion and controlling surfaces for a bell-shaped part.

(continued)

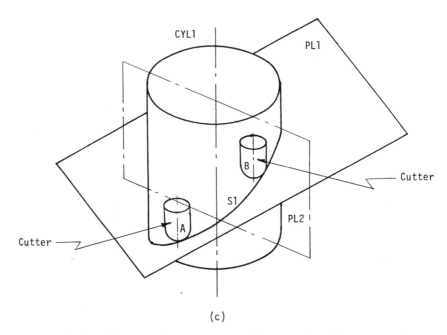

(c)

Figure 5-10(c)   The contouring motion along a helix formed by the intersection of plane PL1 and cylinder CYL1.

The part surface for the motion from C to D is PL4 because it determines the tool motion in the Z direction.

The part given in Fig. 5-10(b) is bell shaped. The tool motion from A to B in the Z direction is controlled by S2, the intersection of surface S1 and plane PL1; the tool motion in the plane perpendicular to the tool axis is controlled by plane PL1. Therefore the surfaces S2, PL1, and PL2 are, respectively, the part, drive, and check surfaces.

In Fig. 5-10(c) the tool is required to move from A to B along a helix, which is a part of the intersection of plane PL1 and cylinder CYL1. The surface that controls the tool movement in the plane perpendicular to the tool axis is cylinder CYL1, and that controlling the axial motion of the tool is PL1. In addition, position B is on the intersection of plane PL2 and cylinder CYL1. Therefore the part, drive, and check surfaces are PL1, CYL1, and PL2, respectively. One may also use curve S1 as the part surface while taking CYL1 as the drive surface, or one may use curve S1 as the drive surface and plane PL1 as the part surface. Therefore the three controlling surfaces can be a surface, a plane, a space curve, or a line.

5.4.2 The Statement Defining Contouring Motion

In APT the three surfaces mentioned above are used as guiding surfaces to define a contouring motion. As a rule, we use the design profile of a given part to determine the three surfaces. A tool position defined by an APT statement is the position of the tool-end center. Since every cutting tool has certain dimensions, moving the tool simply along the three guiding surfaces will generate a profile that is different from

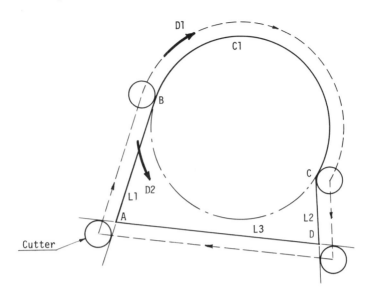

Figure 5-11   The contouring motion for a formed part. The part surface PL1 (not shown) is perpendicular to the tool axis.

the design. Therefore the cutter is required to be positioned on the left- or right-hand side of the design profile to compensate for the cutter radius (Fig. 5-11).

A contouring motion defined by an APT statement is always given with respect to a previous motion. In NC programming, the controller cannot be confused in trying to define the direction of motion because the final destination is defined explicitly by its coordinates and a G code (e.g., G02 or G03), which also indicates motion direction. In APT, however, a motion is defined by three surfaces. In many cases, there exist two possible directions that the tool can follow, although only one of them is correct. In Fig. 5-11, for example, having arrived at position B, the tool can follow either direction D1 or direction D2 in the next step of motion, wherein the cutter motion is defined by circle C1 as the drive surface and line L2 as the check surface. Therefore it becomes necessary to define the direction of the tool motion with respect to the previous one.

Finally, it is also necessary to specify the position of the tool-end center relative to the part surface in a contouring motion statement. This is particularly important in three-dimensional machining.

The general format of a contouring motion statement is as follows:

$$\left[\left[\begin{array}{c} TLON \\ TLLFT \\ TLRGT \end{array}\right]\right]\left[\begin{array}{c} [,TLOFPS] \\ ,TLONPS \end{array}\right], \left(\begin{array}{c} GOLFT \\ GORGT \\ GOFWD \\ GOBACK \\ GOUP \\ GODOWN \end{array}\right)/S1 \left(\begin{array}{c} [,TO] \\ ,ON \\ ,PAST \\ ,TANTO \\ ,PSTAN \end{array}\right),S2[,f]$$

where S1, S2, and *f* are the symbols for the drive surface, check surface, and feedrate, respectively. The four vocabulary words specified in the statement are discussed below.

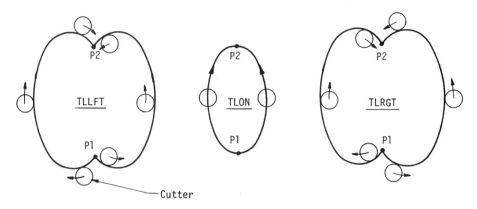

Figure 5-12   The tool positions with respect to the specified drive surfaces. The arrows indicate the motion directions. P1: Starting point; P2: end point.

The first vocabulary word, TLON, TLRGT, or TLLFT, indicates the tool position with respect to the specified *drive surface* or the designed profile (Fig. 5-12). The word TLON indicates that the tool-end center should always be located on the drive surface while it is moving along it. The words TLLFT and TLRGT specify that the tool should be on the left- and right-hand sides of the drive surface, respectively, and always tangent to the surface. This word can be omitted if it is the same as in the preceding statement.

The second word in the statement (i.e., TLONPS or TLOFPS) indicates the relative position of the tool-end center with respect to the *part surface* (Fig. 5-13). If the word TLONPS is specified, then the tool-end center is always on the part surface during the motion. The word TLOFPS indicates (1) that the tool profile is always tangent to or offset from the part surface and (2) that there is only one common point or line of the tool and the part surface, except in the case of tools that are flat ended and perpendicular to the part surface. This is also the default position of the tool if neither of the two words is specified.

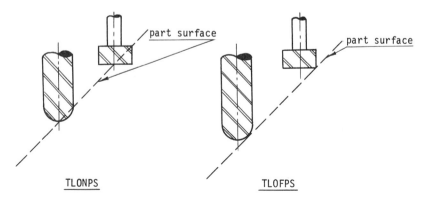

Figure 5-13   The tool positions, specified by words TLONPS and TLOFPS, with respect to the part surfaces.

The third word, GOLFT, ..., or GODOWN, specifies the direction of the tool motion defined by this statement with respect to the *preceding* motion direction. The words GOLFT and GORGT mean that the tool should turn to the left and right directions, respectively, after arriving at the end point of the preceding motion. The words GOFWD, GOBACK, GOUP, and GODOWN mean that the tool should go forward, backward, upward, and downward, respectively, after reaching the end point of the preceding motion.

The change of motion direction at the end point of the preceding motion (which is also the starting point of the current motion) can be determined as follows: Draw the tool axis vector from its end to its top and the vector tangent to both the drive and part surfaces of the preceding movement, thereby determining the two directions UP and FORWARD, as shown in Fig. 5-14. The direction RIGHT is determined by the cross product of the vector FORWARD to the vector UP. Then draw vector V1 tangent to both the drive and part surfaces of the current statement. Generally, the motion direction may change in three directions simultaneously. Since we can specify only a change in a single direction in the statement, we should find out the direction in which a major change occurs. Thus, if the tool does not change its position in the tool axis (Z) direction, or changes it very little, we can project vector V1 onto a plane perpendicular to the Z axis. The options GOUP and GODOWN are excluded in this case. One of the remaining words, GOFWD, GOBACK, GOLFT, or GORGT, is then selected. It should be pointed out that each of these four words indicates a general sense of direction, which, as can be seen from Fig. 5-15, covers an angle of 176 degrees. In addition, the directions covered by GOFWD and GOBACK overlap those covered by GOLFT and GORGT. Therefore we suggest that GOFWD and GOBACK not be used when the angle between motion vector V1 and the FORWARD axis, or between vector V1 and the BACKWORD axis, is larger than 45 degrees; we suggest that GOLFT or GORGT be used instead. The

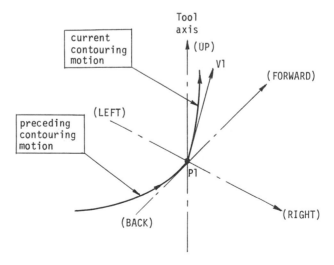

Figure 5-14   The reference axes of motion direction at the end point of a preceding contouring motion.

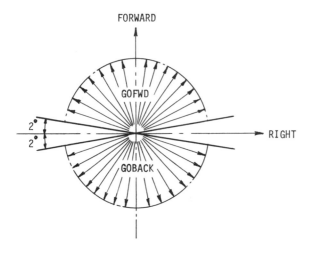

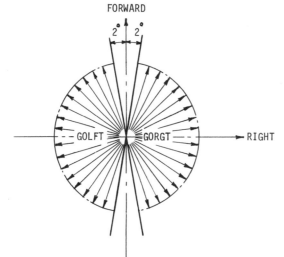

Figure 5-15   The directional ranges covered by various directional words.

word GOUP or GODOWN is selected when the change in tool motion direction is essentially in the vertical (tool axis) direction.

The fourth word in the statement presented above indicates the positional relationship between the tool and the *check surface* at the end of the motion (Fig. 5-16). The words that can be selected are as follows:

TO indicates that the tool stops where it is tangent to the check surface. This word can be omitted (Fig. 5-16[a and d]).

ON indicates that the tool stops where its center is on the check surface (Fig. 5-16 [a and d]).

PAST moves the tool past the check surface and stops it where it is tangent to the surface at the far side (Fig. 5-16[a and d]).

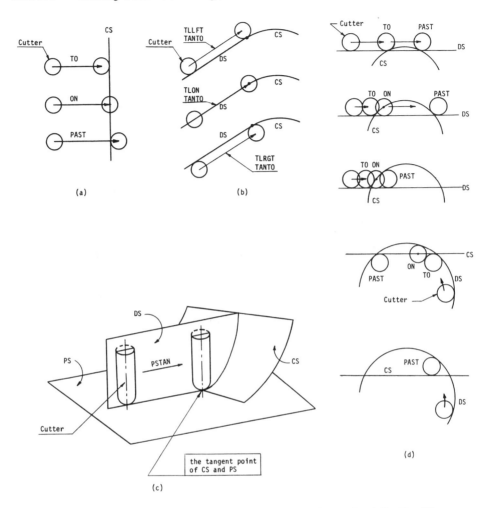

Figure 5-16 The tool positions at the end of the contouring motion defined by different words: (a) TO, ON, and PAST; (b) TANTO; and (c) PSTAN. (d) The possible tool positions resulting from different positional relationships between the drive and check surfaces. DS: Drive surface; PS: part surface; CS: check surface.

TANTO indicates that the drive and check surfaces are tangent to each other. The end point of the tool motion is the tangent point of these two surfaces (Fig. 5-16[b]).

PSTAN indicates that the part and check surfaces are tangent to each other, and the end point of the tool motion is the tangent point of these two surfaces (Fig. 5-16[c]).

The tool positions resulting from different positional relationships between the drive and check surfaces, and the positional modifiers used for defining the motions, are illustrated in Fig. 5-16[d].

The following examples present some applications of the contouring motion statement:

FIGURE	MOTION	PART SURFACE	CONTOURING MOTION STATEMENT
5-10(a)	B to C	PL1	TLRGT,GORGT/PL3,TO,PL4
5-10(a)	C to D	PL4	TLRGT,GOUP/PL5,PL6
5-10(b)	(P1→) A to B	S2	TLON,GOLFT/PL1,ON,PL2
5-10(b)	(P2→) A to B	S2	TLON,GOFWD/PL1,ON,PL2
5-11	B to C	PL1	TLLFT,GOFWD/C1,L2
5-11	C to D	PL1	TLLFT,GOLFT/L2,PAST,L3

It must be pointed out that a contouring motion statement does not include any word indicating the part surface to be used during the motion. As a matter of fact, this surface is defined either in the statement for starting a contouring motion or in a separate statement that defines the part surface.

5.5 STARTING AND TERMINATING A CONTOURING MOTION

As pointed out in Section 5.4, during a contouring motion the deviation of the cutter from the designed profile should be maintained within a specified tolerance. This means that the tool should be positioned at the proper spot where this requirement is satisfied at the beginning of a contouring motion (Fig. 5-17). Before a contouring

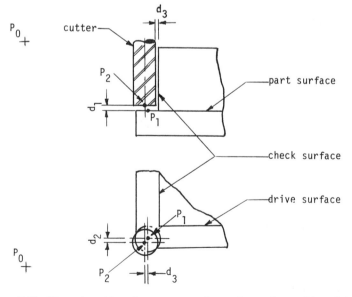

Figure 5-17 The end position of a start-up motion or the starting position of the contouring motion. P_1: Desired end position; P_2: actual end position; P_0: starting position of start-up motion; d_1, d_2, d_3: deviation of tool position from part, drive, and check surfaces, respectively.

motion is started, the tool is away from the designed profile. One or more statements are needed (1) to move the tool to the position where it is ready to start the contouring motion and (2) to define the part surface to be used.

The statement for defining a start-up motion has the following general format:

$$GO/\begin{Bmatrix} [TO,] \\ ON, \\ PAST, \end{Bmatrix} S1, \begin{Bmatrix} [TO,] \\ ON, \\ PAST, \end{Bmatrix} S2, \begin{Bmatrix} [TO,] \\ ON, \\ PAST, \\ TANTO, \end{Bmatrix} S3[,f]$$

where

> GO is a vocabulary word indicating that the defined motion is a start-up one;
>
> S1, S2, and S3 are the supplied drive, part, and check surfaces, respectively;
>
> TO, ON, PAST, and TANTO have the same meaning as described before (Each modifier applies only to the surface specified immediately after it.)
>
> f is the feedrate, which can be omitted if it is the same as defined previously.

The various positions of the tool at the end of the start-up motion are shown in Fig. 5-18. This statement is used only to position the tool to the desired position. The NC processor does not calculate or take care of the path of this motion. Therefore the motion specified by this statement is not a contouring motion, although three controlling surfaces are specified. The end position is calculated on the basis of the specified surfaces, modifiers, and tolerance. The cutter path of the start-up motion is determined in the same way as for the GOTO statement, by the specified motion speed (i.e., either a RAPID or a specified feedrate). The part surface defined in this statement will be effective in the following contouring motion until it is changed by another statement.

In some cases the tool is not required to be positioned at a definite point on the drive surface. Then the two-surface start-up motion statement can be used:

$$GO/\begin{Bmatrix} [TO,] \\ ON, \\ PAST, \end{Bmatrix} S1, \begin{Bmatrix} [TO,] \\ ON, \\ PAST, \end{Bmatrix} S2$$

where S1 and S2 are the *drive* and *part* surfaces, respectively. The part surface will be effective in the following contouring motion until a new part surface is defined. Generally, the end position defined by this statement is determined in such a way that the tool will move the shortest distance possible from the starting point to the intersection of the drive and part surfaces (Fig. 5-19[a]).

If a tool is required to move the shortest distance to a single surface that will be used as the drive surface in the next contouring motion statement, then a one-surface start-up motion statement can be used:

$$GO/\begin{Bmatrix} [TO,] \\ ON, \\ PAST, \end{Bmatrix} S1$$

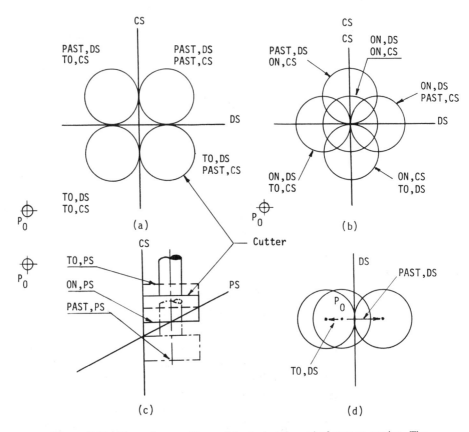

Figure 5-18 The various positions of the tool at the end of start-up motion. The starting position is P_0. (d) Note the end positions of the start-up motion when a control surface interferes with the tool. DS: Drive surface; PS: part surface; CS: check surface.

where S1 is called the *drive* surface in the APT programming manual,[1,2] and the position modifiers TO, ON, and PAST have the same meanings as described above. There are three possible situations:

1. The part surface has been defined in a previous statement. Then the tool motion is similar to a two-surface start-up motion.
2. The part surface has not been defined. Then the X-Y plane is used as the part surface. Again the tool motion is similar to a two-surface start-up motion.
3. An NOPS statement is specified preceding the one-surface start-up motion statement to indicate that no part surface is required for the start-up motion (Fig. 5-19[b]). In such a case the NC processor determines the end point by creating a normal vector passing through the starting point and perpendicular to the specified surface, S1.

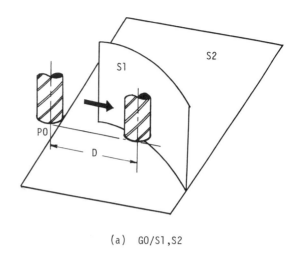

(a)   GO/S1,S2

Figure 5-19 The end positions of start-up motion. (a) Two-surface start-up motion: D is the shortest distance from the starting position to the intersection of the drive surface (S1) and the part surface (S2). (b) One-surface start-up motion with no part surface: the end position is determined on the basis of the intersection point of the drive surface and its normal passing through point P0, the starting position.

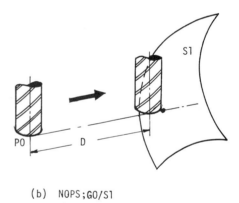

(b)   NOPS;GO/S1

As can be seen from the explanation presented above, surface S1 is actually a check surface, rather than a drive surface, for the start-up motion. The term *drive surface* is used here because S1 is often used as the drive surface in the following contouring motion statement. The discussion also applies to the first surface specified in the two-surface start-up motion statement.

Like all the motion statements, the start-up motion statement defines a motion with respect to a given starting position. This position can be defined by a FROM statement at the beginning of a program or by a previous motion statement in the middle of a program.

Within the NC processor, determining the end point of a start-up motion con- sists in finding a mathematical solution to the given conditions. In many cases the solution is not unique: two or more exist (Fig. 5-20). Therefore additional informa- tion should be provided for the NC processor to select the desired one. The state-

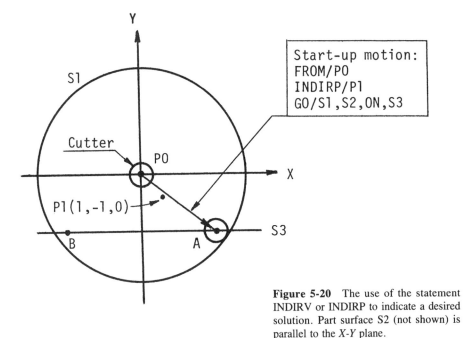

Figure 5-20 The use of the statement INDIRV or INDIRP to indicate a desired solution. Part surface S2 (not shown) is parallel to the X-Y plane.

ments used for this purpose are INDIRV (in the direction of a given vector) and INDIRP (in the direction of a given point), their format being

$$\text{INDIRV/} \begin{Bmatrix} V1 \\ i,j,k \end{Bmatrix}$$

where V1 is the given vector and i, j, and k are its components in the X, Y, and Z directions, respectively; and

$$\text{INDIRP/} \begin{Bmatrix} P1 \\ x,y,z \end{Bmatrix}$$

where P1 is the symbol of the given point and x, y, and z are its coordinates. These statements indicate that the end point of a start-up motion, specified in the immediately following motion statement, is in the direction of V1, starting from the end point of the preceding statement, or in the direction from the end point of the preceding statement to the specified point. For example, if the given point is P1 $(1,-1,0)$ (Fig. 5-20), then the INDIRV or INDIRP statement in the following program indicates the direction in which end point A of the start-up motion is located:

```
.................
.................
FROM/P0
INDIRP/P1$$ OR INDIRV/1,-1,0
GO/S1,S2,ON,S3
.................
```

The start-up motion defined by this program is from point P0 to point A.

It must be pointed out that the given point or vector in the INDIRP or INDIRV statement need not be precise, because these two statements specify a general sense of direction that is used only to distinguish one solution from the other. Therefore, for the case shown in Fig. 5-20, if point A is the desired solution, then point P1 can be any point in the fourth quadrant.

The use of statement INDIRV or INDIRP is not limited to a start-up motion; either can also be used to define a change in motion direction during a contouring motion. For example, in Fig. 5-21, the contouring motion is from A through B and C, to D. The section of the program defining this path is

```
...............................
....(The tool position is B).....
A1)TLLFT,GOLFT/S2,S3
A2) GOLFT/S3,PAST,S4

...............................
...............................
```

An INDIRP statement can be inserted between statements labeled A1 and A2 to indicate the motion direction at point C when the tool starts moving from point C to point D. The program can be modified as follows:

```
...............................
....(The tool position is B).....
A1)TLLFT,GOLFT/S2,S3
 INDIRP/P1
A2) GOFWD/S3,PAST,S4

...............................
...............................
```

The inserted statement indicates a general sense of direction after the tool has arrived at point C. Hence the word GOLFT in the statement labeled A2 should be changed to GOFWD (or GORGT). This indicates that the motion direction word in the contouring motion statement labeled A2 should be selected here, not on the basis of the previous tool motion direction, but on the basis of the direction specified by the statement INDIRP.

When used in a start-up motion, the FROM statement provides a reference point for calculating the end point. Both the direction indicated by the INDIRV or INDIRP statement and the position defined by the start-up motion statement are with respect to the FROM position. A FROM position can be the same as the actual tool position, or different. In the latter case, the FROM position (not the actual tool posi-

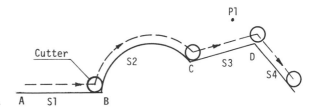

Figure 5-21   A contouring motion.

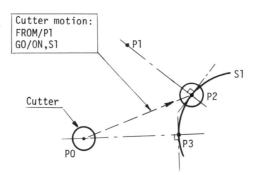

Figure 5-22 The end position of a start-up motion when the FROM position, P1, is different from the actual tool position, P0.

tion) is used to determine the end point of a start-up motion. Thus, for example (Fig. 5-22), assume that the actual tool position is P0 and the following statements are used to start up a contouring motion:

....

FROM/P1
GO/ON,S1

....

The end position of the start-up motion is P2, instead of P3, and the resulting motion is from P0 to P2. If the FROM position is the same as the actual tool position that has been defined by the previous motion statement, then the FROM statement can be omitted.

A contouring motion can be terminated by simply specifying a GOTO statement. The tool will move from the end position of a contouring motion to the position specified by the GOTO statement. Again, the tool path is not necessarily a straight line from the end position of the contouring motion to the position specified by the GOTO statement; it depends on the specified feedrate.

5.6 DEFINING THE PART SURFACE FOR A CONTOURING MOTION

A part surface is necessary for each contouring motion statement because it is one of the three controlling surfaces that define a contouring motion. Once defined, a part surface is effective until it is changed. A part surface can be defined in a two-surface or three-surface start-up motion statement. If a one-surface start-up motion statement is specified with an NOPS statement preceding it, the part surface is undefined. As a result, the following contouring motion will fail because the NC processor cannot determine the tool motion in the Z direction.

Besides the two-surface and three-surface start-up motion statements, two statements are provided in APT to define a part surface, their formats being

PSIS/PL1

where PL1 is a symbol designating a given plane that is defined by this statement as the part surface for the following contouring motion, and

AUTOPS

which defines the plane, being parallel to the *X-Y* plane and having the same *z* co-ordinate as the current tool position, as the part surface for the following contouring motion.

During a contouring motion, the tool is always positioned within the tolerance range with respect to the three controlling surfaces. Therefore, if the part surface in a contouring motion is to be changed, one should check whether the tool position is within the tolerance range with respect to the surface to be used as the new part surface. If so, one can simply use the command

PSIS/PL1

to define PL1 as the part surface in the following contouring motion. Otherwise, a new start-up motion statement should be specified before a new part surface is used. Examples are given in Section 5.8 (examples 2 and 3).

5.7 DEFINING ROUGHING AND FINISHING CONTOURING MOTIONS

It is a well-known machining fact that, generally, a part cannot be cut to the desired dimensional accuracy and surface finish within one pass. One or more cuts are needed to perform the rough and semifinishing cuts, followed by a finishing cut. During the rough cut, a certain amount of allowance should be left over the profile for the finishing cut; in addition, the cutter is to be shifted away from the profile by a distance equal to the desired allowance.

The NC processor allows us to specify allowances to be left over the part, drive, and check surfaces, so that the contouring motion can still be defined on the basis of the designed profile. The statement for specifying allowances is

THICK/a1[,a2[,a3]]

where $a1$, $a2$, and $a3$ are the allowances on the part, drive, and check surfaces, respectively. The last specified allowance applies to all the remaining surfaces. Thus, if $a2$ and $a3$ are omitted, allowance $a1$ applies to the drive, part, and check surfaces. Generally, the chosen allowances for the drive and check surfaces have the same value; thus the tool position is always within the tolerance range with respect to the new drive surface at the end of each motion step.

For example, the cutter path of the rough cut for the part shown in Fig. 5-23 can be defined as follows:

```
. . . . . . . . . . . . . . . . . . . .
FROM/P0
THICK/a1,a2
GO/ON,PL1,S0,S1
INDIRV/0,1,0
TLLFT,GOFWD/S1,PAST,S2
        GORGT/S2,S3
        GOLFT/S3,......
. . . . . . . . . . . . . . . . . . . .
. . . . . . . . . . . . . . . . . . . .
```

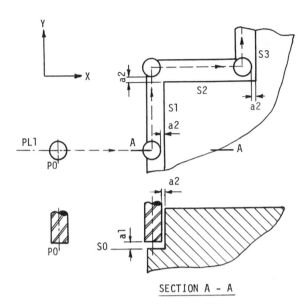

Figure 5-23 The use of statement THICK to specify the allowances ($a1, a2, a3$).

SECTION A - A

In this program the allowances for the drive and check surfaces are the same. If an even allowance is required on all the surfaces, then $a1 = a2 = a3$. The statement can be simplified to

THICK/$a1$

A THICK statement should be specified before the start-up motion statement in such a way that the tool can be correctly positioned at the beginning of a contouring motion. A three-surface or one-surface start-up motion statement without an NOPS statement preceding it is needed to specify the start-up motion after a THICK statement. However, usually a two-surface start-up motion fails in a THICK mode.

A THICK statement is effective for all the following contouring motion statements until it is changed by another THICK statement. Thus a finishing cut can be defined by simply changing the parameters in the THICK statement to zero.

When it is necessary to position a tool on a surface in a start-up or contouring motion (i.e., an ON modifier is specified), the specified allowance for that surface is ignored. For example, the specified allowance will not apply to part surface S2 in the following program:

```
......................
......................
THICK/0.02
GO/S1,ON,S2,PAST,S3
TLLFT,GORGT/S3,.......
......................
```

5.8 PROGRAMMING OF THE CONTOURING MOTION

On the basis of the discussion presented above, the procedure for defining a contouring motion can be summarized as follows:

1. Defining the start-up motion
   a. Define the reference point for the NC processor to calculate the start-up motion. If this position is not the same as the current tool position, a FROM statement should be used to specify it; otherwise, no statement is needed.
   b. Specify the required allowance(s) for various surfaces.
   c. Use a start-up motion statement, together with an INDIRV or INDIRP statement, if necessary, to define the start-up motion.

2. Defining the contouring motion
   a. A contouring motion can be specified immediately after a two- or three-surface start-up motion statement since the part surface has been defined. When the one-surface start-up motion statement is selected with an NOPS statement preceding it, the part surface must be defined before a contouring motion statement is specified.
   b. Since the tool has been positioned within the tolerance range with respect to the three controlling surfaces after a start-up motion, either its drive surface or its check surface can be used as the drive surface in the following contouring motion statement.
   c. If a direction of motion is difficult for the NC processor to detect during a contouring motion, the statement INDIRV or INDIRP can be added. Once it is added, the direction of motion in the next contouring motion is determined with respect to the direction indicated by the statement INDIRV or INDIRP.
   d. The drive surface of a contouring motion statement can be the drive surface or check surface in the preceding contouring motion statement.
   e. If the part surface is to be changed during the contouring motion, make sure that the tool is positioned within the tolerance range with respect to that surface.

3. Terminating a contouring motion. A contouring motion can be ended either by a GOTO statement if a point-to-point movement is allowed or by a contouring motion statement if a designated path is to be followed by the tool to move away from the part profile.

The following are some examples illustrating the use of contouring motion statements.

Example 1
The part given in Fig. 5-24 shows a profile that consists of lines and curves in the X-Y plane. Attention should be paid to programming the cutter path from P2 to P4. The APT program for the rough cut and the finishing cut may appear as follows:

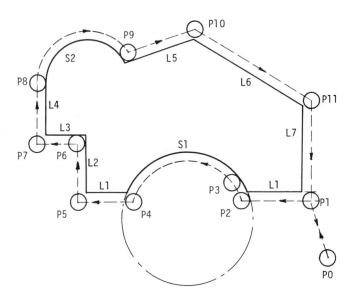

Figure 5-24 A part profile consisting of lines and curves. The part surface, PS (not shown), is parallel to the X-Y plane.

```
. . . . . . . . . . . . . . . . . . . . .
. . . . . . . . . . . . . . . . . . . . .
FROM/P0
THICK/0.01
GO/L7,PS,L1                          $$THE END POINT IS P1
1A)  TLLFT,GOLFT/L1,PAST,S1          $$THE END POINT IS P2
          GORGT/S1,PAST,L1           $$THE END POINT IS P3
          GOFWD/S1,PAST,L1           $$THE END POINT IS P4
          GORGT/L1,PAST,L2           $$THE END POINT IS P5
          GORGT/L2,L3                $$THE END POINT IS P6
          GOLFT/L3,PAST,L4           $$THE END POINT IS P7
          GORGT/L4,TANTO,S2          $$THE END POINT IS P8
          GOFWD/S2,L5                $$THE END POINT IS P9
          GOLFT/L5,PAST,L6           $$THE END POINT IS P10
          GORGT/L6,PAST,L7           $$THE END POINT IS P11
2A)       GORGT/L7,PAST,L1           $$THE END POINT IS P1
THICK/0
GO/L7,PS,L1
. . . . . . . . . . . . . . . . . . .
. . . . . . . . . . . . . . . . . . .      (Repeat the statements labeled 1A through 2A)
. . . . . . . . . . . . . . . . . . .
GOTO/P0
. . . . . . . . . . . . . . . . . . . . .
. . . . . . . . . . . . . . . . . . . . .
```

The starting and end points are P0.

Example 2

The cutter path (Fig. 5-25[a]) defined by the program shown in Fig. 5-25(b) includes a change in the part surface. Since the tool has been positioned within the tolerance range with respect to plane PL2 after point P2 has been reached, it is not necessary to have a

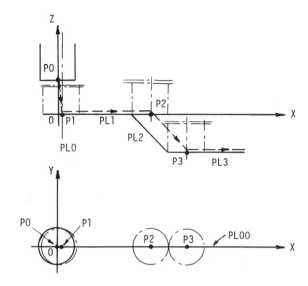

(a)

IBM S/370 APT-AC N/C PROGRAM VERSION=1.3.0000

... BEGIN TRANSLATION PHASE... (SECTION 1) ...

```
ISN 00001 PARTNO TEST GO STATEMENT
ISN 00002 MACHIN/GN5CC,9,OPTION,2,0
ISN 00003 CLPRNT
ISN 00004 CUTTER/1.0
ISN 00005 FEDRAT/1.0
ISN 00006 P0=POINT/0,0,1
ISN 00007 PL0=PLANE/1,0,0,0.1
ISN 00008 PL00=PLANE/0,1,0,0
ISN 00009 PL1=PLANE/0,0,1,0
ISN 00010 PL2=PLANE/1,0,1,2
ISN 00011 PL3=PLANE/0,0,1,-1
ISN 00012 FROM/P0
ISN 00013 GO/ON,PL00,PL1,ON,PL0
ISN 00014 INDIRV/1,0,0
ISN 00015 TLON,GOFWD/PL00,PAST,PL2
ISN 00016 PSIS/PL2
ISN 00017 GOFWD/PL00,TO,PL3
ISN 00018 PSIS/PL3
ISN 00019 GOFWD/PL00,TO,(PL4=PLANE/1,0,0,5)
ISN 00020 END
ISN 00021 FINI
```

NO DIAGNOSTICS ELICITED DURING TRANSLATION PHASE
17 N/C SOURCE RECORDS (SYSIN)

(b)

SECTION 1 ELAPSED CPU TIME IN MIN/SEC IS 0000/00.1333
SECTION 2 ELAPSED CPU TIME IN MIN/SEC IS 0000/00.0966

Figure 5-25 (a) A cutter path including change of the part surface. (b) The program defining the cutter path shown in (a).

start-up motion at point P2. The internal sequence number (ISN) for each statement is assigned by the APT-AC NC processor during processing and does not belong to the statement itself.

Example 3
The cutter path (Fig. 5-26[a]) defined in the program shown in Fig. 5-26(b) also includes a change in the part surface at point C. The statement labeled ISN 00018, which is used to change the part surface and which moves the tool downward, can be replaced by the following two statements:

```
GODLTA/−1.0
AUTOPS
```

5.9 PROCESSING OF CONTOURING MOTION STATEMENTS

In this section, we consider how a series of consecutive contouring motion statements are processed by the NC processor.

Mathematically speaking, to process contouring motion statements is to find the consecutive tool positions of a contouring motion. The given conditions for such a calculation are as follows:

1. The designed or specified profile
2. The specified allowances for drive, part, and check surfaces
3. The specified tolerance(s)*
4. The tool dimensions and its direction in space*

For a start-up motion, the end position of the tool is found on the basis of the given conditions, and the result of the calculation is a single output that is a set of coordinates defining the position of the tool-end center at the end of start-up motion.

For a contouring motion, the processing result depends on the locus of the cutter path (linear or nonlinear). If the contouring motion is linear, the result is the same as for a start-up motion or point-to-point motion statement: only the end position is calculated and output. The processing of a nonlinear contouring motion is more complicated because the path can be described only by a series of consecutive points (or sets of coordinates). For example, the processing of the curved path shown in Fig. 5-27 yields points P1, P2, P3, . . . , and their coordinates are output in the CLDATA. Actually, the tool path is the linear interpolation of the designed profile based on these calculated points. It is evident that the number of points that describe the contouring motion of a given profile varies with the specified tolerance. The tighter the tolerance, the larger is the number of points or motion steps required to maintain the tool position within the specified tolerance range. The maximum number of steps that can be processed by the NC processor for one single contouring motion statement is 400, which is usually enough. In some cases the required number of steps exceeds this limit. In such a case, the statement

NUMPTS/n

can be used to specify a new limit, n.

---

*Tolerance and tool dimension specifications are specified by separate statements (discussed in Chapter 6). For three-axis NC machines, the tool axis is always in the Z direction.

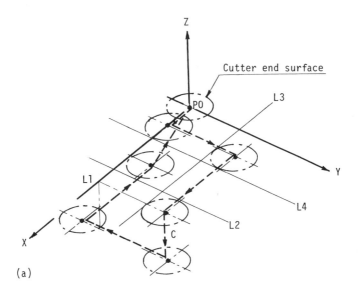

(a)

IBM S/370 APT-AC N/C PROGRAM VERSION=1.3.0000

... BEGIN TRANSLATION PHASE... (SECTION 1) ...

```
ISN 00001 PARTNO TEST GO STATEMENT
ISN 00002 MACHIN/GN5CC,9,OPTION,2,0
ISN 00003 CLPRNT
ISN 00004 CUTTER/1.0
ISN 00005 FEDRAT/1.0
ISN 00006 P0=POINT/0,0,0
ISN 00007 L1=LINE/XAXIS
ISN 00008 L2=LINE/(2,0),(2,1)
ISN 00009 L3=LINE/(2,1),(0,1)
ISN 00010 L4=LINE/(1,1),(1,0)
ISN 00011 PL0=PLANE/0,0,1,0
ISN 00012 PL1=PLANE/0,0,1,-1
ISN 00013 FROM/P0
ISN 00014 GO/L4
ISN 00015 AUTOPS
ISN 00016 TLLFT,GOLFT/L4,PAST,L3
ISN 00017 GORGT/L3,PAST,L2
ISN 00018 GO/L2,PL1,L3
ISN 00019 GORGT/L2,PAST,L1
ISN 00020 GORGT/L1,TO,L4
ISN 00021 GOTO/P0
ISN 00022 END
ISN 00023 FINI
```

    NO DIAGNOSTICS ELICITED DURING TRANSLATION PHASE
    23 N/C SOURCE RECORDS (SYSIN)

                              SECTION 1 ELAPSED CPU TIME IN MIN/SEC IS 0000/00.1499
                              SECTION 2 ELAPSED CPU TIME IN MIN/SEC IS 0000/00.0899
        (b)

Figure 5-26 (a) The cutter path defined by the APT program in (b). (b) An APT program showing a change of the part surface during contouring motion.

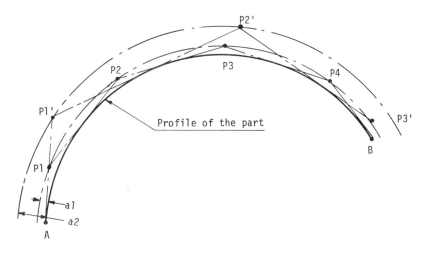

Figure 5-27   The consecutive cutter positions in a contouring motion. The number of points generated by the NC processor depends on the specified tolerance ($a1$ or $a2$).

In any case, the result of processing a nonlinear contouring motion is a series of points. The cutter motion is the linear movement from one point to the next (because a feedrate other than RAPID is usually required). Figure 5-28(a and b) presents the program that defines a circular path (one fourth of a circle) and the CLDATA output. The number of points is 39 for the specified tolerance of 0.0005 in. If we reduce the tolerance to 0.0004 in., the number of points is increased to 43 (Fig. 5-28[c]). The statement CLPRNT in the listed program instructs the NC processor to print the CLDATA, which is often used to check a programmed cutter path. Further discussion on CLDATA is given in Chapters 8 and 11.

PROBLEMS

5.1 Define each of the two groups of points in Fig. P5-1 as a pattern, using as concise a PATERN statement as possible. The order of the points is specified in the figure. All the points are on the plane $z = 0$.

5.2 Assume that the points in the ARC and LINEAR patterns in Problem 5-1 represent holes on levels $z = 1.0$ in. and $z = 2.0$ in., respectively. The depth of the holes in the arc pattern is 0.5 in., whereas that in the linear pattern is 0.6 in. The drilling feedrate is 2.0 in./min. The starting and end points are at (0,0,3). Write a segment of APT program that defines

  a. These two groups of points as one pattern

  b. The drilling operation

5.3 Assume that a point pattern is defined as

PT0=PATERN/(1,1),(0,1),(−1,0),(0,−1)

IBM S/370 APT-AC N/C PROGRAM VERSION=1.3.0000   DATE=08/05/86 TIME=15:04:24 0
                         ... BEGIN TRANSLATION PHASE... (SECTION 1) ...

ISN 00001 PARTNO TEST GO STATEMENT
ISN 00002 MACHIN/GN5CC,9,OPTION,2,0
ISN 00003 CLPRNT
ISN 00004 CUTTER/0.5
ISN 00005 FEDRAT/2.0
ISN 00006 TOLER/0.0005
ISN 00007 P0=POINT/0,0,0
ISN 00008 C1=CIRCLE/3,0,0,2
ISN 00009 L1=LINE/(3,0,0),(3,1,0)
ISN 00010 FROM/P0
ISN 00011 GO/C1
ISN 00012 AUTOPS
ISN 00013 TLLFT,GOLFT/C1,ON,L1
ISN 00014 END
ISN 00015 FINI

NO DIAGNOSTICS ELICITED DURING TRANSLATION PHASE
15 N/C SOURCE RECORDS (SYSIN)

SECTION 1 ELAPSED CPU TIME IN MIN/SEC IS 0000/00.0999
SECTION 2 ELAPSED CPU TIME IN MIN/SEC IS 0000/00.0899

(a)

Figure 5-28 (a) An APT program defining a circular contouring motion. (continued)

253

....SECTION 3....

ISN

```
0001 PARTNO/ TEST GO STATEMENT
0002 MACHIN/              GN5CC        9.00000000           OPTION        2.00000000
                 0.0
0004 CUTTER/     0.50000000
0005 FEDRAT/     2.00000000
0006 OUTTOL/     0.00050000
0006  INTOL/     0.0
0010  FROM/               P0
                 0.0               0.0                 0.0
0011  GOTO/               C1
                 0.75000000        0.0                 0.0
0013
0013 SURFACE             C1                            CIRCLE   DS(IMP-TO)
                 3.00000000        0.0                 0.0
                 0.0               0.0                 1.00000000     2.00000000
0013  GOTO/               C1
                 0.74998067        0.04651086          0.0
                 0.75382479        0.13945312          0.0
                 0.76150646        0.23215713          0.0
                 0.77301256        0.32446450          0.0
                 0.78832344        0.41621752          0.0
                 0.80741292        0.50725945          0.0
                 0.83024840        0.59743473          0.0
                 0.85679087        0.68658931          0.0
                 0.88699497        0.77457086          0.0
                 0.92080911        0.86122906          0.0
                 0.95817551        0.94641588          0.0
                 0.99903034        1.02998575          0.0
                 1.04330378        1.11179592          0.0
                 1.09092021        1.19170659          0.0
                 1.14179827        1.26958126          0.0
                 1.19585104        1.34528687          0.0
                 1.25298616        1.41869408          0.0
                 1.31310603        1.48967748          0.0
                 1.37610793        1.55811578          0.0
                 1.44188422        1.62389207          0.0
                 1.51032252        1.68689397          0.0
                 1.58130592        1.74701384          0.0
                 1.65471313        1.80414896          0.0
                 1.73041874        1.85820173          0.0
                 1.80829341        1.90907979          0.0
                 1.88820408        1.95669622          0.0
                 1.97001425        2.00096966          0.0
                 2.05358412        2.04182449          0.0
                 2.13877094        2.07919089          0.0
                 2.22542914        2.11300503          0.0
                 2.31341069        2.14320913          0.0
                 2.40256527        2.16975160          0.0
                 2.49274055        2.19258708          0.0
                 2.58378248        2.21167656          0.0
                 2.67553550        2.22698744          0.0
                 2.76784287        2.23849354          0.0
                 2.86054688        2.24617521          0.0
                 2.95348914        2.25001933          0.0
                 3.00000000        2.25000000          0.0
```

```
0014 END
0015 ***** FINI *****
....END OF SECTION 3....
```

SECTION 3 ELAPSED CPU TIME IN MIN/SEC IS 0000/00.0733

Figure 5-28   (b) The computer output (CLDATA) of the program shown in (a).   (continued)

```
....SECTION 3....
ISN
0001 PARTNO/ TEST GO STATEMENT
0002 MACHIN/          GN5CC          9.00000000          OPTION     2.00000000
                 0.0
0004 CUTTER/     0.50000000
0005 FEDRAT/     2.00000000
0006 OUTTOL/     0.00040000
0006  INTOL/     0.0
0010   FROM/          P0
                 0.0                 0.0                 0.0
0011   GOTO/          C1
                 0.75000000          0.0                 0.0
0013
0013 SURFACE          C1                                 CIRCLE     DS(IMP-TO)
                 3.00000000          0.0                 0.0
                 0.0                 0.0                 1.00000000          2.00000000
0013   GOTO/          C1
                 0.74999346          0.04207993          0.0
                 0.75314030          0.12618093          0.0
                 0.75942957          0.21010546          0.0
                 0.76885248          0.29373614          0.0
                 0.78139586          0.37695600          0.0
                 0.79704215          0.45964866          0.0
                 0.81576948          0.54169845          0.0
                 0.83755166          0.62299063          0.0
                 0.86235821          0.70341151          0.0
                 0.89015445          0.78284859          0.0
                 0.92090150          0.86119080          0.0
                 0.95455636          0.93832855          0.0
                 0.99107195          1.01415396          0.0
                 1.03039721          1.08856098          0.0
                 1.07247714          1.16144556          0.0
                 1.11725288          1.23270575          0.0
                 1.16466182          1.30224188          0.0
                 1.21463764          1.36995672          0.0
                 1.26711046          1.43575555          0.0
                 1.32200687          1.49954634          0.0
                 1.37925011          1.56123988          0.0
                 1.43876012          1.62074989          0.0
                 1.50045366          1.67799313          0.0
                 1.56424445          1.73288954          0.0
                 1.63004328          1.78536236          0.0
                 1.69775812          1.83533818          0.0
                 1.76729425          1.88274712          0.0
                 1.83855444          1.92752286          0.0
                 1.91143902          1.96960279          0.0
                 1.98584604          2.00892805          0.0
                 2.06167145          2.04544364          0.0
                 2.13880920          2.07909850          0.0
                 2.21715141          2.10984555          0.0
                 2.29658849          2.13764179          0.0
                 2.37700937          2.16244834          0.0
                 2.45830155          2.18423052          0.0
                 2.54035134          2.20295785          0.0
                 2.62304400          2.21860414          0.0
                 2.70626386          2.23114752          0.0
                 2.78989454          2.24057043          0.0
                 2.87381907          2.24685970          0.0
                 2.95792007          2.25000654          0.0
                 3.00000000          2.25000000          0.0
0014 END
0015 ***** FINI *****
....END OF SECTION 3....
                                SECTION 3 ELAPSED CPU TIME IN MIN/SEC IS 0000/00.0833
```

Figure 5-28 (c) The CLDATA of the APT program in (a), with tolerance changed to 0.0004 in.

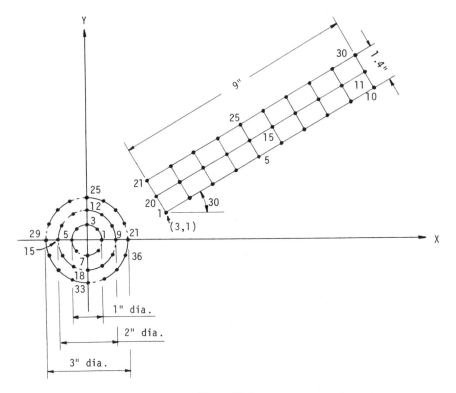

Figure P5-1

and that P0 is the symbol for the point (0,0,0). Determine the positions and the order of the points defined in the following patterns and verify them by means of CLDATA output:

a. PT1=PATERN/PT0,ATTACH,P0,TRANPT,(3,3)

b. PT2=PATERN/PT0,ATTACH,P0,TRANPT,(3,3,1)

c. ZSURF/1.0; PT3=PATERN/PT0,ATTACH,P0,TRANPT,(3,3)

d. PT4=PATERN/PT1,ATTACH,P0,INVX,ATTACH,P0,INVY

e. PT5=PATERN/PT1,ATTACH,P0,INVX,INVY

f. PT6=PATERN/PT0,PT1

g. PT7=PATERN/PT0,INVERS,PT1

h. PT8=PATERN/PT0,PT1,INVERS

Note: To obtain the CLDATA output, add the statement CLPRNT before the motion statement.

5.4 The various formats of the pattern statements can be concatenated to define a complex point pattern. The following are some of the statements that can be used to define the point pattern PAT1, which consists of points P1, P2, P3, ..., P8, shown in Fig. P5-4:

a. PT1=PATERN/(3,1.5),(4,1.5)

  PT2=PATERN/(4,2.5),(4,3.5)

  PT3=PATERN/AT,(4,3.5),RTHETA,CONST,3,0,27,54,81

  PAT1=PATERN/PT1,PT2,PT3

b. PAT1=PATERN/AT,(1,1.5),DELTA,XCOORD,2,1,YCOORD,1,1,AT,$

              (4,3.5),RTHETA,CONST.3,0,27,54,81

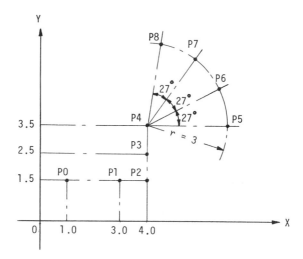

Figure P5-4

c. PAT1=PATERN/AT,(1,1.5),DELTA,XCOORD,2,1,YCOORD,1,1,$
        RTHETA,CONST,3,0,27,54,81
d. PAT1=PATERN/AT,(1,1.5),ABSLTE,XCOORD,3,4,YCOORD,2.5,3.5,ARC,$
        (CIRCLE/4,3.5,3),ATANGL,0,CCLW,81,NUMPTS,4
e. PAT1=PATERN/(3,1.5),(4,1.5),DELTA,YCOORD,1,1,ARC,$
        (CIRCLE/4,3.5,3),ATANGL,0,CCLW,81,NUMPTS,4
    Verify these statements.

5.5 Define the start-up motions given in Fig. P5-5(a through g), using APT statements. Assume that the starting point is P1 and the geometric entities with given symbols have been defined.

5.6 Define the contouring motion given in Fig. P5-6(a through f), using APT statements. Assume that the tool has been correctly placed to position P1, required for starting the contouring motion. Assume that the geometric entities with given symbols have been defined. Assume also that the positive X direction is from the left side of the figure to the right.

5.7 Define the cutter path for machining the profile of the template given in Problem 4.2. Assume that the starting and end positions of the tool are (6,3,1) and that the thickness of the template is 0.15 in. It is required that the origin of the coordinate system is as shown in Fig. P4-2. A feedrate of 2 in./min can be assumed for the cutting operation.

5.8 Define the cutter path for machining the part given in Problem 4.4, whose profile consists of line, circle, tabulated cylinders, and/or LCONICS. The starting position of the tool is (−5,5,1) and the thickness of the part is 0.5 in. The coordinate system is defined as shown in Fig. P4-4, with its origin set on the surface of the part. A feedrate of 2 in./min is used.
    Note: The definition of an LCONIC, or tabulated cylinder, based on its tangency to another geometric entity, often does not ensure that a TANTO modifier can be successfully used, in the contouring motion statement, to move the tool from an LCONIC, or tabulated cylinder, to the geometric entity tangent to it or vice versa. The reason is that the radii of curvature at the tangent point for the two geometric entities can be different even though their slopes are the same at that point. One solution is to define a line that

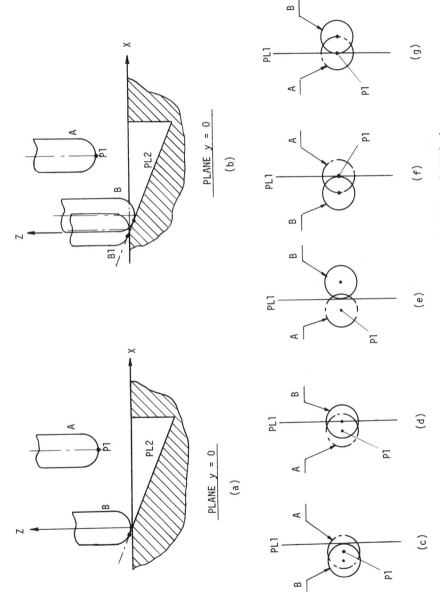

PLANE y = 0

(a)

PLANE y = 0

(b)

(c)

(d)

(e)

(f)

(g)

Figure P5-5 Start-up motions. The positive X direction in (c) through (g) is the same as that shown in (a). A: the starting position of the cutter; B and B1: the end positions of the start-up motion.

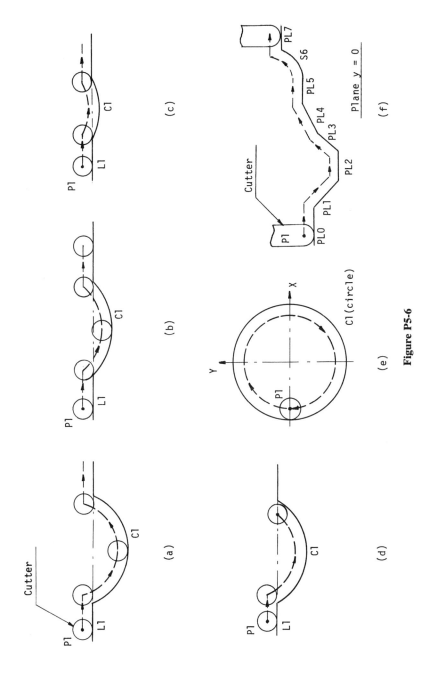

Figure P5-6

259

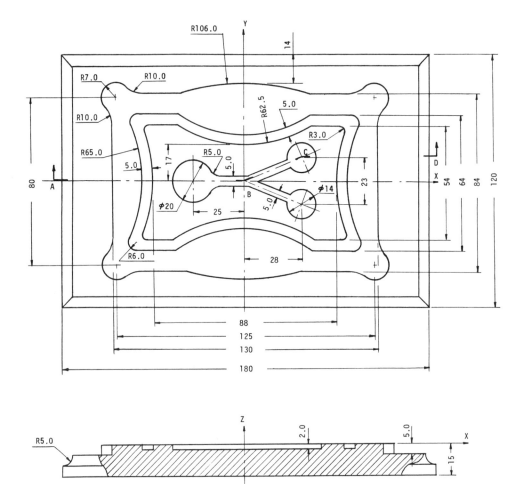

Figure P5-11

passes through the tangent point and is perpendicular to the common tangent line. The line is then used as the check surface, together with the modifier ON, to position the tool to the tangent point of the two geometric entities. Sometimes, because of the severe change in radii of curvature at the tangent point from one curve to another, the contouring motion cannot continue without issuing a start-up motion statement, between the two contouring motion statements, to move the tool to the position required for the next contouring motion.

5.9 If the milling operation for machining the part discussed in Problem 5.8 should be divided into two passes, one for rough cut and the other for finishing, write an APT program segment in such a way that the finishing cut immediately follows the rough cut without detouring. The finishing allowance is 0.02 in.

5.10 The cavity of the matrix shown in Fig. P4-8 is to be cut on a three-axis milling machine. The cutter path is designed in such a way that the vertical planes parallel to section planes B-C and C-D are used as the controlling surface to direct the tool motion from the left end to the right or vice versa, with an incremental motion, from one set of vertical planes to the next, along the end surfaces (e.g., the motion from D to E). The resulting tool motion is B → C → D → E → F → G .... Write an APT program segment that defines the start-up motion and the contouring motion B → C → D → E → F → G. The starting and end positions of the tool are (10,5,3), and the feedrate is 2.0 in./min. Assume that a ball-end mill is used.

5.11 An APT program is needed for machining the part given in Fig. P5-11. The dimensions of the stock material are 180 × 120 × 15 mm. Three cutters are used:

a. A 10 mm ball-end mill for cutting the trimming edge (R = 5 mm)

b. An 8 mm flat-end mill for cutting the exterior profile (with its depth equal to 5 mm)

c. A 4 mm flat-end mill for cutting the groove and circular cavities (with a depth of 2 mm)

Define the geometric entities and write three program segments, each defining the cutter motion for one tool. The starting and end points are at (200,150,30), and the feedrate is 50 mm/min.

chapter 6

Machining Specifications and Miscellaneous Statements

From an APT programmer's point of view, a machining process basically includes two major aspects, namely, the path that the tool(s) should follow and the machining specifications required to produce the part. Machining specifications include, for example, tolerance(s), allowance(s), cutting speed(s), feedrates, cutting tool(s), and coolant. An APT program is a description of the machining process written in computer language, which must be processed by a computer to obtain an NC program. Thus such an APT program should also include instructions on how it is to be processed. In this chapter, we discuss the statements that define the machining specifications and computer operations. Some of these statements depend also on the postprocessor design; in addition, their statement formats and parameters vary with the selected postprocessor. Therefore, to write a correct program, a programmer must have some knowledge of the postprocessor to be used.

6.1 STATEMENTS SPECIFYING TOLERANCES

The cutter motion on an NC machine is controlled by the pulses sent by the NC controller to the motors that generate the motions in the X, Y, and Z directions. At any moment, the tool movement in each direction is determined by the rate at which pulses are sent to the motors, and it is linear even when the desired cutter path is a curve. Similarly, the NC processor calculates the cutter path for a contoured surface on the basis of linear interpolation (Fig. 5-27). Thus some deviation of the cutter path from the desired one is unavoidable. A programmer should specify an acceptable limit to this deviation according to the part design; this limit is called *tolerance*. In APT a tolerance can be specified by the following statements:

INTOL/*t*

where *t*, in millimeters or inches, is the maximum allowable inward deviation of the cutter position from the designed profile (Fig. 6-1[a]);

OUTTOL/*t*

where *t* is the maximum allowable outward deviation of the cutter position from the designed profile (Fig. 6-1[b]); and

TOLER/*t*

where *t* represents the outward tolerance, and the inward tolerance is zero (Fig. 6-1[b]).

There are three possible situations that can arise from a part design: the allowed deviation of the actual profile from the designed one is inward, outward, or in both directions. An inward tolerance is specified by the statement INTOL; an outward tolerance, by the statements OUTTOL or TOLER. Both the INTOL and OUT-TOL statements should be specified to define the tolerances in both directions.

The specified tolerance, *t*, in the statements shown above applies equally to the drive, part, and check surfaces. If different tolerances are needed for the three controlling surfaces, then the following statements can be used:

$$\begin{Bmatrix} \text{INTOL} \\ \text{OUTTOL} \\ \text{TOLER} \end{Bmatrix} /t1\,[,t2[,t3]]$$

where *t*1, *t*2, and *t*3 are the tolerances on the part, drive, and check surfaces, respectively. If *t*3 is omitted, then the tolerance on the drive surface, *t*2, is taken as that on the check surface.

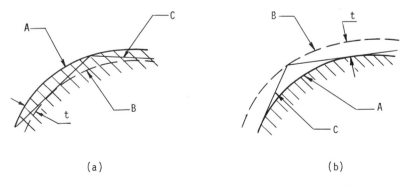

(a)   (b)

Figure 6-1   (a) The cutter path (C) and the acceptable profile (B) specified by the statement INTOL/*t*. (b) The cutter path (C) and the acceptable profile (B) specified by the statement OUTTOL/*t* or TOLER/*t*. A: Designed part profile.

The tolerance commands listed above are modal; that is, they are effective until changed by another tolerance statement.

6.2 THE STATEMENT SPECIFYING THE UNIT OF MEASUREMENT

Two basic units of measurement are used in engineering drawings to specify a dimension: millimeter (mm) and inch (in.). The statement

$$UNITS/ \begin{Bmatrix} [INCHES] \\ MM \end{Bmatrix}$$

can be specified to indicate the unit in use. Inch is the default unit and can be omitted. If more than one UNITS statement is specified in a program, each cancels the previous one and affects the section of the program following it.

It should be pointed out that the conversion of measurement units applies only to geometric entities and cutter positions. The scalar variables remain unaffected. For example, in the following program

```
. . . . . . . . . . . . . . . . . .
UNITS/MM
I=1
P1=POINT/3,4,5
UNITS/INCHES
P2=POINT/I,(I+1),(I+2)
. . . . . . . . . . . . . . . . . .
A)   GOTO/P1
B)   GOTO/P2
. . . . . . . . . . . . . . . . . .
. . . . . . . . . . . . . . . . . .
```

point P1 is defined in millimeters and its x, y, and z coordinates are 3, 4, and 5 mm. The letter I represents a scalar and is unaffected by the statement UNITS/MM. Thus the x, y, and z coordinates of point P2 are 1, 2, and 3 in., respectively. The cutter path in this program segment is defined after the UNIT/INCHES statement, and the cutter position defined by the statements labeled A and B are (3/25.4, 4/25.4, 5/25.4 in.) and (1, 2, 3 in.), respectively.

As a matter of fact, the APT system is dimensionless. If we do not intend to use different measurement systems in a program to define the geometric entities or cutter path, we need not specify the unit of measurement. It is useful only when the geometric entities in a program are required or intended to be defined with different unit systems to save calculation work.

6.3 STATEMENTS SPECIFYING THE CUTTER

6.3.1 The CUTTER Statement

As pointed out previously, the cutter location data (CLDATA) processed by the NC processor is an orderly list of sets of coordinates that represent the cutter center-line positions or, more specifically, the end-center position for a milling cutter or the position of the tool nose center for a turning cutter (Fig. 6-2). For point-to-point mo-

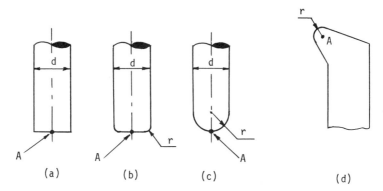

Figure 6-2 The point, A, represents the position of a cutter. (a) Flat-end mill;
(b) filleted-end mill; (c) ball-end mill; (d) turning tool.

tion, the geometric shape of a cutter is of no importance because the cutter
center-line position has been defined in the point-to-point motion statement. However, as discussed in Chapter 5, a contouring motion is defined on the basis of the
part profile, which is the envelope of the cutter profile at the consecutive positions
that it occupies during a contouring motion. The NC processor calculates the cutter
center-line positions according to the part profile, the specified tolerance, and the
geometric form of the cutter. For a milling operation in the plane perpendicular to
the tool axis (i.e., the X-Y plane), with an end mill, only the tool diameter, d, and its
corner radius, r, are needed to calculate the tool center position (Fig. 6-2). The following statement specifies end mills:

CUTTER/$d[,r]$

where d and r are the diameter and corner radius of the cutter, respectively (Fig. 6-2).
They must be positive. The parameter r can be omitted if it is zero (as in a flat-end
mill). For a turning tool, the cutting operation is performed by the tool nose. Therefore it can also be considered as a flat-end mill with a diameter of $2r$ (Fig. 6-2).

   For three-dimensional machining, a part profile can be any combination of the
surfaces allowed in APT. The calculation of the cutter center-line position requires
the three-dimensional contour of the cutter. Therefore additional parameters are
needed to define a cutter. The statement defining a cutter for three-dimensional machining is as follows (Fig. 6-3):

CUTTER/d,r,e,f,a,b,h

where

   d = the diameter of the circle generated by the intersection of the end and side
        surfaces of the cutter
   r = the radius of the corner circle
   e = the distance from the corner circle center to the cutter center line
   f = the distance from the corner circle center to the plane passing through the
        cutter end center and perpendicular to the cutter center line, or axis

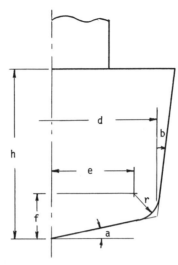

Figure 6-3 The parameters (d, r, e, f, a, b, h) defining a cutter.

a = the angle, in degrees, between the cutter end surface and the plane perpendicular to the cutter axis (It lies in the range $0 \leqslant a < 90°$.)

b = the angle, in degrees, between the cutter side surface and the cutter axis (It lies in the range $-90° < b < 90°$, with positive direction as shown in the figure.)

h = the cutter height measured from the tool end center

Note that the corner circle should be tangent to the end and side surfaces. In addition, all the parameters, except b, should be positive.* Usually, the values of r, e, a, b, and h are known. The parameters d and f should be calculated by the computer according to the following formulas[1,2]:

$$d = 2\{e + r \cdot \cos[45 + (b - a)/2]/\cos[45 - (a + b)/2]\}$$
$$f = r \cdot \sin[45 + (b - a)/2]/\cos[45 - (b + a)/2] + (d/2) \cdot \tan(a)$$

Two calculation statements based on these formulas should be specified before the cutter definition statement is specified.

Besides the limitation mentioned above, there are several restrictions on the values of the parameters in a CUTTER statement:

1. $(a + b) < 90°$
2. $h \geqslant (d/2) \tan(a)$
3. $f \geqslant (d/2) \tan(a)$
4. The corner circle should not extend below the end surface when it is not tangent to the end surface.

When the statement

CUTTER/d[,r]

*These two conditions are listed in the APT programming manuals.[1,2] However, parameter e can also be negative (i.e., the corner circle and its center are on different sides of the center line). In addition, the corner circle can be not tangent to the end and side surfaces.

is used for three-dimensional machining, the following default values are used by the NC processor to define a cutter:

$$a = 0$$
$$b = 0$$
$$f = \begin{cases} r \\ 0 & \text{(if } r \text{ is not given)} \end{cases}$$
$$e = (d/2) - r$$
$$h = \begin{cases} 5 & \text{(if } r \leqslant 5 \text{ or } r \text{ is not given)} \\ r & \text{(if } r > 5) \end{cases}$$

Figure 6-4 gives several examples of the use of the CUTTER statement.

6.3.2 The Effect of the Cutter Top Surface on the Cutter Path

The cutter path of a contouring motion is determined by the interaction of the tool-end and tool-side surfaces with the three controlling surfaces, namely, the part, drive, and check surfaces. This is true when any one of the three controlling surfaces does not become a limiting surface on the upper side of the tool. Generally, the three controlling surfaces are the limiting surfaces for the complete profile of a tool, including its upper surface. Therefore, when a geometric entity that is part of the part profile has more than one intersection point with the tool axis, the tool height should also be taken into consideration in defining a cutter path. For example, the part A-B-C-D, shown in Fig. 6-5, is composed of several circular arcs, each of them being defined as a cylinder. The following statements are used to define the desired cutter path P0-A-C:

```
. . . . . . . . . . . . . . . . . . . . . .
CUTTER/1,0,0,0,0,0,3   $$ D = 1; H = 3
. . . . . . . . . . . . . . . . . . . . .
FROM/P0
INDIRV/0,0,-1
GO/ON,PL1,CYL1,ON,PL2 $$PL2 IS X-Z PLANE AND PL1 IS PLANE X = T
INDIRV/-1,0,0
TLON,GOFWD/PL2,PSTAN,CYL2
. . . . . . . . . . . . . . . . . . . . . .
. . . . . . . . . . . . . . . . . . . . .
```

However, the tool is not able to reach point C (where surface CYL1 is tangent to surface CYL2) because the upper part of the cylinder, CYL1, takes effect after the tool reaches point B. The cutter path will be P0-A-B-C' instead of P0-A-C. Surfaces, such as elliptical cones and hyperbolic cylinders, with more than one sheet in the Z direction will sometimes also cause a problem in defining tool position (Fig. 6-6). Therefore, when defining a cutter path, we should consider the complete outline of the geometric entities that constitute the part profile so that possible interferences of the tool with the upper part of the surfaces can be discovered.

There are several remedies for the problem described above. The first one is to redefine the surface as a geometric entity having only one intersection with the tool axis. For example, if sections AC and CD of cylinders CYL1 and CYL2 are defined as a tabulated cylinder, the tool can move from A to C because these redefined geo-

Cutter	Statement	
	Simple format	Expanded format
	CUTTER/d or CUTTER/d,0	CUTTER/d,0,d/2,0,0,0,h
	CUTTER/d,r	CUTTER/d,r,e,f,0,0,h
		CUTTER/d,0,d/2,0,0,b,h (b > 0)
		CUTTER/d,r,e,f,0,b,h (b < 0)

Figure 6-4 Examples of the CUTTER statement.

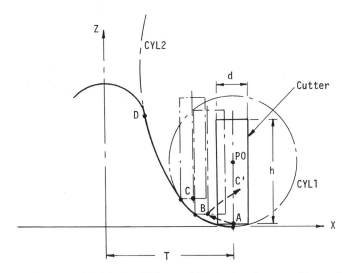

Figure 6-5 The tool height should be considered when the controlling surfaces have more than one intersection with the Z (tool) axis.

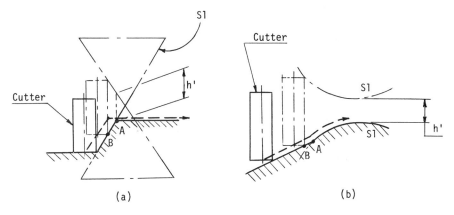

(a)   (b)

Figure 6-6 Surfaces with two sheets in the Z direction might cause a problem in positioning the tool. S1: Elliptical cone (a) or hyperbolic cylinder (b). B: The tool interferes with the surface at this point; —————→: desired cutter path.

metric entities do not have more than one intersection with the tool axis. Another approach is hypothetically to reduce the tool height during calculation of the cutter path. For example, in Fig. 6-6(a), if the tool height is less than h', then the tool can reach point A without interfering with the upper sheet of surface S1. Of course, the CUTTER statement can be used to change the tool height temporarily, but this might still create a problem when the tool height must be reduced to a value that is unacceptable to the CUTTER statement. An additional statement is provided in APT to solve this problem, its format being

$$\text{CUTTER/OPTION,} \begin{Bmatrix} 1 \\ 2 \end{Bmatrix}, \begin{Bmatrix} r,h \\ \text{OFF} \end{Bmatrix}$$

This statement uses two parameters, r and h, to define the tangent (or contact) point of the tool contour to the part profile (Fig. 6-7). A hypothetical disk cutter, with zero height and radius r, is assumed at that position to replace the real cutter. Since the distance between the hypothetical disk cutter and the end center of the real tool is h, the actual tool position can be calculated by the NC processor on the basis of the position of the hypothetical tool. The parameters 1 and 2 specify that the hypothetical tool applies to the part and drive surfaces, respectively. The hypothetical tool can be canceled by another CUTTER/OPTION statement or by the statement

CUTTER/OPTION, $\begin{Bmatrix} 1 \\ 2 \end{Bmatrix}$,OFF

when the real tool takes effect again.

It is evident from the description presented above that the CUTTER/OPTION statement essentially defines the contact point between the tool and the part profile. The position of the contact point on the tool does not change if the tool has a flat end (Fig. 6-8[a]) or if the part profile is a plane or a straight line (Fig. 6-8[b]). Therefore

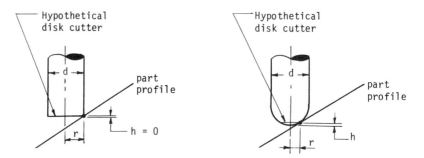

Figure 6-7   The contact point, or hypothetical disk cutter, defined by the CUTTER/OPTION statement.

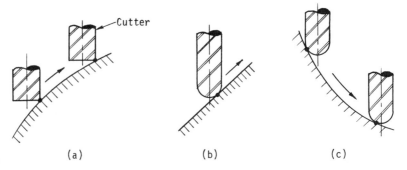

(a)   (b)   (c)

Figure 6-8   The contact point between the tool and the part profile. (a) The contact point of a flat-end tool does not change its position on the tool during machining. (b) The contact point of a profiled-end tool does not change its position on the tool during machining if the part profile is a straight line or plane. (c) The contact point changes its position on the tool if the part profile is nonlinear.

the CUTTER/OPTION statement does not cause a problem when a flat-end tool is used or a linear profile is machined. However, when cutters other than a flat-end one are used to cut a nonlinear profile, the tool positions of a contouring motion calculated on the basis of the hypothetical tool might not be correct (Fig. 6-8[c]).

Another possible approach that may be used to avoid interference of the cutter's upper surface with the part or drive surface consists in defining the cutter path in the *X-Y* plane and then transforming it into the desired one. This method is discussed in Chapter 7.

Two examples are given below to explain the usage of the CUTTER/OPTION statement.

Example 1

The program given previously for machining the part shown in Fig. 6-5 can be used if the tool height is reduced, as in the following program segment:

```
. . . . . . . . . . . . . . . . . . . . .
. . . . . . . . . . . . . . . . . . . . .
CUTTER/1,0,0,0,0,0,3
FROM/P0
INDIRV/0,0,-1
GO/ON,PL1,CYL1,ON,PL2 $$ CYL1 IS THE PART SURFACE
CUTTER/OPTION,1,0.5,0
INDIRV/-1,0,0
TLON,GOFWD/PL2,PSTAN,CYL2
CUTTER/OPTION,1,OFF
. . . . . . . . . . . . . . . . . . . . .
. . . . . . . . . . . . . . . . . . . . .
```

The cutter path for this section of the program is P0-*A*-*C*. The parameter 1 is selected in the CUTTER/OPTION statement because the surface interfering with the tool is a part surface (CYL1).

Example 2

The application of a hypothetical tool to the drive surface can be seen in Fig. 6-9. In this case, the following statements can be used to move the tool from point *A* through point *B* to point *C*:

```
. . . . . . . . . . . . . . . . . . . . $$TOOL IS AT POINT A; PART SURFACE IS PL2
CUTTER/OPTION,2,0.5,0
INDIRV/-0.2,-1,0
TLRGT,GOFWD/CYL1,PAST,PL1
CUTTER/OPTION,2,OFF
PSIS/PL1
INDIRV/0,-1,0
TLLFT,GOFWD/PL2,. . . . .
. . . . . . . . . . . . . . . . . . . . .
```

Here, a hypothetical tool is used to avoid interference of the tool's upper surface with cylinder CYL1.

6.4 STATEMENTS DEFINING FEEDRATE

The feedrate is the speed of the tool moving along the part profile or from one point to another. It is defined as the distance (in inches or millimeters) that the tool moves in 1 minute or in one revolution of the machine tool spindle. Normally, the feedrate

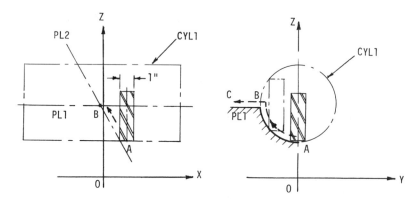

Figure 6-9   The upper surface of the cutter interferes with the drive surface, CYL1.

is given in inches or millimeters per minute for milling machines, and in inches or millimeters per revolution for lathes.

A feedrate can be specified by using the statement

$$\text{FEDRAT/} \begin{Bmatrix} \text{[IPM,]} \\ \text{IPR,} \end{Bmatrix} f$$

where

IPM indicates that the specified feedrate, f, is measured in inches or millimeters per minute, depending on the unit specified by the UNITS statement;

IPR indicates that the specified feedrate, f, is measured in inches or millimeters per revolution.

A rapid tool speed can be specified by the statement

RAPID

A feedrate statement should be specified before the motion statement. The FEDRAT statement is modal; it remains in effect until changed by another FEDRAT statement. However, the effectiveness of a RAPID statement depends on the postprocessor design. Usually, a postprocessor is designed in such a way that the RAPID statement is effective only until the next motion statement. This programming rule is followed throughout this book.

An example is given in Fig. 6-10 to explain the usage of the statements FEDRAT and RAPID.

```
. . . . . . . . . . . . . . . . . . . . . . . .
FEDRAT/2.0
. . . . . . . . . . . . . . . . . . . . . . . . . . .   ⎫
. . . . . . . . . . . . . . . . . . . . . . . . . . .   ⎬    (Feedrate is 2.0)
RAPID                                                   ⎭
GOTO/P1 $$ RAPID FEEDRATE
GOTO/P2 $$ FEEDRATE IS 2.0.
FEDRAT/3.0
GO/. . . . . . . . . . . . . . . . . . . . . .          ⎫
. . . . . . . . . . . . . . . . . . . . . . . . . . .   ⎬    (Feedrate is 3.0)
. . . . . . . . . . . . . . . . . . . . . . . . .       ⎭
```

(Feedrate is 3.0)   **Figure 6-10**   The use of the statements RAPID and FEDRAT.

6.5 STATEMENTS SPECIFYING THE SPINDLE ROTATION MOTION

The spindle rotation motion is defined by the following parameters:

1. ON or OFF, which is used to start or stop the turning of the spindle
2. CLW or CCLW, which specifies the direction of spindle rotation as clockwise or counterclockwise
3. s, which specifies the speed of spindle rotation in revolutions per minute

The statement formats are as follows:

$$\text{SPINDL}/s, \begin{Bmatrix} \text{CLW} \\ \text{CCLW} \end{Bmatrix}$$

$$\text{SPINDL/OFF}$$

$$\text{SPINDL/ON}, \begin{Bmatrix} \text{CLW} \\ \text{CCLW} \end{Bmatrix}$$

The first format is used to define, at the beginning of a program, a spindle speed, s, and the rotation direction; in addition, it is used to turn on the spindle. The second format can be used in the middle or at the end of a program to turn off the spindle. The third format is used in the middle of a program to turn the spindle on again, with the same rotation speed as defined previously. The first format can also be used to change both the spindle speed and its direction. If a change in rotation direction is needed, it is necessary, first, to stop running the spindle. However, many postprocessors are designed in such a way that a spindle stop code (e.g., M05) is automatically generated if a reversal of the spindle rotation direction is experienced. In such a case the SPINDL/OFF statement can be omitted.

6.6 THE STATEMENT FOR CHANGING THE CUTTING TOOL

Most NC machine tools can automatically change tools during a machining operation. On lathes, different tools have different offset values in the X and Z directions of the machine; on milling machines the tool axis is always the spindle axis (except when a cranked or jointed tool is used), and the differences in tool dimensions for different tools appear in the tool length and diameter only. The cutter diameter is taken care of by the CUTTER statement; therefore a statement is needed to identify the required tool and to specify its offsets. The format of this statement is

$$\text{LOADTL}/n \left[,\text{SETOOL}, \begin{Bmatrix} d,e,f \\ d,e \\ f \end{Bmatrix} \right]$$

where n is the tool identification code and d, e, and f are the offsets, in the X, Y, and Z directions, of the APT coordinate system, respectively, from the reference point of the tool to the reference point of the tool holder (Fig. 6-11).

Note that the APT coordinate system differs from the machine tool coordinate system for lathes. This will be explained in Chapter 8.

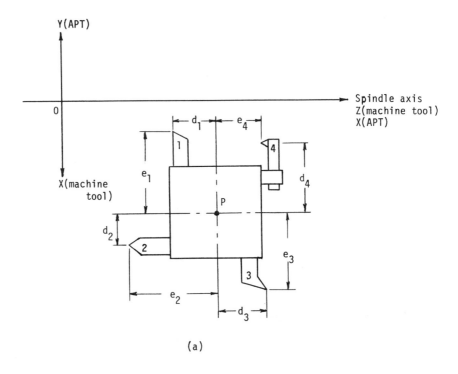

(a)

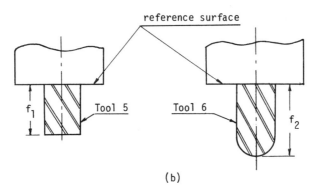

(b)

Figure 6-11   The offsets of the tools. (a) Lathe tools (P: reference point of the turret). (b) Mills.

Example

The LOADTL statements for tools 1, 3, 5, and 6, shown in Fig. 6-11, are, respectively, as follows:

LOADTL/1,SETOOL,$d1$,$e1$,0
LOADTL/4,SETOOL,$d4$,$e4$,0
LOADTL/5,SETOOL,0,0,$f1$ (or LOADTL/5,SETOOL,$f1$)
LOADTL/6,SETOOL,0,0,$f2$ (or LOADTL/6,SETOOL,$f2$)

Note that the offsets of tool 4 are determined on the basis of its working position (i.e., the position that the tool occupies after the turret has turned counterclockwise by 90 degrees). The selection of a tool in the tool magazine is programmed through use of the SELCTL statement, which is described in Section 13.4.1.1.

6.7 STATEMENTS SPECIFYING THE COOLANT

On some NC machines, selected forms of coolant can differ, depending on machining requirements. Coolant can be provided as running flood (FLOOD), tapping coolant (TAPKUL), or mist (MIST). The statements for selecting and turning the coolant on or off are as follows:

$$COOLNT/\begin{Bmatrix} FLOOD \\ MIST \\ TAPKUL \end{Bmatrix}$$

$$COOLNT/\begin{Bmatrix} ON \\ OFF \end{Bmatrix}$$

The first statement is used to select, change, and turn on the coolant. The statement COOLNT/OFF is used to turn off the coolant in the middle or at the end of a program; the statement COOLNT/ON can be used to turn on the coolant.

6.8 THE STATEMENT SPECIFYING A PAUSE DURING MACHINING OPERATION

Sometimes a pause is needed during machining to produce a sharp corner or a better surface finish at the bottom of a hole. A pause for a certain period can be specified by using the following statement:

DELAY/t

where t is the pause length of time in seconds.

Example
The DELAY statement in the following program generates a pause at the bottom of the hole during a drilling operation:

```
....................
....................
FEDRAT/2.0
GODLTA/−1.0 $$ DRILL MOVES DOWN AT SPEED 2.0 IN/MIN
DELAY/0.5    $$ DRILL STOPS AT THE BOTTOM FOR 0.5 SEC
GODLTA/1.0  $$ DRILL MOVES UP
....................
....................
```

6.9 THE STATEMENT SPECIFYING A CIRCULAR MOVEMENT

As explained in Chapter 5, a contouring cutter path is linearly interpolated by the NC processor as a series of consecutive linear motion steps. A number of coordinate sets, corresponding to the end points of these motion steps, will be generated. When these data are processed by a postprocessor, a corresponding number of G01 codes (or linear motion steps) will be generated. Therefore, for a single circular movement, there might be tens or hundreds of NC statements. In fact, a circular movement on an NC machine can be defined by one statement only. A postprocessor can be designed in such a way that it uses the coordinates of the end position and the center point of the circular movement to generate one G02 or G03 code for the complete circular motion. The information indicating that the movement is circular should therefore be passed to the NC processor. The following statement signifies circular movement:

$$\text{ARCSLP/} \begin{Bmatrix} \text{ON} \\ \text{OFF} \end{Bmatrix}$$

The statement ARCSLP/ON indicates that the motion defined in the following statements can be processed by the circular interpolation routine in the postprocessing section of the NC processor. The ARCSLP/OFF statement calls off the circular interpolation routine, and the motion data are handled by the linear interpolation routine. If the statement ARCSLP/ON is specified, the circular interpolation processing is effective only when the drive surface in the motion statement is a circle or cylinder; as a result, a circular interpolation NC code G02 or G03 is generated. For motion statements with a drive surface other than a circle or cylinder, the circular interpolation routine is not invoked, and the linear interpolation routine is used instead. Thus, in the program

```
.....................
.....................
ARCSLP/ON
A1) TLRGT,GORGT/C1,L1 $$ C1 IS A CIRCLE
A2) GORGT/L1,L2          $$ L1 IS NOT A CIRCLE OR CYLINDER
.....................
.....................
```

the statement labeled A1 is processed as a circular interpolation motion by the postprocessing section of the NC processor, whereas the statement labeled A2 is processed as a linear interpolation motion.

As pointed out previously, the number of motion steps for a curved cutter path is determined by the specified tolerance. The processing of a circular motion by outputting a G02 or G03 code, together with the end point coordinates, means that the accuracy in positioning the tool during the circular movement on the NC machine is no longer controlled by the NC processor but by the NC controller. If an accuracy higher than that attainable by the NC controller is required, the circular interpolation routine should not be used, and the CLDATA should be processed by the postprocessor as a linear movement.

6.10 THE STATEMENT SIGNIFYING THE END OF A MACHINING PROCESS

The statement that terminates a machining process is

END

The corresponding code generated by the postprocessor is M02 or M30, depending on the postprocessor design.

6.11 THE STATEMENT DEFINING THE MACHINE COORDINATE SYSTEM

There are cases in which the machine-tool coordinate system and the coordinate system for defining the part geometry and the cutter path are not the same, or in which two identical parts in different positions are to be cut on an NC machine (Fig. 6-12). In the first case, a translation of the coordinate system is needed. In the second case, the programs for cutting the two parts can be the same, provided that the origins of the machine coordinate systems for the two parts are set correctly. A statement is provided in APT to define the origin of the machine-tool coordinate system with respect to the coordinate system used in the APT program. The format of this statement is

ORIGIN/x,y,z

where x, y, and z are the coordinates of the origin of the machine-tool coordinate system within the APT coordinate system, specifically, the coordinate system adopted by the APT programmer to define the part geometry or cutter path. For example, let the APT coordinate system be X-Y-Z in Fig. 6-12; part A and the cutter path are defined in APT with respect to it. If the origin of the machine-tool coordinate system, X'-Y'-Z', is at point $P0(-1,-1,0)$, and the same machining program is

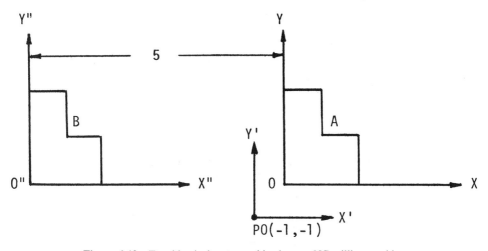

Figure 6-12   Two identical parts machined on an NC milling machine.

to be used for cutting part A, then we have to move the origin from O to $P0$. Thus, for definition of the origin of the machine coordinate system with respect to the APT coordinate system, the statement

ORIGIN/−1,−1,0

should be specified before the statements that define the cutter path. The CLDATA is then translated during the postprocessing stage by 1, 1, and 0 in the X, Y, and Z directions, respectively, so that the part can be machined while positioned, as shown in the figure, at point O. Thus the effect of the ORIGIN statement is to translate the CLDATA by $-x$, $-y$, and $-z$ in the X, Y, and Z directions, respectively.

The following example shows another application of the ORIGIN statement. Suppose that part A is positioned at point O and the part and machine coordinate systems are set as X-Y-Z in Fig. 6-12 (i.e., they are the same). If part B, which is identical to part A but placed at a different position, O'', on the machine table, is to be machined by using the same program segment and the same machine coordinate system, X-Y-Z, for machining part A, then an ORIGIN statement should be specified to move the cutter path by -5 in. in the X direction. Thus the program for cutting these two parts may appear as follows:

```
.....................   (Statements for machining part A. The part and
.....................   machine coordinate systems are the same [i.e.,
.....................   X-Y-Z].)
ORIGIN/5,0,0
.....................   (The same statements as for machining part A.
.....................   The cutter path is translated by -5 in. in the
.....................   X direction, so that part B can be cut while
.....................   the original machine coordinate system X-Y-Z
.....................   is maintained.)
```

6.12 PROCESSOR CONTROL WORDS

The vocabulary words described in this section are used to define or alter the way the NC processor processes a particular APT program. They include PARTNO, MACHIN, NOPOST, CLPRNT, and FINI.

6.12.1 The PARTNO Statement

The format of this statement is as follows:

PARTNO ....(alphanumeric string).........

The word PARTNO should be specified in columns 1 through 6, followed by an alphanumeric string, specified in columns 7 to 72, that serves as a comment or label of the program. This statement should be specified as the first statement in an APT program. The comment will be printed in the CLDATA file and used by the postprocessor to label the postprocessor printed and punched output. Depending on the postprocessor design, the comment can also be punched as man-readable text on the punched tape.

Example

The following is the first statement of the APT program for cutting a shaft:

PARTNO SHAFT-CUTTING PROGRAM

6.12.2 The MACHIN Statement

This statement is used to specify the postprocessor to be used and the output format. It should be specified right after the first statement, PARTNO. The format of this statement* is

MACHIN/$ppname$,$n1$,OPTION,$n2$,$n3$,$n4$

where

$ppname$ = the name of the postprocessor to be used

$n1$ = a number indicating the specific processing subroutine (called *machine tool module*) corresponding to the machine tool in use and the program initializing the values of a series of variables to be used by the postprocessor

$n2$ = a parameter specifying the form and format of the postprocessor output

$n3$, $n4$ = parameters reserved for use by the user (These two parameters can be used to transfer information from an APT program to the postprocessor; therefore they are postprocessor-dependent.)

6.12.3 The NOPOST Statement

This statement contains a single word, NOPOST. It instructs the NC processor that no postprocessing is required after this statement. It can be specified on a line by itself, as

NOPOST

or in conjunction with another statement, with a comma or semicolon separating them; for example:

NOPOST,FEDRAT/2.0

or

NOPOST;FEDRAT/2.0

*See Section 13.4.1.1 and Table 13-5 of Chapter 13 for further details.

6.12.4 The CLPRNT Statement

The processing of an APT program by the NC processor yields a common interface code, called CLDATA, which is the input file to a selected postprocessor. The CLDATA consists of the following:

1. The coordinates of the consecutive positions of the tool as it is directed by the program to produce the part
2. The postprocessor commands or statements specifying machining specifications
3. The information that should be passed to the postprocessor regarding, for example, the record type and record sequence number

Once the CLPRNT statement, which consists of the single word CLPRNT, is specified, the CLDATA after this statement is printed in a tabular format. Detailed explanation of the CLDATA is given in Chapters 8 and 11.

6.12.5 The FINI Statement

This statement consists of the single word FINI. It defines the end of a part machining program and should be the last record (statement) of a program.

6.13 ADDITIONAL STATEMENTS SPECIFYING COMMENTARY TEXT

In addition to the PARTNO statement, there are several other statements that can be used to provide commentary to an APT program, the CLDATA file, and the postprocessor output. Three of these statements, discussed here, are

```
REMARK...........(alphanumeric string)...................
PPRINT...........(alphanumeric string)...................
INSERT...........(alphanumeric string)...................
```

These three vocabulary words should be specified in columns 1 through 6, followed by a comment that can be specified as an alphanumeric string in columns 7 to 72, if needed.

The REMARK statement is used to provide commentary to a part program. This statement will be printed in the program listing file. Normally, it is used to give some explanation of the content of a program or a particular section of a program. It does not have any effect on the processing of a part program.

Similarly, PPRINT is a postprocessor word. The comment following it will be printed in the postprocessor output listing file.

The alphanumeric string in an INSERT statement is transferred directly to CLFILE, which is then used by the postprocessor, without further processing, as direct input to the NC machine controller. Thus the comment in an INSERT statement is directly printed on the NC output and punched on the NC punched tape. Therefore we can make use of the INSERT statement to insert an NC statement, which cannot be handled by the NC processor or postprocessor, in the output NC program.

PROBLEMS

6.1 What is the purpose of specifying a tolerance in an APT program? How does it affect the cutter path or the part profile? What is the difference between tolerance and allowance?

6.2 How does the NC processor determine the directions of the tolerance, namely, inward and outward, for a part profile or geometric entity?

6.3 Define the cutters shown in Fig. P6-3.

6.4 Define the cutter path *A-B-C-D-E*, shown in Fig. P6-4. A flat-end mill of 0.5 in. in diameter and 1.5 in. in length is used.

6.5 A turning process is to be carried out in the following sequence:
  a. Rapidly move the tool to the safe position P0.
  b. Rotate the turret and put the No. 2 tool in working position.
  c. Start running the spindle at 700 rpm in the clockwise direction and rapidly move the tool to point P1 near the part profile.
  d. Turn on the coolant and slowly move the tool to the position required for starting to cut the profile, with a feedrate of 0.006 in./rev.
  e. Cut the first part of the profile with a feedrate of 0.005 in./rev.
  f. Change the spindle speed to 900 rpm.
  g. Cut the second part of the profile with a feedrate of 0.003 in./rev.
  h. Rapidly move the tool to point P0.
  i. Rotate the turret and put the No. 4 tool in working position.
  j. Rapidly go to point P2.
  k. Cut the third part of the profile with a feedrate of 0.002 in./rev and a spindle speed of 500 rpm.
  l. Rapidly move the tool back to P0, stop the spindle, and turn off the coolant.
Define this process in APT.

6.6 If the machine coordinate system is set on the basis of the nose position of the No. 1 tool, what are the offset values to be specified in the LOADTL statements for the No. 1, 2, and 3 tools in Fig. P6-6?

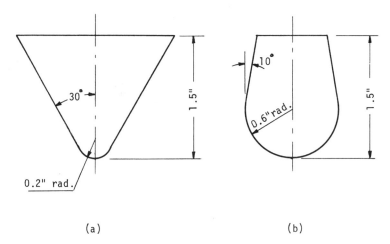

(a)   (b)

Figure P6-3

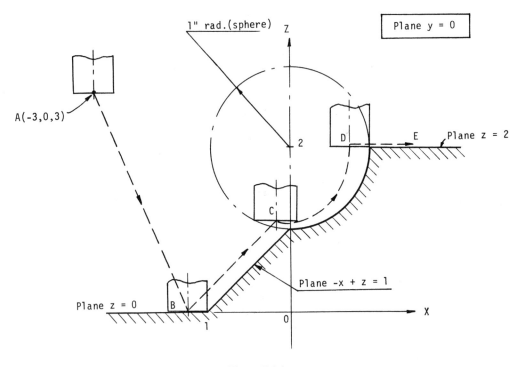

Figure P6-4

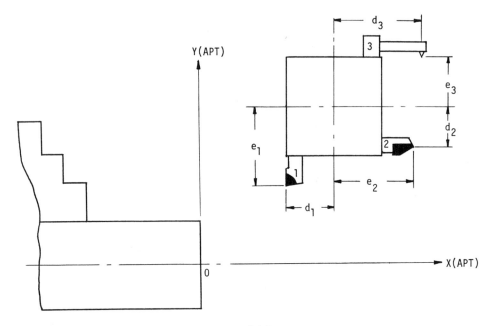

Figure P6-6

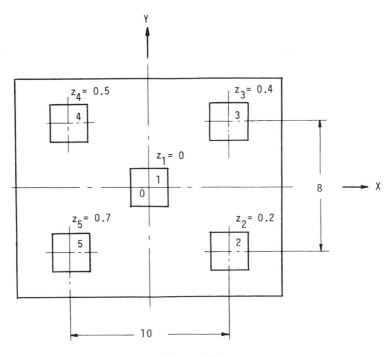

Figure P6-7

6.7 There are five contours in areas 1 through 5 on the part shown in Fig. P6-7. They are of the same form but on different levels, z_1 through z_5, respectively. Write an APT program and use the ORIGIN statement to direct the tool motion for machining contours in areas 2, 3, 4, and 5 by means of the program segment for machining the contour in area 1. Assume that the starting and end points of the tool are at (0,0,3).

6.8 Suppose that the multiple threading cycle motion on the lathe and the corresponding code G76 cannot be processed and generated by the NC processor and the postprocessor. How do we specify the threading cycle in an APT program so that we can still obtain the following NC block

G76X20.752Z−40.0K1.624D0.5F3.0A60*

in the NC output?

chapter 7

Definitions of Geometric Entities and Cutter Motion by Means of More Sophisticated Statements

In many cases a program can be simplified and made much shorter by means of loop-programming, subprogramming, and transformation of the coordinate system or cutter path. The statements for realizing these means are available in the APT programming language. The use of mathematical calculation statements, together with loops and subprograms, can often make easier both the calculation and the definition of geometric entities and the cutter path in a program. APT also provides statements that simplify the definition of some routine machining, such as pocketing. Statements are included in the APT language for defining a start-up or a contouring motion that is difficult or clumsy to define by means of the statements described in Chapter 5. These are the subjects discussed in this chapter.

7.1 SUBSCRIPTED VARIABLES AND CONTROL-BRANCHING STATEMENTS

Subscripted variables are normally used to define a series of similar variables, geometric entities, and subprograms. These variables carry the same symbol and constitute an array. For example, L(1),......, L1(10) can be used to represent 10 parallel lines or 10 different subprograms. The dimension of an array should be defined before any one element of the array is specified in a program. This is realized by the following statement:

RESERV/A1,n1,A2,n2,...

where

$\qquad$ A1, A2, ... = symbols for different arrays

$\qquad$ n1, n2, ... = the dimensions of the arrays A1, A2, ..., respectively (Their values should be within the range 1 to 32767.)

The RESERV statement can be executed only once; therefore it cannot be included in a loop.

The subscript, or index, of a subscripted variable can be specified as a scalar or a mathematical expression. For example, if I = 3 and J = 5, then the subscripted variable A(I+J) is equivalent to A(8), and A(I∗J), equivalent to A(15). The subscript should be an integer. If a scalar other than an integer is specified, it is truncated and only its integer part is used.

In some cases, we need to specify a series of orderly listed subscripted variables in one statement (e.g., in the PATERN statement). When the number of these variables is large, it is tedious to specify all of them individually. To resolve this problem, APT provides inclusive subscripts of different formats, as shown below:

$$A\left(\left\{\begin{matrix} a,\text{THRU,}\begin{Bmatrix} b \\ \text{ALL} \end{Bmatrix} \\ \text{ALL} \end{matrix}\right\}\left[,\begin{Bmatrix} \text{INCR} \\ \text{DECR} \end{Bmatrix},c\right]\right)$$

which can be specified in a statement to represent a series of orderly listed elements of array A. The scalars a and b represent meaningful indices in the array, and the scalar c is the incremental or decremental value of the index running from one subscripted variable to the next. If parameter c and modifier INCR or DECR preceding it are omitted, they are considered by the NC processor as INCR,1. These three parameters can be specified in any form acceptable to the APT language. The meanings of the variables with inclusive subscripts in different formats are explained below.

$\qquad$ The subscripted variable

A(a,THRU,b,INCR,c)

specifies the following elements of array A, in listed order, in the statement

A(a), A(a+c), A(a+2∗c), ..., A(a+n∗c)

where
$$a+n{\ast}c \leqslant b < a+(n+1){\ast}c, \qquad \text{if } c > 0 \text{ and } a < b;$$
$$a+n{\ast}c \geqslant b > a+(n+1){\ast}c, \qquad \text{if } c < 0 \text{ and } a > b.$$

The subscripted variable

A(a,THRU,b,DECR,c)

specifies the following elements of array A, in listed order, in the statement

A(a), A(a−c), A(a−2*c), ..., A(a−n*c)

where

$$a-n*c \geqslant b > a-(n+1)*c, \qquad \text{if } c > 0 \text{ and } a > b;$$
$$a-n*c \leqslant b < a-(n+1)*c, \qquad \text{if } c < 0 \text{ and } a < b.$$

The subscripted variable

$$A\left(a, \text{THRU}, \text{ALL}, \begin{Bmatrix} \text{INCR} \\ \text{DECR} \end{Bmatrix}, c\right)$$

has the same meaning as those of the two formats shown above, except that parameter b is replaced by the first or last index of the array, depending on the resulting sign of the index increment. Finally, the subscripted variable

$$A\left(\text{ALL}, \begin{Bmatrix} \text{INCR} \\ \text{DECR} \end{Bmatrix}, c\right)$$

specifies a list of elements in array A, starting from its first or last index and ending with its last or first index, depending on the specified modifier INCR or DECR and the sign of parameter c. The following examples illustrate their usage.

Example 1

If the array P(1), P(2), ... , P(100), representing 100 points, has been defined, then the subscripted variable

P(3,THRU,25,INCR,5)

specifies the following elements of array P in listed order:

P(3), P(8), P(13), P(18), P(23)

An LCONIC defined by the five points listed above can be specified by any one of the statements listed below:

LCONIC/5PT,P(3,THRU,25,INCR,5)
LCONIC/5PT,P(3,THRU,24,INCR,5)
LCONIC/5PT,P(3,THRU,23,INCR,5)

Example 2

For the array given above, the meanings of subscripted variables in different formats are as follows:

SUBSCRIPTED VARIABLE	SPECIFIED ELEMENTS
P(75,THRU,54,INCR,−7)	P(75), P(68), P(61), P(54) in listed order
P(75,THRU,54,DECR,7)	Same as above
P(97,THRU,ALL)	P(97), P(98), P(99), P(100) in listed order
P(35,THRU,ALL,DECR,10)	P(35), P(25), P(15), P(5) in listed order
P(ALL,INCR,−25)	P(100), P(75), P(50), P(25) in listed order

Two statements in the APT language can be used to realize branching of control within a loop or a subprogram. *Unconditional branching* to an APT statement labeled K1 is accomplished by the statement

JUMPTO/K1

Conditional branching uses the IF and JUMPTO statements. Two different formats are available for the IF statement. The first one is

IF(s)K1,K2,K3

which directs control according to the value of s. The designations K1, K2, and K3 in this statement are, respectively, the labels of the three statements specified in the loop or subprogram. If s < 0, control goes to the statement labeled K1; if s = 0, to the statement labeled K2; and if s < 0, to the statement labeled K3 (Fig. 7-1[a]). The parameter s can be a scalar variable or an arithmetical expression.

The second format is

IF(s), APT statement

where s is a relational expression containing relational operators, such as 'GE'(greater than or equal to), 'EQ'(equal to), 'GT'(greater than), 'LT'(less than), and 'LE'(less than or equal to). The meaning of this statement is that if the relational expression s is true, then the APT statement in this statement is executed; otherwise, it will be skipped and control goes to the next statement. For example, in the following APT program

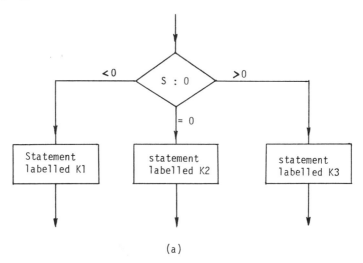

(a)

Figure 7-1 (a) The control flowchart of the statement IF(S)K1,K2,K3. (b) The control flowchart of the statement JUMPTO/(A1, A2, . . . , An), I. (c) The control flowchart of the statement JUMPTO/(A1, A2, A3, A4, A2, . . . , An), I.

(continued)

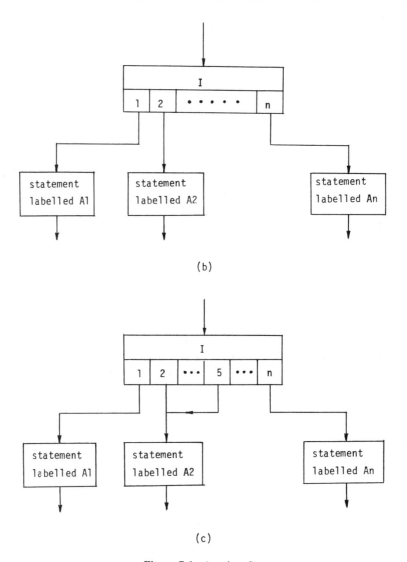

(b)

(c)

Figure 7-1   (continued)

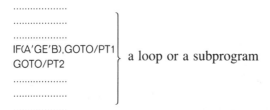

if A ⩾ B, then the tool will go to point PT1 first and then to point PT2; otherwise, it
will go directly to point PT2.

Sometimes the JUMPTO statement is selected as the APT statement in an IF statement to transfer control conditionally to a desired section. For instance, the IF statement in the following APT program transfers control to the statement labeled K1 when A is greater than or equal to B:

```
.................
.................
.................
IF(A'GE'B), JUMPTO/K1
GOTO/PT2                     } a loop or a subprogram
.................
.................
K1)GOTO/PT1
.................
.................
.................
```

Another conditional branching statement is JUMPTO, which has the following format:

JUMPTO/(A1,A2,...,An),I

where A1, . . . , An are the labels of statements within the loop or subprogram, and I is an integer variable or index. If the variable I assumes the value, for example, of i, then control is sent to the statement labeled Ai, which is the ith statement label listed in this statement. The control flowchart for this statement is given in Fig. 7-1(b). It can readily be seen that the value of the index should be in the range $1 \leq I \leq n$. If a real number is used for this variable, the fractional part will be truncated. Thus the value 1.6 for the index is treated as 1 in this statement. The statement labels listed in this statement can differ, or some of them can be the same (Fig. 7-1[c]).

Compared with the IF statement, the conditional branching statement JUMPTO allows control to be branched to more than three different directions. In addition, the index, I, can be a nested mathematical expression. For example, the statement JUMPTO and the statement for calculating the index in the program

```
.....................
I=M+4*tan(C)
JUMPTO/(A1,A2,A3,A4),I
.....................
.....................
```

can be replaced by one JUMPTO statement with a nested calculation expression as the index:

JUMPTO/(A1,A2,A3,A4),(M+4*tan(C))

As for the unconditional branching statement JUMPTO, this statement is used to direct control within a loop or subprogram.

7.2 LOOP-PROGRAMMING

Very often, we need to define a group of similar geometric entities or a series of similar cutter paths. Loop-programming allows us to execute a group of instructions repeatedly. As in FORTRAN, this can be realized by using the DO-loop or by transfer of control within the loop defined by the LOOPST and LOOPND statements. A loop can also be used to make a decision among alternatives.

7.2.1 DO-loop

The format and structure of a DO-loop are shown in Table 7-1. The first statement, DO/..., and the last statement, labeled K, define a DO-loop. They instruct the computer to execute the statements repeatedly within the loop for a given number of times. When the DO-loop is entered, the index, I, is assigned the value $n1$. After each iteration, its value is increased by $n3$. The iteration continues as long as I is less than or equal to $n2$. When I is greater than $n2$, the statement after the one labeled K is executed. The flowchart of the DO-loop is also included in Table 7-1.

A dummy executable statement, CONTIN, is sometimes used as the last statement of a DO-loop (labeled K in Table 7-1), its function being the same as that of

TABLE 7-1 THE FORMAT AND FLOWCHART OF DO-LOOP

Format	Flowchart

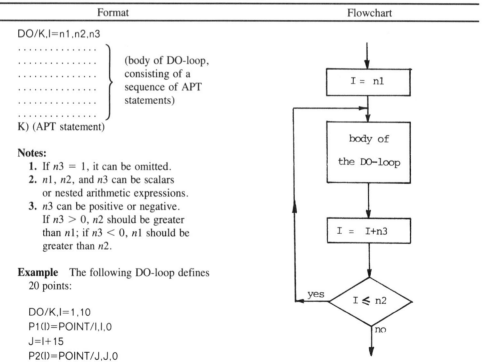

DO/K,I=n1,n2,n3

```
. . . . . . . . . . . . . .
. . . . . . . . . . . . . .        (body of DO-loop,
. . . . . . . . . . . . . .        consisting of a
. . . . . . . . . . . . . .        sequence of APT
. . . . . . . . . . . . . .        statements)
. . . . . . . . . . . . . .
K) (APT statement)
```

Notes:
  1. If $n3 = 1$, it can be omitted.
  2. $n1$, $n2$, and $n3$ can be scalars or nested arithmetic expressions.
  3. $n3$ can be positive or negative. If $n3 > 0$, $n2$ should be greater than $n1$; if $n3 < 0$, $n1$ should be greater than $n2$.

Example The following DO-loop defines 20 points:

```
DO/K,I=1,10
P1(I)=POINT/I,I,0
J=I+15
P2(I)=POINT/J,J,0
K)CONTIN
```

the statement CONTINUE in FORTRAN (i.e., it instructs the computer to continue execution of the program).

One of the functions of the last statement in a loop is to send control back to the beginning of the loop. Hence those statements that send control to other statements or that define a new loop or a subprogram cannot be used as the last statement. Accordingly, the last statement of a DO-loop cannot be any one of the following: CALL, IF, JUMPTO, FINI, MACRO, TERMAC, or another DO statement.

The statements contained in the body of a DO-loop are executed repeatedly T times, where $T = INTGF((n2-n1)/n3) + 1$ (see Table 7-1). Nevertheless, conditions can be set in a DO-loop to transfer control out of the range of a DO-loop and thereby to terminate the execution of a DO-loop prematurely. An IF statement can be used in conjunction with the JUMPTO statement within a DO-loop to obtain such transfer of control. For example, in the following APT program, the IF statement will terminate execution of the DO-loop when the condition $(I - R1) > 25$ is true; control then goes to the statement labeled K2, which is outside the DO-loop but inside the loop nesting the DO-loop:

```
. . . . . . . . . . . . . . . . . .
LOOPST
. . . . . . . . . . . . . . . . . .
. . . . . . . . . . . . . . . . . .
DO/K1,I=1,S,2 $$THE VALUE OF S MUST HAVE BEEN ASSIGNED BY A PREVIOUS
              $$STATEMENT
. . . . . . . . . . . . . . . . . .
. . . . . . . . . . . . . . . . . .
IF((I-R1)'GT'25),JUMPTO/K2
. . . . . . . . . . . . . . . . . .
K1)CONTIN
. . . . . . . . . . . . . . . . . .
K2). . . . . . . . . . . . . . .
. . . . . . . . . . . . . . . . . .
LOOPND
. . . . . . . . . . . . . . . . . .
```

The transfer of control outside the DO-loop will fail if the statement labeled K2 is not in a loop or in a subprogram nesting the DO-loop.

A DO-loop cannot be entered at any point except through its initial DO-statement. DO-loops can be nested; the maximum level of nesting is 31. However, they should not overlap one other because overlapping results in logical mistakes. For example, the program segment

```
DO/K,I=n1,n2,n3        ⌐
. . . . . . . . . . . .
. . . . . . . . . . . .
DO/K1,J=n4,n5,n6    ⌐
. . . . . . . . . . . .
. . . . . . . . . . . .
K1)CONTIN          ⌐
. . . . . . . . . . . .
. . . . . . . . . . . .
K)CONTIN               ⌐
```

has two DO-loops, one of which is nested in the other. The first DO-loop instructs the computer to execute the statements within it INTGF$((n2-n1)/n3)+1$ times. Each time the DO-loop is executed, the nested DO-loop is executed INTGF$((n5-n4)/n6)+1$ times. On the other hand, the program segment

```
DO/K,I=n1,n2,n3
.................
.................
DO/K1,I=n4,n5,n6
.................
.................
K)CONTIN
.................
.................
K1)CONTIN
```

contains two DO-loops that overlap. The first DO-loop instructs the computer to execute the statements within the loop INTGF$((n2 - n1)/n3) + 1$ times. When the first statement of the second DO-loop is executed, it orders the computer to execute INTGF$((n5 - n4)/n6) + 1$ times the statements within the second loop, which includes those out of the range of the first DO-loop. The instructions given by these two DO-loops contradict each other. As a result, the program is not executable.

A DO-loop can be included in a subprogram (MACRO) or in a loop defined by the LOOPST and LOOPND statements. Statements transferring control out of the range of a DO-loop, such as those calling a subprogram defined outside the DO-loop, can also be included in a DO-loop. In this case, control returns to the statement after the one calling the subprogram in the DO-loop after the subprogram is executed.

A subprogram (MACRO) cannot be defined repeatedly; therefore the MACRO statement should not be included in a DO-loop.

Two examples are given below to explain the use of the DO-loop.

Example 1

A quasi cone-shaped part with a horizontal circular cross section on the top and an elliptical one at the bottom is shown in Fig. 7-2(a). The horizontal cross section of this part, at any z level between $z = 0$ and $z = 3$, is defined by the equation

$$x^2/(3 - z/1.5)^2 + y^2/(2.5 - z/3)^2 = 1$$

As can be seen from this equation, all the cross sections are ellipses with the exception of the top one, which is a circle with a 2 in. radius. The DO-loop for defining 11 horizontal cross sections, including the top and bottom ones, with equal distance (0.15 in.) between any two adjacent ones, can be as follows:

```
......................
RESERV/CROS,11 $$ CROS(I) IS THE SYMBOL OF A HORIZONTAL CROSS SECTION
......................
DO/A1,I=1,11
     Z1=1.5-0.15(I-1)
     IF(I'GT'1),JUMPTO/A2
     CROS(I)=CIRCLE/0,0,1.5,2
     JUMPTO/A1
A2)  CROS(I)=ELLIPS/CENTER,(0,0,Z1),(3-Z1/1.5),(2.5-Z1/3),0
A1)CONTIN
......................
......................
```

Example 2

A flat surface is to be cut by a facing mill in three passes with the cutter path in the X-Y plane after the pattern shown in Fig. 7-2(b). The depth of cut is 0.2 in. in the first two passes and 0.05 in. in the last one. Assume that the starting and end points are at $(-1.3,1,1)$. The DO-loop for defining the cutter motion can be as follows:

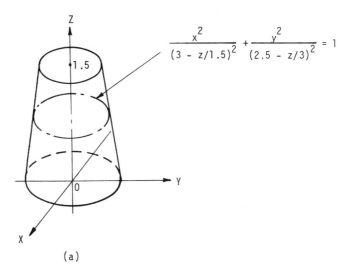

$$\frac{x^2}{(3 - z/1.5)^2} + \frac{y^2}{(2.5 - z/3)^2} = 1$$

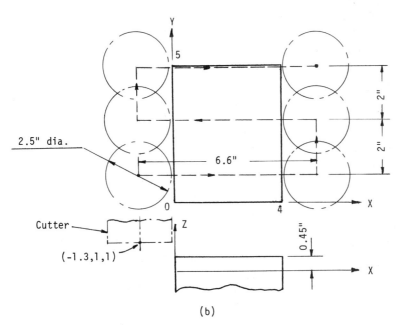

Figure 7-2   (a) A quasi cone-shaped part with a horizontal crosssection defined by an elliptical function. (b) The cutter path of a facing operation.

```
FROM/-1.3,1,1
GODLTA/-0.45
DO/A4,I=1,3
JUMPTO/(A1,A1,A2),I
A1)   GODLTA/-0.3
      JUMPTO/A3
A2)   GODLTA/-0.15
A3)   GODLTA/6.6,0,0
      GODLTA/0,2,0
      GODLTA/-6.6,0,0
      GODLTA/0,2,0
      GODLTA/6.6,0,0
      GODLTA/0.1
      GODLTA/-6.6,-4,0
A4)CONTIN
GOTO/-1.3,1,1
```

7.2.2 Loop Defined by the LOOPST and LOOPND Statements

A loop can also be defined by using a loop starting statement, LOOPST, and a loop ending statement, LOOPND (Table 7-2); between these two statements is the body of the loop. Since there is no DO-statement, as in the DO-loop, to count the index and to direct control, the loop index should be defined properly by additional statements. A LOOPST-LOOPND loop can be invoked only through its beginning statement (i.e., LOOPST). Furthermore, it can be executed only once in a program. The IF and JUMPTO statements are often used to direct control within the LOOPST-LOOPND loop. Table 7-2 also shows two ways to realize repeated execution (N times) of the APT statements within a loop. In the first flowchart, the loop is controlled by a conditional branching statement

IF(I-N)A1,A1,A2

and, in the second one, by a conditional IF statement and an unconditional branching statement, JUMPTO. Transfer of control from a statement other than LOOPND to outside the loop is not permissible.

The statement labels used in a LOOPST-LOOPND loop are effective within the loop only. They can be used elsewhere in another loop or in a subprogram (MACRO). The APT statements that cannot be used within a LOOPST-LOOPND loop are those which cannot be executed repeatedly, such as FINI, MACRO, and RESERV.

In comparison with the DO-loop, one of the usages of the LOOPST-LOOPND loop is to reiterate a sequence of operations when the number of iterations is difficult to predict. For example, suppose that the slope surface shown in Fig. 7-3(a) is to be machined on a three-axis milling machine. The cutter moves in the zigzag way shown in the figure. The allowable scallop height is 0.005 in. For definition of the vertical planes PL(i), $i = 1, 2, \ldots$, which are used as drive surfaces for the tool

TABLE 7-2   THE FORMAT OF A LOOPST-LOOPND LOOP AND THE TWO WAYS TO REALIZE
CONTROL IN A LOOP

Format	Examples	

Format	Example 1	Example 2
. . . .	**Example 1**	**Example 2**
LOOPST	. . . .	. . . .
. . . .	LOOPST	LOOPST
(body of the loop)	I=1	I=1
. . . .	A1).	A1). . . . .
. . . .	(APT statements)	(APT statements)
LOOPND	. . . .	. . . .
. . . .	I=I+1	IF(I−(N−1))B1,B1,B2
	IF(I−N)A1,A1,A2	B1)I=I+1
	A2)LOOPND	JUMPTO/A1
	. . . .	B2)LOOPND
		. . . .

Flowchart — Example 1:

```
   LOOPST
     │
   I = 1
     │
 APT statements
 in the body
 of the loop
     │
   I = I + 1
     │
   I : N   ──≤──> (loop back to APT statements)
     │
     │ >
   LOOPND
```

Flowchart — Example 2:

```
   LOOPST
     │
   I = 1
     │
 APT statements
 in the body
 of the loop
     │
   I : (N-1)  ──>──> (loop back to APT statements)
     │ ≤
   I = I + 1
     │
  JUMPTO/A1
     │
   LOOPND
```

motion on different levels, the tool center-line positions $x(i)$ at consecutive levels are to be calculated on the basis of the allowable scallop height, H. The calculation should be repeated for a number of times until the tool touches surface B. A

LOOPST-LOOPND loop can be designed to realize this calculation process and to define planes PL(i), $i = 1, 2, \ldots$ The loop may appear as follows:

```
.....................
PHI=ATANF(1.5)
R=3/8
H=0.005
X(1)=R*COSF(90-PHI)
D1=2*SQRTF(2*R*H-H*H)
D=D1*COSF(PHI)
X1=1+R*TANF(PHI/2)
PL(1)=PLANE/1,0,0,X(1)
LOOPST
        I=2
A0)     X(I)=X(I-1)+D
        IF(X(I)'GE'X1),JUMPTO/A1
        PL(I)=PLANE/1,0,0,X(I)
        I=I+1
        JUMPTO/A0
A1)     X(I)=X1
        PL(I)=PLANE/1,0,0,X1
LOOPND
```

The LOOPST-LOOPND loop can also be used to make a decision in a complex situation. For example, in Fig. 7-3(b), the tool is going to cut the profile composed of lines L1, L2, L3, and L4. At the corner formed by lines L1, L2, and L3, it is difficult to predict which one of the two lines, L2 and L3, will be in contact with the cutter when the tool moves toward the left. The LOOPST-LOOPND loop can be used to realize the selection of the correct check surface and to define the correct tool movement.* The loop may appear as follows:

```
.....................
L1=LINE/XAXIS
L2=LINE/0,0,ATANGL,95
L3=LINE/(-1.2*TANF(5)),1.2,ATANGL,70
L11=LINE/PARLEL,L1,YLARGE,1
L21=LINE/PARLEL,L2,XLARGE,1
L31=LINE/PARLEL,L3,XLARGE,1
P1=POINT/INTOF,L31,L11; OBTAIN,POINT/P1,X1,Y1,Z1
P2=POINT/INTOF,L21,L11; OBTAIN,POINT/P2,X2,Y2,Z2

.....................
.....................
(THE TOOL HAS MOVED TO POINT A)
LOOPST
INDIRV/-1,0,0
IF(X1-X2)A1,A1,A2
A1)     TLRGT,GOFWD/L1,L3
        JUMPTO/A3
A2)     TLRGT,GOFWD/L1,L2
            GORGT/L2,L3
A3)         GORGT/L3,PAST,L4
LOOPND
```

*This example is presented here only to explain the usage of loop programming. In fact, the APT language does provide a multiple check surface statement to solve this problem. See Section 7.8.2 for details.

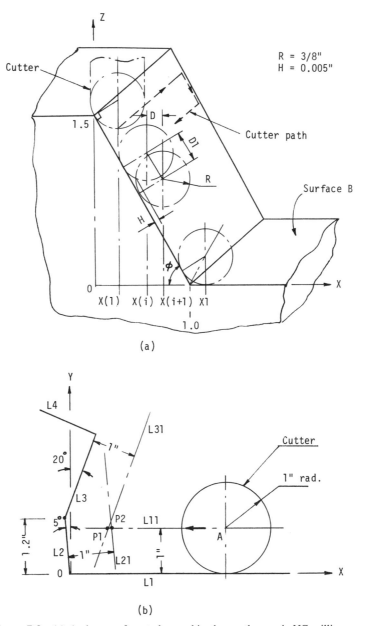

Figure 7-3   (a) A slope surface to be machined on a three-axis NC milling machine. (b) The part profile, L1-L2-L3-L4, and the milling cutter. A calculation is needed to determine which one of the two surfaces, L2 or L3, first comes in contact with the cutter.

A subprogram can also be used in this case to realize the function of the LOOPST-LOOPND loop, with the advantage that the decision-making process can be generalized to consider the variations of tool radius and corner profile.

7.3 PROGRAMMING WITH SUBPROGRAMS (MACRO)

In APT, a subprogram is a part of the part machining program that is read and stored in the computer memory and is not executed until it is invoked by a calling statement. A subprogram should begin with a MACRO statement and end with the TERMAC statement. Between these two statements, a sequence of the instructions that are to be executed when called is defined. A subprogram or MACRO must have a name or symbol, by which the subprogram can be referenced and executed through the use of the CALL statement.

The format of the MACRO statement is

symbol=MACRO/M1,M2,....,Mn

where

$\qquad$ symbol = any combination of alphanumeric characters with the following limitations:
$\qquad\qquad$ 1. The maximum number of characters and digits cannot exceed 6.
$\qquad\qquad$ 2. The first character must be alphabetical.
$\qquad$ M1, M2, . . . , Mn = subprogram or MACRO variables, which are used within the subprogram and can be assigned any values or words when the MACRO is called. If no MACRO variable is included in a subprogram, the statement is simplified to

$\qquad\qquad$ symbol=MACRO

The last statement of a subprogram or MACRO must be

TERMAC

which denotes the end of the MACRO. A MACRO is invoked by the CALL statement, whose format is

CALL/symbol,M1=b1,M2=b2,....,Mn=bn

where the element "symbol" is the name of the MACRO to be called; and b1 through bn are the values or words assigned to MACRO variables M1 through Mn, respectively. The following is an example of an APT program that makes use of a subprogram to define or change the feedrate, type of coolant, and cutter motion:

```
. . . . . . . . . . . . . . . . . . . . .
. . . . . . . . . . . . . . . . . . . . .
M1=MACRO/A,B,C
FEDRAT/A
COOLNT/B
GODLTA/0,0,C
TERMAC
. . . . . . . . . . . . . . . . . . . . .
. . . . . . . . . . . . . . . . . . . . .
CALL/M1,A1=2.0,B=ON,C=1.0
. . . . . . . . . . . . . . . . . . . . .
. . . . . . . . . . . . . . . . . . . . .
```

Note that a subprogram or MACRO must be defined *before* its CALL statement. The computer will not execute a subprogram until the CALL statement is reached. The following vocabulary words cannot be assigned to a MACRO variable: IF, JUMPTO, CALL, MACRO, TERMAC, FINI, LOOPST, and LOOPND.

Other rules concerning MACRO programming are as follows:

1. A subprogram can be defined only once. Thus it cannot be defined within a loop or another subprogram.

2. It is recommended that MACROs be kept as small as possible. Statements that can be specified outside a MACRO should not be included in a subprogram, thus avoiding unnecessary repetitive executions.

3. A LOOPST-LOOPND loop cannot be defined within a MACRO. Furthermore, a MACRO should not contain a FINI statement.

4. A MACRO can be called and executed by a CALL statement specified in another MACRO.

5. It is not allowed to use a MACRO variable to represent a statement label in a JUMPTO statement.

Usually a subprogram is used to define a sequence of instructions that are to be executed more than once. Besides, it is often necessary to change the values of certain parameters in these instructions according to the processing requirement each time the MACRO is called. For example, the program segment

```
. . . . . . . . . . . . . . . . . . . . . .
A1=MACRO/R1,R2,N1,N2
DO/B1,I=N1,N2
        IF(I'GT'5),JUMPTO/B2
        A=72*(I-1)
        JUMPTO/B3
B2)     A=90*(I-6)
B3)     C(I)=CIRCLE/(R1*COSF(A)),(R1*SINF(A)),R2
B1)CONTIN
TERMAC
CALL/A1,R1=2,N1=1,N2=5,R2=0.5   $$DEFINING THE FIRST SET OF CIRCLES
$$ C(1 THRU 5) EQUALLY DISTRIBUTED ON THE CIRCLE OF 2" RADIUS
CALL/A1,R1=3.0,N1=6,N2=9,R2=0.25   $$DEFINING THE 2ND SET OF CIRCLES
$$ C(6 THRU 9) EQUALLY DISTRIBUTED ON THE CIRCLE OF 2.5" RADIUS
. . . . . . . . . . . . . . . . . . . . . .
```

defines two sets of circles. The first one consists of five circles with a radius of 0.5 in. They are equally distributed on the circle with a 2 in. radius and center at (0,0). The same subprogram is also used to define the second set of circles, consisting of four circles with a radius of 0.25 in. that are equally distributed on another circle with a radius equal to 3.0 in. and center at (0,0). The parameters, or MACRO variables, are thus assigned the values corresponding to different sets of circles.

A subprogram is executed once when it is called. However, if the calling statement is within a loop and is executed a number of times, then the subprogram can also be invoked the same number of times. For example, the program segment listed above for defining the two sets of circles can also be replaced by the following program segment

. . . .
```
A1=MACRO
C(I)=CIRCLE/(R1*COSF(A)),(R1*SINF(A)),R2
TERMAC
DO/B1,I=1,5
     A=72*(I-1)
     CALL/A1,R1=2,R2=0.5
B1)CONTIN
D)/B2,I=6,10
     A=90*(I-6)
     CALL/A1,R1=2.5,R2=0.25
B2)CONTIN
TERMAC
```
. . . .

In this example, it is not necessary to have a subprogram because the subprogram consists of only one statement, which can be included in the DO-loop. However, when a number of geometric definition statements are involved, it is evident that a subprogram is needed. Generally, a logical combination of loops and subprograms can solve problems that require complex, repeated calculation and decision making.

7.4 COMMON PROGRAMMING ERRORS IN LOOPS AND SUBPROGRAMS

Apart from logical mistakes, there are several errors often committed by beginners, which are discussed in this section.

One of the general rules of APT programming is that in an APT program, a symbol representing a geometric entity cannot be redefined either as another geometric entity or even as the same geometric entity itself. For example, in the following program

```
.................
.................
P1=POINT/0,0,0
.................
.................
P1=POINT/0,1,0
.................
.................
```

the symbol P1, representing point $(0,0,0)$, has been redefined as point $(0,1,0)$. In this case, an error message, numbered 1321,* will be issued. If geometric definition statements are specified in a loop or a subprogram, the symbols for geometric entities will be redefined when the loop is executed or the subprogram is repeatedly invoked. Thus the same error (No. 1321) results.

There are two ways to solve this problem. The first is to specify these geometric definition statements outside the loop or MACRO, if possible or convenient. As an alternative, one can use these symbols as the names of arrays, and the index can be changed every time a geometric entity is defined. If programming requires redefi-

---

*See Section 8.7 for error messages.

nition of symbols for geometric entities, the statement CANON/ON can be specified before the geometric definition statements. This statement allows symbols of geometric entities to be redefined in subsequent statements. The statement CANON/OFF, which means that a symbol is not allowed to be redefined once it is assigned to a geometric entity, represents the default mode of the APT-AC NC processor.

When contouring motion statements are defined in a loop or a MACRO, they are often executed repeatedly. In such a case, it is necessary to check the cutter starting position at the beginning of each iteration to ensure that the tool has been placed at the correct start-up position. This can be illustrated by the following example. Figure 7-4 shows the APT program that directs an end mill to cut the inner profile. During the rough cut, an allowance of 0.02 in. is left on drive and check surfaces by specifying the statement

THICK/0,0.02,0.02

In the finishing cut, the allowance is removed by the statement

THICK/0,0,0

However, after finishing the rough cut, the tool is 0.02 in. away from the drive surface specified in the statement labeled A1. This means that the tool has not been placed at the correct start-up position and is not ready to start the contouring motion. Therefore the APT-AC NC processor issues the No. 2209 error message, which indicates that the cutter is out of tolerance with respect to the drive surface at the start of the motion sequence. For the solution to this problem, a start-up motion statement

GO/C1,PL1,L41

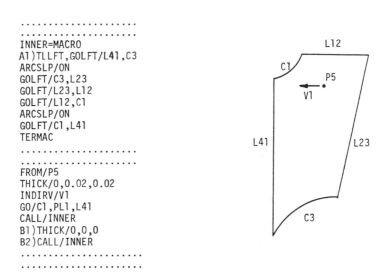

```
. . . . . . . . . . . . . . . . . . .
. . . . . . . . . . . . . . . . . .
INNER=MACRO
A1)TLLFT,GOLFT/L41,C3
ARCSLP/ON
GOLFT/C3,L23
GOLFT/L23,L12
GOLFT/L12,C1
ARCSLP/ON
GOLFT/C1,L41
TERMAC
. . . . . . . . . . . . . . . . . . .
. . . . . . . . . . . . . . . . . .
FROM/P5
THICK/0,0.02,0.02
INDIRV/V1
GO/C1,PL1,L41
CALL/INNER
B1)THICK/0,0,0
B2)CALL/INNER
. . . . . . . . . . . . . . . . . . .
. . . . . . . . . . . . . . . . . .
```

Figure 7-4   An APT program for cutting the inner profile of a part. The part surface PL1 (not shown) is perpendicular to the cutter axis. The finishing cutter path in this program is incorrectly defined, because it does not have a proper starting position.

can be inserted between the statements labeled B1 and B2. The tool will then be placed in the position required to start the contouring motion.

7.5 THE USE OF MATHEMATICAL CALCULATION STATEMENTS TOGETHER WITH LOOPS AND/OR SUBPROGRAMS TO DEFINE GEOMETRIC ENTITIES AND THE CUTTER PATH

Compared with programming in the NC code, programming in APT also has the advantage of allowing mathematical calculation statements in the program. Although a great variety of geometric entities are allowed in APT, there are still many others that cannot be defined. For those not allowed in APT, we can first calculate the coordinates of discrete points on their outline and then define them as LCONIC or tabulated cylinder. Usually, these calculations must be carried out by the computer so that the profile and slope continuity can be maintained. There are also cases in which the cutter paths are similar to one another but have, for example, some changes in depth of cut and in dimension of the profile. The mathematical calculation routine can be specified in loops to define such variation.

In Chapter 4, we illustrated through an example the use of mathematical calculation statements to define an epitrochoid (the profile of a Wankel engine chamber). We found that approximately 100 statements are needed for even as large an incremental angle as 10 degrees to define the profile. Two examples are given below to indicate the use of mathematical calculation statements, together with loop(s) or MACRO(s), thereby reducing significantly the size of a program.

Example 1

In Section 4.13 of Chapter 4, the profile of a Wankel engine chamber is defined, by approximately 100 statements, as nine overlapped loft conics. The program segment can be simplified through loop programming as follows:

```
. . . . . . . . . . . . . . . . . . . .
. . . . . . . . . . . . . . . . . . . .
RESERV/X,36,Y,36,E,9,P,36
. . . . . . . . . . . . . . . . . . . .
. . . . . . . . . . . . . . . . . . . .
DO/A1,I=1,36
        A=10*I
        B=3*A
        X(I)=60*COSF(A)+10*COSF(B)
        Y(I)=60*SINF(A)+10*SINF(B)
A1)    P(I)=POINT/X(I),Y(I)
DO/A2,J=1,8
A2)    E(J)=LCONIC/5PT,P(J*4-3),P(J*4-2),P(J*4-1),P(J*4),P(J*4+1)
E(9)=LCONIC/5PT,P(33),P(34),P(35),P(36),P(1)
. . . . . . . . . . . . . . . . . . . .
. . . . . . . . . . . . . . . . . . . .
```

This section of the program, which consists of only 10 statements, replaces the 117 statements in the program shown in Section 4.13 (Chapter 4).

Example 2

Figure 7-5 shows a part with five through holes of different depths, located along the X axis. A program that directs an NC drilling machine to drill the five holes may appear as follows:

```
. . . . . . . . . . . . . . . . . . . . .
. . . . . . . . . . . . . . . . . . . . .
D1=MACRO/Z1
RAPID;GODLTA/2.0,0,0
GODLTA/-Z1
RAPID;GODLTA/Z1
TERMAC
. . . . . . . . . . . . . . . . . . . . .
. . . . . . . . . . . . . . . . . . . . .
FROM/-5.0,4.0,3.0 $$ THE STARTING POSITION IS (-5,4,3).
RAPID;GOTO/(-2,0,2.1)
FEDRAT/2.0
DO/A1,I=1,5
      Z1=0.1649*X**1.2+1
      CALL/D1
A1) CONTIN
RAPID;GOTO/-5,4,3
. . . . . . . . . . . . . . . . . . . . .
. . . . . . . . . . . . . . . . . . . . .
```

The program shown above can also be written as follows:

```
. . . . . . . . . . . . . . . . . . . . .
. . . . . . . . . . . . . . . . . . . . .
FROM/-5.0,4.0,3.0
RAPID;GOTO/(-2,0,2.1)
FEDRAT/2.0
DO/A1,I=1,5
      Z1=0.1649*X**1.2+1
      RAPID;GODLTA/2.0,0,0
      GODLTA/-Z1
      RAPID;GODLTA/Z1
A1) CONTIN
RAPID;GOTO/-5,4,3
. . . . . . . . . . . . . . . . . . . . .
. . . . . . . . . . . . . . . . . . . . .
```

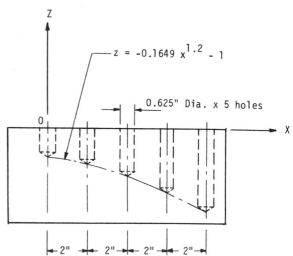

Figure 7-5   A part with five holes of different depths.

7.6 STATEMENTS DEFINING TRANSFORMATION MATRICES

7.6.1 The Matrix Representation of a Transformation

Generally, a geometric entity at a given spatial position and in a given spatial orientation can be placed at any spatial position and in any spatial orientation by both a translation (parallel shift) and a rotation. By the term *transformation*, we mean the moving of a geometric entity, a cutter path, or a coordinate system from one position and orientation to another. Translations and rotations are the two basic types of transformations. The transformation of a geometric entity may include a translation, a rotation, or both. A point can be placed at any position simply by translation. Generally, however, both types of transformations are needed to bring a geometric entity, other than a point, to a desired spatial position and direction.

A geometric entity is considered to be a rigid body, and every point on it goes through the same rotation and translation during a transformation. Therefore its transformation can be described by that of a point on it.

A point, P, in space (Fig. 7-6) can be described by its three coordinates

$$(x, y, z)$$

or, with the use of the matrix representation, by a column matrix

$$[P] = \begin{bmatrix} x \\ y \\ z \end{bmatrix}$$

The translation of point P to point P1 (x', y', z') can be described as

$$[P1] = \begin{bmatrix} x' \\ y' \\ z' \end{bmatrix} = \begin{bmatrix} x \\ y \\ z \end{bmatrix} + \begin{bmatrix} d1 \\ d2 \\ d3 \end{bmatrix} = [P] + [D] \qquad (1)$$

where $[D]$ is the translation vector and $d1$, $d2$, and $d3$ are the incremental distances in the X, Y, and Z directions, respectively.

Mathematically, the rotation of a point, P (x, y, z), about the origin of the coordinate system by an angle, a, is equivalent to the rotation of the coordinate system by an angle, $-a$, while the position of the point remains unchanged. Let the reference coordinate system be X-Y-Z and the unit vectors in the X, Y, and Z directions be

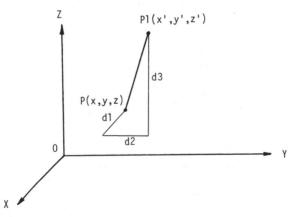

Figure 7-6   The translation of point P to point P1.

i, j, and k, respectively. After rotation, a new coordinate system, X'-Y'-Z', is generated with the direction cosines of the unit vectors i', j', and k' equal to

$$C11 = \cos(X, X') \qquad C12 = \cos(X, Y') \qquad C13 = \cos(X, Z')$$
$$C21 = \cos(Y, X') \qquad C22 = \cos(Y, Y') \qquad C23 = \cos(Y, Z')$$
$$C31 = \cos(Z, X') \qquad C32 = \cos(Z, Y') \qquad C33 = \cos(Z, Z')$$

where (X, X'), (X, Y'), $\ldots$, (Z, Z') represent, respectively, the angles from axis X to axis X', from axis X to axis Y', $\ldots$, and from axis Z to axis Z', with positive angles in the counterclockwise direction. Note that C11, C21, C31, $\ldots$, C33 are also the projections of the unit vector i', j', and k' on the X, Y, and Z axes, respectively.

The equation defining a rotational transformation is*

$$\begin{bmatrix} x' \\ y' \\ z' \end{bmatrix} = \begin{bmatrix} C11 & C21 & C31 \\ C12 & C22 & C32 \\ C13 & C23 & C33 \end{bmatrix} \begin{bmatrix} x \\ y \\ z \end{bmatrix} = [T] \begin{bmatrix} x \\ y \\ z \end{bmatrix} \qquad (2)$$

where (x, y, z) and (x', y', z') are the coordinates of the same point, say P, in coordinate systems X-Y-Z and X'-Y'-Z', respectively, and $[T]$ is the transformation matrix. Thus, given the rotational transformation of the coordinate system and the coordinates (x, y, z) of a point in the original coordinate system, X-Y-Z, the coordinates of the same point in the transformed coordinate system, X'-Y'-Z', can be determined by equation 2, above. The same equation can be viewed differently. It can also be considered as an equation to calculate the new position (x', y', z') of point P after rotation while the coordinate system remains unchanged. As indicated above, the rotation of the point in this case is the reverse rotation of the coordinate system from X-Y-Z to X'-Y'-Z'; the coefficients in the transformation matrix $[T]$ should be changed accordingly.

The rotations of a point by an angle, a, about the Z, X, and Y axes are given, respectively, by the following transformation matrices (Fig. 7-7):

$$\begin{bmatrix} \cos(a) & -\sin(a) & 0 \\ \sin(a) & \cos(a) & 0 \\ 0 & 0 & 1 \end{bmatrix} \qquad (3)$$

$$\begin{bmatrix} 1 & 0 & 0 \\ 0 & \cos(a) & -\sin(a) \\ 0 & \sin(a) & \cos(a) \end{bmatrix} \qquad (4)$$

$$\begin{bmatrix} \cos(a) & 0 & \sin(a) \\ 0 & 1 & 0 \\ -\sin(a) & 0 & \cos(a) \end{bmatrix} \qquad (5)$$

A general transformation that includes both a translation and a rotation can be defined by the expression

$$\begin{bmatrix} x' \\ y' \\ z' \end{bmatrix} = \begin{bmatrix} C11 & C21 & C31 \\ C12 & C22 & C32 \\ C13 & C23 & C33 \end{bmatrix} \begin{bmatrix} x \\ y \\ z \end{bmatrix} + \begin{bmatrix} d1 \\ d2 \\ d3 \end{bmatrix} \qquad (6)$$

*Gellert, W. et al., *The VNR Concise Encyclopedia of Mathematics*, pp. 533-535. New York: Van Nostrand Reinhold Co., 1975.

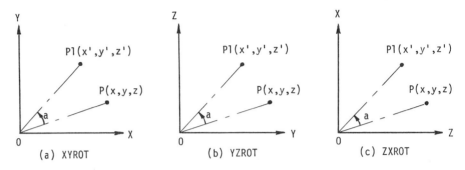

Figure 7-7   The rotation of a point, P, by an angle, a, about the Z (a), X (b), and Y (c) axes, respectively.

which is a combination of formulas 1 and 2, above, and can be represented in matrix form as

$$
\begin{bmatrix} x' \\ y' \\ z' \\ 1 \end{bmatrix} = \begin{bmatrix} C11 & C21 & C31 & d1 \\ C12 & C22 & C32 & d2 \\ C13 & C23 & C33 & d3 \\ 0 & 0 & 0 & 1 \end{bmatrix} \begin{bmatrix} x \\ y \\ z \\ 1 \end{bmatrix} \tag{7}
$$

or

$$
[P1] = [T_a][P] \tag{8}
$$

where $[T_a]$ is called an *affine* or *homogeneous* transformation. Note that the submatrices in the transformation matrix $[T_a]$

$$
\begin{bmatrix} C11 & C21 & C31 \\ C12 & C22 & C32 \\ C13 & C23 & C33 \end{bmatrix} \quad \text{and} \quad \begin{bmatrix} d1 \\ d2 \\ d3 \end{bmatrix}
$$

represent rotational and translational transformations, respectively.

7.6.2 Matrix Definitions in APT

In APT, a transformation matrix is defined by the statement

MATRIX/......

It is stored as a geometric entity in the NC processor. Therefore the matrix statement cannot be repeatedly executed, and the symbol for a matrix cannot be redefined unless a statement

CANON/ON

is specified.

When several transformation matrices are required for the transformation of a geometric entity, the order in which these matrices apply is very important since the commutation law does not apply to the product of matrices. The formats and usages of the MATRIX statement are listed in Table 7-3. MATRIX statements 1 through 8 are generally used to define the transformation of a geometric entity or cutter path from one position into another within a single coordinate system. Statements No. 9

and No. 10 are used to redefine, in the base coordinate system, a geometric entity or cutter path that has been defined in the local coordinate system specified by the MATRIX statement.

It must be pointed out that in the APT-AC NC processor, an APT MATRIX statement is only a mathematical definition of the matrix itself. It does not produce a transformation of the geometric entities and cutter path until it is used in certain APT statements, such as PATERN and those transformation statements described in Section 7.7, below.

TABLE 7-3   THE FORMATS AND USAGES OF MATRIX STATEMENTS

No.	Statement Format	Mathematical Expression

1 Translation matrix

MATRIX/TRANSL,$d1$,$d2$[,$d3$]

$$\begin{bmatrix} 1 & 0 & 0 & d1 \\ 0 & 1 & 0 & d2 \\ 0 & 0 & 1 & d3 \\ 0 & 0 & 0 & 1 \end{bmatrix}$$

Comment   The quantities $d1$, $d2$, and $d3$ are the incremental distances, in the X, Y, and Z directions, respectively, from the initial position, P, of a geometric entity (or cutter path) to the transformed position, P1 (see Fig. 7-6). The distance, $d3$, is taken as zero if omitted.

Examples
M1=MATRIX/TRANSL,1,2,3
or
M1=MATRIX/TRANSL,$d1$,$d2$,3

if $d1$ and $d2$ have been assigned the values 1 and 2, respectively, in the previous statements.

2 Rotation matrix

MATRIX/$\left\{\begin{matrix} \text{XYROT} \\ \text{YZROT} \\ \text{ZXROT} \end{matrix}\right\}$,$a$

$$\begin{bmatrix} \cos(a) & -\sin(a) & 0 & 0 \\ \sin(a) & \cos(a) & 0 & 0 \\ 0 & 0 & 1 & 0 \\ 0 & 0 & 0 & 1 \end{bmatrix} \quad \text{(for XYROT)}$$

$$\begin{bmatrix} 1 & 0 & 0 & 0 \\ 0 & \cos(a) & -\sin(a) & 0 \\ 0 & \sin(a) & \cos(a) & 0 \\ 0 & 0 & 0 & 1 \end{bmatrix} \quad \text{(for YZROT)}$$

$$\begin{bmatrix} \cos(a) & 0 & \sin(a) & 0 \\ 0 & 1 & 0 & 0 \\ -\sin(a) & 0 & \cos(a) & 0 \\ 0 & 0 & 0 & 1 \end{bmatrix} \quad \text{(for ZXROT)}$$

Comment   The matrix defines a rotation of the geometric entity about the coordinate axis, Z (for XYROT), X (for YZROT), or Y (for ZXROT), by an angle, a (Fig. 7-7). The positive direction of the angle follows the right-hand rule.

Example
M1=MATRIX/XYROT,45

(continued)

TABLE 7-3 THE FORMATS AND USAGES OF MATRIX STATEMENTS (continued)

No.	Statement Format	Mathematical Expression

3 Scaling matrix

MATRIX/SCALE,s

$$\begin{bmatrix} s & 0 & 0 & 0 \\ 0 & s & 0 & 0 \\ 0 & 0 & s & 0 \\ 0 & 0 & 0 & 1 \end{bmatrix}$$

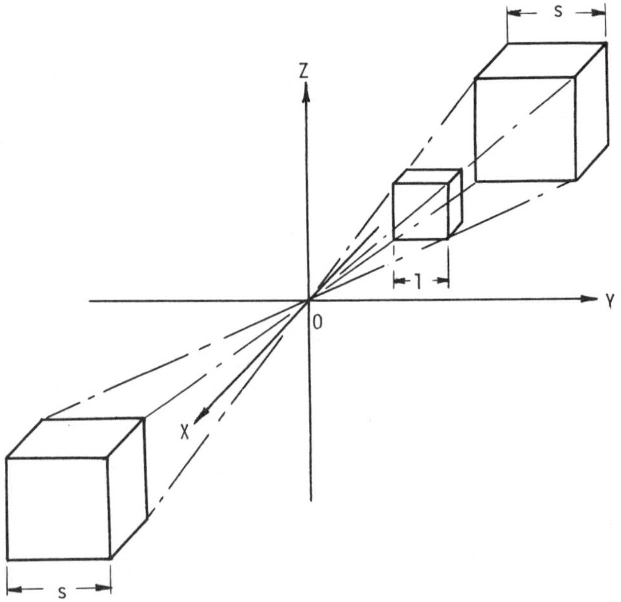

Comment This matrix magnifies or reduces a geometric entity or cutter path by factor s. The magnification (s greater than 1) or reduction (s less than 1) is with respect to the origin of the coordinate system, as can be seen from the figure, above. The transformation is a reflection through the origin when $s = -1$. When s is less than zero, we have

$$s=(-1) \cdot |s|$$

Hence, for any s value less than zero, the transformation defined by a scaling matrix is a reflection, together with a magnification (by a factor $|s|$), through the origin of the coordinate system.

4 A mirror-image transformation about one or more coordinate planes

MATRIX/MIRROR, $\left\{\begin{matrix} \text{XYPLAN} \\ \text{YZPLAN} \\ \text{ZXPLAN} \end{matrix}\right\}$\$

$\left[, \left\{\begin{matrix} \text{XYPLAN} \\ \text{YZPLAN} \\ \text{ZXPLAN} \end{matrix}\right\}\left[, \left\{\begin{matrix} \text{XYPLAN} \\ \text{YZPLAN} \\ \text{ZXPLAN} \end{matrix}\right\}\right]\right]$

$$\begin{bmatrix} 1 & 0 & 0 & 0 \\ 0 & 1 & 0 & 0 \\ 0 & 0 & -1 & 0 \\ 0 & 0 & 0 & 1 \end{bmatrix} \text{(MIRROR,XYPLAN)}$$

$$\begin{bmatrix} -1 & 0 & 0 & 0 \\ 0 & 1 & 0 & 0 \\ 0 & 0 & 1 & 0 \\ 0 & 0 & 0 & 1 \end{bmatrix} \text{(MIRROR,YZPLAN)}$$

$$\begin{bmatrix} 1 & 0 & 0 & 0 \\ 0 & -1 & 0 & 0 \\ 0 & 0 & 1 & 0 \\ 0 & 0 & 0 & 1 \end{bmatrix} \text{(MIRROR,ZXPLAN)}$$

TABLE 7-3   THE FORMATS AND USAGES OF MATRIX STATEMENTS   (continued)

No.	Statement Format	Mathematical Expression

$$\begin{bmatrix} 1 & 0 & 0 & 0 \\ 0 & -1 & 0 & 0 \\ 0 & 0 & -1 & 0 \\ 0 & 0 & 0 & 1 \end{bmatrix}$$ (MIRROR,XYPLAN,ZXPLAN or MIRROR,ZXPLAN,XYPLAN)

$$\begin{bmatrix} -1 & 0 & 0 & 0 \\ 0 & 1 & 0 & 0 \\ 0 & 0 & -1 & 0 \\ 0 & 0 & 0 & 1 \end{bmatrix}$$ (MIRROR,XYPLAN,YZPLAN or MIRROR,YZPLAN,XYPLAN)

$$\begin{bmatrix} -1 & 0 & 0 & 0 \\ 0 & -1 & 0 & 0 \\ 0 & 0 & 1 & 0 \\ 0 & 0 & 0 & 1 \end{bmatrix}$$ (MIRROR,YZPLAN,ZXPLAN or MIRROR,ZXPLAN,YZPLAN)

$$\begin{bmatrix} -1 & 0 & 0 & 0 \\ 0 & -1 & 0 & 0 \\ 0 & 0 & -1 & 0 \\ 0 & 0 & 0 & 1 \end{bmatrix}$$ (MIRROR,XYPLAN,YZPLAN,\$ ZXPLAN)

Comment   A mirror transformation with respect to a coordinate plane consists of changing the sign of the coordinate in the direction of the axis perpendicular to that plane. For example, a point (a, b, c) can be bought to position $(-a, b, c)$, $(a, -b, c)$, and $(a, b, -c)$ by the mirror transformations with respect to *Y-Z*, *Z-X* and *X-Y* planes, respectively. More than one mirror operation can be concatenated in this statement, and their order in this statement has no effect on the result.

5 General transformation matrix

MATRIX/$c11,c21,c31,d1,c12,c22,\$$
$c32,d2,c13,c23,c33,d3$

$$\begin{bmatrix} c11 & c21 & c31 & d1 \\ c12 & c22 & c32 & d2 \\ c13 & c23 & c33 & d3 \\ 0 & 0 & 0 & 1 \end{bmatrix}$$

6 A mirror-image transformation about a line (L1) or a plane (PL1)

MATRIX/MIRROR, $\begin{Bmatrix} L1 \\ PL1 \end{Bmatrix}$

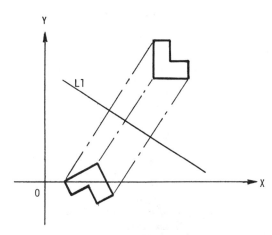

(continued)

TABLE 7-3 THE FORMATS AND USAGES OF MATRIX STATEMENTS (continued)

No.	Statement Format	Mathematical Expression

Comment   A line in APT is a plane passing through it and perpendicular to the X-Y plane. Therefore this statement essentially defines a mirror-image transformation about that plane.

7 Combined transformation matrix

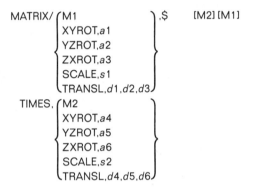

MATRIX/ (M1   ),$ [M2] [M1]
         XYROT,$a1$
         YZROT,$a2$
         ZXROT,$a3$
         SCALE,$s1$
         TRANSL,$d1,d2,d3$
   TIMES, (M2
         XYROT,$a4$
         YZROT,$a5$
         ZXROT,$a6$
         SCALE,$s2$
         TRANSL,$d4,d5,d6$

Comment   M1 and M2 are symbols for two defined matrices. The matrix operation after the word TIMES always applies first.

Examples
(1) The transformation defined by the matrix

M1=MATRIX/XYROT,$a1$,TIMES,TRANSL,$d1,d2,d3$

is a translation $(d1, d2, d3)$, followed by a rotation, by an angle, $a1$, about the Z axis.
(2) The transformation defined by the matrix.

M2=MATRIX/TRANSL,$d1,d2,d3$,TIMES,XYROT,$a1$

is a rotation followed by a translation. It is different from that defined by the matrix M1 in example 1.

8 Inverse transformation matrix

MATRIX/INVERS,M1

If

$$M1 = \begin{bmatrix} c11 & c21 & c31 & d1 \\ c12 & c22 & c32 & d2 \\ c13 & c23 & c33 & d3 \\ 0 & 0 & 0 & 1 \end{bmatrix}$$

the inverse matrix of M1 is

$$M1^{-1} = \begin{bmatrix} c11 & c12 & c13 & -(c11{\cdot}d1 + c12{\cdot}d2 + c12{\cdot}d3) \\ c21 & c22 & c23 & -(c21{\cdot}d1 + c22{\cdot}d2 + c23{\cdot}d3) \\ c31 & c32 & c33 & -(c31{\cdot}d1 + c32{\cdot}d2 + c33{\cdot}d3) \\ 0 & 0 & 0 & 1 \end{bmatrix}$$

Comment   If matrix M1 transforms a point from position A to position B, then the inverse transformation defined by the above statement transforms a point at position B to position A. The inversion of matrix M1, which is the combined matrix of several matrices, can be described by the following formulas:

TABLE 7-3 THE FORMATS AND USAGES OF MATRIX STATEMENTS (continued)

No.	Statement Format	Mathematical Expression

$(M1)^{-1} = (M2')^{-1} \cdot (M1')^{-1}$ if $M1 = (M1') \cdot (M2')$

$(M1)^{-1} = (M3')^{-1} \cdot (M2')^{-1} \cdot (M1')^{-1}$ if $M1 = (M1') \cdot (M2') \cdot (M3')$

$(M1)^{-1} = (Mn')^{-1} \cdot \ldots \cdot (M2')^{-1} \cdot (M1')^{-1}$ if $M1 = (M1') \cdot (M2') \cdot \ldots \cdot (Mn')$

9 Matrix defined by three
planes, PL1, PL2, and PL3

MATRIX/PL1,PL2,PL3

If planes PL1, PL2, and PL3 are
defined by equations

$(a1)x + (b1)y + (c1)z - d1 = 0$
$(a2)x + (b2)y + (c2)z - d2 = 0$
$(a3)x + (b3)y + (c3)z - d3 = 0$

the defined matrix is

$$\begin{bmatrix} a1 & a2 & a3 & (a1 \cdot d1 + a2 \cdot d2 + a3 \cdot d3) \\ b1 & b2 & b3 & (b1 \cdot d1 + b2 \cdot d2 + b3 \cdot d3) \\ c1 & c2 & c3 & (c1 \cdot d1 + c2 \cdot d2 + c3 \cdot d3) \\ 0 & 0 & 0 & 1 \end{bmatrix}$$

N: normal to the planes

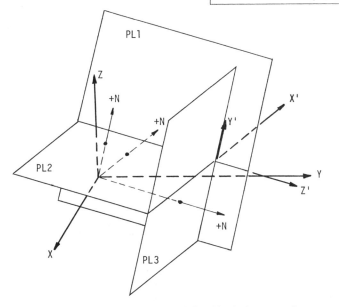

Comment   The three planes PL1, PL2, and PL3, defined in the base coordinate system X-Y-Z, are the coordinate planes Y'-Z', Z'-X', and X'-Y' of the local coordinate system X'-Y'-Z', respectively. A geometric entity or cutter path is supposed to be defined in the coordinate system X'-Y'-Z'. The defined matrix transforms the geometric entity into one defined in coordinate system X-Y-Z. Note that the coordinate system for defining the geometric entity in this case has been changed from the local to the base system, whereas the position of the geometric entity in space has not. The geometric entity after transformation is described in the base coordinate system. Note also that the X', Y', and Z' axes are in the positive direction of the normal to the three given planes, respectively. Thus the definition of the three planes should be such that their normals will form a right-hand coordinate system.

(continued)

TABLE 7-3 THE FORMATS AND USAGES OF MATRIX STATEMENTS (continued)

No.	Statement Format	Mathematical Expression
10	Matrix defined by a point, P1, and two vectors, V1 and V2 MATRIX/P1,V1,V2	

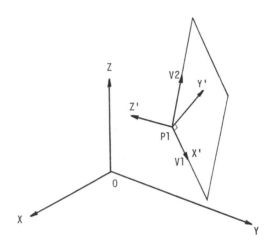

Comment This matrix defines a local coordinate system, X'-Y'-Z', whose origin is at point P1 and whose X' axis is defined by vector V1. The Y' axis of the local system is perpendicular to axis X' and on the plane defined by vectors V1 and V2. The angle included between vectors V2 and Y' should be less than 90 degrees. A transformation of the coordinate system (from the local to the base system X-Y-Z) can be realized through the specification of this statement. It has the same usage as statement No. 9, above.

7.7 DEFINING GEOMETRIC ENTITIES AND CUTTER PATHS BY MEANS OF A TRANSFORMATION MATRIX

Three statements can be used to manipulate the geometric entities or cutter path defined in an APT program: REFSYS, TRACUT, and COPY. The first statement, REFSYS, allows us either to use the most convenient coordinate system to define a *geometric entity,* or to transform a *geometric entity* defined in the base coordinate system. The second statement, TRACUT, is used to transform the CLDATA or *cutter path*. The third one allows us to copy and transform a *cutter path* as well. A detailed description and usages of these statements follow.

7.7.1 Transformation of Geometric Entitites by Means of the REFSYS Statement

Sometimes a geometric entity may be difficult to define in the base coordinate system (the coordinate system used in the machining process) but easy to define in a selected coordinate system. We can first define such a geometric entity in the selected coordinate system and then make use of the REFSYS statement to redefine it in the base coordinate system. The REFSYS statement can also be used simply to transform geometric entities that are defined in the base coordinate system.

The format, meaning, and usages of the statement REFSYS are shown in Table 7-4. As can be seen, the statements REFSYS/(matrix symbol) and REFSYS/NOMORE define the effective range of the transformation in a program. The geometric entities to be transformed should be defined between these two statements. When an APT program is being processed and the REFSYS/ statements are read, the matrix and the geometric definition statements are stored in the memory by the NC processor, and no transformation is carried out at this stage. The transformation is carried out only when these geometric entities are referenced in the statements after the statement REFSYS/NOMORE. The result of the transformation is to redefine, in the base coordinate system, the geometric entities that have been defined in the local coordinate system. If the geometric entities have been defined in the base coordinate system, they are simply transformed according to the matrix specified in the REFSYS statement.

For example, the APT program shown in Fig. 7-8 defines three circles, C1, C2, and C3, with the same set of parameters. C2 and C3 can be considered as geo-

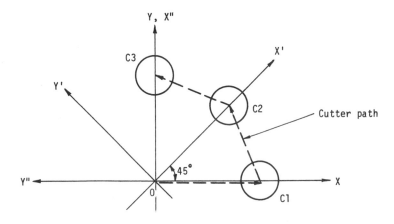

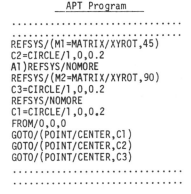

```
              APT Program

..........................
..........................
REFSYS/(M1=MATRIX/XYROT,45)
C2=CIRCLE/1,0,0.2
A1)REFSYS/NOMORE
REFSYS/(M2=MATRIX/XYROT,90)
C3=CIRCLE/1,0,0.2
REFSYS/NOMORE
C1=CIRCLE/1,0,0.2
FROM/0,0,0
GOTO/(POINT/CENTER,C1)
GOTO/(POINT/CENTER,C2)
GOTO/(POINT/CENTER,C3)
..........................
..........................
```

Figure 7-8   The application of the statement REFSYS.

TABLE 7-4 THE FORMAT, MEANING, AND USAGE OF THE REFSYS STATEMENT

Format	Meaning and Usage
. . . .	There are two different usages:
REFSYS/M1$$ M1 IS A MATRIX	(1) The matrix M1 defines the transformation of the base coordinate system X-Y-Z into the local coordinate system X'-Y'-Z'.
. . . . (statements . . . . defining . . . . geometric . . . . entities)	Geometric entities specified between statements REFSYS/M1 and REFSYS/NOMORE are defined in the local coordinate system.
. . . . REFSYS/NOMORE	Whenever any of these geometric entities is referred to in a statement after REFSYS/NOMORE, it is redefined in the base coordinate system.
. . . .	(2) The matrix defines a transformation of the geometric entities in the base coordinate system.
	Geometric entities specified between the two REFSYS statements are defined in the base coordinate system.
	Whenever any of these geometric entities is referred to in a statement after REFSYS/NOMORE, it is transformed by the matrix, M1, in the base coordinate system.
	Note: The geometric entities defined before statement REFSYS/M1 and after statement REFSYS/NOMORE are not transformed.

metric entities defined in the transformed coordinate systems X'-Y' and X''-Y'', respectively. When circles C2 and C3 are referenced in the statements after REFSYS/NOMORE, they are redefined in base coordinate system X-Y through the transformations of the coordinate systems, inverse to the specified transformations M1 and M2, respectively. Thus the statement REFSYS allows us to use a selected convenient coordinate system to define geometric entities. On the contrary, we may regard circles C2 and C3 as geometric entities defined in the base coordinate system. When referred to, they are rotated by 45 and 90 degrees, respectively, in the base coordinate system.

Note that in the APT program shown in Fig. 7-8, the statement labeled A1 can be omitted because the next statement, REFSYS/M2, specifies the conclusion of the former transformation, REFSYS/M1, and the beginning of a new transformation, REFSYS/M2. Both M1 and M2 should be defined with respect to the base coordinate system, *X-Y-Z*. Figure 7-9 presents the computer output of the program shown in Fig. 7-8. The reader should check the CLDATA to verify the transformation.

The geometric entities that can be affected by the REFSYS statement are as follows: point, vector, PNTVCT, SPALIN, pattern, line, plane, circle, ellipse, hyperbola, GCONIC, LCONIC, sphere, cone, cylinder, torus, RLDSRF, and the surfaces defined by the QADRIC statement.

7.7.2 Transformation of a Cutter Path by Means of the TRACUT Statement

As in the case of the REFSYS statement, the TRACUT statement should also be used in pairs: TRACUT/(matrix symbol) and TRACUT/NOMORE. The usage of this statement is as follows:

TRACUT/(symbol of matrix)

....

....

.... } (motion statements)

....

....

TRACUT/NOMORE

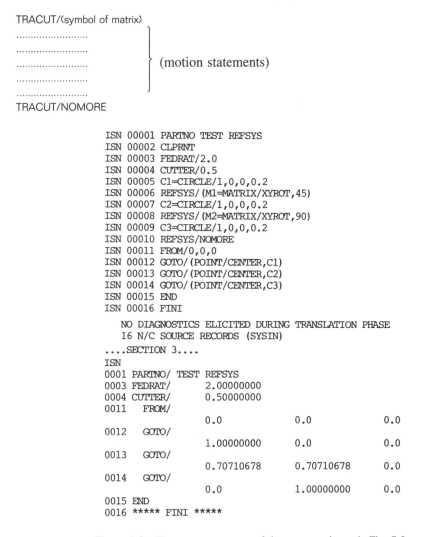

```
ISN 00001 PARTNO TEST REFSYS
ISN 00002 CLPRNT
ISN 00003 FEDRAT/2.0
ISN 00004 CUTTER/0.5
ISN 00005 C1=CIRCLE/1,0,0,0.2
ISN 00006 REFSYS/(M1=MATRIX/XYROT,45)
ISN 00007 C2=CIRCLE/1,0,0,0.2
ISN 00008 REFSYS/(M2=MATRIX/XYROT,90)
ISN 00009 C3=CIRCLE/1,0,0,0.2
ISN 00010 REFSYS/NOMORE
ISN 00011 FROM/0,0,0
ISN 00012 GOTO/(POINT/CENTER,C1)
ISN 00013 GOTO/(POINT/CENTER,C2)
ISN 00014 GOTO/(POINT/CENTER,C3)
ISN 00015 END
ISN 00016 FINI

    NO DIAGNOSTICS ELICITED DURING TRANSLATION PHASE
    16 N/C SOURCE RECORDS (SYSIN)
....SECTION 3....
ISN
0001 PARTNO/ TEST REFSYS
0003 FEDRAT/      2.00000000
0004 CUTTER/      0.50000000
0011    FROM/
                  0.0              0.0              0.0
0012    GOTO/
                  1.00000000       0.0              0.0
0013    GOTO/
                  0.70710678       0.70710678       0.0
0014    GOTO/
                  0.0              1.00000000       0.0
0015 END
0016 ***** FINI *****
```

Figure 7-9 The computer output of the program shown in Fig. 7-8.

The cutter path specified between the two TRACUT statements is defined in the base coordinate system and is transformed in the same coordinate system according to the matrix specified in the TRACUT statement. For example (Fig. 7-10), if we add the statement

TRACUT/(M2=MATRIX/XYROT,90)

in the APT program given in Fig. 7-9, the cutter path will be rotated counterclockwise by 90 degrees (see the CLDATA in Fig. 7-10). The TRACUT/NOMORE statement is used to cancel a TRACUT transformation specified previously.

If a TRACUT/M1 statement (M1 is the symbol of a matrix) is followed by a TRACUT/M2 statement without an intervening TRACUT/NOMORE statement,

```
ISN 00001 PARTNO TEST REFSYS1
ISN 00002 CLPRNT
ISN 00003 FEDRAT/2.0
ISN 00004 CUTTER/0.5
ISN 00005 C1=CIRCLE/1,0,0,0.2
ISN 00006 REFSYS/(M1=MATRIX/XYROT,45)
ISN 00007 C2=CIRCLE/1,0,0,0.2
ISN 00008 REFSYS/(M2=MATRIX/XYROT,90)
ISN 00009 C3=CIRCLE/1,0,0,0.2
ISN 00010 REFSYS/NOMORE
ISN 00011 FROM/0,0,0;TRACUT/(MATRIX/XYROT,90)
ISN 00013 GOTO/(POINT/CENTER,C1)
ISN 00014 GOTO/(POINT/CENTER,C2)
ISN 00015 GOTO/(POINT/CENTER,C3);TRACUT/NOMORE
ISN 00017 END
ISN 00018 FINI

    NO DIAGNOSTICS ELICITED DURING TRANSLATION PHASE
    16 N/C SOURCE RECORDS (SYSIN)

....SECTION 3....

ISN
0001 PARTNO/ TEST REFSYS1
0003 FEDRAT/      2.00000000
0004 CUTTER/      0.50000000
0011    FROM/
                 0.0               0.0               0.0
0012 TRACUT/
                 0.0          -1.00000000        0.0          0.0
                 1.00000000        0.0          0.0          0.0
                 0.0               0.0          1.00000000   0.0
0013    GOTO/
                 0.0           1.00000000        0.0
0014    GOTO/
                -0.70710678    0.70710678        0.0
0015    GOTO/
                -1.00000000        0.0           0.0
0016 TRACUT/            NOMORE
0017 END
0018 ***** FINI *****
....END OF SECTION 3....
```

Figure 7-10   An APT program showing the use of the TRACUT statement. This program is the same as that in Fig. 7-9, with the addition of two TRACUT statements.

the cutter path, or CLDATA, appearing between the two TRACUT statements is transformed by M1 and the CLDATA after the statement TRACUT/M2 is transformed by M2. This means that the second TRACUT statement automatically cancels the first one.

A transformation defined in a TRACUT statement can also be added to another by inserting the modifier LAST between the slash and the matrix symbol:

TRACUT/LAST,M1

An example is given in the following APT program:

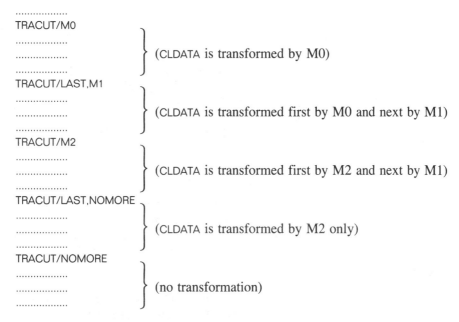

Note that the transformation specified in a TRACUT/LAST statement is added to the transformation defined by the TRACUT statement specified either before or after the TRACUT/LAST statement. A TRACUT statement with a LAST modifier can be canceled only by a succeeding TRACUT statement with a modifier LAST (either TRACUT/LAST,[symbol of matrix] or TRACUT/LAST,NOMORE). In addition, a TRACUT statement with the modifier LAST cannot cancel a previous TRACUT statement without the modifier LAST.

It should be pointed out that the cutter path defined by the transformation statement TRACUT is realized in two stages. The cutter path defined between the TRACUT and TRACUT/NOMORE statements is first calculated in the calculation section* of the NC processor without transformation. Then it is transformed in the edit section of the NC processor and output as the CLDATA according to the matrix specified in the TRACUT statement. However, the NC processor uses the last tool position generated or calculated by the calculation section as the reference point for calculation of the following tool motion. Therefore the reference position for deter-

*See Section 8.5 for the structure of the IBM APT-AC NC Processor.

mining the tool motion immediately after the transformation is the last untransformed tool position. This can be explained by the following example. The program

```
PARTNO
L1=LINE/XAXIS,1
L2=LINE/YAXIS,-1
M1=MATRIX/XYROT,45
CUTTER/1.0
FEDRAT/1.0
FROM/1,0,0
TRACUT/M1
     GO/ON,L1
TRACUT/NOMORE
GO/ON,L2
END
FINI
```

defines the tool motion from point (1,0,0) to point (0,1.4142,0) after execution of
the statement GO/ON,L1. However, in the NC processor, the untransformed tool
position defined by statement GO/ON,L1, that is, (1,1,0), is considered as the last
tool position and is used as the reference point to determine the following motion.
As a result, the end position defined by the statement GO/ON,L2 is (−1,1,0) instead
of (−1,1.4142,0).

7.7.3 Defining a Cutter Path by Means of the COPY Statement

As its name suggests, the COPY statement can be used to copy, or to copy and
transform as well, a previously defined *cutter path*. The cutter path to be copied
should be specified between statements INDEX/i and INDEX/i,NOMORE. A
COPY statement coming after the statement INDEX/i,NOMORE, with the same index number, i, instructs the APT-AC NC processor to transform and copy that part
of the cutter path.

There are a number of modifiers in a COPY statement that can be chosen to
define the transformation. The syntax of the COPY statement is

$$
\text{COPY}/i, \left\{
\begin{array}{l}
\text{SAME} \\
\text{TRANSL},x,y,z \\
\text{XYROT},a \\
\text{YZROT},a \\
\text{ZXROT},a \\
\text{MODIFY},M1 \\
\text{SCALE},s \\
\text{P1}, \left\{ \begin{array}{l} \text{ZXROT} \\ \text{XYROT} \\ \text{YZROT} \end{array} \right\} ,a[,\text{ROTREF}]
\end{array}
\right\} ,n
$$

where *n* defines the number of additional copies of the cutter path to be generated. It
should be noted that the first copy is made on the basis of the defined cutter path,
and the *n*th copy, on the basis of the (*n*−1)th copy. The usages and meanings of
these modifiers and parameters are as follows:

SAME

This modifier specifies that the cutter path is to be copied only and that no transformation takes place.

Example

```
. . . . . . . . . . . . . . . . .
. . . . . . . . . . . . . . . .
INDEX/2
GOTO/0,1,0
GOTO/0,2,0
. . . . . . . . . . . . . . . . .
GOTO/3,4,5
INDEX/2,NOMORE
. . . . . . . . . . . . . . . . .
. . . . . . . . . . . . . . . .   } (A)
. . . . . . . . . . . . . . . . .
COPY/2,SAME,1
. . . . . . . . . . . . . . . . .
. . . . . . . . . . . . . . . .
```

The cutter path between INDEX statements is copied once and placed after the cutter path defined by the statements in section A. This option is useful when a cutter path is the same as the one defined previously between the two INDEX statements.

TRANSL,x,y,z

This modifier specifies that the cutter path data is to be translated by x, y, and z in the X, Y, and Z directions, respectively, as it is copied.

Example   (Fig. 7-11)

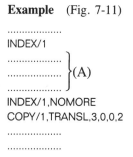

```
. . . . . . . . . . . . .
INDEX/1
. . . . . . . . . . . . .
. . . . . . . . . . . . .   } (A)
. . . . . . . . . . . . .
INDEX/1,NOMORE
COPY/1,TRANSL,3,0,0,2
. . . . . . . . . . . . .
. . . . . . . . . . . . .
```

As can be seen in Fig. 7-11, each transformation is carried out on the basis of the immediately preceding cutter path, that is, the first copy is made on the basis of the original cutter path and the specified translation. Then the second copy is made on the basis of the cutter path of the first copy and the specified translation.

XYROT,a; YZROT,a; ZXROT,a

These modifiers define a rotation, by an angle, a, about the Z, X, and Y axes, respectively. If multiple copies are required, the transformation follows the same principles as those described above.

MODIFY,M1; SCALE,s

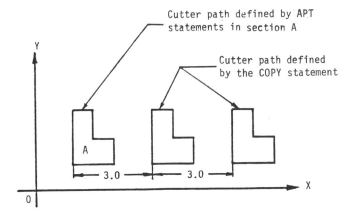

Figure 7-11   The cutter path generated by the COPY statement.

These modifiers specify, respectively, a transformation based on matrix M1 (MODIFY,M1) and a magnification or reduction with respect to the origin of the co-ordinate system based on scaling factor s (SCALE,s).

P1,XYROT,a[,ROTREF],n; P1,YZROT,a[,ROTREF],n; P1,ZXROT,a[,ROTREF],n

The transformations defined by these modifiers are as follows. Before the transformation is made, the point (0,0,0) (i.e., the origin of the coordinate system) is considered to be attached to the cutter path to be transformed. The transformation consists of two steps: the attached point is first moved to the specified point, P1, and then the cutter path is rotated by an angle, a, about the axis that passes through point P1 and is parallel to the Z (for XYROT), X (for YZROT), or Y (for ZXROT) axis. If n is greater than 1, the first copy is made by translating the original cutter path to point P1 and then rotating it by an angle, a. The second copy is made by translating the original cutter path to the same point, P1, and then rotating it by an angle, $2a$, and so on. If the copy is made through the rotation of the cutter path about the speci-fied point, P1, without translation, the optional word ROTREF should be specified.

We stated previously that the copy statement should be used together with the pair of statements INDEX/i and INDEX/i,NOMORE. In fact, this is not always necessary; indeed, a COPY statement can be specified after an INDEX/i state-ment without an intervening INDEX/i,NOMORE statement. In such a case, the COPY statement has the functions of both statements, COPY and INDEX/i,NOMORE. COPY statements can also be nested, as illustrated by the example shown in Fig. 7-12.

7.7.4 Some Comments Regarding the REFSYS, COPY, and TRACUT Statements

REFSYS, COPY, and TRACUT are the three statements used to define a transfor-mation either for geometric entities or for cutter motion. Statement REFSYS is used to transform geometric entities, whereas the other two transform cutter paths. A cut-ter path, being similar to one defined previously, can be defined more easily either by transforming the geometric entities first and then directing the tool along the

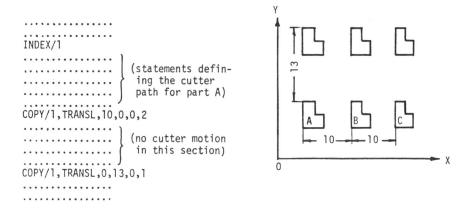

```
. . . . . . . . . . . . . . .
. . . . . . . . . . . . . . .
INDEX/1
. . . . . . . . . . . . . . .
. . . . . . . . . . . . . . .    (statements defin-
. . . . . . . . . . . . . . .     ing the cutter
. . . . . . . . . . . . . . .     path for part A)
. . . . . . . . . . . . . . .
COPY/1,TRANSL,10,0,0,2
. . . . . . . . . . . . . . .
. . . . . . . . . . . . . . .    (no cutter motion
. . . . . . . . . . . . . . .     in this section)
. . . . . . . . . . . . . . .
COPY/1,TRANSL,0,13,0,1
. . . . . . . . . . . . . . .
. . . . . . . . . . . . . . .
```

Figure 7-12   An example showing the effect of nested COPY statements. The first COPY statement makes two copies of the cutter path at position A, thereby obtaining the cutter path B and C. Since the first COPY cycle is nested in the second one, the three cutter paths A, B, and C, are all transformed and copied.

transformed surfaces or by directly transforming the cutter path, which is defined on the basis of nontransformed geometric entities.

As pointed out previously, the CLDATA is generated in different ways as different transformation statements are used. The calculation processes in the NC processor, corresponding to the three transformation statements, are given in Table 7-5.

It should be noted that in the transformation defined by statement REFSYS, the geometric entities have been transformed when referred to, whereas in the transformation defined by statements TRACUT or COPY, they remain unchanged. Moreover, the cutter positions calculated and stored in the calculation section of the NC processor are not changed by the COPY or TRACUT transformations. Since the last tool position calculated in the calculation section is used as the reference point to determine the following motion, the last tool position given by the edit section (i.e., the tool position in the CLDATA) is not the reference point used to calculate the next tool motion. Such a difference may result in incorrect definition of the cutter motion when the TRACUT and COPY statements are repeatedly used or when a cutter motion is defined after either of these two statements. When both TRACUT and COPY statements are used to define a cutter motion, their transformations are applied in the order indicated in Table 7-5. The examples given below explain the importance of the calculation process in the definition of the cutter path.

Fig. 7-13 gives an example of the use of the modifier LAST in a TRACUT statement that is specified between the pair of INDEX statements. The purpose of the COPY statement is to copy the cutter path O-A-B-C and to obtain the cutter path O_1-A_1-B_1-C_1 (Fig. 7-14). However, the cutter path obtained by executing the copy statement is O_1-A_1-B_2-C_1. The reason is that the cutter-path transformation statement TRACUT, with the modifier LAST, will not be executed until the copying has been finished. Therefore the cutter location defined as B' (i.e., the center of circle CC2) is first rotated by 90 degrees to point B'', then transformed and copied to position $B3$, and finally rotated by 180 degrees to position $B2$.

TABLE 7-5   THE CALCULATION PROCESSES WITHIN THE NC PROCESSOR FOR THE THREE TRANSFORMATION STATEMENTS

Transformation Statement	Calculation Process
REFSYS	1. Stores the transformation matrix and the geometric entities to be transformed (carried out by the translation section of the APT-AC NC processor) 2. Transforms the geometric entities when they are referred to (by the calculation section) 3. Calculates the cutter position on the basis of the transformed geometric entities (by the calculation section) 4. Generates the CLDATA according to the calculation results (by the edit section)
TRACUT	1. Stores the geometric entities and the transformation matrix specified by the TRACUT statement (by the translation section) 2. Calculates the cutter path defined between the TRACUT statements without transformation (by the calculation section) 3. Transforms the calculated cutter position data according to the matrices specified in the TRACUT statements and then output it as CLDATA (by the edit section)

COPY	1. Stores the geometric entities and the transformation matrix defined by the COPY statement (by the translation section)
	2. Calculates the cutter path defined between INDEX statements, no transformation carried out (by the calculation section)
	3. Cutter path transformed and copied as well (by the edit section)
	Note: When both the COPY and TRACUT statements are used to define a cutter path, the transformations are applied in the following order:
	    TRACUT (first precedence)
	    COPY (second precedence)
	    TRACUT/LAST (third precedence)

The second example is a gear with 12 teeth whose profile is approximated by circles (Fig. 7-15). The listed program A directs the end mill to go along the right-hand side of the profile. The cutter path is PTA → 1 → 2 → 3 → 4 → 5 →......→ 1 after execution of the statement

COPY/1,XYROT,30,11

It is intended to direct the tool from position 1 to position 2 by using the statement labeled K1. However, the tool goes from position 1 to position W for the following reason. The data concerning the tool movement from position 1, through positions

```
ISN 00001 PARTNO TEST MODIFIER LAST
ISN 00002 CLPRNT
ISN 00003 FEDRAT/2.0
ISN 00004 CUTTER/0.5
ISN 00005 C1=CIRCLE/1,0,0,0.2
ISN 00006 REFSYS/(M1=MATRIX/XYROT,45)
ISN 00007 C2=CIRCLE/1,0,0,0.2
ISN 00008 REFSYS/(M2=MATRIX/XYROT,90)
ISN 00009 C3=CIRCLE/1,0,0,0.2
ISN 00010 REFSYS/NOMORE;INDEX/1
ISN 00012 FROM/0,0,0;TRACUT/(MATRIX/XYROT,90)
ISN 00014 GOTO/(POINT/CENTER,C1);TRACUT/LAST,(MATRIX/XYROT,180)
ISN 00016 GOTO/(POINT/CENTER,C2);TRACUT/LAST,NOMORE
ISN 00018 GOTO/(POINT/CENTER,C3);TRACUT/NOMORE;INDEX/1,NOMORE
ISN 00021 COPY/1,TRANSL,3,0,0,1
ISN 00022 END
ISN 00023 FINI
```

```
     NO DIAGNOSTICS ELICITED DURING TRANSLATION PHASE
     17 N/C SOURCE RECORDS (SYSIN)

ISN
0001 PARTNO/ TEST MODIFIER LAST
0003 FEDRAT/      2.00000000
0004 CUTTER/      0.50000000
0011 INDEX /      1.00000000
0012    FROM/
                  0.0              0.0              0.0
0013 TRACUT/
                  0.0             -1.00000000       0.0              0.0
                  1.00000000       0.0              0.0              0.0
                  0.0              0.0              1.00000000       0.0
0014    GOTO/
                  0.0              1.00000000       0.0
0015 TRACUT/      LAST
                 -1.00000000       0.0              0.0              0.0
                  0.0             -1.00000000       0.0              0.0
                  0.0              0.0              1.00000000       0.0
0016    GOTO/
                  0.70710678      -0.70710678       0.0
0017 TRACUT/      LAST             NOMORE
0018    GOTO/
                 -1.00000000       0.0              0.0
0019 TRACUT/      NOMORE
0020 INDEX /      1.00000000       NOMORE
0021 COPY  /      1.00000000       TRANSL           1.00000000
                  3.00000000       0.0              0.0
                 *** THIS STARTS COPY/    1   PASS 1 ***
0021    FROM/
                  3.00000000       0.0              0.0
0013 TRACUT/
                  0.0             -1.00000000       0.0              0.0
                  1.00000000       0.0              0.0              0.0
                  0.0              0.0              1.00000000       0.0
0014    GOTO/
                  3.00000000       1.00000000       0.0
0015 TRACUT/      LAST
                 -1.00000000       0.0              0.0              0.0
                  0.0             -1.00000000       0.0              0.0
                  0.0              0.0              1.00000000       0.0
```

Figure 7-13 A computer program showing the effect of the modifier LAST in a TRACUT statement when a copy statement is used to transform and copy the cutter path defined between the TRACUT statements.   (continued)

```
0016    GOTO/
                -2.29289322          -0.70710678          0.0
0017  TRACUT/              LAST                NOMORE
0018    GOTO/
                 2.00000000           0.0                 0.0
0019  TRACUT/             NOMORE
0020  INDEX /             1.00000000                      NOMORE
                      *** THIS COMPLETES COPY/        1 ***
0020  COPY  /            1.00000000           TRANSL      1.00000000
                        3.00000000           0.0          0.0

0022  END
0023 ***** FINI *****
....END OF SECTION 3....
```

Figure 7-13   (continued)

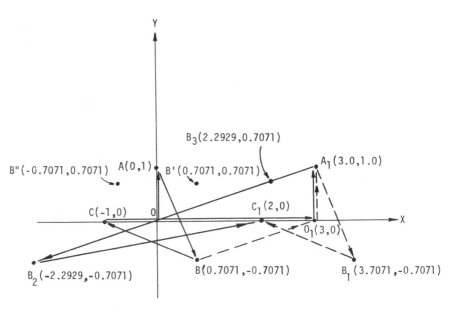

Figure 7-14   The cutter path for the program shown in Fig. 7-13.

2, 3, and 4, to position 5, specified between statements INDEX/1 and INDEX/ 1,NOMORE, are calculated and stored in memory. They are rotated each time they are copied, but the original data, in the memory, regarding the last tool position (point 5) and geometric entities (CD and C1) have not been changed. As the statement labeled K1 is executed, the initial position of the tool is still considered by the processor to be the one defined by the statement labeled K0 (i.e., point 5). As a result, the tool moves from point 5 along circle CD until check surface C1 is reached. Consequently, the end position of the statement labeled K1 must be point W.

Care must also be taken to specify a point-to-point movement in incremental coordinate(s) immediately after the COPY statements. The starting position is not the position transformed by the COPY statement; it is the untransformed position defined by the last of the statements between INDEX statements. Therefore, in pro-

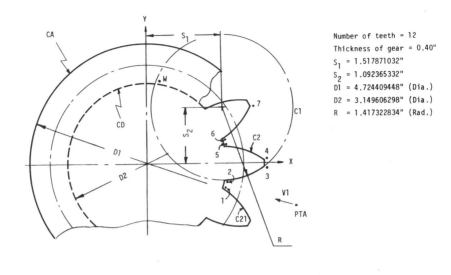

Number of teeth = 12
Thickness of gear = 0.40"
S_1 = 1.517871032"
S_2 = 1.092365332"
D1 = 4.724409448" (Dia.)
D2 = 3.149606298" (Dia.)
R = 1.417322834" (Rad.)

```
            Program A                              Program B
    . . . . . . . . . . . . . . . .          . . . . . . . . . . . . . . . .
    FROM/PTA                                 FROM/PTA
    INDIRV/V1                                INDIRV/V1
    GO/CD,PL1,C21                            GO/CD,PL1,C21
    $$ PL1 IS THE PART SURFACE              $$ PL1 IS THE PART SURFACE
    AUTOPS                                   AUTOPS
    INDEX/1; ARCSLP/ON                       INDEX/1;ARCSLP/ON
    TLRGT,GORGT/CD,C1                        TLRGT,GORGT/CD,C1
          GORGT/C1,PAST,CA                         GORGT/C1,PAST,CA
          GOLFT/CA,PAST,C2                         GOLFT/CA,PAST,C2
    KO)   GOLFT/C2,CD                        GOLFT/C2,CD
    INDEX/1,NOMORE                           INDEX/1,NOMORE
    COPY/1,XYROT,30,11                       COPY/1,XYROT,30,11
    ARCSLP/ON                                K)GODLTA/1.0,0,0
    K1)TLRGT,GORGT/CD,C1                     . . . . . . . . . . . . . . . .
    . . . . . . . . . . . . . . . .          . . . . . . . . . . . . . . . .
    . . . . . . . . . . . . . . . .          . . . . . . . . . . . . . . . .
    . . . . . . . . . . . . . . . .
```

Figure 7-15   A gear with a simplified profile. Program A: The statement labeled K1 specifies a cutter path that is not really intended. Program B explains the use of the GODLTA statement specified immediately after the COPY statement.

gram B listed in Fig. 7-15, the starting position of the tool for the statement labeled K (i.e., GODLTA/1.0,0,0) is point 5. If the coordinates of point 5 are (x_5, y_5, z_5), the end position will be $(x_5 + 1, y_5, z_5)$.

7.8 FURTHER STATEMENTS DEFINING START-UP AND CONTOURING MOTION

Sometimes a start-up or contouring motion is difficult to define by means of the statements given in the previous sections. The statements described in this section might prove useful in such cases.

7.8.1 The Delete-Output Statements (DNTCUT and CUT)

The pair of statements DNTCUT and CUT is used when it is difficult to position a tool from its current position directly to the desired position. In the example given in Fig. 7-16, a start-up motion is to be defined to position the tool from point PTA directly to point PTB, where the tool is tangent to S1 and S3. This can be realized by using the DNTCUT and CUT statements as shown in the figure. The meaning of the DNTCUT statement is to instruct the NC processor to execute the succeeding commands without outputting the resulting cutter points (locations). The CUT command indicates that the processor is to resume outputting the cutter location(s). When a CUT command is specified, the current cutter location is output by the processor. Therefore a start-up motion will be generated from PTA directly to PTB.

7.8.2 The Contouring Motion Statement with Multiple Potential Check Surfaces

In some situations, any one of several surfaces might be taken as the check surface of a motion statement, and the actual check surface cannot be determined unless detailed calculations are made (Fig. 7-17). In such cases the NC processor can determine which surface will be encountered first and then transfer control to the correct motion statement. This is realized by using a multiple-check-surface motion statement together with the TRANTO statement.

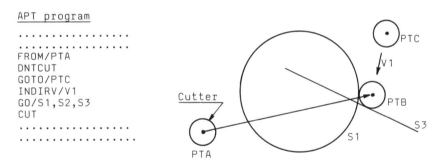

```
APT program

. . . . . . . . . . . . . . .
. . . . . . . . . . . . . . .
FROM/PTA
DNTCUT
GOTO/PTC
INDIRV/V1
GO/S1,S2,S3
CUT
. . . . . . . . . . . . . . .
. . . . . . . . . . . . . . .
```

Figure 7-16   The use of statements DNTCUT and CUT. S2 is the part surface (not shown).

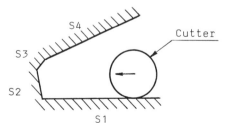

Figure 7-17   A typical case for the use of the contouring motion statement with multiple potential check surfaces.

The format of a multiple-check-surface motion statement is

$$\left[\left[\begin{matrix}\text{TLLFT}\\\text{TLRGT}\\\text{TLON}\end{matrix}\right],\right]\left(\begin{matrix}\text{GOLFT}\\\text{GORGT}\\\text{GOFWD}\\\text{GOBACK}\\\text{GOUP}\\\text{GODOWN}\end{matrix}\right)/\text{S1,}\left\{\begin{matrix}[\text{TO}]\\\text{ON}\\\text{PAST}\\\text{TANTO}\end{matrix}\right\}\text{,S2,L1,}\left\{\begin{matrix}[\text{TO}]\\\text{ON}\\\text{PAST}\\\text{TANTO}\end{matrix}\right\}\text{,S3,L2,}\left\{\begin{matrix}[\text{TO}]\\\text{ON}\\\text{PAST}\\\text{TANTO}\end{matrix}\right\}\text{,S4,L3[,}f\text{]}$$

where

$$S1 = \text{the drive surface}$$
$$S2,\ S3,\ S4 = \text{the first, second, and third check surfaces, respectively}$$
$$L1,\ L2,\ L3 = \text{the labels of the statements to be branched to when the first,}$$
second, and third check surfaces take effect, respectively
$$f = \text{the feedrate}$$

The first vocabulary word in this statement can be omitted if it is the same as in the preceding statement. Up to three check surfaces can be specified in this statement with corresponding labels of statements and modifiers. The multiple-check-surface statement is always used together with one or more TRANTO statements, which is an unconditional branching statement directing control to the appropriate motion statement(s) corresponding to these possible situations. All the motion statements are processed by the translation section of the APT-AC NC processor and stored, but the actual branching is deferred until the cutter path is computed in the calculation section.

A program segment that uses the multiple-check-surface motion statement to define the cutter motion in Fig. 7-17 follows:

```
. . . . . . . . . . . . . . . . . . . . . . . .
(cutter being at the position shown in Figure 7-17)
TLRGT,GOFWD/S1,S2,K2,S3,K3,S4,K4 $$ FORWARD DIRECTION IS AS SHOWN IN
                                 $$ THE FIGURE
K2)    GORGT/S2,S3
K3)    GORGT/S3,S4
       GORGT/S4,..........
TRANTO/K5  $$  THIS STATEMENT DIRECTS CONTROL TO THE ONE LABELED K5
K4)    GOBACK/S4,......
K5).....(the statement specifying the motion along the surface
         following S4).....
```

Fig. 7-18 shows an example of the usage of these statements. In this machining operation, the same profile is to be machined with two cutters of different sizes — a rough-cut cutter of larger size and a finishing cutter of smaller size. The fillet, defined as S2, requires that multiple check surfaces and branching of control be specified in the program. The APT program segments for directing the small and large cutters, and both cutters, are given. The program segment for both tools can be included in a MACRO statement, together with the THICK statement. Then the MACRO (subprogram) can be called to define both the roughing and finishing operations.

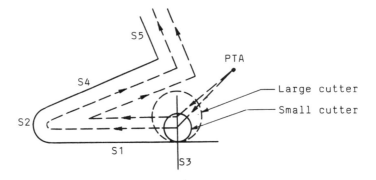

APT statements for small cutter:

. . . .
```
FROM/PTA
GO/ON,S3,PS1,S1
TLRGT,GORGT/S1,TANTO,S2
GOFWD/S2,TANTO,S4
GOFWD/S4,PAST,S5
```
. . . .

APT statements for large cutter:

. . . .
```
FROM/PTA
GO/ON,S3,PS1,S1
TLRGT,GORGT/S1,S4
GORGT/S4,PAST,S5
```
. . . .

APT statements for both large and small cutters:

. . . .
```
FROM/PTA
GO/ON,S3,PS1,S1
TLRGT,GORGT/S1,TANTO,S2,$
    ID1,TO,S4,ID2
ID1)GOFWD/S2,TANTO,S4
GOFWD/S4,PAST,S5
TRANTO/ID3
ID2)GOBACK/S4,PAST,S5
ID3)GOLFT/S5,......
```
. . . .

Figure 7-18 An example of the use of the multiple-check-surface statement. Part surface PS1 (not shown) is perpendicular to the tool axis.

7.8.3 Motion Statement Specifying a Selected Intersection of the Drive Surface with the Check Surface

When a drive surface has several intersections with a check surface, the tool will change its motion direction at the first intersection if the contouring motion statement described in Chapter 5 is used. Several contouring motion statements are then needed to define the tool motion along the same drive surface from the first intersection to the nth one. To simplify programming in such a case, an additional word, IN-TOF, together with a parameter, can be used in the contouring motion statement to indicate the desired end position. Suppose that the desired position for changing tool motion direction is the nth intersection of the drive surface with the check surface; the option "n,INTOF" can then be introduced into the motion statement just ahead of the check surface to indicate the desired turning point of the tool motion. The format is as follows:

$$\begin{bmatrix} \begin{Bmatrix} TLLFT \\ TLRGT \\ TLON \end{Bmatrix}, \end{bmatrix} \begin{Bmatrix} GOLFT \\ GORGT \\ GOUP \\ GODOWN \\ GOFWD \\ GOBACK \end{Bmatrix} /S1, \begin{Bmatrix} [TO] \\ ON \\ PAST \\ TANTO \end{Bmatrix} ,n,INTOF,S2[,f]$$

where S1 is the drive surface; S2, the check surface; and f, the feedrate. The parameter n must be a positive integer. An example is given in Fig. 7-19.

7.8.4 The OFFSET Statement

The OFFSET statement is used to provide additional means for controlling the end position in a start-up motion. The format of this statement is

$$OFFSET/\begin{Bmatrix} [TO] \\ ON \\ PAST \end{Bmatrix},S1\begin{bmatrix} \begin{Bmatrix} [,TO] \\ ,ON \\ ,PAST \end{Bmatrix},S2 \end{bmatrix}[,f]$$

The first surface, S1, in this statement is the drive surface, and S2, the part surface. The brackets mean, as usual, that this part of the statement can be omitted if it is not required.

The OFFSET statement must be preceded by an INDIRV or INDIRP statement. The vector defined by the INDIRV or INDIRP statement is considered to be attached to the current cutter position (PTA); when extended, it intersects with the drive surface (S1) specified in the OFFSET statement (Fig. 7-20). The normal to

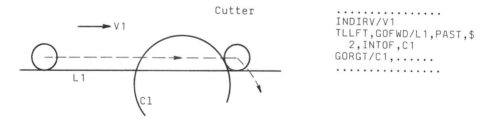

```
                                    . . . . . . . . . . . . . . .
                                    INDIRV/V1
                                    TLLFT,GOFWD/L1,PAST,$
                                       2,INTOF,C1
                                    GORGT/C1,......
                                    . . . . . . . . . . . . . . .
```

Figure 7-19   The motion statement specifying a selected intersection of the drive surface with the check surface.

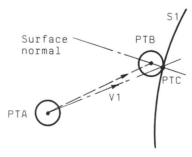

Figure 7-20   A start-up motion defined by an OFFSET statement.

```
FROM/PTA
INDIRV/V1
OFFSET/TO,S1
```

the drive surface at that point is found by the NC processor and is considered to be the check surface of the start-up motion. Thus the tool will move from PTA straight to PTB, which is a valid position corresponding to the specified modifier (in Fig. 7-20 the modifier is TO) on the normal to the drive surface. If the part surface is specified in the OFFSET statement, the tool will move farther along the tool axis TO, ON, or PAST the part surface; otherwise, it will move to the previously defined part surface. These two motions, from PTA to PTB and from PTB to the part surface, say PTD, are to be completed simultaneously; therefore the processor generates only one motion statement, namely, from PTA to PTD.

7.9 THE POCKET STATEMENT

Sometimes we need to cut an area bound by a series of straight lines to a desired depth. The POCKET statement can be used in such a case to save a large number of statements and a great deal of calculations.

The format of the POCKET statement is as follows:

POCKET/$a,b,c,f1,f2,f3,d,e$,PT1,PT2,...,PTn

where

a = effective cutter radius, which is determined by the cutter geometry, the bottom surface of the pocket, and the desired scallop height (Fig. 7-21)

b = step-feed factor (Fig. 7-21 and 7-22) from one cut to the next (The step feed or offset is the cutter radius, a', multiplied by b and is always measured in the part [bottom] surface.)

c = step-feed factor for finishing cut (The step feed for the finishing cut is equal to the cutter radius, a', multiplied by c. Similarly, it is also measured in the part surface.)

$f1$ = feedrate for the initial movement of the tool into the pocket

$f2$ = feedrate for general pocketing machining

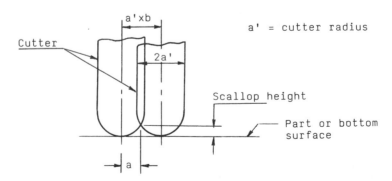

Figure 7-21   The parameters a and b used in the POCKET statement.

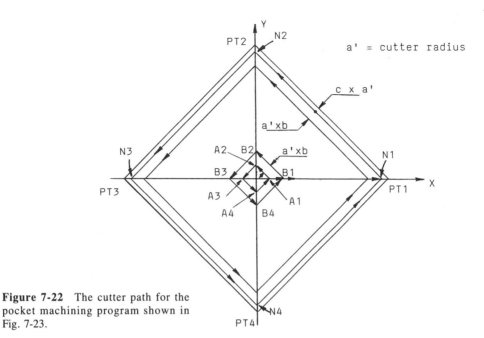

Figure 7-22 The cutter path for the pocket machining program shown in Fig. 7-23.

$f3$ = feedrate for the finishing cut around the pocket

d = 0 (The processor will make a test to ascertain whether the scallop height is less than the allowed value. If not, then new step-feed rates will be set by the processor to replace the specified b and c.)

d = 1 (No test is carried out by the processor. The programmed step feed b and c will be used during pocketing throughout.)

PT1, PT2, ..., Ptn = points defining the boundary vertices of the pocket to be machined

e = parameter indicating the type of points, PT1 through PTn, and the form of the bottom surface (see the following table):

e	BOTTOM OF THE POCKET	CORNER OF THE POCKET
0	The input points are passed through by the cutter-center locus.	

e	BOTTOM OF THE POCKET	CORNER OF THE POCKET

| 1 | The input points, PT1 through PTn, are on a plane, which is the bottom of the pocket. | The input points are the vertices of the pocket. |

2		The input points are passed through by the cutter-center locus.
2	The part surface is used as the bottom of the pocket; the input points, PT1 through PTn, are not necessarily	
3	on a plane.	The input points are the vertices of the pocket.

The order of the input points determines the direction in which the cutter is to travel when cleaning out the pocket.

The POCKET statement is a motion statement. Once it is executed, the tool will go to a starting point (whose z coordinate is the same as that of the pocket bottom) close to the center of the pocket and will clean out the pocket. The starting point is determined by the processor in the following way (Fig. 7-22). The NC processor first calculates the cutter locations for the finishing cut (namely, points PT1, PT2, PT3, and PT4) and then rolls back to calculate those for the final rough cut (namely, N1, N2, N3, and N4) according to factor c. The remainder of the cutter locations are determined by factors a' and b on the basis of the calculation of the last cutter location first and the first cutter location last. Finally, the first point in the first

motion cycle A1-A2-A3-A4 (i.e., point A1) is selected as the starting point
for pocketing.

The cutter motion for a POCKET statement is as follows: the cutter moves
from its current position straight to the starting point in the pocket, and then spirals
in the direction defined by the order of the input points, gradually moving toward the
boundary of the pocket. For example, the POCKET statement in the program given
in Fig. 7-23 will bring the cutter to position A1 (Fig. 7-22). Then the cutter will
move along the path A1-A2-A3-A4-A1......N1-N2-N3-N4-N1-PT1-PT2-PT3-PT4-
PT1. The cutter will move up again at PT1.

The APT-AC NC processor requires that the polygon defined by the input

```
ISN 00001 PARTNO TEST POCKET
ISN 00002 CLPRNT
ISN 00003 CUTTER/0.2
ISN 00004 PT1=POINT/1,0,0
ISN 00005 PT2=POINT/0,1,0
ISN 00006 PT3=POINT/-1,0,0
ISN 00007 PT4=POINT/0,-1,0
ISN 00008 FROM/0,0,0.1
ISN 00009 RAPID
ISN 00010 GOTO/0,0,0
ISN 00011 PSIS/(PLANE/0,0,1,-0.5)
ISN 00012 POCKET/0.1,0.9,0.1,1.0,2.0,1.5,0,2,PT1,PT2,PT3,PT4
ISN 00013 GODLTA/1.5
ISN 00014 END
ISN 00015 FINI


     NO DIAGNOSTICS ELICITED DURING TRANSLATION PHASE
     15 N/C SOURCE RECORDS (SYSIN)


ISN
0001 PARTNO/ TEST POCKET
0003 CUTTER/      0.20000000
0008    FROM/
                  0.0              0.0            0.10000000
0009 RAPID
0010    GOTO/
                  0.0              0.0            0.0
0012
0012 FEDRAT/      1.00000000
0012    GOTO/         POCKET       (    )
                  0.09490332       0.0           -0.50000000
0012 FEDRAT/      2.00000000
0012    GOTO/         POCKET       (    )
                  0.0              0.09490332    -0.50000000
                 -0.09490332       0.0           -0.50000000
                  0.0             -0.09490332    -0.50000000
                  0.09490332       0.0           -0.50000000
                  0.22218254       0.0           -0.50000000
                  0.0              0.22218254    -0.50000000
                 -0.22218254       0.0           -0.50000000
                  0.0             -0.22218254    -0.50000000
                  0.22218254       0.0           -0.50000000
                  0.34946176       0.0           -0.50000000
                  0.0              0.34946176    -0.50000000
                 -0.34946176       0.0           -0.50000000
```

Figure 7-23   An APT pocketing program and its CLDATA.   (continued)

0.0	-0.34946176	-0.50000000
0.34946176	0.0	-0.50000000
0.47674098	0.0	-0.50000000
0.0	0.47674098	-0.50000000
-0.47674098	0.0	-0.50000000
0.0	-0.47674098	-0.50000000
0.47674098	0.0	-0.50000000
0.60402020	0.0	-0.50000000
0.0	0.60402020	-0.50000000
-0.60402020	0.0	-0.50000000
0.0	-0.60402020	-0.50000000
0.60402020	0.0	-0.50000000
0.73129942	0.0	-0.50000000
0.0	0.73129942	-0.50000000
-0.73129942	0.0	-0.50000000
0.0	-0.73129942	-0.50000000
0.73129942	0.0	-0.50000000
0.85857864	0.0	-0.50000000
0.0	0.85857864	-0.50000000
-0.85857864	0.0	-0.50000000
0.0	-0.85857864	-0.50000000
0.85857864	0.0	-0.50000000
0.98585786	0.0	-0.50000000
0.0	0.98585786	-0.50000000
-0.98585786	0.0	-0.50000000
0.0	-0.98585786	-0.50000000
0.98585786	0.0	-0.50000000

```
0012 FEDRAT/    1.50000000
0012   GOTO/      POCKET      (    )
                1.00000000   0.0          -0.50000000
                0.0          1.00000000   -0.50000000
               -1.00000000   0.0          -0.50000000
                0.0         -1.00000000   -0.50000000
                1.00000000   0.0          -0.50000000
0013   GOTO/      TLAXIS
                1.00000000   0.0           1.00000000
0014 END
0015 ***** FINI *****
....END OF SECTION 3....
```

Figure 7-23   (continued)

points in the POCKET statement be convex. This means that the interior angle at the vertex of the polygon should be less than 180 degrees. The error message (numbered 2504)

"POCKET INPUT POINTS DO NOT SPECIFY A CONVEX POLYGON"

will be issued if the point PT1 in the APT program shown in Fig. 7-23 is defined as

PT1=POINT/-0.1,0,0

which means that the interior angle at point PT1 is greater than 180 degrees. A cavity with interior angle(s) of more than 180 degrees between adjacent sides must be divided into two or more pockets with polygons, formed by the sides of the pockets, which are convex only. Overlaps in the machined areas are also required to ensure a complete cleaning out of the cavity (Fig. 7-24).

When the POCKET statement is used for a cavity with an interior angle of less than 90 degrees, step-feed factor b should be adjusted to ensure that no uncut area is

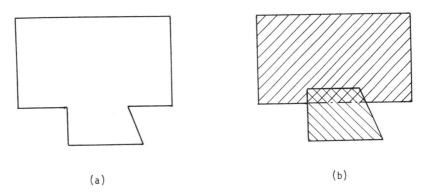

Figure 7-24   A pocket with interior angles greater than 180 degrees (a) should be divided into two pockets (b) overlapping each other.

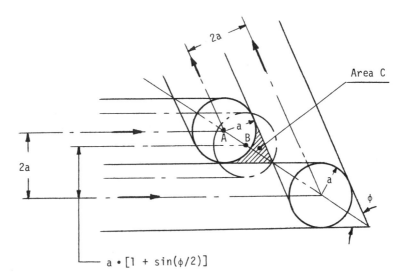

Figure 7-25   The determination of the step feed for a pocket with an interior angle, ϕ, less than 90 degrees. Area C is left uncut if the step feed is two times the effective radius (i.e., $2 \cdot a$). The step feed, $b \cdot a'$, should be reduced to
$$a \cdot [1+\sin(\phi/2)],$$
or the center of the cutter be moved from point A to point B to clean out the area.

left at the corner. The step feed, $b \cdot a'$, should be less than $2 \cdot a$ and can be calculated with the formula (Fig. 7-25)
$$b \cdot a' \leq a \cdot [1 + \sin(\phi/2)].$$

The NC processor also requires that the angle between the bottom plane of the defined pocket and the cutter axis be larger than 1 degree and 10 minutes.

A pocket with a nonlinear boundary should first be simplified as a polygon by selecting several point on or within the boundary. After machining (pocketing) of the polygon pocket, a contouring motion along the curved boundary is needed.

PROBLEMS

7.1 A milling operation is usually divided into two sections, one for the rough cut and the other for the finishing cut. Give the general structure of an APT program that uses the MACRO (subprogram) to define such a milling operation.

7.2 The drilling motion defined by a PATERN statement with the modifier AVOID can have only one motion speed. However, the three motion steps for drilling each hole should be at two different speeds; they are as follows:

a. A rapid motion to the position right above the top surface of the hole

b. A downward motion at a specified feedrate to the depth of the hole

c. A rapid upward retreat motion to the required level

If there are $n + m$ holes on a part, among which n holes have the same depth, $H1$, the other m holes have depth $H2$. The x and y coordinates of the ith hole are $x(i)$ and $y(i)$, respectively. The top surface of these holes are on the same level: $z = z1$. The drilling feedrate is $f1$. The other machining specifications can be assigned arbitrarily. Write as concise an APT program as possible for drilling these holes.

7.3 A type of shafts that we often need to turn is shown in Fig. P7-3. Its basic features are as follows:

a. The profile consists of lines only.

b. Its diameter increases monotonously in the $-X$ direction.

Assume that the origin of the coordinate system is always set at the center point of the end surface. Furthermore, the first line segment is the one after the end surface, that is, L(1), which can be a line with a slope angle between 0 and -90 degrees. The starting position of the tool, P0, for the finishing cut is always selected in such a way that the end position of the start-up motion (i.e., P1) is on the right side of the end surface. Write an APT subprogram or subprograms that can be called to define automatically the finishing cutter path, P0-P1-....-PE, for parts whose profile consists of any desired number of line segments.

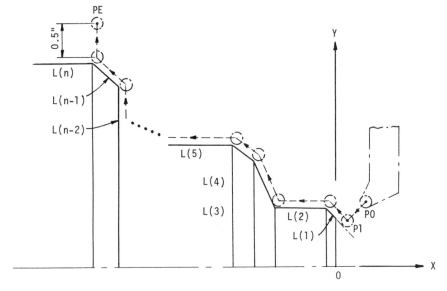

Figure P7-3

7.4 The part shown in Fig. P7-4 is of the same type as that described in Problem 7.3. Its rough turning process consists of a number of motion cycles. Write an APT program, as concisely as possible, for defining such a roughing operation. The following conditions are given:

  a. The part profile consists of line segments only. The first segment, L(1), is parallel to the X axis, and the last one, L(n), is not.

  b. The depth of cut for each pass is D1.

  c. The distance of retreat motion is D2.

  d. The allowance is A1.

  e. The starting and end points are (X0, Y0).

  f. The feedrate is 0.007 in./revolution, and the spindle speed is 500 rpm.

  g. The starting point, P0, is always selected in such a way that Y0 − D1 < R1.

7.5 Give an example to verify the sequence of transformations defined by the statement

MATRIX/M1,TIMES,M2

7.6 Give an example to show that the No. 9 format of the MATRIX statement given in Table 7-3 is used to transform a geometric entity defined in a local coordinate system into one defined in the base coordinate system.

7.7 The gear shown in Fig. 7-15 is to be machined by a flat-end mill of $\frac{1}{8}$ in. diameter in seven passes, each with a depth of cut of 0.06 in. Write an APT program that uses the transformation statement COPY only to transform and define the cutter path. The following conditions are given:

  a. The starting and end points are at the center of the gear and 1 in. above the surface.

  b. Spindle speed = 1000 rpm.

  c. Feedrate = 2 in./min.

7.8 Rework Problem 7.7, using the transformation statement REFSYS only to transform the geometric entities and then to define the cutter path.

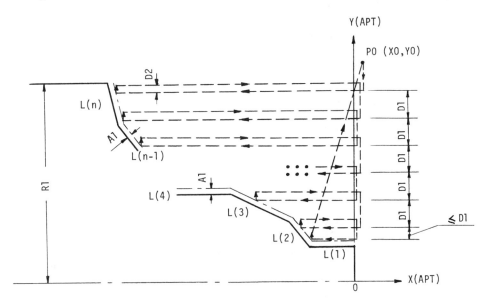

Figure P7-4

7.9   The program listed below uses the TRACUT statement to define the cutter path for machining, on level $z = -0.06$ in., the profile of the gear given in Problem 7.7. The cutter path is first defined as going from point 3, through points 4, 5, and 6, to point 7. Then the cutter path is rotated 11 times, each with an incremental rotational angle of 30 degrees. Does the following program define a correct cutter path for machining the gear profile? If not, explain why not and correct the program.

```
PARTNO PROBLEM 7.9; THE ORIGIN OF THE COORDINATE SYSTEM IS ON THE
$$                       TOP SURFACE
MACHIN/.............
CLPRNT
RESERV/M1,12
C1=CIRCLE/.........
C21=CIRCLE/........
CD=CIRCLE/.........
CA=CIRCLE/.........
C2=CIRCLE/.........
P1=POINT/(2.2*COSF(-15)),(2.2*SINF(-15)),1 $$ A POINT WITHIN THE
$$             AREA BOUNDED BY CIRCLE C1,CD,C21 AND CA
DO/K1,I=1,12
K1) M1(I)=MATRIX/XYROT,(30*(I-1))
REFSYS/M1(2)
      C3=CIRCLE/..(the same parameters as those used to define C1)...
REFSYS/NOMORE
CUTTER/(1/8)
SPINDL/1000,CLW
COOLNT/ON
FROM/0,0,1.0,2.0
RAPID;  GOTO/P1
GODLTA/-1.06
GO/C1,(PLANE/0,0,1,-0.06),PAST,CA
DO/K2,J=1,12
      TRACUT/M1(J)
            ARCSLP/ON;  TLRGT,GOLFT/CA,PAST,C2
                        GOLFT/C2,CD
                        GORGT/CD,C3
                        GORGT/C3,PAST,CA
      TRACUT/NOMORE
K2) CONTIN
RAPID;  GODLTA/1.06
RAPID;  GOTO/0,0,1.0
SPINDL/OFF
COOLNT/OFF
END
FINI
```

7.10   The part shown in Fig. P7-10 is made of cast iron. Its exterior profile has been formed by casting. The interior profile can be divided into three cavities (pockets), one of which has an inclined bottom surface. Suppose that the top surface has been machined to the required surface finish and dimension. Write an APT program for machining the interior profile. The following conditions are given:

    a. Cutter: flat-end mill of $\frac{3}{4}$ in. diameter

    b. Spindle speed: 500 rpm

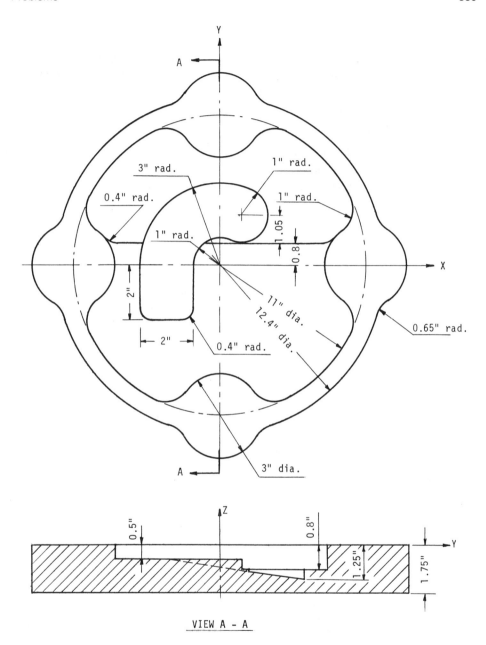

VIEW A - A

Figure P7-10

    c. Feedrate: 3.0 in./min (roughing)

                      2.0 in./min (finishing)

                      1.0 in./min (plunging)

    d. No coolant needed

    e. Finishing allowance: 0.02 in.

f. Starting and end points: (10,5,3)

g. Allowable scallop height: 0.005 in.

Note: For machining a pocket with a bottom plane other than a horizontal one by means of a flat-end mill, the step feed, F, can be determined by using the expressions

$$F = (2H/\sin A) \cdot [R \cdot (2 \sin A/H - 1)]^{\frac{1}{2}} \qquad (0 < A < 90°) \Big\}$$
$$F \leqslant R$$

where

    H = allowable scallop height

    R = cutter radius

    A = angle between bottom plane and horizontal plane

chapter 8

The Construction and Processing of an APT Program

In Chapters 3 through 7, we introduced and described in detail the statements needed to construct a part machining program. Now the question arises as to how to put the necessary statements together to form a correct APT part-machining program and how to process and debug it. Thus, in this chapter, we discuss the composition of an APT program and what, in general, the programmer should know about writing a program. The discussion starts with the difference between programming turning and milling operations, followed by the relationship between a postprocessor and APT statements. Then the general structure of an APT program is discussed. Four examples are given to explain how an APT program is constructed.

An APT program should be processed by the APT-AC NC processor and the postprocessor designed specifically for the NC machine to be used. It is inevitable that a programmer will make mistakes in syntax, programming logic, or both in the program. Usually, he must run (use the NC processor and the postprocessor to process) the program several times to correct the errors according to the error messages issued by the NC processor and the postprocessor. This procedure, known as debugging, might take longer than writing the program for a beginner. Even an experienced APT programmer may find it necessary to run an APT program several times. Therefore, running and debugging a program are necessary procedures for producing a correct and useful program.

In the second part of this chapter, we first introduce the processing of a program by the APT-AC NC processor and describe in detail the procedures and necessary job execution statements for running an APT program. Then we discuss the error messages issued by the NC processor and the postprocessor. Finally, the method of debugging an APT program is described. The discussions in Section 8.6 are based on the IBM VM/CMS (Virtual Machine/Conversational Monitor System) operating system. The procedures for running an APT program under the IBM OS/MVS (Operating System/Multiple Virtual System) environment are similar, and the necessary Job Control Language (JCL) statements are listed in Appendix D.

8.1 THE COORDINATE SYSTEM FOR PROGRAMMING TURNING OPERATIONS

As can be seen in the previous chapters, our discussions are primarily related to milling operation. The APT language provides many statements that handle two-dimensional cases in the X-Y plane. The turning operation on a lathe is also two dimensional; however, it is carried out in the X-Z plane. As a matter of fact, a turning tool is treated, in APT, as a flat-end mill with its radius equal to the nose radius of the turning tool and its axis in the Z direction. Therefore *the cutter path of a turning process should be specified in the* X-Y *plane in an APT program,* instead of in the X-Z plane (Fig. 6-11[a]). A postprocessor for the lathe is usually designed in such a way that the x and y coordinates in the APT coordinate system are, respectively, transformed into z and x coordinates in the NC output or program. For the lathe shown in Fig. 6-11(a), there is still one problem: the cutter is on the negative side of the Y axis of the APT coordinate system, and the NC output in the X direction (of the NC machine coordinate system) is then negative, which is incorrect.

There are two ways to solve this problem. The first is to design the postprocessor in such a way (1) that it will transform a positive y coordinate in the APT program into a negative x coordinate in the NC program and (2) that a cutter path will be defined at its actual position (i.e., on the Y-negative side of the APT coordinate system). However, this is not convenient because all the y coordinates in the APT program become negative. The other way is to specify a cutter path on the positive side of the Y axis of the APT coordinate system (side A' in Fig. 8-1), since, for a turning process, there is no difference in the part profiles resulting from placing a cutting tool on either side of the spindle axis. However, the direction of spindle rotation and the types of tools used should be determined on the basis of the actual tool position (on side A of the spindle axis for the case shown in Fig. 8-1). Many of the lathes or turning centers are designed in such a way that the tools are placed on the upper side, A', of the spindle axis (Fig. 8-1). In this case, it is not necessary to change the tool position in the APT program.

8.2 THE EFFECT OF A POSTPROCESSOR ON APT PROGRAMMING

As indicated in Chapter 1, an APT program is processed by both the APT-AC NC processor and a selected postprocessor. Therefore the formats and usages of statements are determined by the design of these two processors. The majority of APT

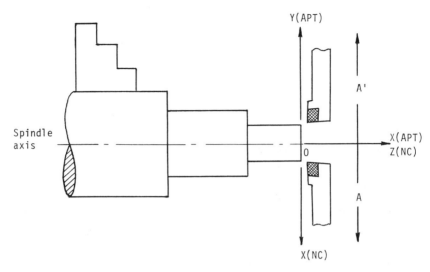

Figure 8-1   The difference in APT and NC coordinate systems. A turning tool is actually on the lower side, A, of the spindle axis. During APT programming, it can be considered as being on the upper side, A′.

statements, including those regarding geometric definitions, tool motions, machining specifications, and computer operations, are determined by the APT-AC NC processor itself. However, there are many statements whose formats are determined by both the NC processor and the selected postprocessor. For example, if there are only two coolant states (ON and OFF) available for a milling machine, the postprocessor might be designed in such a way that it does not accept the minor words other than ON and OFF in the COOLNT statement, thus eliminating the words FLOOD, MIST, and TAPKUL (see Section 6.7). Similarly, the parameter s, regarding spindle speed (see Section 6.5), can be specified in revolutions per minute, as a percentage of maximum attainable speed, or in a speed code, depending on the machine or postprocessor design.

In general, the statements that are affected by the postprocessor are postprocessor words and statements regarding machining specifications. Words that can be accepted by the APT-AC NC processor might not be accepted or processed by the postprocessor. Besides, a word might have different meanings when processed by different postprocessors (typical examples are the FEDRAT and RAPID statements described in Section 6.4 of Chapter 6). Therefore an APT programmer must have a good knowledge of the applicable postprocessor to write a correct APT program; otherwise, he might have difficulty in obtaining a correct NC output.

8.3 THE GENERAL STRUCTURE OF AN APT PROGRAM

An APT program should start with the PARTNO statement and end with the FINI statement. An input file to the NC processor might consist of more than one APT program. These two statements denote the start and end of an APT program to the NC processor.

If an APT program is to be processed by a postprocessor, then the second statement should be

MACHIN/......

which was explained in Section 6.12.2 (Chapter 6).

Starting from the second statement, statements regarding computer operation (e.g., the RESERV statement if subscripted variables are used, or CLPRNT if the CLDATA file is to be printed) and machining specifications (e.g., cutter, coolant, tolerance, feedrate) can be specified, if needed.

The next section of the program defines the geometric entities needed to specify the cutter path. Usually, many geometric entities will be involved; therefore it is suggested that the statements be written in as neat a format as possible for easy review and checking. As a rule, one character is chosen for one type of geometric entity, together with different alphanumeric characters to represent different geometric entities of the same type. For example, we might use the character L as the symbol for line, and L1, L2, ..., L1A, L1B, ... as the first, second, ... lines, respectively. Subscripted variables are often used in programming with loop(s) and subprogram(s), representing geometric entities of the same type and of similar form. The statement

PRINT/3,...

can be specified to check the defined geometric entities.

The cutter path can then be specified after the section defining the geometric entities. If a subprogram is used in defining a cutter path, it should be specified before the statement calling it. A CLPRNT statement can be specified in this section (or in the previous section) to check the cutter paths. Since this statement will usually generate a large CLDATA file, we suggest that the CLPRNT function be turned off by using the statement

CLPRNT/OFF

if the cutter path after it is not to be reviewed. The CLPRNT function can be resumed by specifying the statement

CLPRNT/ON

Usually a machining operation starts with one or several rough cuts and is followed by a finishing cut. Loops and subprograms can be used, if possible, to define similar cutter paths to reduce the length of the program. Machining specifications, such as the feedrate, spindle speed, and tool, can also be changed in this section according to the machining requirement.

The program comes to the end section after having defined the desired cutter path. The tool should move to a safe position, and the coolant and spindle must be stopped.

The END statement should be specified to denote the end of the machining operation. The last statement of an APT program should be FINI, indicating that it is the last record of the current APT program.

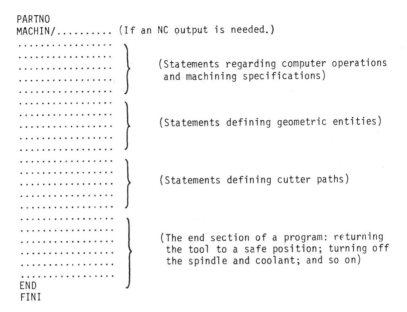

```
PARTNO
MACHIN/......... (If an NC output is needed.)
................
................
................    ⎫
................    ⎬   (Statements regarding computer operations
................    ⎭    and machining specifications)
................
................
................
................    ⎫
................    ⎬   (Statements defining geometric entities)
................
................    ⎭
................
................
................    ⎫
................    ⎬   (Statements defining cutter paths)
................
................    ⎭
................
................
................    ⎫
................    ⎬   (The end section of a program: returning
................    ⎪    the tool to a safe position; turning off
................    ⎭    the spindle and coolant; and so on)
................
END                ⎭
FINI
```

Figure 8-2   The general structure of an APT program.

The general structure of an APT program, as described above, is shown diagrammatically in Fig. 8-2. However, there is always some freedom in specifying many of the statements in different sections of a program, provided the programming logic is correct.

8.4 EXAMPLES OF APT PROGRAMS

The following examples are given to explain the details of APT programming. These programs were processed by the APT-AC NC processor and satisfactory results were obtained. The first example is an APT program that defines the machining operation for the Wankel engine chamber given in Fig. 4-25. This example shows how to use the LCONIC statement and a loop to define the cutter path for a special profile. The second example is a part shaped like a simplified normal distribution. The machining operation of this part is a typical three-dimensional case. The third example shows that a cutter path in the Y-Z plane can be defined through the transformation of an identical one defined in the X-Y plane. Finally, the programming of a turning process is illustrated in example 4.

The program for each example represents one of the many potential approaches defining a machining process. Readers are encouraged to solve these problems in their own way.

Example 1
The profile of the Wankel engine chamber shown in Fig. 4-25 is divided into a number of sections that can be defined as LCONICs or tabulated cylinders. To explain the possible problems related to, and the techniques used in, programming with the LCONIC statement, we decided to use it to define the various sections of the profile. Since the

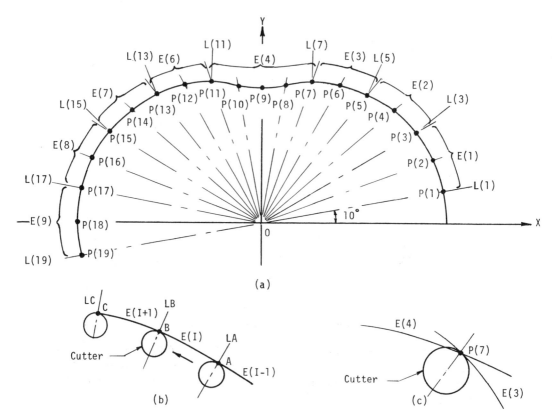

(a)

(b)   (c)

Figure 8-3 The machining of the Wankel engine chamber shown in Fig. 4-25. (a) The chamber half-profile is divided into eight sections, all of them being defined as LCONICs by LCONIC/3PT2SL statements with the exception of E(4), which is defined by the LCONIC/ 5PT statement. Lines L(1) through L(19) are the normals to the profile at points P(1) through P(19), respectively. (b) A normal is defined at the connecting point of two LCONICs. It serves as a check surface in the contouring motion statement. (c) The cutter might be out of tolerance at the start of the contouring motion step, when the tool is moved from point P(7) to point P(11), because of the discontinuity of the slope at point P(7). E(3) is defined by statement LCONIC/3PT2SL, and E(4), by statement LCONIC/5PT.

profile is symmetrical with respect to the X and Y axes, only one half of the cutter path needs be defined; the other half can be duplicated by using a COPY statement. Nineteen points, corresponding to polar angles 10, 20, . . . , 190 degrees on the profile (Fig. 8-3[a]), were chosen (if a higher accuracy is required, more points are needed). Their coordinates are calculated. The slopes of the profile at points $P(i)(i = 2n + 1$, $n = 0, 1, 2, \ldots, 9)$ are also calculated, with the exception of P(9). The half-profile is divided into eight sections: E(1), E(2), E(3), E(4), E(6), E(7), E(8), E(9). All the sections are defined by the statement

LCONIC/3PT2SL,......

with the exception of E(4), which is defined by the statement

LCONIC/5PT,P(7),P(8),P(9),P(10),P(11)

We used the 5PT option, instead of 3PT2SL, to define this section of the profile because of its particular form (i.e., the existence of an inflection point), which causes problems in defining the cutter path from point P(7) to point P(9) if the 3PT2SL option is selected.* Therefore, any two sections of the profile, except for sections E(4) and E(3) and sections E(4) and E(6), are theoretically tangent to each other at the connecting points because the two end slopes are used to define the curves. However, an error message will always be generated for the statement

TL...,GO.../E(I),TANTO,E(I+1)

which moves the tool along one LCONIC, E(I), to the next one, E(I+1), even though the two LCONICs are defined as tangent to each other at the connecting point. This error results from the presence of different radii of curvature of the two LCONICs at the connecting point. The solution is to define a line that is normal to both LCONICs (Fig. 8-3[b]) and to use it as a check surface to position the tool. Thus the cutter motion A-B-C can be defined as

TLLFT,GOFWD/E(I),ON,LB
        GOFWD/E(I+1),ON,LC

The APT program for machining the profile shown in Fig. 4-25 and also in Fig. 8-3(a) may appear as follows:

```
PARTNO EXAMPLE 1
MACHIN/GN5CC,9,OPTION,2,0 $$ GN5CC IS THE POSTPROCESSOR NAME
UNITS/MM
CLPRNT
CUTTER/0.0001
FEDRAT/2.0
CLPRNT
RESERV/X,19,Y,19,S,19,SS,19,E,9,L,19,P,19
DO/A2,I=1,19 $$ CALCULATION AND DEFINITION OF POINTS P(I) AND LINES L(I)
    A=10*I
    B=3*A
    X(I)=60*COSF(A)+10*COSF(B)
    Y(I)=60*SINF(A)+10*SINF(B)
    P(I)=POINT/X(I),Y(I)
    IF((I/2-INTGF(I/2))'EQ'0),JUMPTO/A2
    S(I)=-(COSF(A)+(COSF(B))/2)/(SINF(A)+(SINF(B))/2) $$ THE SLOPE
                                                       $$ OF THE PROFILE
    IF(I'EQ'9),JUMPTO/A1
    SS(I)=-1/S(I)
    L(I)=LINE/P(I),SLOPE,SS(I)
    JUMPTO/A2
A1)  L(I)=LINE/YAXIS
A2)  CONTIN
DO/B1,J=1,9 $$ DEFINITION OF PROFILE SECTIONS E(I)
    JJ=2*J-1
```

*If the 3PT2SL option is selected, the generated equation that defines the section of the profile, P(7)-P(8)-P(9), is

$$0.014585131X^2 - 0.382364792XY + 2.07500844Y^2 +$$
$$19.1182396X - 207.317561Y + 5178.35698 = 0$$

which is a hyperbolic function with its center point located between points P(8) and P(9). Therefore it cannot correctly define the profile between these two points.

```
         IF(J'EQ'4),JUMPTO/B1
         IF(J'EQ'5),JUMPTO/B1
         E(J)=LCONIC/3PT2SL,P(JJ),S(JJ),P(JJ+2),S(JJ+2),P(JJ+1)
B1)    CONTIN
E(4)=LCONIC/5PT,P(7,THRU,11)
SPINDL/700,CLW
FROM/0,0,0
INDIRV/1,0.1,0
GO/ON,L(1),(PLANE/0,0,1,0),E(1)
INDIRV/-0.1,1,0
INDEX/7
   DO/C1,K=1,3
        KK=2*K+1
C1)     TLLFT,GOFWD/E(K),ON,L(KK)
             GOFWD/E(4),ON,L(11)
   DO/D1,K=6,9
        KK=2*K+1
D1)     TLLFT,GOFWD/E(K),ON,L(KK)
INDEX/7,NOMORE
COPY/7,XYROT,180,1
RAPID
GOTO/0,0,0
SPINDL/OFF
END
FINI
```

A hypothetical cutter of very small diameter, 0.0001 mm, is used in this program to simplify verification of the cutter path. A cutter of larger diameter, say, 12.7 mm, should be used for a real machining process. In this case, a problem might appear when the tool moves from point P(7) to point P(11). The cutter might be out of tolerance because the slopes of the two LCONICs, E(3) and E(4), are not the same at point P(7) (Fig. 8-3[c]). Thus a start-up motion statement

GO/L(7),PL1,E(4)

should be inserted between the motion statements

GOFWD/E(3),ON,L(7)

and

GOFWD/E(4),ON,L(11)

to position the tool correctly. Care must be taken to check also the motion direction of the start-up motion statement. If it differs from that of the next statement, then the motion direction should be correctly specified for the next motion. As a matter of fact, we can define the profile as a single tabulated cylinder so that profile and slope continuity are easily maintained.

Example 2

Figure 8-4 shows a part with the profile of a normal distribution in the vertical cross section passing through the Z axis. To simplify the problem, we assume the horizontal cross section at any level to be similar to the base cross section shown in the figure, which is a quasi-ellipse composed of four circular arcs. Moreover, a flat-end tool with

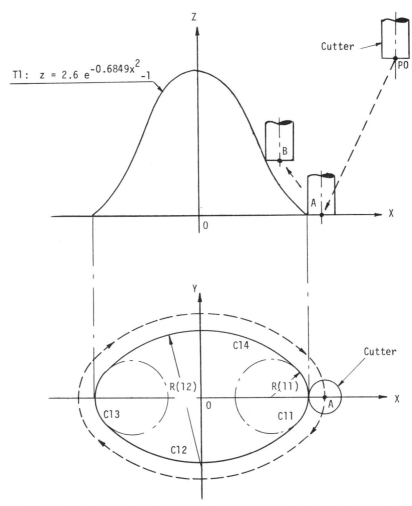

Figure 8-4 A three-dimensional part. Its cross section in the X-Z plane has the shape of a normal distribution, which is defined as a tabulated cylinder, T1. Its cross section in any horizontal plane is similar to the quasi ellipse, which is composed of four circular arcs. R(11) = 0.3937; R(12) = 1.3780.

a diameter of 0.5 in. is used. The starting position of the tool is P0(5,0,2). The cutter path is defined as follows:

1. The tool is positioned at position A, ready to start the contouring motion. The drive, part, and check surfaces for this start-up motion are, respectively, the X-Z plane, the tabulated cylinder, T1, and the circle, C11, which is part of the horizontal cross section.

2. The tool is instructed to move in the horizontal plane, using C11, C12, C13, and C14 as the drive and check surfaces consecutively and going back to position A at the end of this motion.

3. The tool is instructed to move up to position B. The drive, part, and check surfaces in this motion are, respectively, the X-Z plane, T1, and C21 (not shown in

the figure), which is one of the constituent circles of the next horizontal cross section.

This process is repeated until the tool reaches the top. The program for this machining process is as follows:

```
PARTNO EXAMPLE 2
MACHIN/GN5CC,9,OPTION,2,0
CLPRNT
NOPOST
RESERV/X,16,Z,16,R,162,C,164
CUTTER/0.5
FEDRAT/2.0
$$  UNITS/INCHES
DO/A1,I=1,16
      Z(I)=(I-1)*0.1
A1)   X(I)=SQRTF(-LOGF((Z(I)+1)/2.6)/0.6849)
T1=TABCYL/NOY,SPLINE,Z(1),X(1),Z(2),X(2),Z(3),X(3),Z(4),X(4),Z(5),X(5),$
         Z(6),X(6),Z(7),X(7),Z(8),X(8),Z(9),X(9),Z(10),X(10),Z(11),$
         X(11),Z(12),X(12),Z(13),X(13),Z(14),X(14),Z(15),X(15),Z(16),$
         X(16),1.6,0
PL1=PLANE/0,1,0,0    $$ PLANE Y=0
PL2=PLANE/1,0,0,0    $$ PLANE X=0
DO/A2,J=1,16
        J1=10*J+1
        J2=10*J+2
        J3=10*J+3
        J4=10*J+4
        R(J1)=(0.3937/X(1))*X(J)
        R(J2)=(1.3780/x(1))*X(J)
        C(J1)=CIRCLE/(X(J)-R(J1)),0,R(J1)
        C(J3)=CIRCLE/(R(J1)-X(J)),0,R(J1)
        C(J2)=CIRCLE/YLARGE,IN,C(J1),IN,C(J3),RADIUS,R(J2)
A2)     C(J4)=CIRCLE/YSMALL,IN,C(J1),IN,C(J3),RADIUS,R(J2)
SPINDL/600,CLW
FROM/5,0,2
INDIRV/-1,0,-0.3
GO/ON,PL1,T1,TO,C(11)
DO/C1,K=1,16
        K1=10*K+1
        K2=10*K+2
        K3=10*K+3
        K4=10*K+4
        KK1=10*(K+1)+1
        INDIRV/0,-1,0
        AUTOPS
        ARCSLP/ON;   TLLFT,GOFWD/C(K1),TANTO,C(K2)
                     GOFWD/C(K2),TANTO,C(K3)
                     GOFWD/C(K3),TANTO,C(K4)
                     GOFWD/C(K4),TANTO,C(K1)
                     GOFWD/C(K1),ON,PL1
        GO/ON,PL1,T1,C(K1)
        IF(K'EQ'16),JUMPTO/C1
        TLON,GOUP/PL1,C(KK1)
C1)     CONTIN
TLON,GOUP/PL1,ON,PL2
```

```
RAPID
GODLTA/1.0
SPINDL/OFF
END
FINI
```

Example 3

The part shown in Fig. 8-5(a) was designed to illustrate the definition of a cutter path through a coordinate transformation approach. The stock material is a rectangular block with the following dimensions: $7 \times 3.5 \times 2.5$ in. A uniformly reduced undercut with cross sections of circular form in the planes parallel to the X-Z plane is to be cut on a milling machine with a ball-end mill whose diameter is 0.5 in. The centers of the circular cross sections parallel to the X-Z plane are on line L1, with the included angles of the circular arcs equal to 90 degrees. The starting point of the tool is selected at $P0$ (0,0,1.0), and the required cutter path is A_1-A_2-A_3-A_4-...... (the paths A_1-A_2 and A_3-A_4 are circular in the plane parallel to the X-Z plane). Since the circular cross sections are in the X-Z plane, it would be fairly cumbersome to use the CYLNDR or TABCYL statements to define these cross sections. Therefore we defined the geometric entities and cutter path in the X-Y plane and then transformed them to their actual positions (Fig. 8-5[b]). The program is as follows:

```
PARTNO EXAMPLE 3
MACHIN/GN5CC,9,OPTION,2,0
NOPOST
RESERV/R,11,C,11,L,111
CLPRNT
$$ UNITS/INCHES
CUTTER/0.25   $$ THE BALL-END MILL IS IMITATED BY A FLAT-END MILL.
FEDRAT/2.0
L1=LINE/(0,0,0),(1.0,1.0,0)
DO/A1,I=1,11
      I1=I*10+1
      R(I)=(1.5+(I-1)*0.1)*SQRTF(2)
      X=R(I)/SQRTF(2)
      C(I)=CIRCLE/CENTER,X,X,0,RADIUS,R(I)
A1)   L(I1)=LINE/(POINT/CENTER,C(I)),(POINT/(2*X),0,0) $$ SEE FIG. 8-5B
PL3=PLANE/0,0,1,0
M1=MATRIX/TRANSL,0,0,-.125,TIMES,YZROT,90
SPINDL/900,CLW
FROM/0,0,1
DNTCUT
GODLTA/1,1,0
INDIRV/-1,-1,-1
TRACUT/M1
GO/ON,L1,PL3,C(1)
CUT
INDIRV/1,-1,0
ARCSLP/ON
TLLFT,GOFWD/C(1),ON,L(11)
TRACUT/LAST,(MATRIX/TRANSL,0,0.35,0)
INDIRV/1,0,0
GO/ON,L(21),PL3,C(2)
TLRGT,GORGT/C(2),ON,L1
TRACUT/LAST,NOMORE
TRACUT/NOMORE
```

```
RAPID
GOTO/0,0,1
SPINDL/OFF
END
FINI
```

Note that only part of the cutter path, $P0$-$A1$-$A2$-$A3$-$A4$-$P0$, is defined in this program. The reader is encouraged to write a complete program to cut the desired profile.

Example 4

The shaft shown in Fig. 8-6 is machined on a lathe that has only two axes, X and Z. The cutter path is defined on the X-Y plane and is transformed by the postprocessor into an NC program that consists of X and Z codes only. The program for defining the finishing operation may appear as follows:

```
PARTNO EXAMPLE 4
MACHIN/LATH1.........$$ THE LATHE POSTPROCESSOR
CUTTER/0.02  $$ THE CUTTER NOSE RADIUS IS 0.01.
$$ UNITS/INCHES
FEDRAT/IPR,0.005
COOLNT/ON
SPINDL/750,CCLW $$ THE SPINDLE ROTATION DIRECTION DEPENDS ON THE
$$ STRUCTURE OF THE MACHINE TOOL.
LOADTL/1,SETOOL,0,0,0
.............
.............      (geometric definition statement defining L1 through
.............       L11, C1 and starting point P0)
.............
$$ THE FOLLOWING STATEMENTS DEFINES THE FINISHING CUT.
FROM/P0
RAPID
GO/L3,(PLANE/0,0,1,0),L1
TLLFT,GOLFT/L1,TO,L2
TLRGT,GORGT/L2,PAST,L3
      GOLFT/L3,L4
      GORGT/L4,PAST,L5
      GOLFT/L5,L6
      GORGT/L6,PAST,L7
      GOLFT/L7,PAST,L8
      GOLFT/L8,TANTO,C1
      GOFWD/C1,TANTO,L9
      GOFWD/L9,PAST,L10
      GOLFT/L10,PAST,L11
RAPID
GOTO/P0
LOADTL/2,SETOOL,0.1,0.2,0 $$ THE PARAMETERS 0.1, 0.2, AND 0 ARE THE
$$ OFFSETS OF TOOL 2 FROM TOOL 1.
CUTTER/0.01 $$ THE CUTTER NOSE RADIUS IS 0.005.
RAPID
```

```
GO/ON,L5,(PLANE/0,0,1,0),L4
FEDRAT/0.002
TLLFT,GOLFT/L4,L12
RAPID
GODLTA/0,2.0,0
RAPID
GOTO/P0
SPINDL/OFF
COOLNT/OFF
END
FINI
```

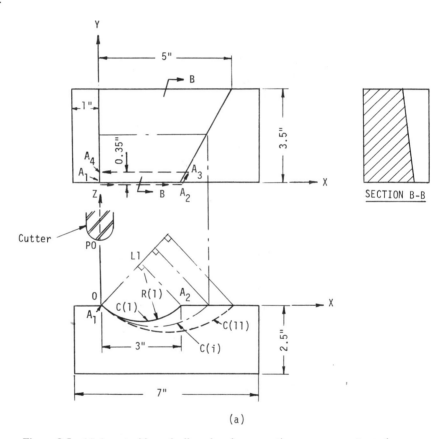

Figure 8-5 (a) A part with gradually reduced cross section. – – – – → cutter path.
(b) (I) The cutter path, C_1-C_2-C_3-C_4, and the geometric entities are first defined in
plane X-Y. (II) A rotation about the X axis by 90 degrees transforms the defined cutter path, C_1-C_2-C_3-C_4, into path B_1-B_2-B_3-B_4. A translation of 0.125 in. is then
needed in the negative Z direction to convert that cutter path into the desired one,
namely, A_1-A_2-A_3-A_4.   (continued)

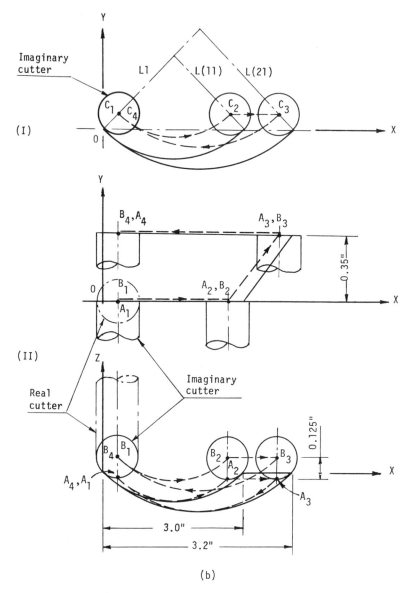

(b)

Figure 8-5   (continued)

8.5 INTRODUCTION TO THE IBM APT-AC NUMERICAL CONTROL PROCESSOR

The IBM APT-AC NC Processor is a program that accepts a user-written APT program as input, processes it, and transforms the processed result into a standard format, or common interface code, known as CLDATA. The NC processor also provides support for creating machine-oriented postprocessors. The CLDATA is fur-

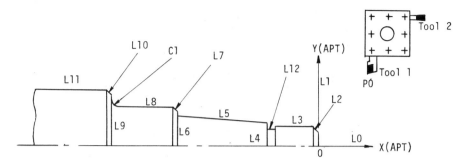

Figure 8-6   A shaft machined on a lathe.

ther processed by the postprocessor as requested by the programmer through the MACHIN statement. The final result is a machine-readable program in NC codes.

The general structure of the APT-AC NC processor is shown in Fig. 8-7; it consists of the following major sections:

1. Control section
2. Translation section
3. Calculation section
4. Edit section
5. Postprocessor section

An APT program is processed successively by the translation, calculation, edit, and postprocessor sections under the control of the control section. A task is assigned to a processing section by the control section. After it has been completed, control returns to the control section, which then assigns the processing task to the next section.

The basic function of the *translation section* is to convert the part geometric entities specified in APT into canonical (standard) forms, which together with other types of APT statements will then be stored as an intermediate file called PROFIL. Note that the geometric entities are not transformed at this stage, even though a transformation described by a REFSYS statement is specified. The matrices defined in REFSYS statements are also stored and passed to the next processing (calculation) stage.

The *calculation section* uses the file PROFIL as input and calculates the successive positions of the cutter-end center. These positions, together with the postprocessor words or commands, constitute a new file called CLFILE. Note that in this section, the transformations specified by the REFSYS statements are performed; however, those defined by TRACUT and COPY statements are postponed until the edit section is invoked. The *edit section* further transforms the cutter-end-center positions, if such transformations are specified, and the result is an updated CLFILE. The printed form of the CLFILE is called CLDATA and consists of data regarding the tool positions required to machine the part and the postprocessor commands specified in the APT program.

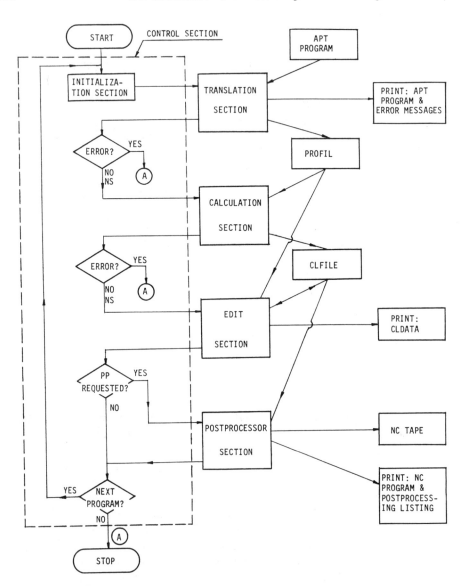

Figure 8-7   The general structure of the IBM APT-AC NC Processor. NOTE: NS = no serious error; PP = postprocessing.

The *postprocessor section* consists of one or several user-created postprocessors. It selects the required postprocessor, according to the MACHIN statement in the APT program, to process the CLDATA. The output of the postprocessor is the required machine-readable NC program, which can be stored in the memory unit of the computer system and output in printed form (the listing file) or as NC tape, depending on the specified output device.

During each processing stage, the corresponding section of the APT processor maintains communication with the control section; control is sent back to the control section prematurely if a serious error has been detected in the program. Then the processing is terminated and the error message(s) is (are) issued according to the types and nature of the errors. If the errors are not serious enough to terminate the processing, then only error messages are issued and the processing continues.

In addition to the functions and capabilities described in Chapters 3 through 7, the IBM APT-AC NC Processor also provides users with the following capabilities:

1. To edit part program files, geometric definition files, and machinability files and to store them in the APT library, which serves as an NC data base and can be used as input or can be called by an APT program. The content of a library can be added to, deleted, or replaced

2. To define a series of logical instructions as a procedure (PROC), to store it in the library, and then to execute it repetitively

3. To define the cutter path for a machining process requiring changes in tool orientation (multiaxis programming)

4. To print, for example, the definitions, part program listings, and symbol cross-reference lists through expanded PRINT statements

In 1985 the IBM Corporation introduced a new release of the APT-AC NC processor (version 1, release 4, modification level 0, program No. 5740-M53), which includes the following additional features[7]:

1. New ways to define the geometric entities accepted by the former release

2. The capability to combine points, line segments, circular arcs, and circles into a single composite curve (or contour) for machining purposes

3. The definition of new geometric entities, such as two-dimensional spline, rational B-spline, three-dimensional bounded line, or circular arc in any special direction

4. The capability to define both the stock and part profiles and then generate the roughing and finishing cutter paths for turning the part

The APT-AC NC processor programs are written in both FORTRAN and Assembler languages. They can be executed on any IBM System/370, 43xx or 30xx computer system under the VM/CMS or OS/MVS operating environment with a minimum virtual storage capacity of 2.5 megabytes.[8] The computer system should have auxiliary storage allocated for the NC processor and its execution requirements. Normally the required capacity of the auxiliary storage is the sum of the spaces allocated to the NC processor itself (about forty-six model 3350 cylinders or the equivalent), the work and library file (two to five model 3350 cylinders or the equivalent), and the input (APT program) and output (NC program and listing output) files, which require about one to three model 3350 cylinders. The total needed capacity of the auxiliary storage is about 49 to 54 IBM model 3350 cylinders.

8.6 THE PROCEDURES FOR RUNNING AN APT PROGRAM

A written APT program should be input into the computer system for processing by the NC processor. The format of the input file should be as follows:

1. Each line (or card) contains up to 80 characters.
2. An APT statement can be typed or punched only within columns 1 through 72.
3. The content of columns 73 through 80 is ignored by the NC processor. These columns can be used to specify alphanumeric characters for identification and sequencing purposes.

A line (or card) may contain more than one APT statement, with a semicolon or semicolons dividing them.

An APT program should be given a name and saved on a direct-access memory unit so that it can be retrieved, edited, and processed.

The detailed procedure for running an APT program on the IBM VM/CMS operating system is as follows[8,9]:

1. Link the disks where the IBM APT-AC NC Processor program, FORTRAN-H Extended program, and their associated libraries reside. Then define the addresses for the linked disks so that they can be accessed.
2. Create the NC processor diagnostic table, which is required for the operation of the NC processor, and save it as a file, say,

LP LP A4

   designations that are, respectively, the symbols for the name, type, and mode of the diagnostic-table file.
3. Store the needed postprocessor programs (or modules) on the A disk of the user's account. If the postprocessor has already been stored in the system, it is better to link the disk storing the postprocessors to the user's account according to procedure 1, above, instead of storing them on the user's A disk.
4. Type in the APT program to be processed and save it. The file name, type, and mode of the saved file might be, for example,

TEST PART A4

   respectively.
5. Type in the following job execution program, which is used to process the APT program:

```
&CONTROL ALL
FI FT05F001 DISK &1 &2 A4 (RECFM F BLOCK 80)
FI ERRFILE DUMMY
FI FT06F001 DISK LT1 LT1 A1
FI FT01F001 DISK F1 F1 (RECFM F BLKSIZE 3228)
FI FT02F001 DISK F2 F2 (RECFM F BLKSIZE 3228)
FI DPOOL DISK SP SP (XTENT 50 RECFM F BLKSIZE 4096)
FI LIBDPOOL DISK LP LP A4 (RECFM F BLKSIZE 4096 DISP MOD)
FI SYSPUNCH DUMMY
FI POSPUNCH DISK LT3 LT3 A1
FI SOURCE DISK PARTPDS PARTPDS A4 (RECFM F BLOCK 80 DISP MOD)
FI 7 TERM
LOADMOD DKWOR6
START * LINELNTH 133 LINECNT 76 POOLSIZE 10 5
```

This job execution program is derived from the ACGOR6 file in the IBM APT-AC NC processor,[8] with the following modifications:

a. The processing listing file is saved as file LT1 LT1 on the A disk.

b. The postprocessor output file POSPUNCH (i.e., the NC program) is saved as file LT3 LT3, also on the A disk.

c. The punched output SYSPUNCH is neglected.

The file FT05F001 is the input file to the NC processor (i.e., the APT program). The two tokens "&1" and "&2" represent the file name and file mode of the APT program. These modifications are made to facilitate debugging and correcting the program. If the punched output is needed after a correct APT program is obtained, the ACGOR6 file in the NC processor can be used instead.

The first two parameters in the last statement, START, can be changed to adjust the format of the output file. The parameter for the LINELNTH option determines the line length of the printed output. It can be either 80 or 133 (print-characters per line). Since some of the outputs cannot be printed in an 80-characters-per-line mode, the parameter 133 is normally used. The parameter for the LINECNT option specifies the maximum number of lines to be printed on each page of the output. It can be any integer between 2 and 132, inclusively. The default value is 55.

The reader should consult the appropriate IBM manuals[7,8] for meanings of the other statements listed above.

These job execution statements can be saved as a job execution program file. Its file name can be abitrary provided it is not the same as the names of other job execution files or as those file names used in the APT system; its file type must be EXEC (indicating that it is a job execution program file); its file mode can be named A, followed by any numerical character other than 0. Thus we may save the above program under, for example, the following file name, type, and mode:

PROAPT EXEC A1

6. Type in

PROAPT TEST PART

to process the APT program. The two tokens in the job execution program are automatically replaced by TEST and PART. Thus the APT program is processed. One or both of files LT1 LT1 A1 and LT3 LT3 A1 will be generated, depending on the extent of the errors. If there is no error or if there are only some minor ones, both files will be generated with or without error messages printed in listing file LT1 LT1 A1. If the error(s) is (are) serious, the NC output LT3 LT3 A1 cannot be generated. Usually a number of error messages are generated in the listing file during the first processing.

7. Correct the program according to the error messages and process it again according to step 6, above. Repeat this procedure until all the errors in the APT program have been corrected.

8.7 DEBUGGING AN APT PROGRAM

Before it has been tested, an APT program usually contains various types of errors that might not have been detected during the reviewing process. However, they can be detected and pointed out by the NC processor and the postprocessor during processing. Error messages issued by the NC processor are in the following form:

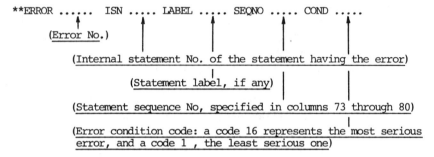

followed by comments regarding the nature of the error, printed in the next line. For example, an error message may appear as follows:

**ERROR 1102 ISN 25 LABEL   SEQNO   COND 04

**** INVALID "CIRCLE' DEF. . . REFERENCED POINTS ARE COLINEAR ***

This means that the error numbered 1102 is located in the twenty-fifth statement of the program, which has no specified statement label or sequence number. The error condition code is 04. The nature of this error is clearly indicated by the comment (i.e., a circle cannot be created on the basis of three colinear points).

Errors can be issued by different sections of the NC processor and by the postprocessor. The section that generates a specific error message can be distinguished through the issued error number (Table 8-1).

The comments explain fairly clearly the type and nature of most of the errors recognized in sections other than the calculation section. Since the location or the statement containing the error is identified by the internal sequence number (ISN) (and also by the statement label and sequence number, if any), these errors can be located and corrected easily. However, an error in a cutter motion statement is not as easy to identify and correct through the issued error messages, especially for a beginner. The reason is that the error might not be in the statement designated by the error message but might be caused by an error in a previous statement. For example, as mentioned in Section 7.4 (Chapter 7), error message No. 2209 will be issued for the statement labeled A1 in the program shown in Fig. 7-4. However, this error is due to the mistake made in the previous statements (i.e., the tool has not been moved to the position required for starting the contouring motion). Thus the correction should be made, not in the statement indicated by the error message, but somewhere in the previous statements (before the CALL statement).

Whenever errors in cutter motion are experienced, the following procedures are suggested to correct them:

1. Check the definitions of the cutter and geometric entities, and make sure that they are defined correctly.

**TABLE 8-1   THE ERROR MESSAGES ISSUED BY DIFFERENT SECTIONS
OF THE APT-AC NC PROCESSOR**

Error No.	Section Issuing Error Messages; Types of Errors
0 to 999	Control section: These errors are related to the job execution program (EXEC program under VM/CMS environment, or JCL statements for OS/MVS operating system) and computer system setup.
1000 to 1999	Translation section: Most of these errors result from incorrect syntax or incorrect parameters provided in a statement.
2000 to 2999	Calculation section: Errors such as cutter motion errors, programming logic errors, and syntax errors in motion statements have been made.
3000 to 3999	Edit section: Errors in transformations of cutter path and input or output errors have been made.
4000 to 4999	Postprocessor dispatcher section: A wrong postprocessor is specified, or the specified postprocessor cannot be found.
6000 to 6999	Postprocessor section: The cutter path, machining specifications, or other elements defined in the APT program do not conform with the specifications of the NC machine or the assumptions made in designing the postprocessor.

2. If the error is experienced during the start-up motion, check to determine whether the three controlling surfaces selected for the start-up motion are specified correctly and can define a correct and meaningful tool position.

3. If the error is in a contouring motion statement, always examine the tool position with respect to the profile, the tool motion direction with respect to the previous motion direction (defined either by the preceding contouring motion statement or by the statement INDIRV or INDIRP), and the starting and end positions of the current motion. Usually, if all the elements specified in a contouring motion statement are correct and the tool has been defined at the correct starting position by the previous motion statement, the error may be due to the fact that the tool motion direction cannot be clearly defined by the motion-direction word or that a surface other than the drive and check surfaces in the preceding motion is used as the drive surface of the current motion. A change in part surface would also cause trouble when a contouring motion is started, if the new part surface is not the drive or check surface in the immediately preceding motion statement and if no start-up motion is specified.

A sound understanding of the definitions of drive, part, and check surfaces and their applications in different formats of start-up and contouring motion statements is the prerequisite for defining a cutter path correctly and for debugging a program. Two examples are given below to illustrate the method for removing errors in cutter motion statements.

Example 1

The error in this example is often found in programs written by beginners. The desired cutter path is shown in Fig. 8-8, and the program might appear as follows:

```
PARTNO TEST ERROR                                  E0001
CLPRNT                                             E0002
UNITS/INCHES                                       E0003
CUTTER/0.5                                         E0004
FEDRAT/2.0                                         E0005
SPINDL/500,CLW                                     E0006
PL1=PLANE/1,0,0,5                                  E0007
PL2=PLANE/0,1,0,0                                  E0008
PL3=PLANE/0,1,0,5                                  E0009
PL4=PLANE/0,0,1,0                                  E0010
C1=CIRCLE/CENTER,5,0,0,RADIUS,2.0                  E0011
A1)   FROM/0,0,0                                   E0012
A2)   GO/PL2,C1                                    E0013
A3)   TLLFT,GOLFT/C1,PL1                           E0014
A4)          GOLFT/PL1,PL3                         E0015
SPINDL/OFF                                         E0016
END                                               E0017
FINI                                              E0018
```

An error message, No. 2200, is issued, after processing by the NC processor for the statement labeled A4. It says

"ÚNABLE TO DETERMINE THE FORWARD DIRECTION OF
CUTTER AT START OF MOTION SEQUENCE. AMBIGUOUS
DIRECTIONAL MODIFIER IS SPECIFIED."

Two points should be noted in this message. First, the error occurs at the beginning of the motion specified by the statement labeled *A4*. Next, it indicates that the NC processor is not able to determine the motion direction at the beginning of the contouring motion and suggests that maybe the directional modifier GOLFT is specified incorrectly. As a matter fact, the statement labeled *A4* is correct with respect to the cutter path O-P1-P2-P3, given in Fig. 8-8. The problem is in the statement labeled *A2*, where the

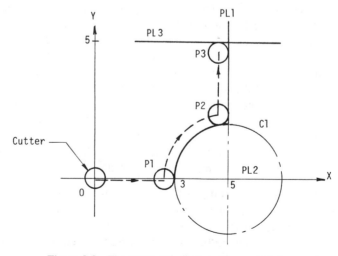

Figure 8-8   The cutter path discussed in example 1.

circle, C1, should be the check surface instead of the part surface. From the CLDATA of this program, we find that the tool moves horizontally ($z = 0$) to circle C1 (or point P1), which is then also used as the drive surface moving the tool to point P2. When processing the statement labeled *A4*, the NC processor found difficulty in determining the tool motion direction because the part surface, C1, was no longer in use. Therefore the correction should be made not in the statement labeled *A4* but in that labeled *A2*. The solution is to replace the statement labeled *A2* by the statement

GO/C1

or

GO/PL2,PL4,C1

Example 2

The profile to be cut is shown in Fig. 8-9(a). For simplification of the error analysis, only part of the cutter path, P0-P1-P2-P3, is discussed. The APT program is as follows:

```
. . . . . . . . . . . . . . . . . .
. . . . . . . . . . . . . . . . . .
CUTTER/0.25
R=0.125
R1=1.9223
C1=CIRCLE/CENTER,1.5625,0,RADIUS,R
C2=CIRCLE/CENTER,0,1.5625,RADIUS,R
C3=CIRCLE/CENTER,-1.5625,0,RADIUS,R
C11=CIRCLE/YLARGE,OUT,C1,OUT,C2,RADIUS,R1
C22=CIRCLE/YLARGE,OUT,C2,OUT,C3,RADIUS,R1
FROM/(P0=POINT/0,0,0)
GO/C11
AUTOPS
TLLFT,GOLFT/C11,TANTO,C2
A1)    GOFWD/C2,TANTO,C22
. . . . . . . . . . . . . . . . . . . . .
. . . . . . . . . . . . . . . . . . . . .
```

Two error messages, namely

ERROR 2210: "WARNING ... CUTTER IS ON THE WRONG SIDE OF DRIVE SURFACE.
                   THE PROCESSING CONTINUES."
ERROR 2619: "UNABLE TO POSITION THE CUTTER INSIDE OF CIRCLE; CHECK
                   CUTTER AND CIRCLE RADII."

are issued for the statement labeled *A1*. These two errors arise because the tool diameter is the same as that of circle C2. As pointed out in Chapter 6, the tool path is always a linear interpolation of the true profile, and certain deviations from the profile at point P2 are unavoidable (Fig. 8-9[b]). The tolerance in this program is the default one, which is +0.0005 in. Therefore the tool will certainly interfere with circle C2 on the other side, and hence it can never be positioned within circle C2 unless the tool position is exactly at the theoretical position at P2, an unlikely occurrence.

The solution is to reduce the cutter diameter. The result can be seen from the following table:

CUTTER DIAMETER (IN.)	GENERATED ERROR(S)
0.2500	2210 and 2619
0.2490	2619 only
0.2480	No error

The two examples presented above clearly show that the solution to an error might not be in the statement where an error is detected. Although it is not possible to list all the possible cases here, the principles introduced in this chapter should prove useful in debugging a program.

8.8 THE PROCESSING LISTING FILE AND THE CLDATA

After finishing the processing of an APT program, the APT-AC NC processor generates the CLDATA file, if requested, and the listing file. A sound understanding of these two files helps in debugging an APT program. The listing file and CLDATA for the program discussed in example 1 in Section 8.7 are listed in Fig. 8-10.

The first part of the listing file is the output of section 1 (translation section). It consists of a list of the APT statements, each of them having been assigned an ISN number. At the end of the statement listing, the NC processor indicates whether any error has been detected during the translation stage. Detailed error messages are issued if they are discovered. If an error is serious, the processing and, accordingly, the listing file are terminated.

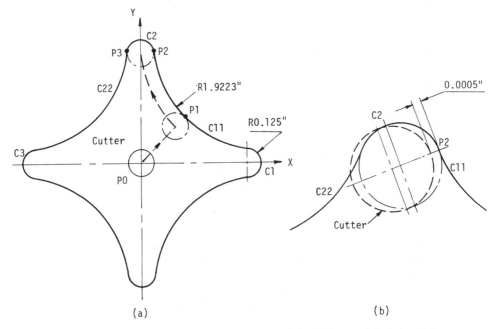

(a)   (b)

Figure 8-9 (a) The cutter path discussed in example 2. (b) A normal deviation of the cutter from the profile.

```
IBM S/370 APT-AC N/C PROGRAM VERSION=1.3.0000                              DATE=09/22/86  TIME=11:32:58

                      ... BEGIN TRANSLATION PHASE...  (SECTION 1) ...

ISN 00001 PARTNO TEST ERROR                                                  E0001
ISN 00002 CLPRNT                                                             E0002
ISN 00003 UNITS/INCHES                                                       E0003
ISN 00004 CUTTER/0.5                                                         E0004
ISN 00005 FEDRAT/2.0                                                         E0005
ISN 00006 SPINDL/500,CLW                                                     E0006
ISN 00007 PL1=PLANE/1,0,0,5                                                  E0007
ISN 00008 PL2=PLANE/0,1,0,0                                                  E0008
ISN 00009 PL3=PLANE/0,1,0,5                                                  E0009
ISN 00010 PL4=PLANE/0,0,1,0                                                  E0010
ISN 00011 C1=CIRCLE/CENTER,5,0,0,RADIUS,2.0                                  E0011
ISN 00012 A1)   FROM/0,0,0                                                   E0012
ISN 00013 A2)   GO/C1                                                        E0013
ISN 00014 A3)   TLLFT,GOLFT/C1,PL1                                           E0014
ISN 00015 A4)        GOLFT/PL1,PL3                                           E0015
ISN 00016 SPINDL/OFF                                                         E0016
ISN 00017 END                                                               E0017
ISN 00018 FINI                                                              E0018

NO DIAGNOSTICS ELICITED DURING TRANSLATION PHASE
18 N/C SOURCE RECORDS (SYSIN)


                             SECTION 1 ELAPSED CPU TIME IN MIN/SEC IS 0000/00.1199
                             SECTION 2 ELAPSED CPU TIME IN MIN/SEC IS 0000/00.0699


....SECTION 3.....

ISN                                                        LABEL   REC M    CARD
0001 PARTNO/ TEST ERROR                                            00002 E0001
0003  UNITS/INCHES                                                 00004 E0003
0004 CUTTER/     0.50000000                                        00006 E0004
0005 FEDRAT/     2.0000000                                         00008 E0005
0006 SPINDL/   500.00000000          CLW                           00010 E0006
0012  FROM/                                                    A1  00012 E0012
                    0.0           0.0           0.0
```

Figure 8-10 The sample program shown in example 1, section 8.7, and its CLDATA.
(continued)

365

```
0013  GOTO/           C1
                      2.75000000        0.0         0.0      CIRCLE       DS(IMP-TO)

0014  SURFACE         C1
                      5.00000000        0.0         0.0      1.00000000   2.00000000
                      0.0               0.0

0014  GOTO/           C1
                      2.74998912        0.04691809  0.0
                      2.75390084        0.14067270  0.0
                      2.76171747        0.23418274  0.0
                      2.77342542        0.32728565  0.0
                      2.78900435        0.41981957  0.0
                      2.80842716        0.51162361  0.0
                      2.83166009        0.60253819  0.0
                      2.85866274        0.69240523  0.0
                      2.88938818        0.78106850  0.0
                      ............      ..........  .......
                      3.98925546        2.01075750  0.0
                      4.07395579        2.05114416  0.0
                      4.16026608        2.08796484  0.0
                      4.24803627        2.12115553  0.0
                      4.33711377        2.15065853  0.0
                      4.42734373        2.17642253  0.0
                      4.51856926        2.19840276  0.0
                      4.61063178        2.21656099  0.0
                      4.70337122        2.23086567  0.0
                      4.75000000        2.23606798  0.0

0015  GOTO/           PL1
                      4.75000000        4.75000000  0.0

0016  SPINDL/         OFF
0017  END
0018  ***** FINI *****
....END OF SECTION 3.....
```

```
A2  00014  E0013

A3  00016  E0014

A3  00017  E0014



A4  00019  E0015

    00021  E0016
    00023  E0017
    00025  E0018
```

SECTION 3 ELAPSED CPU TIME IN MIN/SEC IS 0000/00.0799

TOTAL PART PROGRAM CPU TIME IN MIN/SEC IS 0000/00.2699
**** END OF APT PROCESSING ****

Figure 8-10 (continued)

The calculation and edit sections of the NC processor normally do not yield an output if no error is detected. On request by the statement CLPRNT, the edit section of the NC processor prints the CLDATA in the section 3 (edit section) output.

The structure of the CLDATA is as follows:

1. The contents of the CLDATA consist of the calculated successive cutter center-line positions and the statements other than those regarding computer operations in the APT program. They are listed in the same order as in the APT program.

2. For some of the APT statements, an output line will be generated in the CLDATA; for others, the output may occupy more than one line. A line of CLDATA starts with the ISN number that indicates the parent statement, followed by the contents of the statement. For cross reference, the statement label (LABEL) and the statement sequence number (CARD) are also listed at the end of the line if they are specified in the program. Whenever a MACRO call statement is issued, a character, M, will appear under column M. The record number in the CLDATA is listed in column REC.

3. For a motion statement, the CLDATA starts with the word GOTO, followed by the symbol of the geometric entity specified as the drive surface in the statement. For point-to-point and start-up motion statements, the end point of the motion is then listed in the next line (see ISN 0013 in Fig. 8-10). For contouring motion statements, the drive surface is given in canonical form (see ISN 0014 in Fig. 8-10) before the GOTO word, and then the X, Y, and Z coordinates of the consecutive points based on the linear interpolation calculation are listed, the last set of coordinates representing the end point of this contouring motion.

4. If the COPY and TRACUT statements are specified, they will also be listed in the CLDATA.

The CLDATA is very useful in both debugging a program with errors and verifying a program with no error message issued by the NC processor. A program might be correct in syntax and logic and thus generate no error message after processing by the NC processor, and yet the cutter path may still be wrong because the geometric entities and cutter motion are defined incorrectly.

PROBLEMS

8.1 Write an APT program for the part described in Problem 2.3. The starting and end points are at (200,200,50). The finishing allowance for the hole and outer profile is 0.5 mm; the tolerance is 0.02 mm.

8.2 A gear (Fig. P8-2[a]) with 12 teeth is to be machined on a three-axis NC milling machine. The tooth profile of the gear consists of three parts, namely, an involute curve, LC1 or LC2, a line, L1 or L2, tangent to the involute curve at the intersection of the involute curve with the base circle (with a diameter of 90.2105 mm), and the corner circle (with a radius of 1.5875 mm). An involute curve is the locus of the end point of a

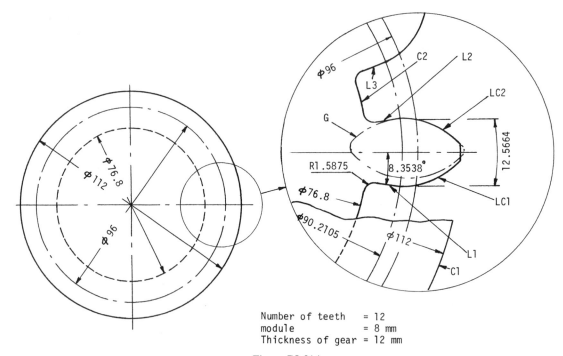

Number of teeth   = 12
module   = 8 mm
Thickness of gear = 12 mm

Figure P8-2(a)

$$x = (d_b/2) \cdot (\cos t + t \cdot \sin t)$$
$$y = (d_b/2) \cdot (\sin t + t \cdot \cos t)$$

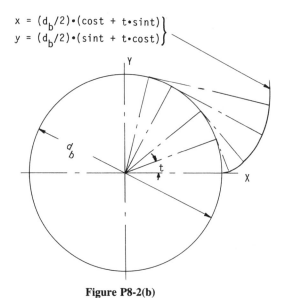

Figure P8-2(b)

string when it is unwinding from a base circle. It is defined by the parametric involute functions (Fig. P8-2[b]):

$$x = (d_b/2) \cdot (\cos t + t \cdot \sin t)$$
$$y = (d_b/2) \cdot (\sin t - t \cdot \cos t)$$

where d_b is the diameter of the base circle. Note that the involute curve is defined outside the base circle only. Lines L1 and L2 pass through the center of the circle. The involute curve can be defined by using statement LCONIC/4PT1SL. Four points on the curve and the slope at the intersection of the curve and the base circle should be calculated. Write an APT program for directing the machining process. The following conditions are given:

a. Cutter $\frac{1''}{8}$ end mill
b. Feedrate   40 mm/min
c. Depth of cut   2 mm
d. Spindle speed   1500 rpm
e. Tolerance   0.01 mm

Note: The reader should pay special attention to the LCONIC function when specifying the contouring motion statement. The actual curve represented by the LCONIC function, which is defined in the way described above, may have the form of curve G, shown in Fig. P8-2(a). Note also that the generated NC program is a file of considerable length.

8.3 Write an APT program for the part described in Problem 2.5, assuming that the NC machine has an automatic tool changer and that the tool-length offset of the $\frac{1}{4}$ in. diameter tool from the other is -0.785 in. In addition, it is assumed that the home position defined by the APT statement GOHOME is the machine position for changing the tool. A tool change can be programmed with the aid of the two statements

SELCTL/n
LOADTL/n,SETOOL,f

where f is the tool-length offset of the No. n tool.
Note: The statement GOHOME orders the tool to go to the home position. See Section 13.4.1.1.

8.4 Write an APT program for directing the finishing cut of the part described in Problem 2.6. The scallop height on the surface should be less than 0.005 in. The starting and end points are at (2,0,2).

8.5 The part shown in Fig. 4-26 is to be machined on a three-axis NC milling machine. The stock material for this part is an aluminum plate with a thickness of 0.16 in. The material is held through the three holes, $\frac{7}{16}$ in. in diameter, located close to the exterior profile. Write an APT program for cutting the interior and exterior profiles and for drilling the holes of different diameters. The following conditions are given:

a. Cutter

TOOL NO.	CUTTER	TOOL-LENGTH OFFSET (in.) WITH RESPECT TO THE NO. 1 CUTTER
1	$\frac{1}{4}''$ end mill	0
2	0.2'' drill	1.47
3	$\frac{3}{8}''$ drill	2.66
4	$\frac{1}{4}''$ drill	1.58
5	$\frac{7}{16}''$ drill	2.90

b. Spindle speed   900 rpm   (milling)
                              400 rpm   (drilling)
c. Feedrate   2.5 in./min (milling)
                              2.0 in./min (drilling)
d. Finishing allowance   0.02 in.

8.6 A turbine impeller with 40 vanes (Fig. P8-6) is to be cut on a three-axis NC milling machine by using a flat-end mill with a $\frac{1}{8}$ in. diameter. Write an APT program for directing the finishing cut of the 40 vanes. The following conditions are given:

    a. Feedrate   1.5 in./min
    b. Spindle speed   2000 rpm
    c. Tolerance   0.0002 in.
    d. Starting and end points   (5,4,1)

8.7 Write an APT program for turning the right end of the shaft described in Problem 2.8. The origin of the coordinate system is set at the center of the finished end surface. The starting and end points are at (0.5,50) in the APT coordinate system. Use the INSERT statement in the APT program to specify the threading cycle in NC codes.

8.8 The impeller of a partial emission pump, shown in Fig. P8-8, is to be machined on a three-axis milling machine with an automatic tool changer. The stock material is first turned on a lathe, and the resulting profile is shown in the figure. Write an APT program for milling the undercut areas and holes, using the following given conditions:

    a. Cutter

TOOL NO.	CUTTER	TOOL LENGTH OFFSET (in.)
1	$\frac{1}{2}''$ end mill with fillet radius of 0.1 in.	0
2	$\frac{1}{4}''$ flat-end mill	−0.32

    b. Spindle speed   800 rpm (for cutter 1)
                                         1500 rpm (for cutter 2)
    c. Feedrate   3 in./min
    d. Allowable scallop height   0.005 in.
    e. Tolerance   0.001 in.
    f. Starting and end points   (3,3,3)

8.9 The part shown in Fig. P8-9 is to be machined on a three-axis NC milling machine. The following conditions are given:

    a. Size of the stock material: 10.2 × 5.2 × 2.0 in.

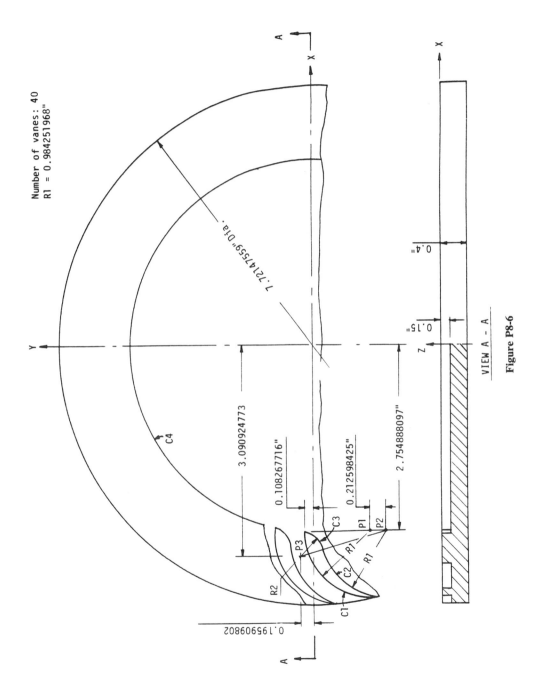

Number of vanes: 40
R1 = 0.984251968"

VIEW A - A

Figure P8-6

371

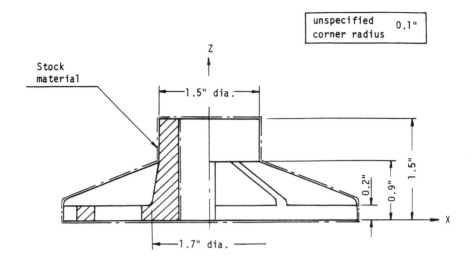

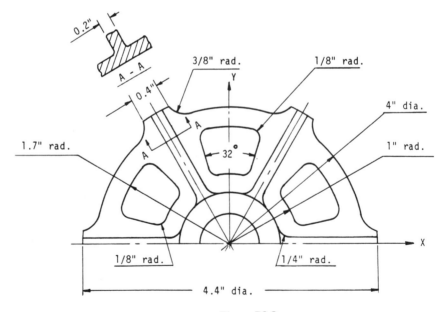

Figure P8-8

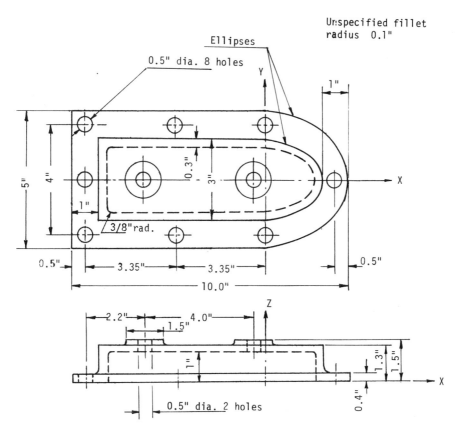

Figure P8-9

b. Cutters, spindle speeds, and feedrates:

CUTTER NO.	CUTTER	TOOL-LENGTH OFFSET (in.)	FEEDRATE (in./min)	SPINDLE SPEED (rpm)
1	$2\frac{1}{2}''$ diameter facing and shoulder milling cutter; nose radius of carbide inserts is 0.1 in.	0	5	800
2	$\frac{1}{2}''$ drill	1.23	2.5	400
3	$\frac{3}{4}''$ end mill with 0.1″ fillet radius	−0.75	3	800

c. Maximum depth of cut for milling operation   0.2 in.
d. Finishing allowance   0.02 in.
e. Tolerance   0.005 in.

First, the profile of the part above the surface $z = 0$ is machined and the holes are drilled. Then the part is turned over, and the bottom surface and the cavity are cut to size. Write two APT programs for directing these two machining processes, respectively.

8.10 Consider the part given in Problem 4.8. Design a method and write an APT program to determine whether the Z-lowest point of the defined polyconic surface, S1, is equal to -0.5. Change the K value in the POLCON statement, run the APT program, and read the corresponding coordinate of the Z-lowest point. Discuss the result.

REFERENCES FOR PART II

1. IBM Manual SH20-1414-1: *System/370 Automatically Programmed Tool—Advanced Contouring (APT-AC) Numerical Control Processor: Program Reference Manual*, 2nd ed. Rye Brook, N.Y.: IBM Corp., 1982.

2. IBM Manual SH20-1414-2: *Automatically Programmed Tool—Advanced Contouring (APT-AC) Numerical Control Processor: Program Reference Manual*, 3rd ed. Rye Brook, N.Y.: IBM Corp., 1985.

3. Chang, Chao-Hwa, *A Guide to APT Programming*. MANE195A course notes, Manufacturing Engineering Program. Los Angeles: University of California at Los Angeles, 1985.

4. Bezier, P., *Numerical Control: Mathematics and Applications*, trans. by A. R. Forrest and A. F. Pankhurst. London: Wiley, 1972.

5. Acherkan, N. S., ed., *Handbook for Mechanical Engineers*, vol. 1, p. 255. Moscow: Mashgiz, 1961. (In Russian.)

6. Korn, G. A. and Korn, R. M., *Mathematical Handbook for Scientists and Engineers*. New York: McGraw-Hill, 1968.

7. IBM Manual SH20-6640-0: *APT-AC NC Processor—Advanced Functions: Program Reference Manual*, 1st ed. Rye Brook, N.Y.: IBM Corp., 1985.

8. IBM Manual SH20-1413-3: *APT-AC NC Processor: Operations Guide*, 4th ed. Rye Brook, N.Y.: IBM Corp., 1985.

9. IBM Manual SC19-6210-2: *CMS User's Guide (Release 3)*, 3rd ed. Rye Brook, N.Y.: IBM Corp., 1983.

part III

The Generation of NC Programs
Through a CAD/CAM System

chapter 9

Principles of NC Programming on CAD/CAM Systems

In Part II we discussed in detail the generation of an NC program by means of the APT language. In APT a cutter path is essentially generated on the basis of part geometry. Although APT provides a great many statements to simplify the definition of part geometry and cutter path, programming in APT depends, to a great extent, on the knowledge and experience of the programmer and on his spatial visualization of the part. On the other hand, computer-aided design (CAD) systems have been widely used in industry for over 20 years to design parts and to draw, store, and modify them. A part is represented in a CAD system by a set of data. If an NC function is incorporated into the system, the very data stored in the memory unit of a CAD system to define a part can also be used to generate the NC machining program.

9.1 AN INTRODUCTION TO CAD/CAM SYSTEMS

All computer-aided design/computer-aided manufacturing (CAD/CAM) systems are based on a technique, called *computer graphics,* that allows the creation and display of an object in graphic form on a cathode-ray tube (CRT) screen through the use of the computer. The object is stored in the computer in the form of mathematical or geometric data, called the *mathematical or CAD model* of the object. By taking advantage of the processing speed of the computer, one can then manipulate the model interactively to obtain a more desirable representation of the object and can further process, analyze, and evaluate the model. Additional information and functions, other than those needed for representing the object graphically, can also be added into the computer graphics system so that the object can be defined in a manner that can meet not only the requirements of design but also those of testing, manufacturing, and management.

A basic CAD system allows the user to generate, display, manipulate, and modify the graphic object. When used together with a plotter, it can produce drawings of consistent quality, store them in a much more compact form (on tape or disk), and retrieve and modify a drawing quickly.

A CAD system can also include other functions that facilitate the design process, such as calculations of centers of gravity and moments of inertia, kinematic analyses, and finite-element analyses. Thus various design options can be tried and tested on the screen, and an optimal design can be reached more easily.

A CAD system can be further upgraded by incorporating software that performs the numerical control function and permits a user to define, on the screen, a cutter path based on part profile and machine operations.* A part machining program, in APT or other higher-level NC programming languages, defining the sequence of machine operations and the cutter motion can then be generated automatically. There are also upgraded CAD systems that include software for facilitating the manufacturing process planning and production simulation. Based on geometric models and an extensive data base, including the necessary data for part classification, process planning, and so forth, these CAD systems can assist the manufacturing engineer in planning and simulating manufacturing processes. Some of the NC-oriented, upgraded CAD systems also include programs that assist the user in generating a customized postprocessor. Such upgraded systems, which support both design and manufacturing, are called *CAD/CAM systems*.

Despite their power and versatility, CAD or CAD/CAM systems cannot, of course, carry out automatically a complete job of design and analysis. Their role is to assist the human engineer during the highly complex design process. Thus the designer and the CAD system work as a team, combining the best characteristics of each. The person conceives a design, inputs it into the system, examines it, modifies it, edits the information, and makes decisions, whereas the CAD system carries out the necessary numerical analyses and organizes, stores, transmits, and accesses the information. Each partner does only the work that he (it) can do best. The result of computer-aided design is a geometric data base† (or CAD model) that defines, in numerical form, the part to be manufactured. The CAM software is then used to generate the information required to manufacture the part on the basis of the geometric data base. During this process, the human being again plays the same role as in CAD. For example, when working on an NC program, the designer must determine the starting point, the sequence of machine operations, the machining specifications, and the cutter path. The CAD/CAM system then carries out the necessary mathe-

---

*As a matter of fact, the earliest CAD system, CADAM, was originally designed to perform such an NC function.

†Although it is called a *data base*, this geometric information is usually more of a special file system than a true data base in the modern sense of the word.

matical calculations and translates the calculated results and the information input by the designer into an NC machining program.

In recent years, the trend has been to incorporate more and more functions performed by human designers into the CAD/CAM system. The reason is the growing shortage of experienced, capable manufacturing planners and NC programmers, while parts, machines, and processes become increasingly complex. Artificial intelligence programs that incorporate the experience and knowledge of experienced designers or engineers in solving specific types of problems are currently being designed for inclusion into CAD/CAM systems.

Existing CAD systems can be divided into three basic types: microprocessor (or microcomputer) -driven dedicated systems, minicomputer-driven systems, and systems based on mainframe computers.[1] Microcomputer-based systems are the smallest and least expensive ones. They are normally provided with basic drafting software and may also include some limited analytical functions. Such systems are mostly used as drafting tools, and their processing speed is the lowest among the three types. Minicomputer-driven systems are provided, in addition to drafting software, with a number of programs that perform various analytical functions used in design or manufacturing, or both. A minicomputer can normally service several graphics terminals. Mainframe-based computer graphics systems are necessary to realize the large-scale data processing required by applications of a complex nature and to maintain the large data bases required for design, manufacturing, and production management.

A CAD/CAM system works on the geometric model of a physical object created in computer memory. The process by which the geometric model of a part is defined is called *geometric modeling*. Three types of three-dimensional models, namely, *wire-frame, surface,* and *solid models,* are used in CAD/CAM systems. A wire-frame model defines a part through its edges and vertices.* The model is stored in the computer as a list of points, lines, and curves representing the vertices and edges of the part. Wire-frame modeling is the earliest and simplest method used in computer graphics for defining a part or an object in three-dimensional space, and many existing systems are still based on wire-frame models. Wire-frame modeling requires the least amount of memory space and much less processing time than do the other three-dimensional models. However, it provides no information regarding the surfaces and volumetric extension of a part. Therefore a wire-frame model is generally ambiguous. For example, the wire-frame model of a cube is displayed on a CRT screen as shown in Fig. 9-1(a). It can be identified as either the object shown in Fig. 9-1(b) or the one in Fig. 9-1(c). Thus this wire-frame model does not represent a unique object. Moreover, the model represents solely the interconnected edges of the boundary surfaces and cannot represent smoothly varying surfaces. Thus the controlling elements for NC application (i.e., part, drive, and check surfaces) must be defined, as lines or curves, for each step of cutter motion.

In surface modeling, an object or a part is defined by a collection of data that specify analytically its boundary surfaces. A surface model of a part can be created by defining boundary surfaces on a wire-frame model. These surfaces can be simple surfaces, such as planes, cylinders, and cones, or complex sculptured surfaces. The

---

*More precisely, a wire-frame model is composed uniquely of the lines and curves joining the points at which the tangents to the surface of the object are discontinuous.

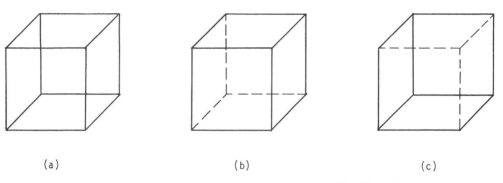

(a)   (b)   (c)

Figure 9-1 The wire-frame model of a cube (a), and the possible objects (b and c) represented by the model.

representation of the surfaces can be improved by attributing to them various characteristics, such as color, texture, transparency, and shading, that affect their visual rendering.[2] A surface model gives a considerable number of details about a part and can be used to generate the contouring cutter motion with respect to any of its surfaces. Thus a surface model can be used to define a surface, rather than a line or curve in the wire-frame model, as the controlling surface of the NC cutter path, thereby simplifying the process needed to define an NC cutter path. For those engineering analyses and applications that require information regarding the part interior, such as volume, weight, and moment of inertia, solid modeling should be used.

Compared with the two approaches described above, solid modeling makes it possible to define the solid nature of an object in the computer, in addition to its vertices, edges, and boundary surfaces. There are two principal representation techniques commonly used in solid modeling, namely, *constructive solid geometry* (CSG) and *boundary representation* (B-rep). CSG uses some primitive solids, such as planes, cubes, cylinders, spheres, and cones, to define an object through Boolean operations (Fig. 9-2). In the computer, the object is defined by a group of data, called a tree, including the types and positions of elementary objects or primitives, the transformations applied to them, and the set of Boolean operations that define how these primitives are combined successively with subtrees. In a boundary representation, an object is represented by its spatial boundary, together with the information that indicates on which side of the boundary the solid is located. A solid model is defined mathematically in the computer as a volume bounded by surfaces; it includes the complete information required to define a part unambiguously and to calculate further data, such as mass and moment of inertia, required for engineering analyses. Automatic generation of NC programs directly from the CAD model or based on the geometric features of a part always involves consideration of its volume (see Section 9.2), thereby requiring that the part be defined unambiguously by the solid model.

Further discussion of modeling techniques falls beyond the scope of this book. The interested reader should consult Besant and Lui[3] and Mortenson.[4]

A number of CAD/CAM systems with the capability of generating NC programs based on design drawings have been developed since the late 1960s. The CADAM system, developed by CADAM INC. and marketed by IBM, the CATIA

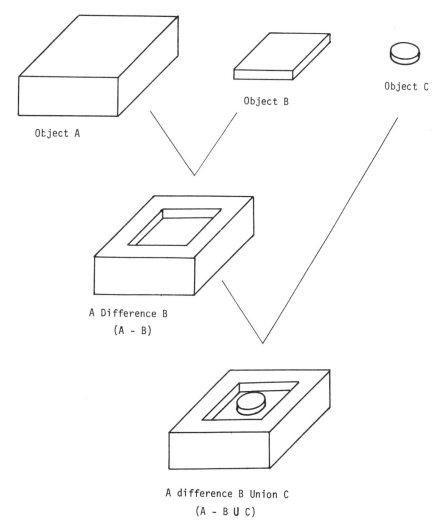

Object A

Object B

Object C

A Difference B
(A - B)

A difference B Union C
(A - B ∪ C)

Figure 9-2   The definition of an object by Boolean operations, starting with primitive solids.

system, developed by Dassault Systems Co. of France and marketed by IBM, and the CDS-CAD/CAM system, developed by Computer Vision Corp., are among the most comprehensive commercial CAD/CAM systems. There are a great many vendors supplying CAD/CAM systems with varying capabilities for design and manufacturing, including NC program generation. Detailed lists of these vendors can be found in references 5 and 6.

One of the pioneer CAD/CAM systems is CADAM, which is mainframe based and allows two- and three-dimensional wire-frame models. It includes a design-drafting software package and a number of additional functions provided to simplify design and analysis, including NC program generation. The system uses a central data base for storing geometric models and data, thus enabling designers and manufacturers to share geometric and alphanumeric data. Two-dimensional, three-dimensional, and multiaxis NC machining programs can be generated on the basis of part geometry in different views. The tool motion can be visualized and replayed on the screen for verification.

CATIA is another mainframe-based, interactive graphics system supported by IBM, with CAD and CAM capabilities. It features a workstation consisting of a graphics console and an optional alphanumeric console, thus enabling users to read a drawing and the design data at the same time. The system provides an integrated data base environment for drafting, three-dimensional design, and surface and solid modeling, as well as for advanced CAD/CAM applications. An APT part machining program can be generated from a geometric model for two- through five-axis NC machining work. The tool path can be displayed, with signs and symbols pointing to errors in cutter positions. The system can also communicate with the CADAM system and retrieve data from the CADAM data base.

9.2 PRINCIPLES OF NC PROGRAM GENERATION BASED ON A CAD MODEL

There are various ways to generate the cutter path from a CAD model. In a wire-frame model, the geometric elements that can be defined are points, lines, and curves (i.e., the edges and vertices of a geometric entity). Thus, in a CAD/CAM system based on the wire-frame model, the cutter path of a contouring motion can be defined only by lines and/or curves representing drive, check, and part surfaces. An algorithm similar to that used in the APT-AC NC processor can be used to calculate the cutter positions of the contouring motion.

When surfaces can be defined in a CAD/CAM system, the cutter path for cutting (profiling) a surface can be generated automatically by specifying the route that the cutting tool should follow. For example, a surface (Fig. 9-3) can be defined by the vector expression

$$\mathbf{r} = \mathbf{r}(u, v)$$

where $\mathbf{r}$ is the position vector, and u and v are the parameters of surface A.[7] If one of the parameters is fixed, say, $u = u_1$ or $v = v_1$, the vector equation

$$\mathbf{r} = \mathbf{r}(u_1, v)$$

or

$$\mathbf{r} = \mathbf{r}(u, v_1)$$

represents a curve on the surface, named the *parametric curve* of surface A. The cutter motion for profiling the surface can be specified so that the cutter moves along the parametric curves. The starting and end points of each cutter motion along a parametric curve are determined by the boundaries of the surface, which are defined by some other equations. The starting point for profiling the surface can also be selected. Thus the cutter profiling motion is completely defined. The cutter moves, on the surface, in a zigzag fashion along the parametric curves (see Fig. 9-3), and the incremental values of u or v can be determined by the allowable scallop height left on the surface and by the tool geometry. An algorithm can be designed according to the techniques described above and incorporated into the CAD/CAM system. A sur-

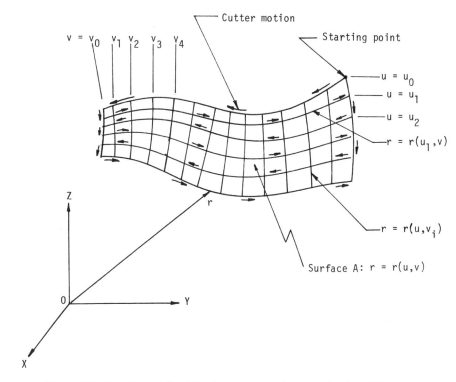

Figure 9-3   A surface, defined by the vector equation $\mathbf{r} = \mathbf{r}(u, v)$, and the cutter motion along the parametric curves, for profiling the surface.

face and its boundary having been defined, the cutter path for profiling the surface can be generated automatically after specification of the starting point of the cutter profiling motion, the tolerance, the allowable scallop height, and the tool geometry. On the basis of these ideas, some commercial CAD/CAM systems, such as CATIA, include the function that permits the generation of tool motion for profiling a complete surface without requiring the user to define the cutter path step by step.

In general, a machined part can be considered as the subtraction of the solid representing the redundant material from the solid representing the stock material. The cutter motion for profiling or cutting the part should sweep over and clean out the volume occupied by the redundant material, which determines the boundary of the cutter profiling motion. Given this volume, one can determine the cutter path by first specifying the route of cutter motion in space. Now the cutter motion can be divided into two parts, one for rough cut, which does not necessarily follow the exact profile of the part, and one for finishing, which is determined by the profile or boundary surfaces of the part. An algorithm for determining the cutter path for a part can be designed and incorporated into a computer graphics system based on the solid model of the part. Thereby an NC cutter path can be generated automatically on the basis of the solid models of the cutting tool, workpiece, and the stock material after definition of the necessary machining specifications, starting and end points, tolerance, and allowable scallop height. Additional information, such as the solids representing fixtures, can also be input into the system so as to avoid collision of the tool with the fixtures. Since solid modeling techniques provide a complete and unambiguous computer representation of real objects, it is an ideal tool for CAM systems that seek to automate NC programming.

9.3 THE DETERMINATION OF CUTTER MOTION WITHIN A CAD/CAM SYSTEM

In this section, we explain how cutter motions are determined within a CAD/CAM system. The cutter motion discussed includes a single step of contouring motion and the cutter motion for profiling a surface. A point-to-point motion is simply determined by the coordinates of the end point; no explanation seems to be needed. Most of the CAD/CAM systems use the geometric data of a CAD model to generate a source machining program (e.g., in APT) and then use an existing NC processor and a suitable postprocessor to generate the NC program.

9.3.1 Simple Contouring Motion

A contouring motion should follow the part profile, and the deviation of the cutter from the profile should always be maintained within a designated tolerance. An NC contouring motion is approximated by a series of linear-interpolation motion steps (see Section 5.9). Two motion elements are needed for defining a single motion step: its spatial direction and its distance (or length). These are determined by the drive and part surfaces and by the tolerances on these surfaces as well. The check surface is used only to determine the end point of the final motion step. The algorithm introduced below is the one used in the APT III and HAPT-3D processors.[7,8]

   Suppose that the cutter has been at position T, ready to start a contouring motion (Fig. 9-4). The distances between the tool profile and the two controlling surfaces, $S1$ and $S2$ (i.e., $P'Q'$ and $P''Q''$), are less than the tolerances d_1 and d_2, respectively. Let $\overline{N}'$ be the common normal to the cutter profile and the drive surface, S_1, and let $\overline{N}''$ be the common normal to the cutter profile and the part surface, S_2. S_1' and S_2' are the intersections of the plane determined by $\overline{N}'$ and $\overline{N}''$ with surfaces S_1 and S_2, respectively. Thus the direction of the linear-interpolation motion step, $\delta\overline{T}$, should be perpendicular to both normals, $\overline{N}'$ and $\overline{N}''$.

   The normals, $\overline{N}'$ and $\overline{N}''$, are calculated iteratively according to the principles[7] described as follows. For tool position T, a tentative point, P, on the tool profile is selected. The normal to the tool profile, $\overline{N}$, the intersection point of $\overline{N}$ and surface S_2, Q, and the normal to surface S_2 at point Q, $\overline{N}_{S2}$, are calculated. Then the angle between $\overline{N}$ and $\overline{N}_{S2}$ is found and compared with a predetermined small angular value, for example, 0.01 radian in APT. The calculated normal, $\overline{N}$, will be accepted as the common normal to both the tool profile and surface S_2 if the calculated angle is smaller than this value; otherwise, a new estimate of point Q is made with the use of the curvature of surface S_2 at point Q, according to the Newton-Raphson method.*[7] This process is repeated until the solution (i.e., $\overline{N}''$) is within the desired accuracy.

---

*The Newton-Raphson method is a simple calculation procedure that yields the numerical solution of polynomial equations to any desired accuracy.

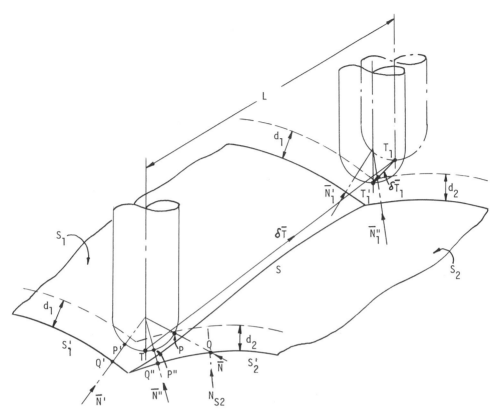

Figure 9-4   The determination of a linear interpolation motion step. d_1 and d_2: Tolerances for surfaces S_1 and S_2.

The common tangent, $\overline{N}'$, is determined similarly. The motion direction, $\delta\overline{T}$ (Fig. 9-4), should be perpendicular to both vectors $\overline{N}'$ and $\overline{N}''$. Thus it can be determined by the vector expression

$$\overline{N}'' \times \overline{N}'/|\overline{N}'' \times \overline{N}'| \tag{9.1}$$

Distance L, or the end point of this motion step, is first estimated according to the radius of curvature of the intersection curve S at point T and the tolerance value (Fig. 9-5). Thus the first tentative position of the end point is at T_1 (Fig. 9-4). However, this cutter position may exceed tolerances d_1 and d_2 for surfaces S_1 and S_2, respectively. It should be further adjusted by moving step $\delta\overline{T}_1$ from position T_1. The end position of this adjustment motion, $\overline{T}_1'$, is determined from the following three conditions:

1. The distance between the tool profile and surface S_1 should be less than or equal to d_1.
2. The distance between the tool profile and surface S_2 should be less than or equal to d_2.
3. The incremental movement $\delta\overline{T}_1$, should be in the plane determined by the two normals, $\overline{N}_1'$ and $\overline{N}_1''$.

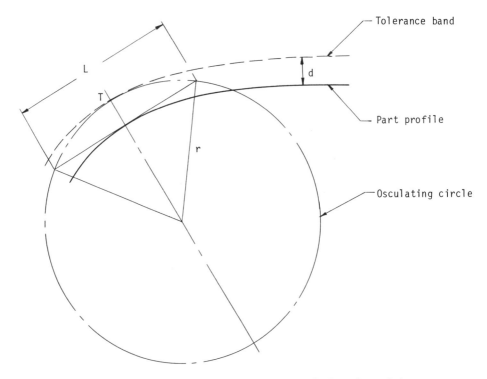

Figure 9-5   The determination of the step length, L, of a linear interpolation motion based on the radius of curvature r and the tolerance value d:

$$L = 2[d \cdot (2r - d)]^{1/2}$$

These three conditions yield the following three vector equations[7]:

$$\overline{N_1}' \cdot \delta\overline{T_1} = D_1 \leqslant d_1 \qquad (9.2)$$

$$\overline{N_1}'' \cdot \delta\overline{T_1} = D_2 \leqslant d_2 \qquad (9.3)$$

and

$$(\overline{N_1}'' \times \overline{N_1}') \cdot \delta\overline{T_1} = 0 \qquad (9.4)$$

where $\overline{N_1}'$ and $\overline{N_1}''$ can be determined by the method described above and $\delta\overline{T_1}$ is adjusted until one of the distances, D_1 or D_2, is equal to the tolerance value and the other is within the tolerance band. The calculation is an iterative process carried out automatically by the computer. The end point of the tool position is finally determined at point T_1'. Then the actual length of the motion step is found and a calculation is made to determine whether any point in this motion step, TT_1', is beyond the tolerance band. If so, the length of the motion step should be reduced; otherwise, point T_1' is accepted. Calculation of the next end point begins after end point T_1' has been determined.

At the end of this contouring motion, the check surface, S_3 (which is not shown in Fig. 9-4), should be used to determine the end position of the last motion step. Thus equation 9.4, above, should be replaced by

$$\overline{N_n} \cdot \delta\overline{T_n} = 0$$

where $\overline{N}_n$ is the normal to check surface S_3 at the end point of the last motion step and δT_n is the adjustment motion at the end point of the last motion step.

An algorithm based on the above method can be designed and incorporated into the CAD/CAM system to calculate the cutter positions in a contouring motion. A series of APT GOTO statements that define the contouring motion can then be generated according to the calculated positions.

9.3.2 The Determination of Cutter Motion for Profiling a Surface

The cutter motion for profiling a surface consists of moving the cutter along the surface in a predetermined way while maintaining the distance between the cutter and surface at less than or equal to the tolerance value. The cutter movement can be designed to follow either the intersection curves of the surface and the planes parallel to the coordinate planes (Fig. 9-6), or the parametric curves shown in Fig. 9-3. Generally, it is simpler to use parametric curves as controlling surfaces since they can be described by a single parametric expression. Thus the calculation routine can be greatly simplified.

One principal motion direction should be chosen for a cutter motion determined by parametric curves. Figure 9-3 shows the principal motion direction as being along the curve

$$\mathbf{r} = \mathbf{r}(u, v)\big|_{u = \text{const}}$$

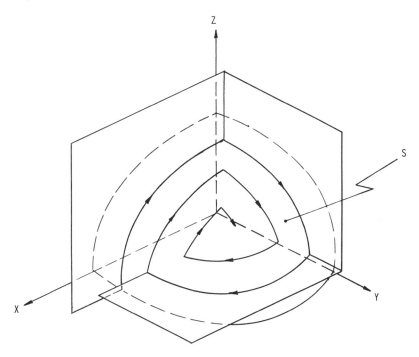

Figure 9-6   The cutter motion for profiling surface S. The route of the cutter motion is determined by the intersection curves of surface S and the planes parallel to the coordinate planes.

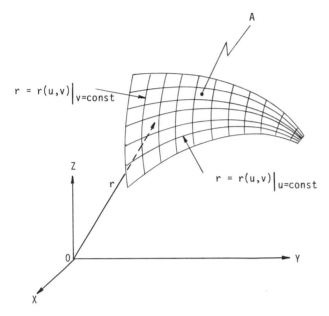

$r = r(u,v)\big|_{v=const}$

$r = r(u,v)\big|_{u=const}$

Figure 9-7 The selection of parametric surface $\mathbf{r} = \mathbf{r}(u, v)|_{v=\text{const}}$ to guide the principal motion.

whereas the step-feed motion, determined by the allowable scallop height, is along the curve

$$\mathbf{r} = \mathbf{r}(u, v)|_{v=\text{const}}$$

In some cases, the selection of a parametric curve as the principal motion curve has a significant effect on the required machining time. In Fig. 9-7, surface A is very narrow on the right-hand side. A principal motion along the curve

$$\mathbf{r} = \mathbf{r}(u, v)|_{u=\text{const}}$$

results in repeated motion on the right-hand side, which wastes time. Therefore the curve

$$\mathbf{r} = \mathbf{r}(u, v)|_{v=\text{const}}$$

should be selected as the controlling surface of the principal motion.

Once the principal motion direction has been defined, determining the profiling motion is reduced to calculating the cutter positions for each step of profiling motion by using the technique described in Section 9.3.1. Thus the profiling motion is completely defined after a starting point is specified.

A calculation routine can be designed according to the method described above, and the APT statements defining the cutter profiling motion for an arbitrary surface can be generated automatically on the basis of the defined surface, tool geometry, starting point, tolerances, feedrate, and allowable scallop height.

9.4 CONSIDERATIONS REGARDING THE AUTOMATIC GENERATION OF NC MACHINING PROGRAMS DIRECTLY FROM A CAD MODEL

One of the recent trends in the development of CAD/CAM systems is the automatic generation of NC machining programs directly from a CAD model. Although interactive CAD/CAM systems do reduce the time for specifying (writing) programming statements, this effect might not be significant in certain cases (see Chapter 10). Besides, a well-designed machining plan should be designed in advance so that the machining steps can be defined interactively. The automatic generation of an NC machining program directly from a CAD model is defined as *an automatic process that designs the machining strategy (plan), determines the route of the cutter motions, and generates the NC program on the basis of the geometric features of a part and the manufacturing data base as well by means of a CAD/CAM system.* Considerable efforts have been under way to develop such systems[9–12]; however, research in this field is still in its initial stage. The purpose of the following description is to introduce briefly the concept and the major problems.

Generally, a complete machining operation includes both rough and finishing machining. The tool movement should remove the redundant material from the stock but should never interfere with the fixtures. The volume of the redundant material is determined by the geometry of the part and the stock material. The tool motion should then be defined on the basis of this volume, the positions and geometry of the fixtures in use, and the tool geometry.

The cutter motion for machining a part consists of roughing and finishing, which should be considered separately. The basic requirements on the cutter motion in a roughing operation are as follows:

1. It should be as simple as possible and preferably consist of the linear type only.
2. It should minimize machining time. In other words, the cutter path should be as short as possible, and the depth of cut and feedrate should be as large as possible.

The cutter finishing motion should follow the part profile as described in Section 9.3.2.

Before a cutter path is designed, the system must first examine which surfaces of the part are machinable at the current holding position. Since the cutter motion in a rough cut is usually a simple linear motion in the X, Y, or Z direction, simple envelope planes, usually parallel to coordinate planes, should be created for those surfaces or profiles that need not be followed exactly during the rough cut (Fig. 9-8). Hypothetical boundary planes, parallel to the coordinate planes, can also be added to those parts of the workpiece that are not machinable. An optimized cutter path for roughing is then defined on the basis of the modified volume of redundant material, obtained by subtracting the volume of the part with added envelope surfaces, boundary planes, or both from that of the stock material.

The major parts of the roughing motion can be considered to be pocketing motion (i.e., the removal of material layer by layer within certain boundaries). Thus patterns of pocketing motion, usually in a zigzag way, can be designed for directing the cutter roughing motion. For example, the pocketing motion can have its principal motion in the X (or Y) direction, with step feed in the Y (or X) direction. After a layer of material has been removed, the tool moves down by an amount determined by the maximum depth of cut and resumes the pocketing on the second level. One major task for the automatic NC program generation system is to divide appropriately the volume representing redundant material into several pocket areas so that pocketing motions can be deduced.

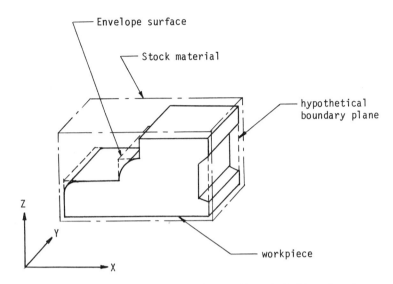

Figure 9-8   The addition of envelope surfaces and hypothetical boundary planes to simplify the rough machining.

The cutter motions in rough machining can be easily determined by the boundaries of the volume representing redundant material, because all the motions are of linear type.

The finishing operation should be further divided into semifinishing and finishing cuts because the completion of the first one results in an even allowance on the part surface for the finishing cut. The cutter(s) should follow the profile during these operations, and the cutter positions for these profiling motions can be derived by using the method described in Section 9.3. As the part might exhibit several different geometric features, the major task of the system is to determine the order in which these features should be finished, the boundary conditions, and the route of the cutter motion for each step of the finishing operation. A source program in APT or some other higher-level NC language can then be generated. Finally, the NC program is obtained as a result of processing the source program by an existing NC processor and a suitable postprocessor.

As can be seen from the description given above, the automatic generation of an NC program based on a CAD model is basically a process consisting of a great deal of decision making and information deduction, which are normally performed by a human programmer or process planner. For both the roughing and finishing operations, two central problems must be resolved: (1) planning the machining process, including determination and modification of areas to be machined, the sequence of machining, the route of the cutter motion, the cutting tools to be used, and machining specifications, and (2) calculating the cutter positions. Since the cutter positions can be calculated by mathematical calculation routines similar to those available in existing NC processors, the major function of the system consists in planning the machine process on the basis of the information provided by the CAD model and the manufacturing data base, which contains data regarding, for example, tools, fixtures, material, and machining data. An artificial intelligence program based on human experience should be incorporated into the system to carry out the planning task. The decisions are made and planning is done essentially on the basis of information provided by the CAD model (preferably a solid model) of the part. Therefore the correct interpretation and extraction of information from a CAD model, equivalent to the work performed by a human manufacturing engineer in obtaining the necessary information by reading engineering drawings, constitute the crucial part of the artificial intelligence program.

REVIEW QUESTIONS

9.1 Explain the major differences among a basic CAD (or computer-aided drafting) system, a CAD system, and a CAD/CAM system.

9.2 What is the role that a human engineer plays in the design and analysis process with a CAD/CAM system?

9.3 What is the CAD or mathematical model of a part?

9.4 What does it mean to create a CAD model in a CAD or CAD/CAM system?

9.5 What are the three basic types of models that are used to represent a part in CAD systems? Compare these models, giving the advantages and disadvantages of each.

9.6 Explain how a CAD/CAM system defines a single step of contouring motion based on the wire-frame model of a part.

9.7 How does a CAD/CAM system define a surface contouring motion on the basis of a defined surface?

9.8 What type(s) of modeling technique(s) is (are) needed for automatic generation of a finishing cutter path on a CAD/CAM system? If both roughing and finishing cutter paths are to be generated automatically by a CAD/CAM system, which modeling technique is considered appropriate? Why?

9.9 What are the basic problems to be resolved in realizing automatic generation of the cutter path for a complete machining process?

9.10 What is the role that artificial intelligence technique plays in automatic generation of NC programs based on a CAD model?

chapter 10

The Generation of NC Programs Through the CADAM System

We shall now use the CADAM system as an example of the way an NC program is generated from a CAD model in the CAD/CAM system.

An NC program usually consists of statements defining the machining specifications and cutter path. In manual programming and computer-aided APT programming, the crucial part of the work is related to the definition of the cutter path. The definition of machining specifications involves selecting the right words in NC or APT and specifying the correct parameters, tasks that are less error prone. When a CAD/CAM system is used to generate an NC program, the situation is basically the same: the definition of the cutter path still represents the major part of the work. However, the specification (writing) of programming statements and their components is replaced by selection of the proper elements on the screen in correct order, thus reducing significantly the possibility of making syntax errors in a program. Moreover, as the defined geometric entities can be vividly seen and checked on the screen, errors are less likely. The general procedure for generating an NC program on the CADAM system is shown diagrammatically in Fig. 10-1. It is essentially the same as in APT programming except that the program is constructed by the CAD/CAM system under the guidance of the user.

Definition of a cutter path on the CRT screen of a CAD/CAM system is, in principle, similar to definition of the cutter motion in APT. A cutter motion is de-

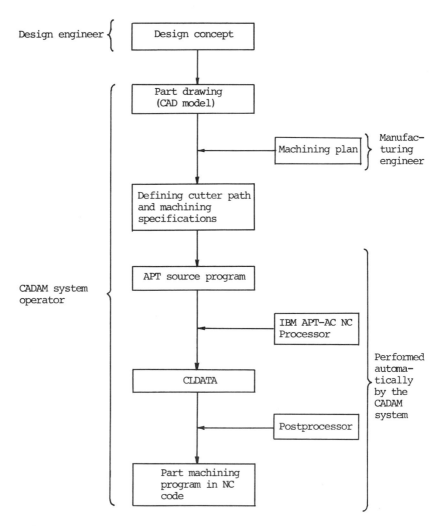

Figure 10-1   The flowchart for generating an NC program on the CADAM system.

fined through selection of the elements of the part profile on the screen in correct order. A continuous cutter path is generated on the basis of a continuous profile. However, a profile that looks continuous on the screen might not actually be continuous because of internal numerical approximations, thus creating problems in generating the NC cutter path. As a result, detailed knowledge of, or rather, skill in the use of, the CADAM NC function plays an important role in defining a cutter path successfully.

In some cases in which the cutter path for a comparatively complicated profile or form is to be defined on the screen of a CAD system, a series of cutter paths may overlap one another. This creates problems in distinguishing a desired tool path from others. In addition, the use of a wire-frame model in CADAM may require a large number of steps to define the necessary geometric entities. As a result, the process becomes very time-consuming for parts with a complicated three-dimensional profile (such as the one shown in example 2 of Section 8.4 [Chapter 8]) and cannot be carried out efficiently. In such cases, a CAD/CAM system based on surface or solid modeling, such as CATIA, or even programming the machining process in APT may well prove to be more efficient.

In general, the generation of an NC program through a CAD/CAM system is simple and straightforward. Furthermore, the cutter path can be displayed and replayed step by step on the screen, thus permitting visual verification and correction of the program.

The following discussion is based on the IBM VM/CMS operating system and CADAM Release 20.1.[13,14]

10.1 PROCEDURES FOR GENERATING AN NC PROGRAM ON THE CADAM SYSTEM AND CONTROL FLOW

As stated previously, the CADAM system provides a common data base for storing and retrieving geometric and NC data. An APT interface module in the CADAM system makes it possible to interface the CADAM NC data with the APT-AC NC processor. The CADAM drawing files are not needed for generating the NC program; therefore they are not accessed during APT processing.

Once a part has been defined on the screen, the CADAM operator specifies the machining procedures and the cutter paths according to the machining plan. The NC data are then saved through the subfunction POST under the CADAM NC function.

The CADAM operator now works on a mainframe terminal under his account. The NC data are retrieved through the APT interface module and then processed by the APT-AC NC processor and a selected postprocessor. The result is a machine-readable NC program.

Before processing the CADAM NC data, the user should ascertain that his mainframe account is privileged to access the NC data in the CADAM data base. This is realized by creating two files in the CADAM system[13] during its installation:

REQUEST IDS A1

and

GROUP OWNER A1

These two files should be stored together with the CADAM modules on one disk, which is accessed as an extension of the user's A disk. The user should also have access to the postprocessor. This is realized through the creation of the file

MACHINE STMTS A1

on the user's A disk. This file is used to supply to the APT program generated by the CADAM system the necessary statements that cannot be defined on the CADAM graphics terminal. Usually, a MACHIN statement can be included in this file for selection of the required postprocessor. The file normally contains several sets of APT statements, each including both a MACHIN statement pertinent to the related postprocessor and other APT statements, and is required for processing of the CADAM NC data aimed at a particular NC machine. The CADAM interface module selects the right set of statements from the file

MACHINE STMTS A1

through the characters specified in columns 73 through 80. The characters in columns 73 through 76 represent the machine to be used, and those in columns 77 through 80 indicate the specific postprocessor to be selected. When there is only one postprocessor for each machine, columns 77 through 80 can be left blank. For example, suppose there are three and four statements to be added for processing the NC data for a lathe and a milling machine, respectively. The "MACHINE STMTS A1" file may appear as follows:

COLUMN 1	COLUMN 73
MACHIN/......	LATH
....	LATH
....	LATH
MACHIN/......	MILL
....	MILL
....	MILL
....	MILL

The characters specified in columns 73 through 76 can be any combination of alphanumeric characters; however, they should be the same for each set of statements. As an example, the characters LATH and MILL are used in the file shown above. Columns 77 through 80 are left blank. For each machine, up to 20 statements can be specified for each set in this file. There is no limit on the number of data sets specified in this file.

Before processing the NC data, the user must link the disks storing the CADAM program and the APT-AC NC processor (or simply, the CADAM disk and the APT disk) to his virtual machine. Necessary procedures should also be carried out to make these disks accessible during processing.

After all the procedures described above have been carried out, the user is in a position to process the CADAM NC data. To do so, he simply types in the job execution command

CADAPT

This command invokes the job execution statements, or program, "CADAPT EXEC" included in the CADAM program product, and the processing is started. The processing consists of three steps.[13] In job step 1 the CADAM system uses as input

1. The NC data stored in the CADAM data base
2. The data set selected from the file "MACHINE STMTS A1"

It generates an APT program that contains the statements taken from the file "MACHINE STMTS A1" and a statement that invokes a module GRAPT in the translation section of the NC processor. It also collects the NC data into a sequential stream so that it can be processed as a unit by the NC processor.

In job step 2, the two data sets in job step 1 are taken as input. Once the APT program has been read in, the GRAPT module is activated. The NC data are read in and converted to standard format. Then the standard APT processing after the translation stage takes effect. In this way, an NC program is generated.

In job step 3, a routine within the CADAM program reinitializes the various data sets in the system so that they are ready for use by the next job.

10.2 METHODS AND PROCEDURES FOR DEFINING MACHINING OPERATIONS AND CUTTER PATHS ON THE SCREEN

As can be seen from the description given above, the generation of an NC program has been reduced to the definition of the cutter path and machining operations on the screen of a computer graphics system. The programming effort is replaced by the selection of the necessary words and geometric elements in correct order on the screen. Thus the skilled use of the NC function of the CAD/CAM system becomes more important. The procedures for defining a cutter path in CADAM are designed around APT programming logic. Thus it is easier for an operator with APT programming knowledge to understand them. In this section, we describe in detail the procedure for defining a part machining process on a CADAM screen.

10.2.1 An Introduction to the CADAM Graphic Display Terminal

A CADAM graphics terminal or console consists of four parts: a CRT display screen, a function keyboard, an alphanumeric keyboard, and a tablet with a cursor or light pen (Fig. 10-2). There is also a control unit through which the input data are processed and sent to the mainframe computer, where the information is further processed and stored in the memory.

The function keyboard (Fig. 10-3) has 32 keys, each of which, when depressed, selects a particular function. Thus, for example, when the LINE function key is depressed, the computer responds to the proper input for defining a line, and the result is displayed on the screen. When the NC function key is depressed, the system is ready to accept input regarding the definition of an NC machining process.

The alphanumeric keyboard permits an operator to enter commands, functions, and data consisting of letters, numbers, and other symbols into the CAD system. An input message is first displayed on the screen (see note 10 in Fig. 10-4) for verification and possible revision before it is entered into main storage. The keyboard also controls the location of a movable cursor on the screen.

The light pen detects a light beam on the CRT screen. It is pen shaped and allows a user to point to a particular position or to an element of the displayed image.

The CRT screen is basically divided into three areas (Fig. 10-4). The central area displays the drawing and geometric elements constructed by the user. The upper part of the screen posts various kinds of messages from the computer, such as drawing scale, selected function, window size, error message, and system acknowledg-

Figure 10-2   The CADAM system workstation. It consists of the following components: a CRT display screen, a function keyboard, an alphanumeric keyboard, and a tablet with cursor or light pen. (Photo courtesy of CADAM INC.)

ment. The lower part of the screen shows the various subfunctions that can be selected under the chosen function and the input data before it is verified and entered by the user into the computer memory.

The control unit acts as a buffer between the high-speed computer and the slower-speed display consoles. A control unit may control several display consoles.

10.2.2 Defining an NC Machining Process under the NC Function[14,15]

Before a machining process is defined, a drawing or geometric model, should be constructed either in one or two views or in multiviews; the reader should consult a CADAM reference manual[16] for details of CADAM computer-aided design procedures. An NC machining process is defined under the NC function of the CADAM system. A description of the procedures follows.

When the NC function key on the function keyboard is depressed, a list of subfunctions is displayed in the lower area of the screen, as follows:

/ENTER/PASS/ERASE NC/PACK/NEST/PLOT/EXIT/

An explanation of these functions is given in Fig. 10-5. A subfunction can then be selected with the light pen. Usually, the function PASS is selected to start defining the machining process.

Figure 10-3 The function keyboard template of the CADAM system. (Courtesy of CADAM INC.)

Once the PASS function is selected, a new menu (list of functions that can be selected) is displayed in the lower area of the screen:

/START/REPLAY/RESTART/POST/ABORT/LEAVE/

An explanation of these functions is given in Fig. 10-6. At this time, a function is selected according to the design requirement. The function START should be selected when one is defining a new machining process. The function RESTART is chosen when an interrupted definition of the machining process is to be resumed. The REPLAY function permits a user to read, step by step, the commands and cutter motion in the designed order on the screen by depressing the YES/NO button successively. If a machining process has been correctly defined, the data can be stored in

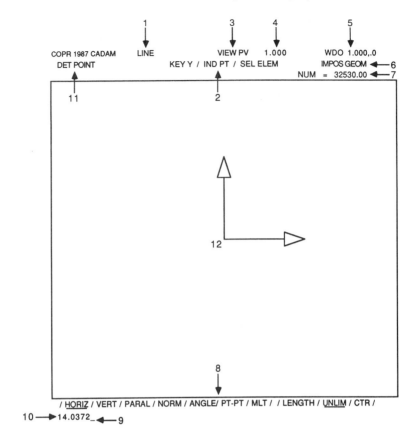

Figure 10-4 The graphics display screen of the CADAM system. Key:

1. The selected function
2. Computer message (listing the options that a user can choose to finish the current operation step)
3. View identifier
4. Scale
5. Window parameters
6. Error message

7. Counter
8. Menu (listing the various subfunctions available under the selected function)
9. Nondestructive cursor
10. Temporary hold register
11. Acknowledgment
12. Origin indicator

(Courtesy of CADAM INC.)

the data base under the function POST. The ABORT function permits deletion of the NC data defined previously. The function LEAVE allows a user to return to the previous menu shown in Fig. 10-5.

10.2.2.1 The Procedures Used to Start Defining a Machining Process.

The START function should be selected to start defining a machining process. This selection is similar to specifying the first statement, PARTNO, in APT programming. The system then prompts the user with the message (Fig. 10-7)

SEL PASS TYPE/KEY CUTTER DIA = ( )

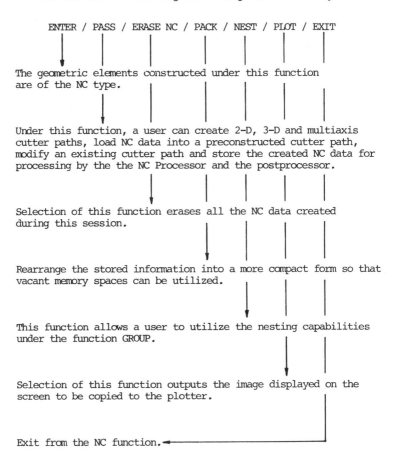

ENTER / PASS / ERASE NC / PACK / NEST / PLOT / EXIT

The geometric elements constructed under this function are of the NC type.

Under this function, a user can create 2-D, 3-D and multiaxis cutter paths, load NC data into a preconstructed cutter path, modify an existing cutter path and store the created NC data for processing by the the NC Processor and the postprocessor.

Selection of this function erases all the NC data created during this session.

Rearrange the stored information into a more compact form so that vacant memory spaces can be utilized.

This function allows a user to utilize the nesting capabilities under the function GROUP.

Selection of this function outputs the image displayed on the screen to be copied to the plotter.

Exit from the NC function.

Figure 10-5   An explanation of the NC subfunctions.

The user should first select the type of NC machining process and then key in the diameter of the cutter. At this time, the menu

/3-AXIS/MULTAX/

is also displayed on the screen, with the 3-AXIS option underlined, meaning that this option is preselected by default. Option 3-AXIS signifies that the machining process will be performed on a three-axis NC machine (two-axis machining, such as turning, can also be defined under this option). A machining process involving a change in the direction of the tool axis during machining should be defined under option MULTAX. We shall consider here only the 3-AXIS case.

If the machining process is of the 3-AXIS type, the user can neglect the first option, SEL PASS TYPE (since the subfunction 3-AXIS is preselected), and type in the cutter diameter value (for the turning tool, it is the tool nose diameter) on the alphanumeric keyboard. Having accepted the diameter input, the system prompts the user with the message

KEY COR-R = ( )

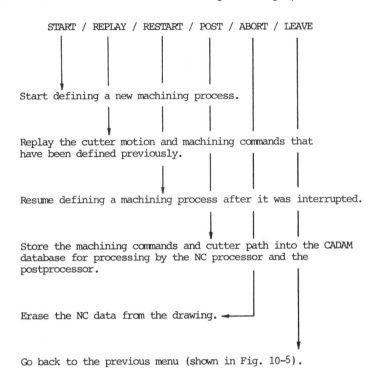

START / REPLAY / RESTART / POST / ABORT / LEAVE

Start defining a new machining process.

Replay the cutter motion and machining commands that
have been defined previously.

Resume defining a machining process after it was interrupted.

Store the machining commands and cutter path into the CADAM
database for processing by the NC processor and the
postprocessor.

Erase the NC data from the drawing.

Go back to the previous menu (shown in Fig. 10-5).

Figure 10-6   The subfunctions under function PASS.

instructing the user to key in the corner radius of the end mill (for turning tools and
flat-end mills, it is zero).

The two steps described above have the same effect as specifying the CUTTER
statement in APT programming.

The system next displays the message

Y/N NEXT AXIS/SEL LOAD PT

The first option, Y/N NEXT AXIS (i.e., using the YES/NO function key to select
the next transformation axis), is used to define the transformation of the CLDATA
(i.e., a TRACUT statement). If no transformation is needed, the procedure SEL
LOAD PT (select a starting point) should be carried out. Here, the user can select
with the light pen a predefined point in the X-Y plane as the starting point; this procedure is similar to defining the starting position in the FROM statement at the beginning of an APT program.

The system then asks for the initial Z position of the tool through the prompt
message

KEY ZL = 0.00000

which means that the default initial tool position in the Z direction is zero and that a
user can change it by keying in the desired value.

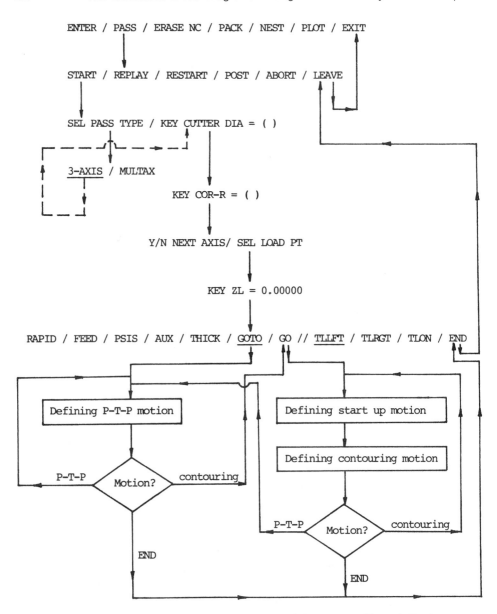

ENTER / PASS / ERASE NC / PACK / NEST / PLOT / EXIT

START / REPLAY / RESTART / POST / ABORT / LEAVE

SEL PASS TYPE / KEY CUTTER DIA = ( )

3-AXIS / MULTAX

KEY COR-R = ( )

Y/N NEXT AXIS/ SEL LOAD PT

KEY ZL = 0.00000

RAPID / FEED / PSIS / AUX / THICK / GOTO / GO // TLLFT / TLRGT / TLON / END

Defining P-T-P motion

P-T-P   Motion?   contouring

END

Defining start up motion

Defining contouring motion

P-T-P   Motion?   contouring

END

Figure 10-7   The procedures for defining a machining process. The underlined functions are preselected by the system when the menu is displayed. P-T-P: point-to-point.

The steps described above constitute the necessary preliminary procedures for starting the definition of a machining process. None of the data defined in these steps can be altered later in the REPLAY mode.

10.2.2.2 The Procedures Used for Defining Point-to-Point and Contouring Motions. On completion of the preliminary procedures described above, the menu for defining the cutter motion and machine operation appears on the screen:

/RAPID/FEED/PSIS/AUX/THICK/<u>GOTO</u>/GO/<u>TLLFT</u>/TLRGT/TLON/END/

The functions GOTO and TLLFT, which have the same meanings as in APT, are preselected. As a matter of fact, the option TLLFT, although preselected, has no effect on the point-to-point motion defined under the option GOTO. The words used in this menu are the same as in APT except for the word AUX, which denotes the selection of APT functions (APT and postprocessor commands). A menu of available APT functions is displayed for selection by the user if the option AUX is selected (Fig. 10-8). The user can specify the APT commands, such as CLPRNT or END, or define machining specifications, such as spindle speed or coolant, by selecting the appropriate APT words in the displayed menu of APT functions by means of the light pen and keying in the necessary parameters.

After the machining specifications have been correctly defined, the user can start defining the cutter motion either in the GOTO mode (point-to-point motion) or in the GO mode (contouring motion).

COPR 1985 CADAM	NUM CON		VIEW PV	SCL 1.000	WDO 1.000,.0
		SEL MAJOR WORD			

CLPRNT		ATANGL	MANUAL	XYPLAN
GOHOME	DELAY/	AUTO	MAXIPM	YZPLAN
OPSTOP	HEAD/	BORE	MAXRPM	ZXPLAN
PPRINT	INSERT	CCLW	MIST	
RAPID	LEADER/	CLW	NEXT	
STOP	LINTOL/	CSINK	NIXIE	
GTOLER/	LOADTL/	DECR	NOW	
MIRX	MCHTOL/	DEEP	OFF	
MIRY	OPSKIP/	DOWN	ON	+
MIROFF	PITCH/	DRILL	RADIUS	
	PREFUN/	FACE	RANGE	
	REWIND/	FLOOD	REAM	
	ROTABL/	HIGH	RIGHT	
ROTATE/	ROTHED/	INCR	ROTREF	
	SELCTL/	IPM	RPM	
	SEQNO/	IPR	SFM	
	SPINDL/	LARGE	SMALL	
AUXFUN/	THREAD/	LEFT	TAP	
COOLNT/	TMARK/	LENGTH	THRU	
CUTCOM/	TRANSZ/	LOCK	TURN	
CYCLE/	TURRET/	LOW	UP	

/ POCKET / INDEX / COPY / RETURN /

Figure 10-8 The APT auxiliary functions that can be selected under menu option AUX in the NC function. (Courtesy of CADAM INC.)

If a point-to-point motion is to be defined, the option GOTO should be selected. The system responds with the message

KEY X,Y/IND PT/SEL ELEM

which suggests that in the GOTO mode, a motion can be defined by typing in the X and Y coordinates of the destination point, by indicating the destination through the use of the light pen, or by selecting a predefined element, either a point, line, or curve. A curve or line selected as the guiding element of the point-to-point motion must have a previously defined point where the motion along the curve begins.

If the next motion is a contouring motion, the option GO should be selected, together with the proper option regarding the tool position with respect to the desired profile. For example, if the tool is required to be on the right-hand side of the profile, the TLRGT option should be selected. The next step is to define the start-up motion.

As in APT programming, there are various ways to define a start-up motion on the CADAM system. Since an engineering drawing may consist of several views, the start-up motion can be initiated either from the primary view (X-Y plane) or the side view (Y-Z or X-Z plane). When the GO function and an appropriate modifier, TLON, TLLFT, or TLRGT, are selected, the system responds with the message

SEL ELEM

indicating to the user that a line or curve should be selected, either in the primary view or in the side view, as the drive surface of the first contouring motion step. The following description applies to the definition of a start-up motion in the primary view (Fig. 10-9[a, b, and c]).

After an element (e.g., S1 in Fig. 10-9[b and c]) has been selected, the message requesting the allowance to be left on surface S1

KEY THICK =

is displayed on the screen. The user should respond by typing in the allowance value. The user can then choose either to define the part surface by selecting the function PSIS on the lower part of the screen, and then to continue defining the start-up motion, or to leave the part surface to be specified after completing the definition of the start-up motion. In either case, the next message displayed by the computer is

IND NORMAL/SEL PT/SEL ELEM

which provides three ways, illustrated in Fig. 10-9(b and c), to specify the end position of the start-up motion. When the first option (IND NORMAL) is selected (Fig. 10-9[b]), the user should use the light pen to indicate a position through which an imaginary normal line, N1, passes on the proper side of the selected drive surface S1. This normal line is used as the check surface to determine the end position of the start-up motion.

If the second option, SEL PT, is chosen (Fig. 10-9[b]), then a predefined point (e.g., P1) should be selected. Again, an imaginary normal, N1, projecting from selected point P1 is used to determine the end position of the tool.

The third option, SEL ELEM, allows the user to define the tool end position on the basis of a selected check surface, S2 (Fig. 10-9[c]). After the second element, S2, is selected, an allowance, t2, on that surface should also be entered in response to the message "KEY THICK." Since the tool can be positioned in any one of the

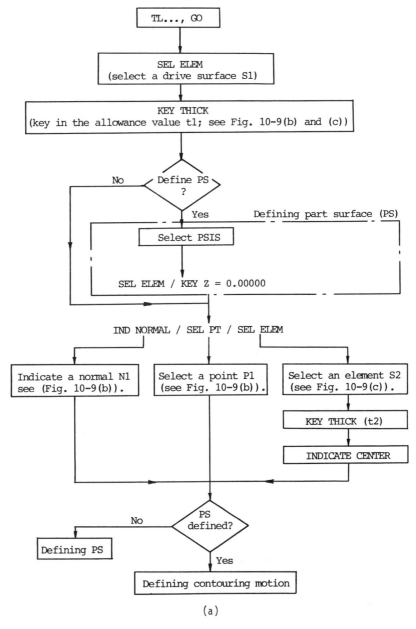

(a)

Figure 10-9(a)   The procedures used for defining start-up motion in the primary view.

(continued)

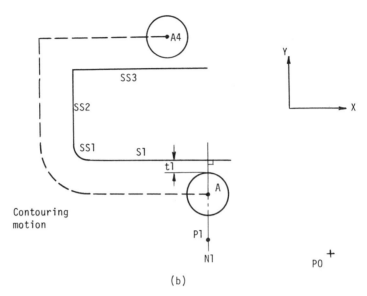

(b)

Figure 10-9(b) Definition of the end position, A, of the start-up motion by selecting option IND NORMAL (indicating the normal N1) or SEL PT (selecting a point, P1). P0: tool starting position.

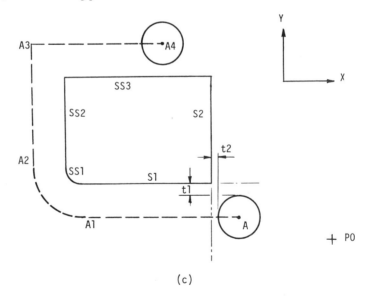

(c)

Figure 10-9(c) Definition of the end position, A, of the start-up motion by selecting option SEL ELEM (selecting a second element, S2, as the check surface), by specifying the second allowance $t2$, and by indicating the position of the tool.

four quadrants formed by lines S1 and S2 if the TLLFT or TLRGT option is chosen, the proper position should be indicated through the use of the light pen and the INDICATE function key. The tool will then be positioned at the proper end position, shown in Fig. 10-9(c).

Once the start-up motion is properly defined, the following operations simply select the consecutive surfaces that constitute the part profile (recall that the first drive surface has been selected in the start-up motion). For example (Fig. 10-9[c]), if the desired profile consists of S1, SS1, SS2, and SS3, and if the end position of the tool is A4, then surfaces SS1, SS2, and SS3 can be selected (surface S1 need not be selected again since it has been specified in the start-up motion). Each selected surface is used as the check surface for the previous motion step and as the drive surface for the next motion step.

To terminate a contouring motion, one simply chooses the GOTO function and selects a proper end position (A4 in this case) on the profile.

During the definition of contouring motion, one can also choose the options listed preceding the GOTO option (i.e., RAPID, FEED, PSIS, AUX, and THICK) to change the machining specifications and part surface.

After all the machining operations have been defined, the option END (Fig. 10-7) is selected. It denotes, as does the APT statement END, the end of the machining process. Then the REPLAY option can be chosen for step-by-step review of the machining process by depressing repeatedly the YES/NO key on the function keyboard. Forward and backward replay options are also available. During the replay, one can also delete or modify a machining specification or part of the cutter path. The POST option is selected when the designed machining process has been verified. Two messages must be input into the system: (1) the program identification, which is similar to the commentary text specified in the statement PARTNO in APT, in response to the computer message

KEY PN,FN / Y/N CURRENT NO

where PN and FN are abbreviations for "part number" and "file number," and the second option accepts the current file name as PN and FN by depressing the YES/NO function key, and (2) the machine code, which should be the same as that specified in columns 73 through 76 in the appropriate set of statements in the file "MACHINE STMTS A1," in response to the computer message

KEY MACHINE =

After completion of the steps described above, the message DONE appears on the upper left corner of the screen, indicating that the NC data have been stored in the CADAM data base.

Figure 10-10 presents an example of the use of the CADAM system to define an NC machining process.

An APT Source Geometry Generator (ASGG) Program is provided in the CADAM system to help an NC programmer generate automatically the APT geometric definition statements based on the created design model. This saves the APT programmer the time-consuming task of defining and labeling geometric entities for a complicated part.

One of the disadvantages of the CADAM system is that a circular cutter motion is defined as a series of point-to-point motions specified by a number of GOTO statements based on linear interpolation. Thus a large number of linear motion statements will be generated in the NC output for circular motion, causing the resulting NC output to be unnecessarily lengthy.

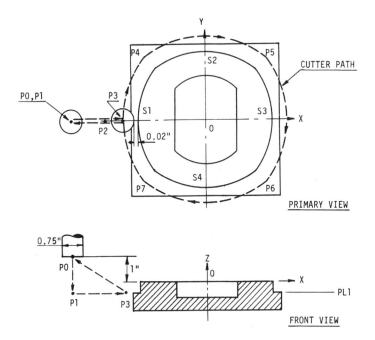

Figure 10-10   The procedures used to define an NC machining process on the CADAM system:

1. Depress function key NC on the function keyboard.
2. Select function PASS.
3. Select function START.
4. Type in the cutter diameter value: 0.75″.
5. Type in the corner radius value: 0.
6. Select P0 in the primary view. (P0 is the starting point).
7. Type in the z coordinate: 1″.
8. Select function AUX and specify spindle speed and direction of rotation, coolant, etc.
9. Select RAPID.
10. Select P1 in the front view first and P1 in the primary view next (remember that the subfunction GOTO has been pre-selected). The tool now comes to point P1.
11. Select FEED and type in the desired feedrate.
12. Select subfunctions GO and TLLFT.
13. Select surface S1 to start defining the start-up motion.
14. Type in the allowance value, e.g., 0.02″.
15. Select function PSIS.
16. Select surface PL1 (the part surface).
17. Select point P2 (the tool is now at point P3).
18. Select S2, S3, S4 and S1 successively.
19. Select function GOTO to terminate the contouring motion.
20. Select point P3 (the tool is now at point P3).
21. Select P0 first in the front view and then in the primary view (the tool is now at point P0).
22. Select function AUX and turn off the spindle and coolant.
23. Select function END.
24. Select function LEAVE.
25. Select function EXIT.

10.3 THE TRANSFER OF AN NC PROGRAM FROM A COMPUTER TO AN NC CONTROLLER

An NC program generated by the APT-AC NC processor or by a CAD/CAM system is stored in the memory of the computer. It must be input into the NC controller to direct the operations of the NC machine. There are several ways to transfer an NC program from the computer to the NC controller:

1. Using manual input
2. Using the paper or magnetic tape preparation unit connected to the computer to punch a paper tape or to record a magnetic tape, and then loading the tape onto the NC controller
3. Sending the program from the computer directly to the machine (The computer can be used to send the NC statements one by one during machining [i.e., the computer directly controls the NC process — called direct numerical control, or DNC for short] or to send a complete NC program to the memory unit of the NC controller [distributed numerical control, or also DNC for short].)

All three of these methods are used in production; however, the selection of a specific method depends on the technical environment (i.e., available facilities, technical expertise of the user, and application). In general, manual input is the slowest and most error-prone process. It should be replaced by the other two if possible. The third method is the most straightforward. In direct numerical control, the computer and the controller communicate with each other during the machining process. The computer sends an NC statement on request by the NC controller. Therefore the NC machine is under direct control of the computer. Generally, a computer on which the APT and CAD/CAM systems can be run is also used by many departments for various purposes in a company. The work load on the computer system may be so heavy that it is not able to control the machine tool on a real-time basis. In addition, the control of a number of NC machines by one central computer system results in job interruptions when the computer system functions improperly, a situation that occurs relatively often.

In modern CNC machines, there is a memory unit in the CNC controller for storing NC programs. A complete NC program can be down-loaded from the computer directly to the CNC controller (i.e., distributed numerical control). Thus a stored program, instead of a computer in direct numerical control, is used to control the NC machining process.

Storing an NC program on a magnetic cassette tape or punched paper tape is also a useful way to run a long program when the memory space of the CNC controller is insufficient. A magnetic tape is generated by the magnetic cassette tape unit connected to the computer, and a paper tape, by the paper tape punch unit.

In recent years, the introduction of the personal computer (PC) with expanded capability has led to the increasing use of PCs to generate NC programs and to control NC machines. A personal computer can be used as a stand-alone system that generates NC programs by using less sophisticated NC software, or it can be used as a satellite computer of a central computer to store the program generated by the latter. In either case, an NC program can be sent from the PC to the CNC controller.

The CNC controller and the PC can be linked through the well-known RS232 inter-face. According to the EIA RS-232-C standard,[18] several parameters should be prop-erly set on both the PC and the CNC controller to ensure satisfactory communication between the two units. On the PC side, the speed at which the data are sent (i.e., the baud* rate), the type of parity checking (odd for an EIA code and even for an ISO code), and the number of the stop bit (usually one) should be correctly defined. On the CNC controller side, the parameters regarding the baud rate (which should be the same as that for the PC), the code type (EIA or ISO), and the input device need to be properly set. The procedures for setting up these parameters on an IBM PC and a FANUC 6MB CNC controller are described in references 19 and 20, respectively.

REFERENCES FOR PART III

1. Kaplinsky, R., *Computer-Aided Design: Electronics, Comparative Advantage and De-velopment*. New York: Macmillan, 1982.

2. Scrivener, S. A. R., ed., *Computer Aided Design and Manufacture: State of the Art Re-port*. Oxford, UK: Pergamon Infotech, 1985.

3. Besant, C. B. and Lui, C. W. K., *Computer-Aided Design and Manufacture,* 3rd ed. Chichester, U.K.: Ellis Horwood, 1986.

4. Mortenson, M. E., *Geometric Modeling*. New York: Wiley, 1985.

5. *CAD/CAM, Productivity Equipment Series,* 2nd ed. Dearborn, Mich.: Society of Manu-facturing Engineers, 1985.

6. *Modern Machine Shop 1986 NC/CAM Guidebook* Cincinnati, Ohio: Gardner Publica-tions, 1986.

7. Faux, J. D. and Pratt, M. J., *Computational Geometry for Design and Manufacture*. Chichester, U.K.: Ellis Horwood, 1979.

8. Hyodo, Y., "HAPT-3D: A Programming System for Numerical Control." In: *Computer Languages for Numerical Control,* J. Hatvany, ed. Amsterdam: North-Holland, 1973.

9. Parkinson, A., "An Automatic NC Data Generation Facility for the BUILD Solid Model-ing System." In: *16th CIRP Intl. Seminar on Mfg. Systems,* Tokyo, 1984.

10. Parkinson, A., "The Use of Solid Models in BUILD as a Database for NC Machining." In: *Software for Descrete Manufacturing,* J. P. Crestin and J. F. McWaters, eds. Amsterdam: North-Holland, 1986.

11. Ferstenberg, R. et al., "Automatic Generation of Optimized 3-Axis NC Programs Using Boundary Files." *Proceedings, 1986 IEEE International Conference on Robotics and Automation,* The Institute of Electrical and Electronics Engineers, 1986, p. 325.

12. Preiss, K., et al., "Automated Part Programming for CNC Milling by Artificial Intelli-gence Techniques," *Journal of Manufacturing Systems,* vol. 4, No. 1, 1985, p. 51.

13. IBM Manual SH20-2095-6: *CADAM APT Interface Installation Guide,* 7th ed. Rye Brook, N.Y.: IBM Corp., 1985.

14. IBM Manual SH20-3026-3: *CADAM Numerical Control (User Reference Manual)*. Rye Brook, N.Y.: IBM Corp., Dec. 1985.

15. IBM Manual SH20-2036-1: *Computer-Graphics Augmented Design and Manufacturing system: NC Supplement to User Training Manual*. Rye Brook, N.Y.: IBM Corp., 1982.

*The term "baud" means "bits per second."

16. IBM Manual SH20-6509-0: *CADAM Interactive User Reference Manual*, vols. 1 and 2. Rye Brook, N.Y.: IBM Corp., 1985.

17. IBM Manual SC19-6210-2: *CMS User's Guide, Release 3*, 3rd ed. Rye Brook, N.Y.: IBM Corp., 1983.

18. EIA Standard RS-232-C: *Interface Between Data Terminal Equipment and Data Communication Equipment Employing Serial Binary Data Interchange*. Washington, D.C.: Electronic Industries Association, Aug. 1969 (reaffirmed 1981).

19. IBM Personal Computer Professional Series: *Asynchronous Communication Support*, Version 2.0, 2nd ed. Rye Brook, N.Y.: IBM Corp., 1982.

20. *GN6MB-2 Maintenance Manual*. Elk Grove Village, Ill.: General Numeric Corp.

part IV

The Design and Implementation of Postprocessors Based on the IBM APT-AC Numerical Control Processor

chapter 11

Introduction to Postprocessors and CLDATA

11.1 THE NECESSITY FOR A POSTPROCESSOR

The need for a postprocessor, in addition to the APT-AC NC processor, for processing APT programs is due to the fact that there are no universal NC codes that can be interpreted by NC controllers made by various manufacturers. Although an international standard (ISO 6983/1)[1] and an American standard (EIA RS-274-D),[2] specifying the NC statement (block) format and the definition of address words, have been proposed, different manufacturers may use different codes to define a function. As a result, NC programs in various formats and codes are needed for machining the same part on NC machines with similar capability but of different makes. Moreover, a machine tool manufacturer may choose to use different codes to define the same function on different types of NC machines. The use of codes G50 and G92 for defining the machine coordinate system in the FANUC 6T (for lathe) and FANUC 6MB (for milling machine) controller systems, respectively, is a typical example. In addition, controllers of the same make and model are often equipped with different options as required by customers; as a result, these machine-controller systems still require different postprocessors, although they have the same type of controller. Therefore an APT program should be processed in the way required by the given NC machine-controller system so that the result is recognizable by the NC machine.

415

There are two ways to solve this problem. The first is to design a processor that can be used to process the APT programs for a specific NC machine only. An APT program can then be processed and directly translated into the format required by the NC machine. Such an approach is unjustifiable because the translation and calculation that must be done within the processor are so complicated that it is very difficult and economically unreasonable for a user to develop a special NC processor for each individual machine. The other approach is to design an NC processor that performs general translations and calculations and that generates a standard output consisting of data regarding tool positions and machine operations. These data, known as cutter location data or cutter center line data (or CLDATA* for short), are then processed and translated into the format required by the NC controller by a computer program, known as a *postprocessor,* designed specifically for the given machine. The postprocessor is a smaller program, in comparison with the NC processor, and its design and implementation are comparatively easier. In this approach, the NC processor carrying out the first stage of processing can be used to process APT programs for use by various types of NC machines, including nonmachining types; a user needs only to provide the postprocessor in order to obtain the NC program for a given machine. This second approach has been universally adopted for processing NC machining programs, not only in APT, but in other higher-level languages as well.

In order for users and machine tool manufacturers to interface with the CLDATA more easily, the International Organization for Standardization (ISO) has developed two standards, namely, ISO 3592-1978[3] and ISO 4343-1978,[4] for the NC processor output. These standards apply to the output of any general-purpose NC processor, including those which process programs written in languages other than APT.

11.2 THE CLFILE

According to the ISO 3592-1978 standard, the CLDATA consists of a sequence of logical records, each representing a postprocessor command or a cutter position calculated by the NC processor. A logical record is composed of a sequence of logical words, each of which represents an integer number, a real number, or six characters. The first three logical words of a record have the same physical size and are integers. The rest of the words also have the same physical size, although not necessarily the same as the first three. The first word represents the sequence number of the logical record; the second represents the record type. The remaining words in a record depend on the record type and are related to the contents of the record.

According to the ISO standard, there are 18 types of records (listed in Table 11-1). They include all the records needed by the calculation stage and are contained in the output file, PROFIL, from the translation section of the NC processor. After processing by the calculation section, some types of records, such as 4000, 7000, 8000, and 12000 types, are no longer needed for defining the cutter

---

*Note that the output of the IBM APT-AC Numerical Control Processor is CLFILE, which contains the data that a postprocessor should process.

motion and machine operation; therefore they are not included in the output file, CLFILE, from the calculation section.

The ISO standard does not define the physical representation of a logical record. Therefore NC processors designed by different vendors may have different record formats of the output file (CLFILE).[5]

The result of the processing by the IBM APT-AC Numerical Control processor is the CLFILE; CLDATA is a term that is commonly used to denote the output from the NC processor (i.e., CLFILE) or its printed form. The physical representation of the CLFILE from the IBM APT-AC NC processor is as follows. The first three words in a record are four-byte words indicating the record sequence number, the record type or major word class, and the major word subclass, respectively. The rest are eight-byte words representing the minor words, parameters, or character strings in a statement. Each logical record can be called for and processed by a postprocessor. For the IBM APT-AC NC processor, only 8 of the 18 types of records defined in the ISO standard (Table 11-1) are used in the CLFILE; they are records of types (or classes in the IBM manual) 1000, 2000, 3000, 5000, 6000, 9000, 14000, and 20000. Further, an additional class of record, class 13000, is included in the CLFILE to represent error diagnostic messages. The contents of a logical record are different for different classes of records; they are listed in Tables 11-2 to 11-9. Accordingly, the number of words in a logical record is different for different classes of records. A record of the CLFILE from the IBM APT-AC NC processor has the following format[5]:

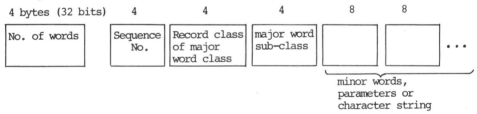

As can be seen, an additional word of four bytes, representing the number of words contained in the current record, is added at the beginning of each record. The APT major and minor postprocessor words are coded by numbers (see the IBM manual[6]). For example, the postprocessor words ARCSLP, MACHIN, and PARTNO are represented by numbers 1029, 1015, and 1045, respectively, and the minor word ON is represented by two numbers, 58 and 71, indicating its class and subclass. Thus, for example, the APT statement

PARTNO NCTEST   A0001

which is a class 2000 record, has the following form in the CLFILE:

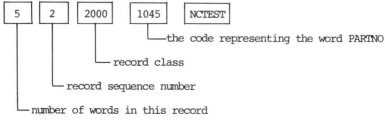

TABLE 11-1   THE TYPES OF RECORDS CONTAINED IN THE OUTPUT OF AN NC PROCESSOR*

Record Type	Record Name	Content
1000	Input sequence	Sequence and identification of statement in original part program
2000	Postprocessor commands	Instructions that are interpreted and processed by postprocessor
3000	Surface data	Canonical form of input geometry
4000	Relative tool position	Relative tool position with respect to drive and part surfaces
5000	Tool position	Information regarding tool position and tool motion vector
6000	Tolerance or cutter information	Information regarding tolerance, cutter, CUT-DNTCUT command, etc., which is needed for calculation section of NC processor
7000	Start-up motion command	Tool position with respect to surface defined in start-up motion
8000	Motion information	Information indicating tool direction with respect to last tool motion
9000	Axis mode; units	Information regarding axis mode (multiaxis or 3-axis) or units
12000	Special program parameter	Information or data for invocation of a special program
14000	FINI record	Termination record of a part program
15000	Unsegmented tool path	Unsegmented information regarding nonlinear tool paths
16000	Workpiece contour description	Parameters or data that describe workpiece contour
17000	Tool description	Record types reserved for tool, material, and machine descriptions, respectively.
18000	Material description	
19000	Machine description	
20000	Literal type of postprocessor command	Literal instructions or postprocessor commands
28000 to 32000	Proprietary records	Record types reserved for private use (will not be standardized)

NOTES:
1. Type 4000, 7000, 8000, and 12000 records do not normally form part of CLDATA files as input to postprocessors.
2. Postprocessor words used in a part program may be represented in the output data in two different ways, namely, as type 2000 or type 20000 records. The data in type 2000 records consist of intermixed strings of integer and real numbers; whereas those in type 20000 records appear as literal strings of characters.

*Modified from ISO Standard 3592-1978.[3] This material is reproduced with permission from the International Organization for Standardization Standard Numerical Control of Machines—NC Processor Output—Logical Structure (and Major Words), ISO 3592-1978, copyrighted by the American National Standards Institute, 1430 Broadway, New York, NY 10018.

A class 1000 record (Table 11-2) is generated for each statement, indicating the internal statement number (ISN), the APT statement label, and the sequencing specified in columns 73 through 80 of the input card. Thus, for the statement given above, two records are generated, namely, the class 1000 record

| 6 | 1 | 1000 | 0001 | 0 | A001 |

and the class 2000 record given previously.

A postprocessor command (or word) and its parameters are represented by a class 2000 record (Table 11-3). The length of the record varies with the APT statement; up to 252 parameters may be contained in a class 2000 record.

TABLE 11-2 THE CONTENTS OF A CLASS 1000 RECORD IN THE CLFILE

Word	Content	Word Length (Bytes)
1st	Record sequence number (integer)	4
2nd	Record-class code (i.e., 1000)	4
3rd	Internal statement number (ISN)	4
4th	Statement label (it is 0 if not assigned)	8
5th	Sequencing specified in columns 73 through 80 of input card in alphanumeric form (it is blank or 0 if none)	8

Modified from IBM Manual (SH20-1414-2): *Automatically Programmed Tool–Advanced Contouring Numerical Control Processor: Program Reference Manual,* 3rd ed. Rye Brook, N.Y.: IBM Corp., 1985. (Courtesy of International Business Machines Corporation.)

TABLE 11-3 THE CONTENTS OF A CLASS 2000 RECORD IN THE CLFILE

Word	Content	Word Length (Bytes)
1st	Record sequence number (integer)	4
2nd	Record-class code (i.e., 2000)	4
3rd	Major postprocessor command (word) code	4
4th	First parameter to right of slash in APT statement	8
5th	Second parameter (if any) to right of slash in APT statement	8
6th	Additional parameter (if any) to right of slash in APT statement	8
	(A maximum of 252 parameters, each with a word length of 8 byte, may occur in this record.)	

Modified from IBM Manual (SH20-1414-2): *Automatically Programmed Tool–Advanced Contouring Numerical Control Processor: Program Reference Manual;* 3rd ed. Rye Brook, N.Y.: IBM Corp., 1985. (Courtesy of International Business Machines Corporation.)

A class 3000 record (Table 11-4) is generated when the drive surface is a circle or cylinder. This record is included in the CLFILE, together with a class 5000 record indicating the tool positions with respect to the part profile so that a circular interpolation NC code, G02 or G03, can be generated when the APT statement ARCSLP/ ON, which appears as a class 2000 record in the CLFILE, is specified.

A class 5000 record (Table 11-5) is related to the cutter position defined in an APT motion statement. For the simple motion defined by statement GOTO/(point) or GODLTA, there is only one cutter position in this record. For the cutter motion specified by a contouring motion statement defining a nonlinear motion or by a point-to-point type statement defining more than one cutter position, for example, GOTO/(pattern), several cutter positions may occur in this record. A contouring motion statement defining the cutter motion along a circle, C1 — for example

TL...,GO.../C1,...

generates a set of records in the CLFILE, consisting of a class 1000 record defining the input card sequence, a class 3000 record specifying the canonical form of circle C1, and a class 5000 record defining the cutter positions.

A class 6000 record (Table 11-6) is related to the cutter specified in the APT program, whereas a class 9000 record (Table 11-7), regarding axis mode, allowance, and measurement unit, is for use by the calculation section. Diagnostic messages are stored as class 13000 records (Table 11-8) in CLFILE, and a class 14000 record (Table 11-9) is reserved solely for the last statement, FINI, of an APT program.

TABLE 11-4 THE CONTENTS OF A CLASS 3000 RECORD IN THE CLFILE

Word	Content	Word Length (Bytes)
1st	Record sequence number (integer)	4
2nd	Record-class code (i.e., 3000)	4
3rd	Subclass code (In CLFILE, this word always has value of 2, indicating that surface defined by this record is a drive surface.)	4
4th	Tool-surface relationship code	8
5th	Surface type of indicator	8
6th	Number of parameters in canonical form	8
7th	Alphanumeric name of surface	8
8th	Subscript value of the surface symbol	8
9th	X coordinate of circle (cylinder) center	8
10th	Y coordinate of circle (cylinder) center	8
11th	Z coordinate of circle (cylinder) center	8
12th	X component of circle (cylinder) axis	8
13th	Y component of circle (cylinder) axis	8
14th	Z component of circle (cylinder) axis	8
15th	Radius of circle (cylinder)	8

Modified from IBM Manual (SH20-1414-2): *Automatically Programmed Tool–Advanced Contouring Numerical Control Processor: Program Reference Manual*, 3rd ed. Rye Brook, N.Y.: IBM Corp., 1985. (Courtesy of International Business Machines Corporation.)

TABLE 11-5 THE CONTENTS OF A CLASS 5000 RECORD IN THE CLFILE

Word	Content	Word Length (Bytes)
1st	Record sequence number (integer)	4
2nd	Record-class code (i.e., 5000)	4
3rd	Subclass code indicating motion type	4
	$= 3$ for statement FROM/	
	4 for statement GODLTA/ (if only one motion vector is specified in this statement)	
	5 for statement GODLTA/ (if more than one motion vector is specified), or for statement GOTO/ or GO . . ./ when one or more than one resultant point is generated	
	6 for the same type of statements as indicated by code 5 with multiple-point record (Under this subclass a series of points* may occur.)	
4th	Symbol for geometric entity specified in statement FROM/, GODLTA/, GOTO/ or GO . . ./ (blank if none)	8
5th	Subscript used in symbol for geometric entity in 4th word	8
6th–8th*	x, y, and z coordinates of point in 4th word or incremental components dx, dy, and dz of motion vector	8
9th–11th*	x, y, and z components of cutter axis vector (These three words appear in this record for each resultant point when multiaxis programming is in effect.)	8

*The maximum number of points allowed in this record is 83 for three-axis programming or 41 for multiaxis programming. After processing, the NC processor outputs a maximum of 81 points for three-axis programming or 41 points for multiaxis programming.

Modified from IBM Manual (SH20-1414-2): *Automatically Programmed Tool–Advanced Contouring Numerical Control Processor: Program Reference Manual*, 3rd ed. Rye Brook, N.Y.: IBM Corp., 1985. (Courtesy of International Business Machines Corporation.)

There may also be a class 20000 record in the CLFILE. It is used as a substitute for a class 2000 record when a special APT statement specifying the format of the CLFILE is given.

Figure 11-1 gives an example of an APT program and its corresponding records in the CLFILE.

11.3 THE GENERAL STRUCTURE OF A POSTPROCESSOR

A CLFILE is transformed into NC codes through a sequence of processing operations by the postprocessor. The processing is carried out as follows (Fig. 11-2): When the postprocessor is first called, a record in the CLFILE is read into the postprocessor. It is then processed by the general processing section of the postprocessor and the result is stored in a buffer array. The data are further processed, checked, and then arranged into a format that is recognizable by the NC controller. Then they are output by the output section of the postprocessor. When the postprocessor has finished processing a record, it reinitializes or updates the buffer arrays, counters, and variables in such a way that they can be used for processing the next record. Then the next record is read in. The postprocessor repeats this cycle until the FINI record is read and processed.

TABLE 11-6 THE CONTENTS OF A CLASS 6000 RECORD IN THE CLFILE

Word	Content	Word Length (Bytes)
1st	Record sequence number (integer)	4
2nd	Record-class code (i.e., 6000)	4
3rd	Record subclass code (In CLFILE, only value 6 is used, which indicates cutter form, dimension, and cutter-controlling surface relationship.)	4

The following words depend on cutter definition:

	CUTTER/A1,A2,A3,A4,A5,A6,A7	CUTTER/OPTION,K, $\left\{ \begin{array}{c} r,h \\ \text{OFF} \end{array} \right\}$	
4th	A1	1 for K=1 2 for K=2 3 for OFF	8
5th	A2	r	8
6th	A3	h	8
7th	A4		8
8th	A5		8
9th	A6		8
10th	A7		8

Modified from IBM Manual (SH20-1414-2): *Automatically Programmed Tool–Advanced Contouring Numerical Control Processor: Program Reference Manual,* 3rd ed. Rye Brook, N.Y.: IBM Corp., 1985. (Courtesy of International Business Machines Corporation.)

TABLE 11-7 THE CONTENTS OF A CLASS 9000 RECORD IN THE CLFILE

Word	Content	Word Length (Bytes)
1st	Record sequence number (integer)	4
2nd	Record-class code (i.e., 9000)	4
3rd	Code with values 2, 5, or 9, defining axis mode, allowance, or measurement unit	4

The following words depend on the code in the 3rd word:

	2	5 (THICK/A1,A2,A3,A4,A5)	9 UNITS/ $\left\{ \begin{array}{c} \text{INCHES} \\ \text{MM} \end{array} \right\}$	
4th	1 for multiaxis programming 0 for nonmultiaxis programming	A1	1.0 for INCHES 25.4 for MM	8
5th		A2		8
6th		A3		8
7th		A4		8
8th		A5		8

Modified from IBM Manual (SH20-1414-2): *Automatically Programmed Tool–Advanced Contouring Numerical Control Processor: Program Reference Manual,* 3rd ed. Rye Brook, N.Y.: IBM Corp., 1985. (Courtesy of International Business Machines Corporation.)

TABLE 11-8   THE CONTENTS OF A CLASS 13000 RECORD IN THE CLFILE

Word	Content		Word Length (Bytes)
1st	Record sequence number (integer)		4
2nd	Record-class code (i.e., 13000)		4
3rd	Code indicating error message from calculation section (Two codes, 3 and 7, are used, indicating that following words represent error message number and textual message, respectively.)		4
	The following words depend on the code in the 3rd word:		
	3	7	
4th	Error message number	Number of characters in textual message	8
5th		First eight characters in textual message	8
6th		9th to 16th characters	8
...		....	8
...		....	8
...		....	8

Modified from IBM Manual (SH20-1414-2): *Automatically Programmed Tool–Advanced Contouring Numerical Control Processor: Program Reference Manual,* 3rd ed. Rye Brook, N.Y.: IBM Corp., 1985. (Courtesy of International Business Machines Corporation.)

TABLE 11-9   THE CONTENTS OF A CLASS 14000 RECORD IN THE CLFILE

Word	Content	Word Length (Bytes)
1st	Record sequence number (integer)	4
2nd	Record-class code (i.e., 14000), indicating that a FINI statement is specified	4
3rd	0	4

Modified from IBM Manual (SH20-1414-2): *Automatically Programmed Tool–Advanced Contouring Numerical Control Processor: Program Reference Manual,* 3rd ed. Rye Brook, N.Y.: IBM Corp., 1985. (Courtesy of International Business Machines Corporation.)

It is evident from the discussion presented above that a postprocessor should generally consist of the following sections:

1. The initialization and housekeeping section. Various variables, counters, and arrays are initialized at the start of the processing operation and updated when a new processing cycle begins.

2. Reader section. The records in the CLFILE are read in, one by one, by this section.

3. General or machine-independent processing section. Two different kinds of processing are performed: calculation of geometric data regarding tool posi-

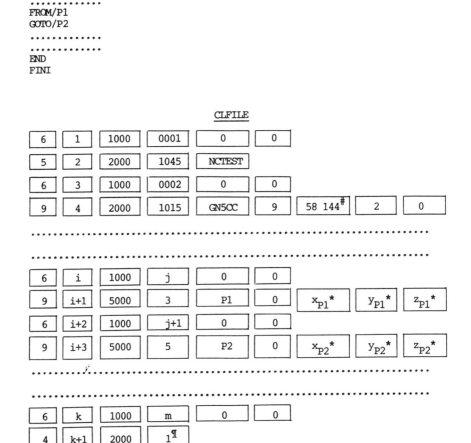

Figure 11-1 An APT program and its corresponding CLFILE. The first four words in a record are of four bytes, the rest are of eight bytes. # = code for the minor word OPTION. * = coordinates for points P1 and P2, respectively. ¶ = code for the postprocessor word END.

tions, and transformation of the machine operation commands into the appropriate numeric codes.

4. Special or machine-dependent processing section. The data are transformed into the format required by the NC controller. A further processing of the cutter path data may also be needed in this section to accommodate machine dynamics (e.g., for control of feedrate to prevent overshoot in corners). In

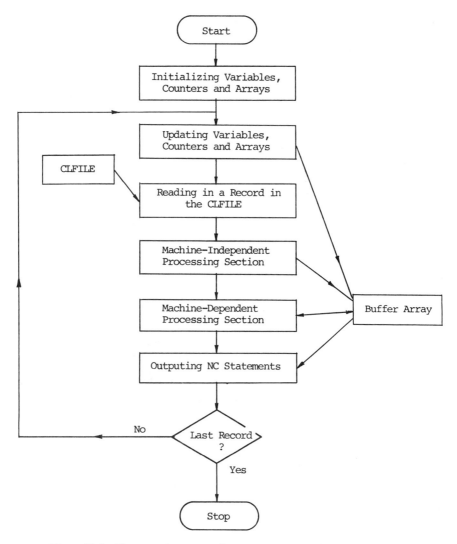

Figure 11-2   The general structure of a postprocessor and its processing flow.

addition, the tool positions, feedrates, and spindle speeds are checked to ensure that they will not exceed permissible limits.

5. Output section. The data are output (printed, punched, or both) as NC codes that can be used as direct input into the NC machine.

The design of a postprocessor is a complex task because a great number of words, parameters, and variables are involved, and communication within the postprocessor is very intricate. However, an analysis of the functions of the five sections indicates that the processing carried out by sections 1, 2, 3, and 5 is not machine related and is generally similar for every postprocessor. Only section 4 performs processing operations that must be designed according to the input requirement of the NC controller and the characteristics of the NC machine. Therefore if one post-processor already exists, the design of a new postprocessor can be restricted to the design of that single section which performs the machine-dependent processing (i.e., section 4). This is the very principle adopted by the designer of the IBM APT-AC NC processor. A special subprogram called *DAPP* (Design Aid for Postprocessor) is included in the NC processor.[7] In the DAPP subprogram, a postprocessor framework has been provided, including the four sections that perform the processing operation applicable to any postprocessor. To design a postprocessor for a specific NC machine, a user need only provide the characteristic parameters regarding that NC machine and to write a subroutine that performs the special processing. Thus the design of a postprocessor is greatly simplified.

In DAPP, the four universal processing sections (i.e., sections 1, 2, 3, and 5, as indicated above) are named the *Initialization and Housekeeping,* the *Reader,* the *Dispatcher,** and the *Output* sections, respectively, whereas the section performing machine-dependent processing is called the *Machine Tool Module* (MTM).

11.4 THE FUNCTIONS OF A POSTPROCESSOR

A DAPP-based postprocessor is a computer program that uses the CLFILE as input, processes it, and then outputs the part-machining program in machine-readable code. Apart from the read-in and output functions, which are common to every postprocessor, a postprocessor should have the following functions[8]:

1. The conversion of a tool position defined in the part coordinate system to that defined in the machine coordinate system

2. The processing of linear interpolation, circular interpolation, higher-order (parabolic) interpolation (if the controller has this function), and cycle motion

3. The generation of correct spindle speed, feedrate, tool, and various machine operation codes

4. Modifying and generating commands that allow for requirements of the NC machine-controller systems

5. Checking or verifying the correctness of, for example, the tool or part travel range, spindle speed, feedrate, and tool motion, and producing diagnostics in the event of errors

6. Setting the output format as required by the NC machine

These functions are briefly explained below.

When writing an APT program, a programmer usually chooses a coordinate system that is most convenient for defining the part geometry and the cutter path. The actual coordinate system selected by a machinist or an NC machine operator during a machining operation might not be the same. The relationship or relative position of these coordinate systems is defined by the ORIGIN statement. A translation

---

*This term is used in the IBM manual.[7] It is introduced here for reference only. In the following discussion, the term "general processing section," which indicates the function of this section more precisely, is used instead.

of tool position is also needed when several tools are used in a machining program. The relative position of one tool to another is defined by the statement

LOADTL/...,SETOOL,......

which was described in Section 6.6. These transformations are not carried out by the NC processor. Since they are independent of machine tools, they can be included in the machine-independent processing section of the postprocessor (Fig. 11-2).

Most NC milling machines and lathes can realize three basic types of motion, namely, point-to-point, linear interpolation, and circular interpolation motion. Some NC machines have only point-to-point and linear interpolation motion, whereas some special NC contouring machines include the function that realizes higher-order interpolation motion. A postprocessor for a milling machine or lathe should usually have the capability to process the three basic types of motion. The processing of these motions is carried out in two steps: general processing, which is not machine related, and special processing, which is machine related.

All postprocessor words and commands regarding machine tool operation should be processed and converted into the appropriate codes. Since the processing is specific to the machine to be used, they should be included in the machine-dependent processing section (MTM).

The overtravel of the tool, resulting from the system gain when its motion direction is changed, affects the accuracy of the workpiece at the corner. The tool should accordingly be decelerated before it changes motion direction. For NC machines of older design, this is realized by the postprocessor by segmenting the motion and assigning a correspondingly slower feedrate to the last segment of the motion. In modern CNC controllers, this function has been included to control automatically the deceleration at the end of one motion and the acceleration at the start of the next motion. The acceleration and deceleration rates can be preset and readjusted according to the required machining accuracy. Therefore the dimensional accuracy can be maintained at the corner of the workpiece without relying on the help of a postprocessor.

Finally, a postprocessor should translate the data in computer code to a code format (EIA or ISO) that can be interpreted by the NC machine controller. The output NC statement (block) should also be set in the block format required by the NC controller. The output should also be generated in various forms, such as paper tape, card, and printed listing, as required by the user.

The realization of these functions will be discussed in detail in the following chapters.

PROBLEMS

11.1 How many types of CLDATA records will be generated for the following statement?

GODLTA/1,2,3,2.5

11.2 How many types of CLDATA records will be generated for the following statement?

GOTO/1,2,3,2.5

11.3   Assume that C1 and L1 are the symbols for a circle and a line, respectively. How many types of CLDATA records will be generated for the following statement?

TLRGT,GORGT/C1,TANTO,L1

11.4   How many types of CLDATA records will be generated for a POCKET statement?

11.5   Assume that a general postprocessor is designed according to the DAPP concept, as described in Section 11.3, and processes the CLDATA for three different NC machines. What is the general structure of this postprocessor? Draw the processing flowchart for this postprocessor on the basis of the concept presented in Fig. 11-2.

11.6   What is the general structure of a postprocessor if it is designed specifically for one NC machine only, without the use of the DAPP routine?

chapter 12

Communication Channels and Major Variables in the DAPP-Based Postprocessor

As we pointed out in Section 11.3, the DAPP is a special subprogram that has been included in the IBM APT-AC NC processor to simplify the design of processors. Both the NC processor and the DAPP subprogram are written in FORTRAN and ASSEMBLER languages.[7] The DAPP provides the complete routines that perform general, machine-independent processing, including the Initialization and Housekeeping sections for initializing and updating variables, counters, and arrays, the Reader Section for retrieving records from the CLFILE, the General Processing Section for machine-independent processing of the CLDATA, the Output Section for outputting the processed results in the required format, and the Control Section for directing the control flow within the DAPP program. The Machine Tool Module (MTM), which performs machine-dependent processing, must be written and provided by the user.

There are a number of variables used in DAPP, some of them representing information from the CLFILE and others used by the DAPP to realize the normal processing function. These variables are specified in the FORTRAN COMMON statement and serve as communication channels within DAPP. Before postprocessing is started, all the variables should be assigned initial values; after postprocessing of a set of CLFILE record, some of the variables should be reinitialized or updated. The initial values of these variables should be assigned by the user according to postprocessor output requirements and the characteristics of the NC machine-controller system to be used.

429

A postprocessor can then be created by linking the MTM and the BLOCK DATA subprogram,* which specifies the initial values of various variables, with the DAPP.

After a set of CLDATA has been read in, it is stored in DAPP as variables of diverse types or classes. These variables are then passed from one section to another as they are processed by the DAPP. They are finally output as an NC paper tape or as a printed output in the desired format.

In this chapter, we first introduce the major variables used in the DAPP. We then describe the method used to collect the information needed by the postprocessor to process the CLFILE. Finally, we discuss the method provided by DAPP for initializing the variables.

12.1 THE BASIC ARRAYS AND VARIABLES USED IN DAPP

Although a large number of variables are used in DAPP, only five arrays and three variables need usually concern the designer of a DAPP-based postprocessor. These are the arrays DATBUF, OUTGRD, LOGCAL, LIMITS, and TIME, the variables OUTCTL, CLKARG, and JLINE, and a few others that will be mentioned later. These variables are initialized to values that allow the postprocessor to perform correct processing, and some of them are changed as a result of processing CLDATA in different sections. Therefore a thorough knowledge of these variables is required to understand the processing that takes place within a DAPP-based postprocessor.

12.1.1 The DATBUF Array

The DATBUF array is used to store the CLDATA information, which is processed by the General Processing Section.[†] These variables are further processed by the MTM* and finally output as NC codes by the Output Section of the DAPP. The DATBUF array is defined as a double-precision, real-number (REAL*8) array. It has a total of 85 elements, DATBUF(1) through DATBUF(85). Positions 1 through 71 in this array are set up as a result of processing by the General Processing Section, the MTM, or both, whereas the rest of the array, that is, DATBUF(72) through DATBUF(85), is reserved for use by the MTM. A full list of the variables in the DATBUF arrays is given in Table 12-1, together with their content and explanation.

DATBUF(1) through DATBUF(9) are used to store information regarding the positions of the cutter-end-center of the mill or the nose-radius-center of the turning tools. For each tool motion statement, positions 1, 2, and 3 contain the destination coordinates in the X, Y, and Z directions, respectively; positions 4, 5, and 6 contain the x, y, and z coordinates of the starting point; and positions 7, 8, and 9 contain the incremental x, y, and z coordinates from the starting point to the end point. The coordinates stored in these positions have been transformed to the machine tool coordinate system so that they can be output as X, Y, and Z codes for use by the NC

---

*A BLOCK DATA subprogram in FORTRAN can be used to initialize values of variables in labeled or unlabeled COMMON areas. It may consist solely of the nonexecutable statements, namely, DATA, COMMON, DIMENSION, and EQUIVALENCE, and any type of declaration.

[†]See Chapters 13 and 14 for the general processing routines and the MTM, respectively.

TABLE 12-1 THE CONTENTS OF DATBUF ARRAY

DATBUF (i)	Content	Explanation
DATBUF (1)	x coordinate of destination point	Coordinate in this position has been transformed and is given with respect to machine tool coordinate system
DATBUF (2)	y coordinate of destination point	Same as above
DATBUF (3)	z coordinate of destination point	Same as above
DATBUF (4)	x coordinate of starting point	Same as above
DATBUF (5)	y coordinate of starting point	Same as above
DATBUF (6)	z coordinate of starting point	Same as above
DATBUF (7)	DATBUF (1) − DATBUF (4)	
DATBUF (8)	DATBUF (2) − DATBUF (5)	
DATBUF (9)	DATBUF (3) − DATBUF (6)	
DATBUF (10)	Programmed machining feedrate, f (units/min) (see note 1)	From FEDRAT/f statement, if statement FEDRAT/IPR.f is specified, then DATBUF (10) $= f \cdot n$, where n is the spindle rotation speed
DATBUF (11)	Rapid motion feedrate (units/min) (see note 2)	Defined by answer to question M05 (see Table 12.6)
DATBUF (12)	Rapid feedrate switch	DATBUF (12) = 0 Programmed feedrate is not rapid 1 Programmed feedrate is rapid
DATBUF (13)	Spindle rotation direction	From SPINDL statement: DATBUF (13) = 1 For clockwise (CLW) direction 2 For counterclockwise (CCLW) direction 3 For spindle stop
DATBUF (14)	Spindle speed, s (rpm)	From statement SPINDL/s,. . . in CLFILE
DATBUF (15)	Tool code, t	From statement LOADTL/t in CLFILE (see note 3)
DATBUF (16)	Tool selection code, t	From statement SELCTL/t in CLFILE (see note 3)
DATBUF (17)	Coolant code	From statement COOLNT/. . . in CLFILE DATBUF (17) = 1 For FLOOD 2 For MIST 3 For TAPKUL (tapping coolant) 4 For ON 5 For OFF
DATBUF (18)	Angular position of rotary table, in number of increments	From ROTABL statement in the CLFILE (see note 4)

(continued)

TABLE 12-1 THE CONTENTS OF DATBUF ARRAY (continued)

DATBUF (i)	Content	Explanation
DATBUF (19)	Dwell time (seconds)	From DELAY or CYCLE statement in CLFILE (see note 4)
DATBUF (20)	Internal sequence number (ISN) of current CLFILE data set	Assigned by NC processor and stored as a class 1000 record in CLFILE
DATBUF (21)	NC block sequence number	Can be assigned by SEQNO statement (see note 4), if specified, or automatically set to 1 for first output block and incremented by 1 for each block if answer to question C21 (see Table 12.7) is other than 0
DATBUF (22)	Rotation direction of rotary table	Set by statement ROTABL:
		DATBUF (22) = 0 No ROTABL statement specified
		1 CLW specified in statement ROTABL
		−1 CCLW specified in statement ROTABL
DATBUF (23)	Programming mode switch (MULTAX/ON or MULTAX/OFF)	DATBUF (23) = 0 When statement MULTAX/OFF is specified (i.e., tool axis is always in **Z** direction)
		1 When statement MULTAX/ON is specified (i.e., tool changes its spatial direction during machining)
DATBUF (24)	Error switch	DATBUF (24) = 0 No postprocessing error
		1 Error detected
DATBUF (25)	Special operation code	DATBUF (25) = ANULL No special operation (see note 5)
		9001 For statement END
		9002 For statement OPSTOP (see note 6)
		9003 For statement STOP (see note 7)
		9004 For statement DELAY
		9005 For statement MCHTOL (see note 4)
		9006 For statement ROTABL
		$+n$ For statement AUXFUN/n (see note 8)
		$-n$ For statement PREFUN/n (see note 8)
DATBUF (26)	Allowable machining error (n) caused by dynamic response of workpiece-machine-controller system	From statement MCHTOL/n in CLFILE (see Table 13-5)

432

| DATBUF (27) | Cutter compensation code | This code assigned by statement CUTCOM/... (see note 4) |

$$\text{DATBUF (27)} = \begin{array}{ll} 0 & \text{For CUTCOM/OFF} \\ -1 & \text{For CUTCOM/LEFT,YZPLAN} \\ -2 & \text{For CUTCOM/LEFT,ZXPLAN} \\ -3 & \text{For CUTCOM/LEFT,XYPLAN} \\ -4 & \text{For CUTCOM/RIGHT,YZPLAN} \\ -5 & \text{For CUTCOM/RIGHT,ZXPLAN} \\ -6 & \text{For CUTCOM/RIGHT,XYPLAN} \\ n & \text{For CUTCOM/}n \end{array}$$

| DATBUF (28) | Type of machining cycle | Set by CYCLE statement (see note 4) in CLFILE; array CYCTYP (INTEGER*4) with two elements is assigned same memory location as that used by DATBUF (28), which is a REAL*8 variable (see note 9) |

$$\text{CYCTYP (2)} = \begin{array}{ll} 82 & \text{For BORE} \\ 151 & \text{For MILL} \\ 153 & \text{For DEEP} \\ 163 & \text{For DRILL} \\ 167 & \text{For REAM} \\ 168 & \text{For TAP} \\ 206 & \text{For BRKCHP (chip-breaking drilling)} \\ 255 & \text{For CSINK (countersinking)} \end{array}$$

DATBUF (29)	z coordinate (z_2) of lowest (feed-in) point in cycle	Coordinate calculated by canned-cycle routine (DLWCYC) (see note 4) in General Processing section
DATBUF (30)	z coordinate (z_3) of tool retreat position in cycle	Same as above
DATBUF (31)	LOGCAL (1) through LOGCAL (4)	LOGCAL array (LOGICAL*1) has six elements; both LOGCAL array and DATBUF (31) have been assigned same memory location by statement EQUIVALENCE (see note 9), for meaning of LOGCAL array, see Section 12.1.3
DATBUF (32)	Clearance drum number	Data needed for NC machines of older design, wherein drums are used to set clearance and depth of drilling cycle
DATBUF (33)	Depth drum number	Same as above
DATBUF (34)	Parameter $n3$ in statement MACHIN (see Sections 6.12.2 and 13.4.1.1)	
DATBUF (35)	Parameter $n4$ in statement MACHIN (see Sections 6.12.2 and 13.4.1.1)	

(continued)

TABLE 12-1 THE CONTENTS OF DATBUF ARRAY (continued)

DATBUF (i)	Content	Explanation
DATBUF (36)	Circular interpolation plane	Assigned by General Processing Section after calculation is made by circular interpolation routine (DLWCIR) (see note 4): DATBUF (36) = ANULL — No circular interpolation requested 1 — Circular interpolation in X-Y plane 2 — Circular interpolation in Y-Z plane 3 — Circular interpolation in Z-X plane
DATBUF (37)	Direction of circular movement	Assigned by General Processing section after calculation is made by circular interpolation routine (DLWCIR): DATBUF (37) = ANULL — For ARCSLP/OFF 1 — For ARCSLP/ON and motion in CLW direction 2 — For ARCSLP/ON and motion in CCLW direction
DATBUF (38)	x coordinate of center of circular arc	Obtained from class 3000 record and then transformed to machine tool coordinate system
DATBUF (39)	y coordinate of center of circular arc	Same as above
DATBUF (40)	z coordinate of center of circular arc	Same as above
DATBUF (41)	Radius of circular arc	Same as in class 3000 record
DATBUF (42) to DATBUF (44)	Incremental distances dx, dy, and dz from starting point to end point in first section of circular movement	Data for first section (quadrant) of circular arc
DATBUF (45) to DATBUF (47)	Incremental distances (I, J, and K) from starting point to circular arc center in X, Y, and Z directions, respectively	Data for first section (quadrant) of circular arc
DATBUF (48) to DATBUF (53)	dx, dy, dz, I, J, and K, respectively	Data for second section (quadrant) of circular arc

DATBUF (54) to DATBUF (59)	Same as above	Data for third section (quadrant) of circular arc
DATBUF (60) to DATBUF (65)	Same as above	Data for fourth section (quadrant) of circular arc
DATBUF (66) to DATBUF (71)	Same as above	Data for fifth section (quadrant) of circular arc
DATBUF (72) to DATBUF (85)	Reserved	Reserved for storing data during processing by MTM

NOTES:
1. The actual unit is defined by the postprocessor designer. See machine tool description question M30 in Section 12.2.1.
2. For modern CNC controllers, the rapid feedrate is adjusted by setting the parameters of the controller. Besides, it is not necessary to specify the feedrate in a rapid motion statement in an NC program.
3. LOADTL is a statement calling for an automatic tool change, whereas the statement SELCTL specifies selection of the tool to be used from the tool magazine and placement of the tool, as well, in position for automatic tool changing.
4. See Chapter 13.
5. In DAPP, the word ANULL is equivalent to 99999.99 and is used as a special symbol representing no input.
6. An APT statement corresponding to the M01 code (optional stop) in the NC program (see Chapter 13).
7. An APT statement corresponding to the M00 code (program stop) in the NC program (see Chapter 13).
8. Statements AUXFUN/n and PREFUN/n are used to generate a miscellaneous function code (M) and a preparatory function code (G), respectively.
9. See the statement EQUIVALENCE in the BLOCK DATA program in Fig. 12.4.

Modified from IBM Manual (SH20-2469-0): *System/370 Automatically Programmed Tool — Intermediate Contouring and Advanced Contouring Numerical Control Processor: Program Reference Manual*. Rye Brook, N.Y.: IBM Corp., 1980. (Courtesy of International Business Machines Corporation.)

machine. DATBUF(1) through DATBUF(9) are used as part of the information needed for defining point-to-point and linear interpolation motion. DATBUF(36) through DATBUF(71) are used, in addition, for circular interpolation motion.

The element DATBUF(37) indicates whether circular interpolation motion is requested (ARCSLP/ON) or not (ARCSLP/OFF). The circular interpolation plane is represented by DATBUF(36). The positions DATBUF(38) through DATBUF(41) contain information regarding the circle-center coordinates and the radius of the circular motion. The APT-AC NC processor divides a circular cutter path into quadrants, with the lines passing through the circle center and parallel to the coordinate axes. For a general case, up to five sections of circular arc will be generated (Fig. 12-1). For each section of a circular motion, the incremental x, y, and z coordinates from the starting point to the end point and the x, y, and z offset values from the starting point to the circle center are stored in positions DATBUF(42) through DATBUF(47) for the first section of the circular path, in positions DATBUF(48) through DATBUF(53) for the second section, and so on.

The positions DATBUF(28) through DATBUF(33) are used to store information regarding cycle motion, which will be explained in detail in Chapter 13. DATBUF(22) and DATBUF(18) contain the position and rotation direction of the rotary table. Machining parameters, such as those regarding feedrate, spindle speed, tool, coolant, and machining tolerance, are stored in DATBUF(10) through DATBUF(19), DATBUF(26), and DATBUF(27). The machining mode MULTAX/ON or MULTAX/OFF is recorded in DATBUF(23). DATBUF array elements 34 and 35 are used to accept the parameters specified in the MACHIN statement in an APT program, whereas elements 20 and 21 contain the record sequence number in the CLDATA and NC output. The variable DATBUF(24) keeps track of the error status during the postprocessing.

The contents of DATBUF(1) through DATBUF(71) are either set by the General Processing Section or first set by the General Processing Section and later processed and reset by the MTM. For those output NC codes that cannot make use of

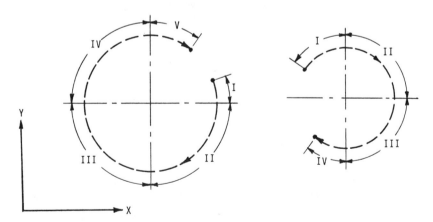

Figure 12-1 Division of circular motion into sections by lines passing through the center and parallel to the coordinate axes. Generally, up to five sections of circular arc result.

the above-mentioned positions of the DATBUF array to store the source data, elements 72 through 85 can be used.

The DATBUF array is designed for use by postprocessors for controllers of various designs, including old hard-wired NC and new CNC controllers. Thus some of its elements may not be used for designing postprocessors for modern CNC machine-controller systems. For example, motion speed is not required to be specified in the statement defining a rapid tool motion for the FANUC 6MB controller; thus DATBUF(11) will not be used. In addition, modern CNC machines do not use the drum to control the tool position in the machining cycle; therefore DATBUF(32) and DATBUF(33) are also not used.

Some of the elements in the DATBUF array are reset by the Housekeeping Section of the DAPP after the processing of a set of CLDATA records has been completed and before the processing of a new set has begun.

12.1.2 The OUTGRD Array

The OUTGRD array is a two-dimensional 20×10 array having a total of 200 elements. These elements are defined as four-byte integers (i.e., INTEGER*4). They can be divided into twenty groups:

OUTGRD$(i,1)$, OUTGRD$(i,2)$, ..., OUTGRD$(i,10)$ $(i = 1, 2, ..., 20)$

Each group has 10 elements. The elements 1 through 9, that is, OUTGRD$(i,1)$, ..., OUTGRD$(i,9)$, in a group are used to define the output format of an NC word; the tenth element, OUTGRD$(i,10)$, indicates the position in the DATBUF array where the data to be output as an NC code in the defined format are stored. Thus, if OUTGRD$(3,10)$ is set as 1, then the data stored in DATBUF(1) will be output in the format defined by OUTGRD$(3,1)$ through OUTGRD$(3,9)$. The content and the use of various positions in a group are listed in Table 12-2. Therefore up to 20 NC words and their formats can be defined in the DAPP.

The values and words contained in elements 1 through 9 in a group are set by the postprocessor designer by answering a series of questions called the *DAPP Questionnaire*. These values remain unchanged during postprocessing. The value of element 10 can be changed and is set by the MTM. Thus the data stored in a specific position of the DATBUF array can be selected and output in the desired format.

The data stored in the DATBUF array can be numeric or alphanumeric. When outputting a numeric value, the Output Section needs the information contained in the nine elements OUTGRD$(i,1)$ through OUTGRD$(i,9)$, to determine the output format. When outputting alphanumeric data, the system needs only the information stored in OUTGRD$(i,4)$ and OUTGRD$(i,8)$.

12.1.3 The LOGCAL Array

The LOGCAL array is a LOGICAL*1 array with six elements. Elements 1 through 4 are used for processing machining cycles; their content is as follows:

LOGCAL(1) = .TRUE.   Tool position change in Z direction occurs in cycle.

TABLE 12-2 THE CONTENT OF EACH GROUP (i) OF OUTGRD ELEMENTS

Element	Content	Used for Defining Punched (PU) or Printed Output (PR)	Assigned According to Answer(s) to Question(s)*
OUTGRD (i, 1)	Letter for word address format (e.g., N, G)	PU and PR	C09
OUTGRD (i, 2)	= 0 No tabulation code preceding each NC word = n n is number of tabulation codes preceding each word	PU and PR	C10
OUTGRD (i, 3)	Number of fractional digits in NC word	PU and PR	C12
OUTGRD (i, 4)	= n Total number of digits (not including decimal point) in NC word = negative value Output being alphanumeric literal string	PU and PR	C11
OUTGRD (i, 5)	= 0 or 1 Decimal point to be omitted in printed output 2 Decimal point to be printed −1 Decimal point to be omitted and plus sign to be printed for positive value −2 Both decimal point and plus sign to be printed	PR	C13 and C15
OUTGRD (i, 6)	Same as above (However, this position is used to define format of punched output.)	PU	C14 and C16
OUTGRD (i, 7)	= 0 Zeros not to be suppressed in output 1 Trailing zeros to be suppressed 2 Leading zeros to be suppressed 3 Both leading and trailing zeros to be suppressed	PU and PR	C17 and C18
OUTGRD (i, 8)	Column number, in printed output, where printing of this word begins	PR	C19
OUTGRD (i, 9)	= 0 This NC word to be printed only 1 This NC word to be printed and punched	PU and PR	C20
OUTGRD (i, 10)	= n Position number in DATBUF array, DATBUF (n) contains data to be output as NC code	PU and PR	Assigned by MTM

Example: For the FANUC 6MB CNC controller, we may define elements OUTGRD $(2, j)$ ($j = 1, \ldots, 9$) for the G code as follows:

OUTGRD $(2, 1)$ = G; OUTGRD $(2, 2)$ = 0; OUTGRD $(2, 3)$ = 0; OUTGRD $(2, 4)$ = 2; OUTGRD $(2, 5)$ = 1

OUTGRD $(2, 6)$ = 1; OUTGRD $(2, 7)$ = 0; OUTGRD $(2, 8)$ = 6; OUTGRD $(2, 9)$ = 1

The defined output format is such that the G code contains only two digits, with no fractional digit. Also, no decimal point is needed, and no plus sign is either printed or punched. Zeros in the G code are not to be suppressed, and the printing of this word begins at column 6. The code is to be printed and punched. Thus if the position OUTGRD $(2, 10)$ is assigned as 73 and DATBUF (73) = 1, then the output G code is G01.

*See Section 12.2.2 and Table 12-7.

Modified from IBM Manual (SH20-2469-0); *System/370 Automatically Programmed Tool—Intermediate Contouring and Advanced Contouring Numerical Control Processor: Program Reference Manual.* Rye Brook, N.Y.: IBM Corp., 1980. (Courtesy of International Business Machines Corporation.)

	.FALSE.	Tool position change in Z direction does not occur in cycle.
LOGCAL(2) =	.TRUE.	Tool change follows immediately.
	.FALSE.	No tool change follows immediately.
LOGCAL(3) =	.TRUE.	Deep-hole drilling cycle (BRKCHP OR DEEP) is effective.
	.FALSE.	BRKCHP or DEEP cycle is not effective.
LOGCAL(4) =	.TRUE.	When in BRKCHP or DEEP cycle, last z position has been reached.
	.FALSE.	When in BRKCHP or DEEP cycle, last z position has not been reached.

If a CYCLE statement* is specified, LOGCAL(1) is set to .TRUE. It is reset to .FALSE. if the statement CYCLE/OFF appears. LOGCAL(2) is set to .TRUE. in a cycle if the LOADTL statement is the next statement. This variable is used by the General Processing Section to adjust the z position of the tool. LOGCAL(3) and LOGCAL(4) are used in the peck drilling cycle of a deep hole. The data in these four positions are set by the DAPP main routine and by the canned-cycle routine, DLWCYC (see Chapter 13), in the General Processing Section. The other two positions are reserved for use by the MTM.

12.1.4 The LIMITS Array

The LIMITS array is a real*8 array with 11 elements (Table 12-3). Eight of them are used to store, respectively, the limits of the tool travel in the X, Y, and Z axes, the maximum feedrate, and the controller resolution (or the minimum incremental

*See Chapter 13.

TABLE 12-3   THE LIMITS ARRAY (REAL*8) WITH ELEVEN ELEMENTS

Limits (i)	Content	Set According to Answer to Question*
LIMITS (1)	Maximum x coordinate	M01
LIMITS (2)	Maximum y coordinate	M02
LIMITS (3)	Maximum z coordinate	M03
LIMITS (4)	Reserved	
LIMITS (5)	Reserved	
LIMITS (6)	Controller resolution or minimum incremental coordinate	C04
LIMITS (7)	Reserved	
LIMITS (8)	Maximum feedrate	M06
LIMITS (9)	Minimum x coordinate	M01
LIMITS (10)	Minimum y coordinate	M02
LIMITS (11)	Minimum z coordinate	M03

*Refer to Sections 12.2.1 and 12.2.2 and to Tables 12-6 and 12-7.

Modified from IBM Manual (SH20-2469-0): *System/370 Automatically Programmed Tool—Intermediate Contouring and Advanced Contouring Numerical Control Processor: Program Reference Manual.* Rye Brook, N.Y.: IBM Corp., 1980. (Courtesy of International Business Machines Corporation.)

motion in each coordinate axis). These limits are specified by the postprocessor designer by answering the corresponding questions in the DAPP Questionnaire. They are used by the General Processing Section and the MTM to determine whether the limit(s) of a motion or speed has (have) been exceeded. If so, an error message will be issued. The element LIMITS(6) is used by the calculation routine in the General Processing Section to round off the results of calculations.

12.1.5 The TIME Array and the Variable CLKARG

The TIME array (REAL*4) has eight elements. The elements TIME(2) through TIME(6) are related to the times needed to perform the necessary machine operations. They should be specified by the postprocessor designer according to the characteristics of the NC machine-controller system. The element TIME(1) contains the time needed to complete the tool motion specified by the current machining statement; it includes also the time for selecting or loading a tool and for dwelling (if a SELCTL, LOADTL, or DELAY statement is specified). TIME(1) is calculated and set by both the Geometric and Auxiliary sections, which are subroutines in the General Processing Section. The elements 7 and 8 of this array are reserved for use by the MTM.

Table 12-4 gives the content of the various elements in the TIME array.

The variable CLKARG is a REAL*4 variable that represents the time required for execution of the current NC block (statement). This variable should be set by the MTM, and its value is the sum of the time required to finish the tool movement and the time needed to execute the machine operation(s) specified in the current NC block. The variable CLKARG is reset to zero after it has been used (output) by the Output Section.

TABLE 12-4   THE TIME ARRAY (REAL*4) WITH EIGHT ELEMENTS

Time (i)	Content	Explanation
TIME (1)	Machining time for current statement	Set by Geometric and Auxiliary sections. Time is calculated by routines processing linear, circular, and cycle machining motions. Then, if DELAY, SELCTL, or LOADTL statement is specified, time in this position will be updated accordingly by Auxiliary section. Refer to respective sections of Chapter 13 for detail.
TIME (2)	Minimum time for executing one statement	Set according to answer to question C05 in DAPP Questionnaire.
TIME (3)	Time to rotate rotary table by 360 degrees	Set according to answer to question M12.
TIME (4)	Time to start or stop spindle	Set according to answer to question M21.
TIME (5)	Time to load a tool	Set according to answer to question M22.
TIME (6)	Time to select a tool	Set according to answer to question M23.
TIME (7)	Reserved for use by MTM	
TIME (8)	Same as TIME (7)	

Modified from IBM Manual (SH20-2469-0): *System/370 Automatically Programmed Tool—Intermediate Contouring and Advanced Contouring Numerical Control Processor: Program Reference Manual.* Rye Brook, N.Y.: IBM Corp., 1980. (Courtesy of International Business Machines Corporation.)

TABLE 12-5   VALUES THAT CAN BE ASSUMED BY THE VARIABLE OUTCTL AND THEIR RESULTING ACTION

OUTCTL	Resulting Action
1	Control is to return to Housekeeping section; then Reader section reads in new set of CLFILE records.
4	MTM is to be called.
5	Print accumulated machining time; print internal sequence number (ISN); output data processed by MTM as one NC block (statement).
6	Print internal sequence number (ISN) and output data processed by MTM as one NC block.
7	Output data processed by MTM as one NC block.
10	Postprocessing is terminated.
11	Punch output.
12	Print "*************" only.

Modified from IBM Manual (SH20-2469-0): *System/370 Automatically Programmed Tool—Intermediate Contouring and Advanced Contouring Numerical Control Processor: Program Reference Manual*. Rye Brook, N.Y.: IBM Corp., 1980. (Courtesy of International Business Machines Corporation.)

12.1.6 The Variable OUTCTL

The variable OUTCTL (INTEGER*4) is used by the MTM to select a required output operation and to direct the control flow within the DAPP. The values that can be assumed by this variable and their meaning are listed in Table 12-5. If different values of OUTCTL are selected during processing by the MTM, different output operations and a different control flow will result. The value of this variable should be selected during the design of the MTM according to output and control requirements in a particular processing stage of the MTM.

12.1.7 The Variable JLINE

An INTEGER*4 variable, JLINE, represents the current line count in the current page of the printed output. It is used to control the number of lines printed on a page, and is automatically incremented by 1 for each line printed by the Output Section of the DAPP. If the postprocessor designer needs to print a message in the MTM without using the Output Section of the DAPP (for example, to use the PRINT statement in FORTRAN to directly print data), then a corresponding statement

    JLINE=JLINE+N

should be added in the MTM to increase the value of JLINE by N for the N lines printed.

12.2 THE ASSIGNMENT OF INITIAL VALUES TO THE VARIABLES IN A DAPP-BASED POSTPROCESSOR: THE DAPP QUESTIONNAIRE

As in every processing program, all the variables should be initialized before the processing begins. Since a postprocessor processes various types of data in the CLFILE, the number of variables that must be initialized according to processing requirements is relatively large. The initial values of these variables contain information regarding the following:

1. The NC machine, such as the maximum programmable feedrate, the rotary table, and the time to select and load a tool
2. The NC controller, such as the types of motion that it can control, the minimum incremental motion in each axis, and the format of the input program
3. The requirements on the postprocessor output, such as the format of paper tape and the printed output

A postprocessor designer usually has to collect all the variables that must be initialized and the information needed to determine their initial values. This task is certainly difficult, especially for a programmer without much experience in NC machine tools and postprocessor design.

To facilitate the collection of information needed to initialize the variables, the DAPP provides a list of questions called the DAPP Questionnaire. By answering these questions, a postprocessor designer can find the variables that should be initialized and assign their initial values as well. The DAPP also provides a program called QUEST that can be used to create the FORTRAN BLOCK DATA subprogram that initializes these variables.

Four groups of questions are listed in the DAPP Questionnaire. They concern the machine tool, the NC or CNC controller, and the requirements for the punched and printed output.

12.2.1 The Machine-Tool Description Questions

The 30 questions in the machine-tool description group (Table 12-6) pertain to the characteristics and parametric setup of the NC machine. The answers to these questions yield the machine tool parameters needed by the postprocessor to process a CLFILE. The questions listed in Table 12-6 are for different types of NC machines, including machines of old or new design. A question can be answered or left unanswered; in the latter case, the answer is provided by the DAPP by default. An answer can be given as one or more numeric values or characters. The respective variables or elements of arrays that will be set by these answers are listed in the last column of Table 12-6. Each question should be studied carefully before it is answered.

For example, the answers to questions M01, M02, and M03 are used to set certain elements of the LIMITS array, namely, LIMITS(1) through (3) and (9) through (11). The Geometric Section of the DAPP will compare the programmed coordinates with these limits. If these limits have been exceeded, an error message will be issued. However, on many NC machines the origin of the machine coordinate system can be set at any point within the working range. Thus the maximum and

TABLE 12-6 THE MACHINE TOOL DESCRIPTION QUESTIONS

Question No. (See Note 1)	Question	Answer (See Note 2)	Default Answer(s)	Unit	Variable(s) Set by Answer(s) (See Note 3)
M01	Values of minimum and maximum attainable x coordinates?	x_{min}† x_{max}†	-99999.99 $+99999.99$	†	LIMITS (9) LIMITS (1)
M02	Values of minimum and maximum attainable y coordinates?	y_{min}† y_{max}†	-99999.99 $+99999.99$	†	LIMITS (10) LIMITS (2)
M03	Values of minimum and maximum attainable z coordinates?	z_{min}† z_{max}†	-99999.99 $+99999.99$	†	LIMITS (11) LIMITS (3)
M04	Value of rapid feedrate?	f†	1.0	in./min or mm/min‡	DATBUF (11)
M05	Should rapid feedrate be in effect until it is canceled?	YES/NO	NO		IRPMOD (INTEGER*4) = 1 YES = 0 NO
M06	Maximum programmable feedrate?	f_1†	99999.99	in./min or mm/min‡	LIMITS (8)
M07	Feedrate that should be assigned to cutter motion if it has not been specified (i.e., default feedrate)?	f_2†	1.0	Same as above	DATBUF (10)
M08	Default spindle speed?	s†	1.0	rpm	DATBUF (14)
M09	Does machine have a rotary table? If answer is YES, enter also number of positions, m, that rotary table can assume.	YES/NO m†	NO 1	degree	TABLGO (see note 4) TPLMT (REAL*8) $= (360/m) \cdot 10^6$

Questions M10 through M14 can be neglected if answer to question M09 is NO.

M10	In which direction can rotary table rotate?	CLW/CCLW/BOTH			TABLGO
M11	Which is positive rotation direction of rotary table?	CLW/CCLW	CLW		TABLGO
M12	How many seconds does rotary table take to rotate 360 degrees?	s†	1.0	second	TIME (3)
M13	What are coordinates of rotary table center with respect to machine coordinate system?	x† y† z†	0 0 0	†	CORTRS (1) (REAL*8) CORTRS (2) CORTRS (3)

(continued)

TABLE 12-6 THE MACHINE TOOL DESCRIPTION QUESTIONS (continued)

Question No. (See Note 1)	Question	Answer (See Note 2)	Default Answer(s)	Unit	Variable(s) Set by Answer(s) (See Note 3)
M14	Should rotary table also return to zero position when GOHOME statement is executed?	YES/NO	NO		IGOTAB (INTEGER*4) = 1 YES 0 NO
M15	Can machine tool move simultaneously in X, Y, and Z axes at rapid rate?	YES/NO	NO		SIMOSW (4) (INTEGER*4) = 1 YES 0 NO
M16	Can machine tool move simultaneously in X and Y axes at rapid rate?	YES/NO	NO		SIMOSW (1) = 1 YES 0 NO
M17	Can machine tool move simultaneously in X and Z axes at rapid rate?	YES/NO	NO		SIMOSW (2) = 1 YES 0 NO
M18	Can machine tool move simultaneously in Y and Z axes at rapid rate?	YES/NO	NO		SIMOSW (3) = 1 YES 0 NO
M19	What is relationship between machine tool axes and their corresponding axes in APT? (See note 5.)	+1 or −1 (for X axis) +1 or −1 (for Y axis) +1 or −1 (for Z axis)	+1 +1 +1		CONVEN (1) (REAL*8) CONVEN (2) CONVEN (3)
M20	"Home" position needed on machine tool? If answer is YES, coordinates of "home" position (x_1, y_1, z_1) should be specified (See note 6.)	YES/NO x_1^+ y_1^+ z_1^+	0 0 0	‡ ‡ ‡	IGOHOM (INTEGER*8) = 1 YES 0 NO HOME (1) (REAL*8) HOME (2) HOME (3)
M21	Time to start and stop spindle?	t_1^+	0	second	TIME (4)
M22	Time to load tool?	t_2^+	0	second	TIME (5)
M23	Time to select tool?	t_3^+	0	second	TIME (6)
M24	Direction of machine spindle: horizontal (H) or vertical (V)?	H or V	V		IZHOR (INTEGER*4) = 1 for H 0 for V

		YES/NO	NO	ZMOD (INTEGER*4) = 0 for NO 1 for YES
M25	Can machine perform machining cycle operation (or so-called canned-cycle operation)?			

Questions M26 through M29 need not be answered if answer to question M25 is NO.

		YES/NO	NO	DRUMZE (INTEGER*4) = 0 for NO 1 for YES R10 (INTEGER*4)
M26	Does machine tool use drums or cams as stops for canned cycle?			
	If answer is YES, enter the starting clearance drum or cam number?	n_1†	0	
M27	Starting depth drum or cam number?	n_2†	0	Z10 (INTEGER*4)
M28	Number of clearance drums or cams?	n_3†	0	CLDRMS (INTEGER*4)
M29	Number of depth drums or cams?	n_4†	0	DPDRMS (INTEGER*4)
M30	Which unit, inches or millimeters, does machine use?	IN or MM		PPUNIT (REAL*8) and QNUNIT (REAL*8) = 1 for IN 25.4 for MM

Notes:

1. On the input card, the question number should be specified in columns 1 through 3.
2. Specified anywhere between columns 5 and 72 of the input card. If the question has more than one answer, they should be input on the same line and be separated by a blank or blanks.
3. If no explanation is given, the variable listed in this column assumes the value of the answer.
4. TABLGO is an INTEGER*4 variable; its value is determined as follows:

TABLGO =	0	No rotary table
	−1	Rotary table with CLW rotation only
	−1	Rotary table with CCLW rotation only
	+3	Rotary table with rotation in both directions, positive direction of rotation is CLW
	−3	Rotary table with rotation in both directions, positive direction of rotation is CCLW

5. For some NC machines, the tool moves in the negative direction for cutting a positive dimension of the part. In such case, the answer for that axis is −1; otherwise, it should be +1.
6. This position will be used by the postprocessor to determine the home position for the GOHOME statement. If the answer is NO, the position specified in the FROM statement will be taken as the home position.

†A number.

‡The unit is defined by the answer to question M30.

Modified from IBM Manual (SH20-2469-0): *System/370 Automatically Programmed Tool—Intermediate Contouring and Advanced Contouring Numerical Control Processor: Program Reference Manual.* Rye Brook, N.Y.: IBM Corp., 1980. (Courtesy of International Business Machines Corporation.)

minimum x, y, and z coordinates that represent the actual limiting positions depend on the position where the origin of the coordinate system is set. The answers to questions M01, M02, and M03 are usually determined on the basis of a practical range in which the origin is set (Fig. 12-2). For instance, suppose that the origin is set at point A and the coordinates of the defined position, B, are (x,y). Then the defined position may exceed the allowable range of tool movement, even though $x <$ x_{max} and $y < y_{max}$. Therefore the answers to these questions generally are not the actual limits on tool motion for a particular machining operation.

The rapid feedrate in question M04 is needed only for those NC machines of old design, for which a rapid feedrate should be specified in a rapid motion statement. As indicated in Section 12.1.1, the answer to this question is not needed for modern CNC controllers.

Questions M09 through M14 are needed to define a rotary table. For NC machines without a rotary table, these questions can be neglected.

Questions M15 through M18 should be answered according to the design characteristics of the NC controller. Most NC machines can move the tool in two directions simultaneously, whereas many of the milling machines with a CNC controller can also have simultaneous movement in three coordinate axes.

An answer "-1" to question M19 for a coordinate axis will cause the Geometric Section to change the sign of the output coordinate in that axis. In most cases, the sense of the coordinate axes in an APT program is the same as that of the machine axes, in which case the default answers should be used. The home position specified in the answer to question M20 will be used to process a GOHOME statement. The answers to questions M21, M22, and M23 are used to calculate the necessary time for starting the spindle and for selecting and loading a tool.

An answer YES to question M25 is needed for NC machines with machine cycle capability. Since modern CNC machines do not use drums to control the tool limiting positions, questions M26 through M29 can be neglected, in which case the default answers are used.

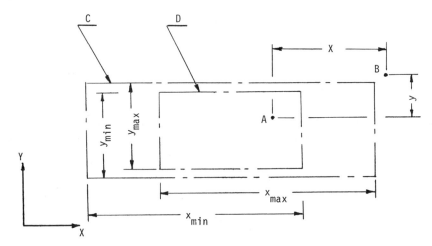

Figure 12-2 Determination of tool movement limits. C: Allowable tool motion area; D: area in which the origin of the machine coordinate system is usually set.

12.2.2 The Controller Description Questions

The motion and operations of an NC machine are controlled by the NC controller. Thus the design of the controller of an NC machine determines the function and motion that the machine can realize. As described in Chapter 2, four kinds of NC block formats have been used for NC machines of different design. In each format, different NC words may be represented in different forms. Therefore the functions of the NC controller and the formats of the used NC words should be defined in the postprocessor.

The DAPP provides 21 questions regarding the characteristics of a controller. These questions are listed in Table 12-7. Questions C06 through C20 are used to define the formats of NC codes to be processed by the postprocessor for printed and punched output. Usually, the rewind-stop code mentioned in question C06 is necessary to stop rewinding of the tape when used with a tape reader with reels. When the tape is used with a tape reader without reels, this code is usually not necessary. Question C07 should be answered according to the operator manual or programming manual of the machine. For the FANUC controller 6MB, an EOB (end of block) code is needed to indicate the start of a program.

The answer to question C08 represents the number of NC codes that a controller can handle and hence determines the number of codes that the postprocessor should process. Since OUTGRD is a 20 × 10 array, the maximum number of codes that can be handled by the DAPP is 20. Questions C09 through C20 should be answered as many times as the number of codes specified in the answer to question C08. Each set of answers to these questions corresponds to one code only and determines the format of that code. To distinguish the answers to these questions from one NC code to another, a letter (A to T inclusive) should be added to each of the question numbers C09 through C20, with A for the first code, B for the second code, and T for the twentieth code. For example, suppose that the answer to question C08 is 12, indicating that the twelve codes N, G, X, Y, Z, I, J, K, M, F, S, and T are to be defined in the postprocessor in listed order. Then the set of questions regarding code N are numbered as C09A, C10A, ..., C20A, those regarding code G, as C09B, C10B, ..., C20B, and so forth, until finally those regarding code T are numbered as C09L, C10L, ..., C20L. Questions other than C09 through C20 should be answered only once. In this example, the code representing block sequence number N is the first defined code; therefore the answer to question C21 should be 1.

The printing position of an NC code in the output listing file is determined by the answer to question C19. Therefore the answers to questions C19 for different NC codes determine the order of outputting NC codes in an NC block.

12.2.3 The Questions Regarding the Formats of Punched and Printed Output

The answers to these two sets of questions provide the information needed to define the formats of punched and printed output. The Punching Questions (Table 12-8) require answers that are needed to define the format of the punched tape. A leader that is 4 to 6 inches in length is normally sufficient. It is required for loading the tape into the tape reader correctly so that the beginning part of the program can be read.

TABLE 12-7 THE CONTROLLER DESCRIPTION QUESTIONS

Question no. (See Note 1)	Question	Answer(s) (See Note 2)	Default Answer	Unit	Variable Set by Answer
C01	Does controller have circular interpolation function?	YES/NO	NO		CNTOUR (INTEGER*4) = 1 NO 2 YES
C02	Is circular interpolation function in effect until it is canceled?	YES/NO	NO		DATBUF (37) = 0 for NO 1 or 2 for YES
C03	Maximum radius for circular interpolation?	$r^\dagger$	99999.99	‡	MAXRAD (REAL*8)
C04	Controller resolution or minimum incremental coordinate?	$d^\dagger$	0.0001	‡	LIMITS (6)
C05	Minimum time for executing one NC statement (block)?	$t^\dagger$	1.0	second	TIME (2)
C06	"Rewind Stop" code needed in NC program before first block of NC data?	YES/NO	NO		IRSTO (LOGICAL) = TRUE for YES FALSE for NO
C07	An EOB (end of block) code needed before first block of NC data?	YES/NO	NO		IEOB (LOGICAL) = TRUE for YES FALSE for NO
C08	Number of NC codes that controller or postprocessor can handle?	$n^\dagger$			OUTGRD (n+1) = 999
C09§	Does NC code have letter address? If YES, enter letter; otherwise enter zero.	letter or 0	0		OUTGRD $(i, 1)$¶
C10§	Is "tab" code precede each NC word needed? If YES, give number also.	YES/NO $n_1^\dagger$	NO		OUTGRD $(i, 2)$¶ = 0 for answer NO n_1 for answer YES and n_1
C11§	Total number of digits in this code? Note that alphabetic letter in word address format and decimal point should not be counted?	$n_2^\dagger$	0		OUTGRD $(i, 3)$¶

448

C12§	Number of fractional digits in this code?	n_3†	0	OUTGRD $(i, 4)$¶
C13§	Decimal point to be printed?	YES/NO	NO	OUTGRD $(i, 5)$¶
C14§	Decimal point to be punched?	YES/NO	NO	OUTGRD $(i, 6)$¶
C15§	Plus sign required for positive numbers in printed listing?	YES/NO	NO	OUTGRD $(i, 5)$¶
C16§	Plus sign required for positive numbers in punched output?	YES/NO	NO	OUTGRD $(i, 6)$¶
C17§	Leading zeros in this NC code to be suppressed?	YES/NO	NO	OUTGRD $(i, 7)$¶
C18§	Trailing zeros in this NC code to be suppressed?	YES/NO	NO	OUTGRD $(i, 7)$¶
C19§	Starting from which column should this code be printed?	n_4†	1	OUTGRD $(i, 8)$¶
C20§	Punched output needed for this code?	YES/NO	NO	OUTGRD $(i, 9)$¶
C21	What is order number of code representing block sequence number (e.g., N) in list of codes defined in questions C09 through C20? If no block sequence number code has been defined in questions C08 through C20, answer is zero.	n_5†	0	DISKID (INTEGER*4)

Notes:
1. Specified in columns 1 through 3 of the input card.
2. Specified anywhere between columns 5 and 72 of the input card.
† A number.
‡ The unit is defined by the answer to question M30.
§ An additional alphabetic character is needed immediately after question number.
¶ The index, i, is equal to the alphabetic order number of the character specified after question numbers C09 through C20. For example,

$i = 1$ for C09A,...,C20A
$i = 2$ for C09B,...,C20B

and

$i = 20$ for C09T,...,C20T

Modified from IBM Manual (SH20-2469-0): *System/370 Automatically Programmed Tool—Intermediate Contouring and Advanced Contouring Numerical Control Processor: Program Reference Manual.* Rye Brook, N.Y.: IBM Corp., 1980. (Courtesy of International Business Machines Corporation.)

TABLE 12-8   THE PUNCHING QUESTIONS

Question No.[†]	Question	Answer[‡]	Default Answer	Variable to Be Set by Answer
P01	Length of NC tape leader?	n (a number)	0[§]	LDRLEN (INTEGER*4)
P02	Should commentary text specified in statement PARTNO be punched in man-readable form before leader part of tape?	YES/NO	NO	IPART (LOGICAL) = TRUE for YES FALSE for NO
P03	Character used in tape leader?	Consult programming manual of NC machine	Blank	LCHAR (1) (REAL*8)
P04	Character representing tape "Rewind-Stop"?	Same as above	%	LCHAR (2) (REAL*8)
P05	Character representing end-of-block code?	Same as above	$	LACHR (3) (REAL*8)
P06	Character representing "tab" code?	Same as above	*	TABCOD (LOGICAL*1)

Notes:
[†]Specified in columns 1 through 3 of the input card.
[‡]Specified anywhere between columns 5 and 72.
[§]The unit of measurement should be consistent with the answer to question M30.

Modified from IBM Manual (SH20-2469-0): *System/370 Automatically Programmed Tool—Intermediate Contouring and Advanced Contouring Numerical Control Processor: Program Reference Manual.* Rye Brook, N.Y.: IBM Corp., 1980. (Courtesy of International Business Machines Corporation.)

An answer YES to question P02 will have the computer punch the commentary text, specified in the PARTNO statement, in human-readable characters, preceding the leader part of the NC tape. The character normally used in the leader part of the tape is a blank. The programming manual of the NC machine should be consulted before questions P03 through P06 are answered. For modern CNC machines, wherein the word address block format is used, question P06 can be neglected.

The answers to Editing Questions E01 and E02 (Table 12-9) define the name of the designed postprocessor and the program number of the Machine Tool Module (MTM). The DAPP can handle up to 10 postprocessors; thus the answer to question E02 can be any numeric value from 1 to 10. The answers to questions E03, E04, and E05 are used to print the heading for each page of the printed output and for the co-ordinate listing. Two input cards (lines) are required for each of the questions. The text input should be listed between columns 11 and 70; therefore a total of 120 char-acters can be entered. They will become, respectively, the title of each page of the printout and the first and second heading lines of the coordinate listing.

12.3 THE PROCEDURES USED TO OBTAIN THE BLOCK DATA SUBPROGRAM

The first step in designing a postprocessor consists of obtaining the necessary infor-mation regarding a given NC machine-controller system and the required output for-mat of the postprocessor. This information is then used to answer the DAPP Questionnaire. These answers should be input according to the format described in Section 12.2. A program called QUEST in DAPP has been included in the APT-AC NC processor to facilitate the generation of the BLOCK DATA program, which ini-tializes the DAPP variables related to the NC machine and the requirements on the postprocessor output. The corresponding job execution programs for generation of the BLOCK DATA subprogram under the VM/CMS and OS/MVS environment have also been provided by the NC processor; they are listed in Appendix D.

In the following, we use the Cadillac NC-100 lathe/FANUC 6T controller sys-tem as an example to explain the procedures and details for generating a BLOCK DATA program. The Cadillac NC-100 lathe is a universal turning machine designed for chucking, bar, and shaft work. It is equipped with a pneumatically operated square turret with four tool positions. There is also a manually movable tail stock that can be used to support the workpiece at the right end. Therefore this machine is similar to a manually operated engine lathe except that the motion of the spindle, saddle, cross slide, and turret is controlled by the CNC controller. Figure 2-2(a) shows a diagram of the structure and coordinate system of the lathe. The general ca-pacity of this machine is as follows:

Swing over bed	$23\frac{1}{4}$ in.
Swing over cross slide	$11\frac{13}{32}$ in.
Distance between center	$39\frac{3}{8}$ in.

The spindle can run in two speed ranges: high gear (70 through 1980 rpm) and low gear (12 through 249 rpm). A spindle speed can be specified as the percentage of the

TABLE 12-9 THE EDITING QUESTIONS

Question No.[+]	Question	Answer[‡]	Default Answer	Variable Set By Answer
E01	Name of postprocessor?	One to six alphanumeric characters, first of which must be alphabetic (Same as used in MACHIN statement)	DAPP	PPCODE (REAL*8)
E02	Program number of Machine Tool Module (MTM)?	Number between 1 and 10	1	PPPARM (REAL*8)
E03[§]	Heading printed at top of each printed page?	Alphanumeric text specified between columns 11 and 70	Blank	ULINE0 (a REAL*8 array with 15 elements)
E04[§]	First line of heading of coordinate listing?	Same as above	Blank	ULINE1 (a REAL*8 array with 15 elements)
E05[§]	Second line of heading of coordinate listing?	Same as above	Blank	ULINE2 (a REAL*8 array with 15 elements)
E06	Accumulated machining time to be printed at end of each page of listing?	YES/NO (If answer is YES, enter in columns 11 through 70 a commentary text to precede printed time.)	NO	ULINE3 (a REAL*4 array with 15 elements)
E07	Accumulated paper tape length to be printed at end of each page of listing? (Unit of length is determined by answer to question M30.)	Same as above	NO	ULINE4 (a REAL*4 array with 15 elements)

Notes:
[+]Specified in columns 1 through 3 of the input card.
[‡]Specified between column 5 and 72 for questions E01 and E02. For questions E03 through E07, the answers should be specified between columns 11 through 70. Two input cards (lines) are required for questions E03, E04, and E05.
[§]Two input cards (lines) are required for this question. The answer will become the heading of the printed output, with a line length of 133 characters.

Modified from IBM Manual (SH20-2469-0): *System/370 Automatically Programmed Tool — Intermediate Contouring and Advanced Contouring Numerical Control Processor: Program Reference Manual.* Rye Brook, N.Y.: IBM Corp., 1980. (Courtesy of International Business Machines Corporation.)

maximum speed of that range with the minimum incremental speed of 1 percent. The specifications and functions of the FANUC 6T controller system are described in Chapter 2.

A complete list of answers to the DAPP Questionnaire, in the required format, is given in Fig. 12-3.

The first three questions, M01, M02, and M03, are concerned with the motion ranges of the NC machine in the X, Y, and Z axes. One specific problem, which was pointed out in Section 8.1, should be considered when answering these questions for

```
M01 -38.0   +38.0
M02 -12.5   +12.5
M03 0.0    0.0
M04 +600.0
M05 NO
M06 +600.0
M07 1.0
M08 12.0
M09 NO
M15 NO
M16 YES
M17 NO
M18 NO
M19 +1 +1 +1
M20 NO
M21 1.0
M22 2.0
M23 0
M24 H
M25 YES
M30 IN
C01 YES
C02 NO
C03 3937.0
C04 0.0001
C05 1.0
C06 YES
C07 YES
C08 12
C09A N
C09B G
C09C X
C09D Z
C09E I
C09F K
C09G F
C09H S
C09I T
C09J M
C09K U
C09L W
C10A NO
C10B NO
C10C NO
C10D NO
C10E NO
C10F NO
```

Figure 12-3   The list of answers, in the required input format, to the DAPP Questionnaire for the Cadillac NC-100 lathe/FANUC 6T controller system.   (continued)

```
C10G NO
C10H NO
C10I NO
C10J NO
C10K NO
C10L NO
C11A 4
C11B 2
C11C 7
C11D 7
C11E 7
C11F 7
C11G 7
C11H 2
C11I 4
C11J 2
C11K 7
C11L 7
C12A 0
C12B 0
C12C 4
C12D 4
C12E 4
C12F 4
C12G 4
C12H 0
C12I 0
C12J 0
C12K 4
C12L 4
C13A NO
C13B NO
C13C YES
C13D YES
C13E YES
C13F YES
C13G YES
C13H NO
C13I NO
C13J NO
C13K YES
C13L YES
C14A NO
C14B NO
C14C YES
C14D YES
C14E YES
C14F YES
C14G YES
C14H NO
C14I NO
C14J NO
C14K YES
C14L YES
C15A NO
C15B NO
C15C NO
C15D NO
C15E NO
C15F NO
C15G NO
C15H NO
```

Figure 12-3 (continued)

```
C15I NO
C15J NO
C15K NO
C15L NO
C16A NO
C16B NO
C16C NO
C16D NO
C16E NO
C16F NO
C16G NO
C16H NO
C16I NO
C16J NO
C16K NO
C16L NO
C17A NO
C17B NO
C17C YES
C17D YES
C17E YES
C17F YES
C17G YES
C17H NO
C17I NO
C17J NO
C17K YES
C17L YES
C18A NO
C18B NO
C18C NO
C18D NO
C18E NO
C18F NO
C18G NO
C18H NO
C18I NO
C18J NO
C18K NO
C18L NO
C19A 1
C19B 7
C19C 11
C19D 20
C19E 30
C19F 39
C19G 67
C19H 75
C19I 80
C19J 86
C19K 48
C19L 57
C20A YES
C20B YES
C20C YES
C20D YES
C20E YES
C20F YES
C20G YES
C20H YES
C20I YES
C20J YES
```

Figure 12-3   (continued)

```
C20K YES
C20L YES
C21 1
P01 6
P02 NO
P03
P04 %
P05 *
P06
E01 LATH2
E02 08
E03         NC-100 LATH/FANUC-6T CONTROLLER SYSTEM
E03         POSTPROCESSOR
E04      N      G   X         Z         I        K        U      W
E04             F       S    T    M                       TIME   ISN
E05
E05
E06 YES     TOTAL MACHINING TIME (IN HR-MIN-SEC) =
E07 YES           TOTAL TAPE LENGTH (IN FEET) =
```

Figure 12-3   (continued)

a lathe, namely, the fact that two-dimensional programming in APT is usually carried out in the X-Y plane. Thus a tool movement in the X-Z plane on the NC lathe has a corresponding programmed movement in the X-Y plane of the APT coordinate system. Since the CLDATA processed by the postprocessor is derived from the APT program, the X and Y axes in the APT program correspond, respectively, to the Z and $-X$ axes of the machine (Fig. 8-1).

If the cutter path in an APT program is defined in the upper half of the X-Y plane (in this case, most of the y coordinates in the program are positive), it is the mirror image of the desired cutter path with respect to the machine axis Z. If the postprocessor translates the $+x$ and $+y$ coordinates in APT to $+z$ and $+x$ coordinates in the NC program, respectively, the cutter path defined by the NC program is still the same as the desired cutter path. However, care must be taken, in writing an APT program, to select the correct types of cutters to be used and the spindle rotation direction. The cutter-nose radius compensation has already been considered during APT programming by the statements TLLFT and TLRGT, and the resulting CLDATA is a series of coordinates of the cutter-nose center. Therefore it is not necessary to consider the cutter radius compensation during transformation of the cutter path in CLFILE format into NC codes.

If the other approach indicated in Section 8.1 is selected, the cutter path is basically designed in the lower half of the X-Y plane during APT programming. Since most of the y coordinates in APT become negative in this case, it is necessary to transform the x and y coordinates in APT to z and $-x$ coordinates in the machine coordinate system.

We selected the first approach because it is easier to program. Then the answers to questions M01 and M02 are the motion ranges in the Z and X axes, respectively. The actual limits of the motion in each axis can be derived from the machine capacity parameters. Since the origin of the machine coordinate system can be set at any point within the range, the upper and lower limits of the motion can be set to be the same. The Z axis of the APT coordinate system is not used; therefore the answer is zero. The answer to question M19 is $+1$ for all three axes accordingly.

The time required for rotating the turret by 90 degrees is about 2.0 seconds; therefore the answer to question M22 is 2.0. Because there is no tool magazine on this lathe, the answer to question M23 is zero. The answers to the other questions are straightforward, and no further explanation is required.

Questions regarding the NC controller can usually be answered according to the specifications of the NC controller and the programming manual of the machine. Basically, the answers to these questions are straightforward. However, question C02 deals with a function that is not controlled solely by the controller. When the answer to this question is NO, it means that DATBUF(37) is zero or that a circular interpolation motion is not requested in APT (i.e., ARCSLP/OFF). When the answer is YES, DATBUF(37) will be initialized as 1 or 2, indicating that circular interpolation motion is requested. Therefore the answer should normally be selected as NO. The answer to question C03 is 3937.0 in., which is the maximum programmable radius for the FANUC 6T controller system. The remaining questions can be answered according to the programming requirement of the FANUC 6T system, which was described in detail in Chapter 2.

With regard to punching questions, the NC tape for this machine requires a 6 in. leader with blank characters only. The tape rewind-stop code is "%" and the end-of-block code is "*". No tab code is needed before each NC word.

The name of the postprocessor is LATH2 and the number of MTM is 8. Questions E01 and E02 must be answered accordingly. Questions E03 to E07 must be answered in accordance with the requirement on the printed output. The reader should compare these answers with the postprocessor listing output given in Chapter 15.

The answers to the DAPP Questionnaire should be saved as a FORTRAN file. For example, in the VM/CMS environment, the file can be saved by file name, type, and mode, as follows:

ANSWER FORTRAN A1

The job execution program QUESTEXE[11] (under VM/CMS operating environment) listed below can then be used to process the answers, generating the FORTRAN BLOCK DATA subprogram (Fig. 12-4):

```
&CONTROL ALL
FILEDEF FT06F001 PRINTER
FILEDEF FT08F001 DISK F1 F1 A4 (XTENT 50 RECFM F BLKSIZE 400)
FILEDEF FT07F001 PUNCH
FILEDEF FT05F001 DISK ANSWER FORTRAN
LOADMOD QUEST
START
```

The Job Control Language (JCL) statements for OS/MVS operating environment are listed in Appendix D. A comparison of the DATA statements in this program, with the respective answers listed in Fig. 12-3, indicates that they are consistent.

```
      BLOCK DATA
C**********************************************************************
C                                                                     *
C        INITIALIZATION OF DAPP COMMON BLOCK QUES08.                   *
C                                                                     *
C**********************************************************************
      IMPLICIT     REAL*8 (A-H,O-Z)
      COMMON   /QUES08/
     A     CORTRS(3)    ,DATBUF(85)  ,HOME(3)     ,LIMITS(11)  ,MAXRAD
     B     ,PPCODE      ,PPPARM      ,TPLIMT      ,ULINE0(15)  ,ULINE1(15)
     C     ,ULINE2(15)  ,XCONVT      ,YCONVT      ,ZCONVT      ,XTRA(2)
     D     ,TIME(8)     ,ULINE3(15)  ,ULINE4(15)
     E     ,CLDRMS      ,CNTOUR      ,DISKID      ,DPDRMS      ,DRUMZE
     F     ,IGOHOM      ,IGOTAB      ,IRPMOD      ,IZHOR       ,LDRLEN
     G     ,LEADSW      ,NLINE       ,OUTGRD(20,10)   ,R10     ,STRTSQ
     H     ,SIMOSW(4)   ,TABLGO      ,ZMOD(2)     ,Z10
     I     ,CLKPRT      ,EOB         ,IEOB        ,IPART       ,IRSTO
     J     ,LCHAR(2)    ,TABCOD      ,TAPPRT      ,TAPMET
      REAL*8           LIMITS      ,MAXRAD
      REAL*4           TIME        ,ULINE3      ,ULINE4
      INTEGER*4        CLDRMS      ,CNTOUR      ,DISKID      ,DPDRMS
     A     ,DRUMZE      ,OUTGRD      ,R10         ,STRTSQ      ,SIMOSW
     B     ,TABLGO      ,ZMOD        ,Z10
      LOGICAL*1        CLKPRT      ,EOB         ,IEOB        ,IPART
     A     ,IRSTO       ,LCHAR       ,TABCOD      ,TAPPRT      ,TAPMET
      EQUIVALENCE      (ZDEPTH,DATBUF(29))     (ZFINAL,DATBUF(30))
      REAL*8           CONVEN(3)
      EQUIVALENCE (CONVEN(1),XCDIG,XCONVT),(YCDIG,YCONVT),(ZCDIG,ZCONVT)
      INTEGER          ASSIGN(2)               ,CYCTYP(2)
      EQUIVALENCE      (ASSIGN(1),DATBUF(20))  ,(CYCTYP(1),DATBUF(28))
      LOGICAL*1        LOGCAL(6)               ,OUTSUX(40)
      EQUIVALENCE      (LOGCAL(1),DATBUF(31))  ,(OUTSUX(1),OUTGRD(1,1))
      DATA  CORTRS /        0.0  D0,       0.0  D0,       0.0  D0/
      DATA  DATBUF /        8H99999.99,    8H99999.99,    8H99999.99,
     A     0.0  D0,         0.0  D0,       0.0  D0,       0.0  D0,
     B     0.0  D0,         0.0  D0,       1.0000D0,      600.0000D0,
     C     0.0  D0,         8H99999.99,    12.0000D0,     8H99999.99,
     D     8H99999.99,      8H99999.99,    8H99999.99,    8H99999.99,
     E     8H99999.99,      0.0  D0,       8H99999.99,    8H99999.99,
     F     0.0  D0,         8H99999.99,    8H99999.99,    8H99999.99,
     G     8H99999.99,      8H99999.99,    8H99999.99,    0.0  D0,
     H     8H99999.99,      8H99999.99,    8H99999.99,    8H99999.99,
     I     8H99999.99,      0.0  D0,   48*8H99999.99/
      DATA  HOME   /        0.0  D0,       0.0  D0,       0.0  D0/
      DATA  LIMITS /        38.0000D0,     12.5000D0,     0.0  D0,
     A     8H99999.99,      8H99999.99,    0.000100D0,    8H99999.99,
     B     600.0000D0,      -38.0000D0,    -12.5000D0,    0.0  D0/
      DATA  MAXRAD /     3937.0000D0/
      DATA  PPCODE /6H  TURN/, PPPARM / 8.0D0/
      DATA  TPLIMT /360000000.0D0/
      DATA  ULINE0 /        8HNC-100 L,8HATH/FANU,8HC-6T CON,8HTROLLER
     A,8HSYSTEM   ,8H         ,8H         ,8H     POST,8HPROCESSO,8HR
     B,8H         ,8H         ,8H         ,8H         ,8H          /
      DATA  ULINE1 /        8HN         G ,8H  X      ,8H    Z    ,8H    I
     A,8H     K  ,8H     U,8H      W,8H       F ,8H S       T
     B,8H    M   ,8H         ,8H         ,8H TIME,8H    ISN/
      DATA  ULINE2 /        8H         ,8H         ,8H         ,8H
     A,8H         ,8H         ,8H         ,8H         ,8H         ,8H
     B,8H         ,8H         ,8H         ,8H         ,8H          /
      DATA  ULINE3 / 4HTOTA,4HL MA,4HCHIN,4HING ,4HTIME,4H(IN ,4HHR-M
```

Figure 12-4 The BLOCK DATA program generated from the answers given in Fig. 12-3.

```
A          ,4HIN-S,4HEC) ,4H=    ,4H     ,4H     ,4H     ,4H     ,4H     /
DATA  ULINE4 / 4H     ,4H     ,4H TOT,4HAL T,4HAPE ,4HLENG,4HTH(I
A          ,4HN FE,4HET) ,4H=    ,4H     ,4H     ,4H     ,4H     ,4H     /
DATA  XCONVT / 1.0D0/,    YCONVT / 1.0D0/,    ZCONVT / 1.0D0/
DATA  XTRA   /              1.0000D0,        1.0000D0/
DATA  TIME   /    0.0    ,   1.0000,     1.0000,     1.0000,
A        2.0000,      0.0    ,      0.0    ,      0.0    /
DATA  CLDRMS /   0/, CNTOUR /    2/, DISKID /   1/, DPDRMS /    0/,
A     DRUMZE /   0/, IGOHOM /    0/, IGOTAB /   0/, IRPMOD /    0/,
B     IZHOR  /   1/, LDRLEN /    6/, LEADSW /   0/, NLINE  /   46/
DATA  OUTGRD / 1HN, 1HG, 1HX, 1HZ, 1HI, 1HK, 1HF, 1HS, 1HT, 1HM,
A             1HU, 1HW, 1HO, 1HO, 1HO, 1HO, 1HO, 1HO, 1HO, 1HO,
B   0,   0,   0,   0,   0,   0,   0,   0,   0,   0,   0,   0,   0,
C   0,   0,   0,   0,   0,   0,   0,   0,   0,   4,   4,   4,   4,
D   4,   0,   0,   0,   4,   4,   0,   0,   0,   0,   0,   0,   0,
E   0,   4,   2,   7,   7,   7,   7,   7,   2,   4,   2,   7,   7,
F   0,   0,   0,   0,   0,   0,   0,   0,   1,   1,   2,   2,   2,
G   2,   2,   1,   1,   1,   2,   2,   0,   0,   0,   0,   0,   0,
H   0,   0,   1,   1,   2,   2,   2,   2,   2,   1,   1,   1,   2,
I   2,   0,   0,   0,   0,   0,   0,   0,   0,   0,   0,   2,   2,
J   2,   2,   2,   0,   0,   0,   2,   2,   0,   0,   0,   0,   0,
K   0,   0,   0,   1,   7,  11,  20,  30,  39,  67,  75,  80,  86,
L  48,  57,   0,   0,   0,   0,   0,   0,   0,   0,   1,   1,   1,
M   1,   1,   1,   1,   1,   1,   1,   1,   1,   0,   0,   0,   0,
N   0,   0,   0,   0,   0,   0,   0,   0,   0,   0,   0,   0,   0,
O   0,   0,   0, 999,   0,   0,   0,   0,   0,   0,   0/
DATA  R10    /   0/, STRTSQ /   1/, SIMOSW /   1,    0,   0,   0/,
A     TABLGO /   0/, ZMOD   /   0,   1/,              Z10    /   0/
DATA  CLKPRT /T/,    EOB    /1H*/,  IEOB   /T/,    IPART  /F/,
A     IRSTO /T/,    LCHAR  /1H ,1H%/,                TABCOD /1H*/,
B     TAPPRT /T/,    TAPMET /F/
END
```

Figure 12-4 (continued)

PROBLEMS

12.1 Give the answers to the DAPP Questionnaire and generate the BLOCK DATA program for a sample NC vertical milling machine with the FANUC 6MB controller based on its specifications as given below:

a. Machine specifications:

(1) Range of programmable machine table movement:
   X direction: 27.5 in.
   Y direction: 17.75 in.
(2) Range of programmable spindle head movement:
   Z direction: 4.0 in.
   spindle speed: 10–3000 rpm (four-digit direct command)
(3) Programmable simultaneous linear and point-to-point motion in X, Y, and Z axes
(4) Programmable circular interpolation motion in three coordinate planes
(5) Programmable cycle motion (as described in Chapter 2)
(6) Rapid traverse feedrate (not programmable):   480 in./min
(7) Programmable machining feedrate: 0.01 to 150 in./min
(8) Manual tool change

b. Controller specifications:

    (1) NC codes used: N, G, X, Y, Z, R, F, S, M, Q, P (time)
    (2) Formats of NC words: see Table 2-3
    (3) Length of tape leader: 6 in.
    (4) Character used in tape leader: blank
    (5) Character representing "end-of-block" code: ";"
    (6) Character representing "rewind-stop" code: "%"

The postprocessor name is SAMPLE; and the program number of the machine-dependent processing routine (MTM) is 8.

12.2 Besides the characteristics of the NC machine-controller system and the requirements on the postprocessor output, which are needed to answer the DAPP Questionnaire, what kinds of information are further needed to design a DAPP-based postprocessor?

chapter 13

The Machine-Independent Processing Routines of a DAPP-Based Postprocessor

As pointed out previously, the machine-independent processing routines are generally the same for every postprocessor, and they have been incorporated into the DAPP. A postprocessor designer normally does not need to take care of these routines during the design of a DAPP-based postprocessor. However, a sound understanding of these routines is necessary to write the machine-dependent processing routine, MTM, and hence to design a postprocessor.

The machine-independent processing routines discussed in this chapter can be divided into the following sections:

1. Initialization Section
2. Housekeeping Section
3. Reader Section
4. General Processing Section
5. Output Section

For each section, we shall explain the functions of the routine, the changes in variables resulting from execution of the routine, and the control flow.

13.1 THE INITIALIZATION SECTION

In this section, initial values are assigned to the variables specified in the DAPP COMMON statements. The machine-related variables are initialized either by the DATA statements in the BLOCK DATA program (in other words, they are determined by the answers to the DAPP Questionnaire) or by the MTM subroutine provided by the user. In the DAPP, the variables that are initialized according to the answers to the Questionnaire are specified in the COMMON statement labeled QUEST. In the BLOCK DATA program generated with the use of the method given in Chapter 12, the COMMON statement is labeled QUESnn, where the last two digits, "nn," are the program number of the corresponding MTM. A special routine named LDCOM in this section moves the COMMON block labeled QUESnn in the BLOCK DATA program to the COMMON block, labeled QUEST, of the DAPP, thus realizing initialization of these variables.

The other machine-related variables in the DAPP are defined in the COMMON statements labeled DBFPRT and SEGCOM. The range of the DATBUF elements to be printed on the listing file is defined by the variables in the COMMON statement labeled DBFPRT

COMMON/DBFPRT/N1,ISTART(5),LENGTH(5)

where

$$N1 = \text{the number of ranges of the DATBUF locations to be printed}$$
$$\text{ISTART}(i) = \text{a number defining a DATBUF element as the starting position}$$
$$\text{of the } i\text{th range } (i = 1, \ldots, 5)$$
$$\text{LENGTH}(i) = \text{a number specifying the total number of DATBUF elements, in}$$
$$\text{the } i\text{th range, to be printed } (i = 1, \ldots, 5)$$

These variables are the parameters of the DLWDBF routine which, when called, prints the specified elements of the DATBUF array processed by the machine-independent processing section. Since the size of the arrays ISTART and LENGTH is 5, the maximum number of ranges should also be 5. These variables are initialized as follows:

$$N1 = 1$$
$$\text{LENGTH}(i) = \begin{cases} 85 & (i = 1) \\ 1 & (i = 2, \ldots, 5) \end{cases}$$
$$\text{ISTART}(j) = 1 \quad (j = 1, \ldots, 5)$$

which indicate that all 85 DATBUF elements will be considered as one range and printed. These variables and the common storage area labeled DBFPRT should be defined in the MTM. Then the printing routine DLWDBF can be called to print the specified DATBUF elements. For example, if the following statements

COMMON/DBFPRT/N1,ISTART(5),LENGTH(5)
· · · · · · · · · · ·
· · · · · · · · · · ·
N1 = 2
ISTART(1)=1
ISTART(2)=36
LENGTH(1)=3
LENGTH(2)=36
CALL DLWDBF

are specified in the MTM, elements 1 through 3 and 36 through 71 of the DATBUF array will be printed together with their names and values.*

The variables defined in the COMMON statement labeled SEGCOM are used to break a cutter path into two or more sections. They are initialized also in this section and can be redefined in the MTM if needed. An explanation of these variables is given in Chapter 14.

The machine-independent variables are internal to the DAPP. Their values are initialized by the DAPP and generally should not be changed unless necessary. A list of these variables can be found in the IBM program reference manual,[7] together with their description.

During the processing of a CLFILE, the initialization routine is executed only once, at the beginning. Then control is sent to the Housekeeping Section.

13.2 THE HOUSEKEEPING SECTION

Some of the data in the DATBUF array should be updated or set to zero or ANULL[†] after a set of CLDATA has been processed and output, and before a new set of CLDATA record is read. This housekeeping operation is evidently necessary for correct processing of CLDATA. For example, after a set of CLDATA has been processed and output, the DATBUF elements that represent the x, y, and z coordinates of the destination point, that is, DATBUF(1), (2), and (3), should be set to ANULL, which means that no input regarding the destination point has been recorded. If, in the next set of CLDATA, the tool position changes in only one coordinate axis, say X, then the DATBUF position corresponding to that axis, DATBUF(1), will be updated by the Reader Section, and the values of the other two positions, DATBUF(2) and (3), will remain unchanged (i.e., ANULL). Thus the tool movement in these two directions will not be generated. Had the three positions not been set to ANULL by the Housekeeping Section, an incorrect tool movement in Y and Z axes would have been generated. A study of the 85 elements of the DATBUF array reveals that the elements that should not be reset in the Housekeeping Section are as follows:

DATBUF(i) i = 10, 11, 13–17, 20, 23, 26–30, 34, 35

When control enters the Housekeeping Section, a test of the value of LOGCAL(3) is first made to determine whether a peck-drilling cycle[‡] is in effect. If the value of the LOGCAL(3) is true, which means that several drilling subcycles are to

---

*See the example in Chapter 15.
[†]See footnote 5 in Table 12-1.
[‡]See Section 13.4.2.3.

be generated and that the DATBUF array should not be updated, control is then by-passed to the canned-cycle routine DLWCYC to process the next step of the drilling cycle. Otherwise, the housekeeping routine is executed.

Table 13-1 lists the values of various DATBUF positions after the housekeeping routine is executed.

The Housekeeping Section is called under two conditions:

1. At the beginning of processing a CLFILE, the Housekeeping Section is entered after the initialization routine is executed

2. On completion of processing a set of CLDATA record, control is sent by the DAPP main routine to the Housekeeping Section

In either case, the DATBUF array is updated as described in Table 13-1, and then control is sent to the Reader Section.

TABLE 13-1   THE VALUES OF THE DATBUF ELEMENTS RESET BY THE HOUSEKEEPING SECTION

i	Value of DATBUF (i) After Execution of Housekeeping Routine
1, 2, 3	ANULL
4	DATBUF (4) + DATBUF (7)
5	DATBUF (5) + DATBUF (8)
6	DATBUF (6) + DATBUF (9)
7, 8, 9	0.0
10, 11	*
12	0.0 If answer to question M05 is NO
	* If answer to question M05 is YES
13–17	*
18, 19	ANULL
20	*
21, 22	0.0
23	*
24	0.0
25	ANULL
26–30	*
31	LOGCAL (j) = FALSE $j = 1, 2, 4$
32–33	ANULL
34, 35	*
36	Reset as ANULL if NC machine has circular-interpolation function; otherwise, this element is not reset.
37	*
38–41	Same as DATBUF (36)
42–71	ANULL
72–85	*

*This element(s) is (are) not reset by the Housekeeping Section.

Modified from IBM Manual (SH20-2469-0): *System/370 Automatically Programmed Tool — Intermediate Contouring and Advanced Contouring Numerical Control Processor: Program Reference Manual*. Rye Brook, N.Y.: IBM Corp., 1980. (Courtesy of International Business Machines Corporation.)

13.3 THE READER SECTION

The data in an APT program are passed from the CLFILE to the DAPP through the Reader Section, which includes a routine, named CLREAD, for retrieving data from the CLFILE. When a set of CLDATA record is read (or retrieved) by CLREAD, the information in the record is transformed into the parameters listed below:

IRECNO (INTEGER) = record number of current set of CLDATA record

INNBUF = REAL*8 array with 252 positions, which are used as a temporary buffer area for storing data read from current set of CLDATA record

IWORDS (INTEGER) = number of words transferred from routine CLREAD to array INNBUF

CONTRL (INTEGER) = record class of current set of CLDATA (see Table 13-2)

NSUBCL (INTEGER) = parameter that further defines current set of CLDATA (For different classes of records, this parameter is determined in different ways.[7])

For example, assume that the 11th record of the CLDATA is a postprocessor command, say CLPRNT, with statement label A1 and sequence number 00000020 in columns 73 through 80 of the input card. As indicated in Chapter 11, two records, one of class 1000 and one of class 2000, are generated from this statement. For the class 1000 record, the parameters passed by the CLREAD routine are as follows:

IRECNO = 11

INNBUF(1) = 00000020 (INNBUF(i), $i = 2, \ldots,$ or 252, contains no record)

IWORDS = 1

CONTRL = 6 (The class number of a class 1000 record is 6. See Table 13-2.)

NSUBCL = internal sequence number (ISN)

The class 2000 record (postprocessor command) is transformed into the following parameters:

IRECNO = 11

INNBUF(i) = contains no record since there is no parameter after postprocessor word CLPRNT (For postprocessor words with minor words and parameters, this array contains the minor words and parameters specified after the major word.) (i = 1 to 252)

IWORDS = 0

CONTRL = 1 (The class number of a postprocessor word without parameters is 1, see Table 13-2.)

NSUBCL = position number of this postprocessor word in CLSETP array[7] (This position number has been set up in DAPP to contain those major postprocessor words to be processed by the DAPP.)

TABLE 13-2   THE TYPES OF DATA THAT ARE PROCESSED BY THE DAPP

Record Class	CONTRL	Description of Record
1000	6	Internal sequence number (ISN) of read-in CLDATA record
2000	1	Postprocessor word without parameter(s) and/or minor word(s)
	2	Postprocessor word with parameter(s) and/or minor word(s)
3000	7	Circular drive surface
5000	3	Tool position(s) defined by statement FROM, GODLTA, or GOTO when tool axis is always in Z direction (MULTAX/OFF)
	5	Same as above with MULTAX/ON (tool axis changes its spatial direction during machining)
9000	9	Tool axis orientation data (MULTAX/ON or OFF)
	10	Measurement unit data
13000	8	Error indication
14000	4	FINI record signifying end of CLFILE

For a class 5000 record, the number of tool positions processed and generated by the NC processor can be greater than 1. Thus the INNBUF array contains the x, y, and z coordinates of those points.

Before a postprocessor command is transformed into parameters, the major postprocessor word in the command is compared with the array CLSETP, which contains the major postprocessor words that are allowed by the DAPP; the minor postprocessor words are compared with the array containing the minor postprocessor words that are accepted by the DAPP. If a postprocessor word is listed in one of the arrays, its position in the array will be stored as NSUBCL for the major word, whereas the minor word(s) and parameter(s) will be saved as INNBUF(i); otherwise, it is ignored by the CLREAD routine.

Ten types of data can be handled by the DAPP; they are listed in Table 13-2. The reader should consult the appropriate IBM manual[7] for details of the parameters that are transferred by the CLREAD routine for different types of records.

Once a CLDATA record has been read in, it is sent to the processing routine in the General Processing Section, corresponding to the record class indicated by the parameter CONTRL. The data is further decoded, processed, or both and then stored in the appropriate arrays (e.g., DATBUF) or as variables.

13.4 THE GENERAL PROCESSING SECTION

The General Processing Section carries out the machine-independent processing of the CLDATA. One of its major functions is to transform the tool position data. The tool positions given in the CLDATA are defined with respect to the part coordinate system; furthermore, the offset (or the position difference) of one tool from another is not considered during the processing of an APT program by the NC processor. However, the tool positions in the postprocessor output (the NC program) must be defined in the machine coordinate system. Correction of tool positions according to the tool offset is also necessary if a tool change has been made. Therefore the General Processing Section should be able to transform the tool position coordinates from the CLDATA on the basis of the information contained in the postprocessor commands regarding, for example, the machine coordinate system and the tool offset.

Another major function of the General Processing Section is to decode the postprocessor commands (words) and to store the translated information as appropriate variables so that they can be easily retrieved and used by the machine-dependent processing section, MTM.

The General Processing Section can be divided into 10 different subsections, called decode areas in DAPP, that process, respectively, the 10 types of records listed in Table 13-2. Records other than postprocessor words and tool position data are processed in the corresponding decode areas and are then sent either to the machine-dependent processing section, MTM, or directly to the DAPP Output Section. The processing of postprocessor words and tool-position data is carried out with the aid of the subroutines given in Table 13-3.

For an explanation of the way data are processed by the General Processing Section, a discussion of these routines follows.

13.4.1 The Auxiliary Section

The Auxiliary Section consists of six processing routines, four of which, DLWPP1, DLWUP1, DLWPP2, and DLWUP2, handle, respectively, the postprocessor words with and without parameter(s). Two calculation routines, DLWROT and DLWORG, are provided to process, respectively, the statements ROTABL, which defines the position of the rotary table, and ORIGIN, which defines the origin of the machine coordinate system with respect to the part coordinate system. ROTABL and ORIGIN are also postprocessor words (commands).

A postprocessor command is one that is passed by the APT-AC NC processor to the CLFILE without being processed. It is then processed by the postprocessor. The number of postprocessor commands or words that can be processed

TABLE 13-3 THE SUBROUTINES IN THE GENERAL PROCESSING SECTION

Section	Subroutine	Function
Auxiliary	DLWPP1	Processing postprocessor words without parameter(s)
	DLWPP2	Processing postprocessor words with parameter(s)
	DLWUP1	Processing postprocessor words without parameter(s), which are not handled by routine DLWPP1
	DLWUP2	Processing postprocessor words with parameter(s), which are not handled by routine DLWPP2
	DLWORG	Processing postprocessor command ORIGIN
	DLWROT	Processing postprocessor command ROTABL
Geometric	DLWLIN	Processing linear-interpolation motion
	DLWCIR	Processing circular-interpolation motion
	DLWCYC	Processing machining cycle motion

by the DAPP Auxiliary Section is limited. However, they are adequate for most applications.

13.4.1.1 The Routines DLWPP1, DLWUP1, DLWPP2, and DLWUP2.

Table 13-4 lists the postprocessor words without parameter(s) that can be processed by the routine DLWPP1 and the processing results. These words are used relatively often in APT programming. Postprocessor words without parameter(s), which are not processed by the routine DLWPP1, are nonstandard words. They are transferred by the routine DLWPP1 to the routine DLWUP1. Nevertheless, no processing function is provided in routine DLWUP1, which issues only an error message, numbered 6117, after receiving the postprocessor word. The word is then ignored and control returns to routine DLWPP1.

As can be seen from Table 13-4, after the DLWPP1 and DLWUP1 routines complete the processing of a postprocessor word without parameters, control may go to one of three possible sections: the Reader Section, Geometric Section, or MTM.

Postprocessor words with parameter(s) are processed by the routines DLWPP2 and DLWUP2. There are 24 words that can be processed by routine DLWPP2; they are listed in Table 13-5. The words that cannot be processed by DLWPP2 are sent to DLWUP2. Again, no processing function is provided in the DLWUP2 routine, which issues merely an error message (numbered 6118) and then sends control back to DLWPP2.

After the DLWPP2 and DLWUP2 routines finish processing a postprocessor word, control goes to the Reader Section, which reads in another CLDATA record, to the MTM, or to the Output Section, depending on the processed word (Table 13-5).

Routines DLWUP1 and DLWUP2 are included in this section to make it possible for DAPP to accept additional postprocessor words. These words are nonstandard, and each word might need different processing routines for different applications. Therefore only a skeleton of the processing routine is provided to DLWUP1 and DLWUP2. If a user needs to process one or several of those additional postprocessor words, he should modify the corresponding program DLWUP1 and/or DLWUP2 by adding appropriate processing routines. For details, the reader should consult the appropriate IBM manual.[7]

13.4.1.2 The ROTABL Statement and Its Processing Routine DLWROT.

The command ROTABL specifies a rotation of the rotary table. It has two different formats. The first format of this statement is

$$\text{ROTABL}/ \begin{Bmatrix} \text{INCR} \\ \text{ATANGL} \end{Bmatrix} ,a, \begin{Bmatrix} \text{CLW} \\ \text{CCLW} \end{Bmatrix}$$

where

a = rotation angle of the rotary table

INCR = indication that the specified angle of rotation is an incremental value from the current position

ATANGL = indication that the specified angle of rotation is the absolute angle from the zero angular position of the rotary table

TABLE 13-4 THE POSTPROCESSOR WORDS WITHOUT PARAMETER(S), PROCESSED BY ROUTINE DLWPP1

Postprocessor Word	Meaning	Result of Processing This Record
END	This statement should be specified at end of APT program to signify termination of part machining program. If not specified, it will automatically be generated by DAPP when FINI record is read. Corresponding NC code generated by postprocessor is M02 or M30.	1. DATBUF (25) = 9001. 2. After this record is processed, control goes to MTM.
OPSTOP	This command is used to specify or generate optional stop code M01 in NC program.	1. DATBUF (25) = 9002. 2. Control goes to MTM after this record is processed.
STOP	This command is used to specify or generate program interruption code M00 in NC program.	1. DATBUF (25) = 9003. 2. Control goes to MTM after this record is processed.
RAPID	This command specifies rapid feedrate so that G00 code can be generated in NC program. It is effective either to next motion statement only if answer to question M05 is NO, or until canceled by statement defining feedrate if answer to question M05 is YES.	1. DATBUF (12) = 1. 2. Control goes to MTM after this record is processed.

GOHOME	This statement orders tool to go at rapid feedrate to home position defined either by answer to question M20 or in FROM statement when answer to question M20 is NO.	1. a. DATBUF (1), (2), and (3) are equal to x, y, and z coordinates of home or FROM position, respectively.
		   b. DATBUF (12) = 1.
		   c. If answer to question M14 is YES, an additional statement that orders rotary table to return to zero position is generated.
		2. Control goes to routine processing linear interpolation motion (DLWLIN) by way of Reader Section.

(continued)

TABLE 13-4 THE POSTPROCESSOR WORDS WITHOUT PARAMETER(S), PROCESSED BY ROUTINE DLWPP1 (continued)

Postprocessor Word	Meaning	Result of Processing This Record
GOCLER	This command should be specified after postprocessor command CLEARP/... to order tool to move at rapid feedrate up to clearance level, defined by CLEARP command, in machining cycle. See discussion on routine DLWCYC (Section 13.4.2.3) for detail.	1. a. DATBUF (1), (2), and (3) will be set according to specified clearance plane defined by statement CLEARP/... b. DATBUF (12) = 1. 2. Error message 6112 is issued if no CLEARP/... statement is specified preceding this statement. 3. Control goes to routine DLWLIN via Reader Section if no error occurs.
RETRCT	This statement should be specified after postprocessor command CLRSRF/..., which defines clearance plane parallel to X-Y plane. It will cause tool to move up to clearance plane. See Section 13.4.2.3 for detail.	1. a. DATBUF (3) is set according to defined clearance plane. b. DATBUF (12) = 1. 2. Error message numbered 6113 is issued if no preceding CLRSRF statement is specified. 3. Control goes to routine DLWLIN if no error occurs.
RESET	This statement forces DAPP to punch NC tape even when there is serious error in CLDATA or during postprocessing.	Whenever this statement is processed, output condition is reset so that Output Section will still punch tape even if serious error occurs. Control returns to Reader Section after this record is processed.

Modified from IBM Manual (SH20-2469-0): *System 370/Automatically Programmed Tool — Intermediate Contouring and Advanced Contouring Numerical Control Processor: Program Reference Manual.* Rye Brook, N.Y.: IBM Corp., 1980. (Courtesy of International Business Machines Corporation.)

TABLE 13-5 THE POSTPROCESSOR COMMANDS WITH PARAMETER(S), PROCESSED BY THE DLWPP2 ROUTINE

Postprocessor Command	Meaning	Result of Processing This Command
ARCSLP/$\begin{Bmatrix} \text{ON} \\ \text{OFF} \end{Bmatrix}$	To control use of circular interpolation routine DLWCIR to process tool position data. When option ON is selected, DLWCIR is called; otherwise, linear interpolation routine DLWLIN will be used instead.	1. DATBUF (37) = 1 for ON; 0 for OFF. 2. Error message 6001 is issued if there is syntax error. 3. After this command is processed, control goes back to Reader Section to read in class 5000, GOTO/ (multiple points) data.
AUXFUN/n n: a scalar	This command is used to generate M code in NC program.	1. DATBUF (25) = n. 2. Error message 6002 is issued if there is syntax error. 3. Control goes to MTM after this statement is processed.
CLEARP/$\begin{Bmatrix} \text{XYPLAN} \\ \text{YZPLAN} \\ \text{ZXPLAN} \end{Bmatrix}$, $d1$ $\left[, \begin{Bmatrix} \text{XYPLAN} \\ \text{YZPLAN} \\ \text{ZXPLAN} \end{Bmatrix} \right.$ $\left[, \begin{Bmatrix} \text{XYPLAN} \\ \text{YZPLAN} \\ \text{ZXPLAN} \end{Bmatrix}, d3 \right]$, $d2\text{S}$	To define clearance plane, line, or point. This statement is used, together with statement GOCLER, to define retreat motion in machining cycle. See Section 13.4.2.3 for detail.	1. Clearance plane, line, or point is set and stored internally. 2. Error message 6003 is issued in case of syntax error. 3. Control returns to Reader Section after this statement is processed.
$d1$ (only): Distance between defined clearance plane and specified parallel coordinate plane		
$d1$, $d2$: Distances between defined clearance line and specified parallel coordinate planes, respectively (Note that two specified coordinate planes should not be the same.)		
$d1$, $d2$, $d3$: Distance between defined clearance point and specified coordinate planes, respectively (Any two of three coordinate planes should not be the same.)		

(continued)

471

TABLE 13-5 THE POSTPROCESSOR COMMANDS WITH PARAMETER(S), PROCESSED BY THE DLWPP2 ROUTINE (continued)

Postprocessor Command	Meaning	Result of Processing This Command
CLRSRF/*d* *d*: Distance between defined clearance plane and *X-Y* plane	To define clearance plane as parallel to *X-Y* plane. This statement should be used together with statement RETRCT to define retract motion in machining cycle.	1. Clearance plane parallel to *X-Y* plane is set and stored internally. 2. Error message 6021 is issued in case of syntax error. 3. Control returns to Reader Section.
COOLNT/ $\left\{\begin{array}{l}\text{FLOOD}\\\text{MIST}\\\text{TAPKUL}\\\text{ON}\\\text{OFF}\end{array}\right\}$	To define coolant type and to turn on coolant.	1. DATBUF (17) is set (see Table 12-1). 2. Error message 6004 is issued in case of syntax error. 3. Control returns to Reader Section.
CUTCOM/ $\left\{\begin{array}{l}\left\{\begin{array}{l}\text{LEFT}\\\text{RIGHT}\end{array}\right\},\left\{\begin{array}{l}\text{XYPLAN}\\\text{YZPLAN}\\\text{ZXPLAN}\end{array}\right\}\\\text{OFF}\\n\end{array}\right\}$ *n*: Positive scalar	To define side of cutter radius compensation in specified *X-Y*, *Y-Z*, or *Z-X* plane. [†]	1. DATBUF (27) is set (see Table 12-1). 2. Error message 6005 is issued in case of syntax error. 3. Control returns to Reader Section.
CYCLE/..... See discussion on routine DLWCYC in Section 13.4.2.3 for format of this statement.	To specify parameters needed to define drilling, tapping, or milling cycle. Parameters in this statement include type of cycle, clearance height and depth of machining, feedrate, and dwelling time.	1. a. DATBUF (31) or the LOGCAL array is set according to specified cycle type (see Sections 12.1.3 and 13.4.2.3). b. CYCTYP (2) or DATBUF (28) is set according to specified cycle type. c. The z coordinate of clearance plane, depth of drilling, or milling are stored internally. d. DATBUF (19) is set according to specified dwelling time. 2. Error message 6008 is issued in case of syntax error. 3. Control returns to Reader Section.

DELAY/t

To order tool to dwell at current position for t seconds.

1. a. DATBUF (19) = t.
   b. DATBUF (25) = 9004.
   c. TIME (1) = TIME (1) + t.
2. Error message 6006 is issued in case of syntax error.
3. Control goes to MTM.

FEDRAT/ $\begin{cases} \text{[IPM,]}f1 \\ \text{[IPR,}f2\text{[,MAXIPM,}f3\text{]} \end{cases}$

$f1$, $f2$, and $f3$: must be greater than zero
IPM: inches per minute
IPR: inches per revolution of spindle or tool
MAXIPM: maximum feedrate in inches per minute

To specify feedrate for tool motion defined in following statements. Option "MAXIPM,$f3$" is specified whenever change of maximum allowable feedrate, f_{max}, defined by answer to question M06, to feedrate $f3$ is required.

1. a. DATBUF (12) = 0.
   b. DATBUF (10) = $f1$ for IPM, $f1$;
      $f2 \cdot n$ for IPR, $f2$
      (n = spindle speed).
      (DATBUF (10) is set to f_{max} when $f1$ or $f2$ is greater than maximum allowable feedrate, f_{max}.)
2. Error message 6107 is issued if there is syntax error, or if $f1$ or $f2$ is greater than f_{max}.
3. Control returns to Reader Section.

INSERT......

... : NC statement or comment

To order postprocessor to output specified comment or NC statement in output NC program.

DAPP directly transfers specified NC statement or comment to output section without examining its content. Then control goes to Output Section, which punches and/or prints specified statement or comment.

LEADER/n

n: positive scalar

To define length of tape leader, in units specified by answer to question M30. This statement overrides answer to question P01.

1. Information (scalar) is stored internally and is used when Output Section punches NC tape.
2. Error message 6009 is issued if there is syntax error.
3. Control returns to Reader Section.

LOADTL/t[,SETOOL,d,e,f]

See section 6.6 for parameters.

To load No. t cutting tool, which has offsets of d, e, and f in X, Y, and Z directions, respectively, from reference point on tool holder to tool end center (for mill) or tool nose center (for turning tool).

1. a. DATBUF (15) = t.
   b. Offsets d, e, and f are stored internally and transferred to Geometric Section to carry out transformation of tool position.

(continued)

473

TABLE 13-5 THE POSTPROCESSOR COMMANDS WITH PARAMETER(S), PROCESSED BY THE DLWPP2 ROUTINE (continued)

Postprocessor Command	Meaning	Result of Processing This Command
		2. Error message 6020 is issued if syntax error is detected. 3. Control goes to MTM.
MACHIN/*ppname*,*n1*,OPTION,*n2*[,*n3*[,*n4*]] See section 6.12.2 for meaning of parameters.	To specify postprocessor (i.e., MTM and BLOCK DATA program) to be used to process CLDATA, and to pass information to postprocessor.	1. DAPP calls MTM and BLOCK DATA program, specified by parameters *ppname* and *n1*, when they are needed. Appropriate output routine is selected according to parameter *n2*. Also, following variables are set: $\quad$ DATBUF (34) = *n3* $\quad$ DATBUF (35) = *n4* 2. Error message 6010 is issued whenever syntax error is detected. 3. Control returns to Reader Section.
MCHTOL/ $\left\{\begin{array}{l} \text{ON} \\ \text{OFF} \\ \text{HIGH} \\ \text{LOW} \\ n \end{array}\right\}$ *n*: a scalar	To specify machining tolerance that is required by MTM for processing acceleration and deceleration at corner (see Section 14.3.3.1, Chapter 14, for detail).	1. a. DATBUF (25) = 9005. $\quad$ b. DATBUF (26) = 1 $\quad$ for ON; $\qquad\qquad\qquad\qquad$ 2 $\quad$ for OFF; $\qquad\qquad\qquad\qquad$ 3 $\quad$ for HIGH; $\qquad\qquad\qquad\qquad$ 4 $\quad$ for LOW; $\qquad\qquad\qquad\qquad$ *n* $\quad$ for *n*. 2. Error message 6011 is issued in case of syntax error. 3. Control goes to MTM after this statement is processed.
OPSKIP/ $\left\{\begin{array}{l} \text{ON} \\ \text{OFF} \end{array}\right\}$	Statement OPSKIP/ON orders postprocessor to generate slash sign at beginning of each NC statement until it is canceled by statement OPSKIP/OFF. Beginning slash in NC statement means that statement is ignored by NC controller when optional-stop button of machine is depressed.	1. Corresponding internal parameter is set to order Output Section to place slash sign at beginning of each output NC statement. 2. Error message 6012 is issued in case of syntax error. 3. Control returns to Reader Section.

ORIGIN/x,y,z.[a]

See Section 13.4.1.3 for its usage and processing result.

PARTNO.......

Specified commentary text is used as identification of processed part machining program.

1. Text is stored internally. It will be printed as heading on each page of listing file and, if answer to question P03 is YES, punched as man-readable text on punched tape preceding leader.
2. Control returns to Reader Section.

PLABEL/n1[,n2[,...[,n6]]]]]]

n1,...,n6: positive scalars consisting of 1 to 6 digits.

To specify man-readable codes n1,...,n6 on tape. These codes are preceded, respectively, by two-character prefixes—TN (tape No.), PN (part No.), OP (operator No.), EC (engineering change), MC (master change) and MG (machining group)—to indicate their meaning.

1. These scalars are stored internally.
2. Error message 6014 is issued in case of syntax error.
3. Control returns to Reader Section.

PPRINT/.......

...: commentary text (maximum length = 66 characters)

To order postprocessor to print commentary text on output listing file.

1. After receiving this command, DLWPP2 calls immediately Output Section, which prints text and line of asterisks on output listing file.
2. Control returns to Reader Section.

PREFUN/n

n: positive scalar

To specify preparatory function code (G code) in output NC program.

1. DATBUF (25) = −n.
2. Error message 6015 is issued in case of syntax error.
3. Control goes to MTM.

ROTABL/$\left\{ \begin{matrix} \text{INCR} \\ \text{ATANGL} \end{matrix} \right\}$,a,$\left\{ \begin{matrix} \text{CLW} \\ \text{CCLW} \end{matrix} \right\}$[,ROTREF]

(See Section 13.4.1.2 for its usages and processing result.)

SELCTL/n

n: positive scalar

To move tool, numbered n, in tool magazine to the position where it can be picked up by automatic tool changer.

1. a. DATBUF (16) = n.
   b. TIME (1) = TIME (1) + TIME (6).
2. Error message 6017 is issued if there is syntax error.
3. Control returns to Reader Section.

(continued)

475

TABLE 13-5 THE POSTPROCESSOR COMMANDS WITH PARAMETER(S), PROCESSED BY THE DLWPP2 ROUTINE (continued)

Postprocessor Command	Meaning	Result of Processing This Command
SEQNO/n1,[,INCR,n2,n3] 1. n1, n2, n3: positive integers 2. n2 = n3 = 1 if not specified 3. If this statement is not specified, statement SEQNO/1 is assumed instead.	Defining sequence number (n1) for current block. Sequence number is increased by n2 for every n3 blocks after block with sequence number n1.	1. Number n1 is first stored in element DATBUF (21), which is updated according to condition set by parameters n2 and n3. 2. Error message 6018 is issued in case of syntax error. 3. Control returns to Reader Section.
SPINDL/ $\begin{Bmatrix} \text{ON} \begin{Bmatrix} [,\text{CLW}] \\ \{,\text{CCLW}\} \end{Bmatrix} \\ \text{OFF} \\ s \begin{Bmatrix} ,\text{CLW} \\ ,\text{CCLW} \end{Bmatrix} \end{Bmatrix}$	To define rotation speed and direction of spindle and to turn spindle on or off.	1. DATBUF (13) and (14) are set (see Table 12-1). If parameter s is not specified, spindle speed is one specified in preceding SPINDL statement or default spindle speed given by answer to question M08 if no spindle speed is specified in program. 2. Error message 6019 is issued in case of syntax error. 3. Control returns to Reader Section.

†In fact, the tool positions in the CLFILE have already been compensated for the cutter radius. Therefore it is not necessary to use this statement to define the direction (left or right) of a cutter radius compensation. However, this statement can be used to define the circular interpolation plane or to transfer useful information to the postprocessor by specifying a scalar, n, in the statement (an example is given in Chapter 15).

Modified from IBM Manual (SH20-2469-0): *System/370 Automatically Programmed Tool—Intermediate Contouring and Advanced Contouring Numerical Control Processor: Program Reference Manual*. Rye Brook, N.Y.: IBM Corp., 1980. (Courtesy of International Business Machines Corporation.)

CLW = indication that the angle is measured in the clockwise direction

CCLW = indication that the angle is measured in the counterclockwise direction

This statement is used to define simply a rotation of the rotary table, which can be considered as a kind of machine motion such as those in the X, Y, and Z directions. It has no effect on the part or machine coordinate system.

The second format of the ROTABL statement can be used to define a rotation of the origin of the part coordinate system together with the rotary table. The format is

ROTABL/$\begin{Bmatrix} \text{INCR} \\ \text{ATANGL} \end{Bmatrix}$,a,$\begin{Bmatrix} \text{CLW} \\ \text{CCLW} \end{Bmatrix}$,ROTREF

where the minor word ROTREF indicates that the origin of the part coordinate system is to be rotated together with the rotary table and about the center of the rotary table by angle a, for option "INCR,a," or to the angular position, a, for option "ATANGL,a," in the direction defined by the word CLW or CCLW.

It should be pointed out that in the DAPP, the axis of the rotary table is assumed to be parallel to the Y axis and the spatial orientation of the part coordinate system remains unchanged after rotation. If the axis of the rotary table is in a direction other than Y, the COPY or TRACUT statement should be used to transform the cutter path.

Figure 13-1 shows diagrammatically the difference in meaning of these two statements. In Fig. 13-1(a), a hole on surface B is to be drilled after the one on surface A is finished. A rotation of 45 degrees is required for the rotary table to bring surface B into position (Fig. 13-1[b]). In this case, the part coordinate system remains unchanged. The effect of the ROTABL statement with the word ROTREF is shown in Fig. 13-1(c). After execution of the statement

ROTABL/INCR,90,CCLW,ROTREF

both the part and the origin of the part coordinate system have been rotated by 90 degrees; however, the spatial orientation of the coordinate axes remains unchanged.

The ROTABL statement is processed by the routine DLWROT. The processing is carried out in the designated sequence. When the ROTABL statement is read, it is sent to the decode area that processes the postprocessor word with parameters. The routine DLWROT (Fig. 13-2) is called once the word ROTABL has been identified. The routine first checks the statement format and issues error message 6016 if there is a syntax error. Control is then sent back to the Reader Section. Otherwise, processing begins. DATBUF(22) is set according to the specified rotation direction. If the word ROTREF is read, the routine calculates the X, Y, and Z displacements of the coordinate system origin. Otherwise, control is bypassed to the next section, which calculates the time and checks the rotation direction for the rotary table to reach the programmed position. Then the variables DATBUF(18), DATBUF(25), and TIME(1) are set on the basis of the calculation shown above. Finally, the total corrections of the tool position coordinates for the table rotation, the position of the rotary table in the machine coordinate system (defined by the answers to question

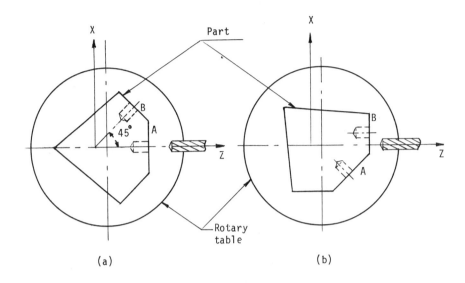

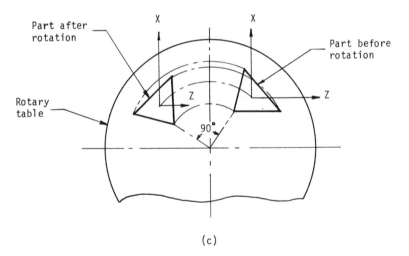

(c)

Figure 13-1   The part and the part coordinate system (a) before and (b) after execution of the statement ROTABL/INCR,45,CLW. (c) The rotation of the origin of the part coordinate system Z-X, together with the rotary table, is defined by the statement ROTABL/INCR,90,CCLW,ROTREF.

M13), and the tool offsets (defined by the LOADTL statement) are calculated and stored internally to enable the DAPP to calculate the correct coordinates for each tool position. Finally, control goes to the MTM after processing.

13.4.1.3 The Routine DLWORG.   Routine DLWORG is used to process the statement

ORIGIN/x,y,z,[a]

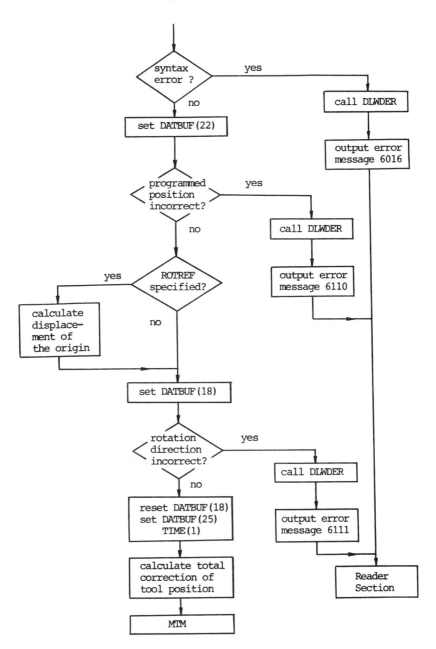

Figure 13-2 The flowchart of the DLWROT routine. Control is entered from the decode area processing postprocessor words without parameters.

which is used to define (1) the origin of the machine coordinate system with respect to the part coordinate system and (2) the zero angular position of tne rotary table with respect to the Z axis. If the machine does not have a rotary table, the parameter *a* is not needed. In this case, the statement has the same meaning as described in

Section 6.11 (i.e., it defines the origin of the machine coordinate system in the part coordinate system).

When an NC machine has a rotary table, its axis of rotation is assumed to be parallel to the Y axis. This statement then defines, in the part coordinate system, the machine coordinate system whose center is assumed to be at the center of the rotary table (Fig. 13-3[a]). Thus parameters x, y, and z represent the coordinates of the machine coordinate system or the rotary table center in the part coordinate system. Parameter a is the angle that defines the zero angular position of the rotary table and is measured from the $+Z$ axis in the direction defined by the answers to question M10.

If the center of the rotary table is not at the origin of the machine coordinate system, this statement relates the rotary table center to the part coordinate system (Fig. 13-3[b]). Therefore the answers to question M13 should be used to further transform the tool position data.

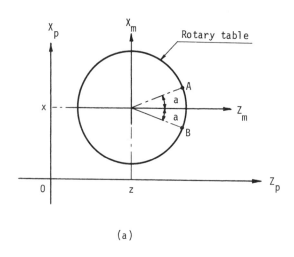

(a)

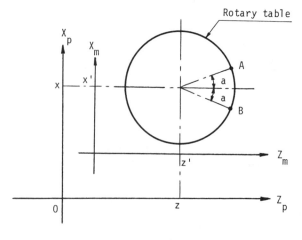

(b)

Figure 13-3   The application of the ORIGIN statement. The origin of the machine coordinate system Z_m-X_m and the center of the rotary table (a) are or (b) are not the same point.

NOTE: Z_p-X_p = part coordinate system; A = zero angular position of the rotary table defined by the ORIGIN statement when it can rotate only in the CCLW direction; and B = zero position of the rotary table defined by the ORIGIN statement when it can rotate in the CLW direction only or in both directions.

The processing in this routine is simple. First, the syntax is checked, and if any errors are found, an error message numbered 6013 is issued. Otherwise, the position of the rotary table with respect to the part coordinate system is calculated in polar coordinates, which are saved for use by the DLWROT routine. The corrections in the X, Y, and Z directions for the tool position, as required by the change in coordinate system, are $-x$, $-y$, and $-z$ (see Section 6.11). Then the total corrections in the X, Y, and Z directions are calculated as follows:

$$k_x = p_x \cdot (-x + x' + x'')$$
$$k_y = p_y \cdot (-y + y' + y'')$$
$$k_z = p_z \cdot (-z + z' + z'')$$

where

$x, y, z =$ parameters in the ORIGIN statement

$x', y', z' =$ coordinates of the rotary table in the machine coordinate system (i.e., the answers to question M13)

$x'', y'', z'' =$ compensations for tool offsets in the X, Y, and Z directions, respectively

$p_x, p_y, p_z = +1$ or -1, depending on the answers to question M19

Then these values are saved internally for use by the Geometric Section of the DAPP to calculate the tool position with respect to the machine coordinate system.

Control finally returns to the Reader Section after processing.

13.4.2 The Geometric Section

There are three processing routines in the Geometric Section:

1. Routine DLWLIN, which processes the data regarding linear interpolation motion
2. Routine DLWCIR, which processes the data regarding circular interpolation motion
3. Routine DLWCYC, which processes the data regarding a machining cycle

Apart from the incremental motion defined by statement GODLTA, a tool motion is defined either by the record GOTO/(single point) or by the record GOTO/(multiple points) in the CLDATA. Usually, routine DLWLIN is used by the DAPP to process the motion record. However, routines DLWCIR and DLWCYC are called when statements ARCSLP/ON and CYCLE/..., respectively, are specified. The respective variables are set during processing by this section, and control is then sent to the MTM after processing.

13.4.2.1 The Linear Interpolation Routine DLWLIN.
As one of the basic routines for processing motion records, routine DLWLIN can be called whenever a GOTO or GODLTA record is read. The function of this routine is to calculate the parameters needed for defining a linear interpolation motion, namely, the absolute coordinates of the destination point, the displacement from the starting point to the destination, the feedrate, and the time required for moving the tool.

The absolute coordinates of the destination point, which are described in the part coordinate system in the CLDATA, should first be transformed into those with respect to the machine coordinate system. This transformation is based on the calculation results from the DLWORG and DLWROT routines (i.e., the total corrections of the tool position coordinates). The transformed coordinates of the destination point are stored in DATBUF(1), (2), and (3). In addition, the displacements in the X, Y, and Z directions, specified in the GODLTA statement, are set as DATBUF(7), (8), and (9), respectively. Then the DLWLIN routine recalculates the feedrate if a programmed feedrate is expressed in inches per revolution. Otherwise, the specified feedrate is set directly as DATBUF(10). Finally, the routine calculates the time needed for moving the tool from the starting point to the destination. In this stage, if the motion is specified in a cycle with a programmed feedrate other than rapid, and if the machine has the canned cycle function, control is bypassed to the DLWCYC routine to calculate the pertinent Z positions for the cycle. Then control returns to this routine to finish calculation of the machining time, which is saved as TIME(1) in the DAPP.

In summary, the DLWLIN routine is called from the decode areas that process the GODLTA and GOTO records. The data passed to this routine by the DAPP consist of the following:

1. DATBUF(4), (5), and (6), representing the starting point
2. The x, y, and z coordinates of the destination point specified in the GOTO or GODLTA statement
3. The corrections of the tool position accounting for the difference between the part and machine coordinate systems, the offset of the tool, and the rotation of the rotary table (These are calculated by routines DLWORG and DLWROT.)
4. DATBUF(10) and DATBUF(14)

This routine also receives the limits on the tool movement, the resolution of the NC controller, and the allowed simultaneous motions in different axes. These are defined, respectively, by the answers to questions M01 through M03, C04, and M15 through M18. These data are used by DLWLIN to determine whether the specified tool motion exceeds the limits, to round off the calculation results according to the resolution of the NC controller, and to calculate the machining time.

The calculation results are used to reset the following variables:

1. DATBUF(1), (2), and (3), which define the destination point in the machine coordinate system
2. DATBUF(7), (8), and (9), which define the displacement from the starting point to the destination point
3. DATBUF(10), which is the transformed feedrate, expressed in inches per minute, if the feedrate is programmed in inches per revolution (Otherwise, it is the same as the input value.)
4. TIME(1), which is the time for moving the tool from the starting point to the destination

Control is sent to various sections in the DAPP according to the situation. If the processed record is GODLTA or GOTO/(single point), control goes to the MTM after completion of the processing; the Output Section is then called to print or punch the result. Finally, control returns to the Reader Section to retrieve a new set of records. If a GOTO/(multiple points) record is processed, control goes to the MTM after the data for the first point are processed and then to the Output Section. After the data are output, control returns to routine DLWLIN to process the data for the next point.

If the specified destination point is the same as the starting one, control returns to the Reader Section. However, if the processed record consists of more than one point, control is sent back to routine DLWLIN again by the DAPP main routine.

If the processed record is within a machining cycle, the calculation of machining time is interrupted and control is bypassed to the canned cycle routine DLWCYC, which returns control to routine DLWLIN after having determined the appropriate z coordinates for the cycle motion.

During processing by routine DLWLIN, several tests are made to determine whether the motion and feedrate are out of defined ranges. If the tool motion exceeds the limits on X, Y, or Z motion, a corresponding error message, numbered 6101, 6102, or 6103, is issued. The error message numbered 6116 or 6107 is generated for a programmed feedrate of zero or exceeding the maximum value. The postprocessing is not terminated because of these errors.

13.4.2.2 The Circular Interpolation Routine DLWCIR.   As suggested by its name, routine DLWCIR processes CLDATA records concerning circular cutter paths. A circular motion defined in APT has the following corresponding set of class 3000 and 5000 CLDATA records:

SURFACE   (symbol for circular drive surface)
                (parameters in canonical form of circle)
GOTO/   (symbol for circular drive surface)
                (x, y, and z coordinates of first point)
                (x, y, and z coordinates of second point)
                ....
                ....
                (x, y, and z coordinates of last point)

The points in the GOTO/(multiple points) record are calculated by the calculation section of the NC processor; in other words, a circular path defined in APT is approximated by a series of linear interpolation motion vectors, which are calculated according to the specified tolerance (see Section 5.9). Routine DLWCIR determines the parameters needed for generating the circular interpolation code G02 or G03 in the output NC program and stores them as the corresponding variables, that is, DATBUF(36) through (71), so that they can be retrieved easily when needed. If the CLDATA of a circular cutter path is processed as a series of linear motions without using the circular interpolation routine DLWCIR, the resulting part profile is still within the tolerance; however, the resulting NC program segment for the circular motion usually consists of a large number of G01 codes and is indeed much longer than a program that defines the same cutter path with only a G02 or G03 code.

The parameters needed for defining circular motion in NC codes are the coordinates of the destination point, the motion direction (i.e., clockwise or counterclockwise), and the parameter(s) necessary for defining the center (either the radius or the offsets I, J, and K of the center from the starting point). If the statement ARCSLP/ON is specified, routine DLWCIR will be called to process the CLDATA records. The linear interpolation routine DLWLIN will be used instead of DLWCIR to process circular motion CLDATA records when

1. No ARCSLP/ON statement has been specified preceding the circular motion statement
2. Specified feedrate is RAPID
3. Circular movement is not in the coordinate planes allowed by the NC controller
4. Number of points in the GOTO/(multiple points) record is less than three
5. Radius of the circular motion exceeds the maximum allowable radius defined by the answer to question C03 or is smaller than the controller resolution specified by the answer to question C04

When the CLDATA records concerning circular motion have been read, control goes to the corresponding decode area, and if the ARCSLP/ON statement is specified, routine DLWCIR is called. Processing in DLWCIR begins by checking for the five conditions listed above. If any of them are detected, control is sent to the linear interpolation routine DLWLIN*; otherwise, the processing continues. The coordinates of the point in the GOTO record are transformed on the basis of the result calculated in the DLWORG and DLWROT routines and hence are redefined in the machine coordinate system. The X, Y, and Z offsets of the circle center from the starting point are also determined. The routine does not stop at this stage but determines whether there is a further point in the GOTO record. If so, control is sent back to the Reader Section to retrieve the record of the next point. The process is repeated until the last point of the first quadrant is reached. Then the displacements and offsets in the X, Y, and Z directions are calculated and stored as DATBUF(42) through (47). The coordinates of the last point are also compared with the machine limits; error messages 6101, 6102, and 6103 are issued if the tool position exceeds the X, Y, and Z limits, respectively. In the same way, the DLWCIR routine calculates the displacements and center offsets in the X, Y, and Z directions for the following quadrants, and the variables DATBUF(48) through (71) are set. This calculation is terminated whenever the last point in the GOTO/(multiple points) data is processed. Finally, the time required for machining the circular arc is determined. Control goes to the MTM after the processing has been finished.

The data needed by the DLWCIR routine consist of the following:

1. DATBUF(4) through (6) (the coordinates of the starting position)
2. Class 3000 CLDATA record (circle center coordinates and radius)

*Error messages 6119 and 6122 are issued, respectively, for situations 3 and 4, described above, and error messages 6120 and 6121, when the radius of the circular motion exceeds the maximum value or is less than the minimum value, respectively.

3. Class 5000 CLDATA record: GOTO/(multiple points)
4. X, Y, and Z corrections of the tool position, calculated by the DLWROT and DLWORG routines
5. Postprocessor command ARCSLP/ON
6. LIMIT(1) through (3), (6), and (8) through (11) (the limits on machine movement, radius of circular movement, and feedrate)
7. Feedrate specified by statement FEDRAT: can be the variable DATBUF(10) if the specified feedrate is in inches per minute, or another variable if the specified feedrate is in inches per revolution
8. DATBUF(14) (the spindle speed specified by the statement SPINDL)

The variables calculated and set by the DLWCIR routine are as follows:

1. DATBUF(1) through (3) (the coordinates of the destination point of the circular movement)
2. DATBUF(7) through (9) (the total displacements in the X, Y, and Z directions)
3. DATBUF(36) (the circular interpolation plane)
4. DATBUF(37) (the direction of circular movement)
5. DATBUF(38) through (40) (the center coordinates)
6. DATBUF(41) (the radius)
7. DATBUF(42) through (47) (the X, Y, and Z displacements and center offsets for the first quadrant)
8. DATBUF(48) through (71) (the X, Y, and Z displacements and center offsets for the following quadrants)
9. DATBUF(10) (the same as the input value of DATBUF(10) if the feedrate is in inches per minute; otherwise calculated according to the feedrate, in inches per revolution, and the spindle speed)
10. TIME(1) (the time required for moving the tool from the starting point to the destination point)

It should be pointed out that there is a mistake in the routine (of the APT-AC NC processor, version 1, modification level 3) that performs the calculation of circle center offsets, namely, DATBUF(45) through (47), (51) through (53), (57) through (59), (63) through (65), and (69) through (71).[9] The problem is that these calculated values are always positive even if one or several of them should be negative. This error can be corrected by incorporating a special calculation routine in the MTM to recalculate the values of the DATBUF positions mentioned above (see Section 14.3.3.2).

13.4.2.3 The Machining (Canned) Cycle Routine DLWCYC.   When a machining process involves drilling a number of holes, it is necessary to define the upward and downward motions at each hole position. This procedure is tedious, especially when the number of holes is large. In Chapter 5, we described a method that uses the PATERN statement, together with the minor word AVOID, to define the drilling operation of multiple holes. An APT programmer needs to define the point pattern and the avoidance parameters only, the upward and downward motions of the tool being generated by the NC processor. In many cases the depth/diameter ratio of the holes to be drilled is so large that the drill should be moved up periodically to bring out the chips. This type of motion is called a *peck-drilling cycle* and cannot be defined by the PATERN statement. Further, it is preferable to generate one statement with the G83 code for the drilling cycle instead of several statements with the G01 code. A postprocessor command CYCLE/... is available to simplify the definition of the drilling, peck-drilling, tapping, and milling operations.

The format of the CYCLE statement defining a drilling or tapping operation is as follows (Fig. 13-4):

$$\text{CYCLE/}\begin{Bmatrix} \text{DRILL} \\ \text{TAP} \end{Bmatrix}, z_d, \begin{Bmatrix} \text{IPM} \\ \text{IPR} \end{Bmatrix}, f, z_c [, \text{DWL}, t]$$

where

> DRILL = postprocessor word defining a drilling cycle
> TAP = postprocessor word defining a tapping cycle
> z_d = depth of drilling below points PTB and PTC specified in cycle
> IPM = postprocessor word indicating that the feedrate, f, specified in this statement is in inches per minute

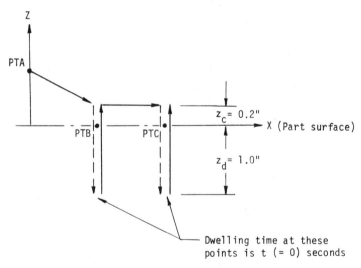

Figure 13-4   The cutter motion defined by the program segment
　　　　　　　FROM/PTA
　　　　　　　CYCLE/DRILL,1.0,IPM,2.0,0.2
　　　　　　　GOTO/PTB
　　　　　　　GOTO/PTC
　　　　　　　CYCLE/OFF
Points PTB and PTC are on the surface of the part.
- - - - → cutter motion at programmed feedrate 2.0;
———→ cutter motion at rapid feedrate.

IPR = postprocessor word indicating that the feedrate, f, specified in this statement is in inches per revolution

f = feedrate

z_c = clearance height above specified points PTB and PTC;

DWL = postprocessor word specifying a pause at the bottom of the hole for t seconds

t = dwelling time in seconds

A CYCLE/OFF statement is needed to terminate the defined CYCLE function, whereas a CYCLE/ON statement can be specified to resume a terminated machining cycle. For example, the cutter motion for the cycle shown in Fig. 13-4 can be defined in APT as follows:

```
..............................
...(the tool is at point PTA)...
CYCLE/DRILL,1.0,IPM,2.0,0.2,DWL,0.1
GOTO/PTB
GOTO/PTC
CYCLE/OFF
```

The following program segment, which defines the same cutter motion, indicates the usage of statements CYCLE/ON and CYCLE/OFF:

```
..............................
.....(the tool is at point PTA)...
CYCLE/DRILL,1.0,IPM,2.0,0.2,DWL,0.1
GOTO/PTB
CYCLE/OFF
CYCLE/ON
GOTO/PTC
CYCLE/OFF
```

When the points defined in a cycle are not on the same level, the retreat motion at one point depends also on the z coordinate of the next point. As can be seen from Fig. 13-5, if $z_{PTC} > z_{PTB}$, then the z coordinate of the retreat motion at point PTB is

$$z_3 = z_{PTC} + z_c$$

otherwise

$$z_3 = z_1$$

Thus the function of the CYCLE statement is to define an additional Z motion at the points specified in the cycle. The absolute z coordinates of the starting, depth, and final positions at one point, that is, z_1, z_2, and z_3 (Fig. 13-5), are calculated by routine DLWCYC and stored as DATBUF(3), (29), and (30), respectively.

The statements GODLTA/z and GOHOME and the statement pairs GOCLER–CLEARP and RETRCT–CLRSRF can also be specified in the cycle. The purpose of using these statements in a cycle is to define a cutter retraction motion to avoid collision of the tool with an obstacle between two holes. A comparison of the calculation results for cycles with different retraction statements is shown in Table 13-6. As can be seen from the listed results, the tool position data DATBUF(29) and (30), corresponding to the additional statements for Z retraction (i.e., GOHOME, GOCLER,

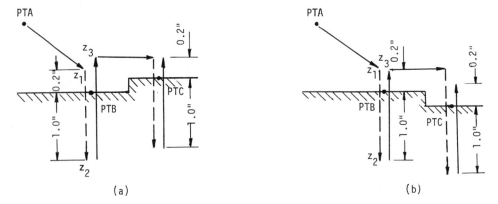

(a)   (b)

Figure 13-5   The cutter motion defined by the program segment
FROM/PTA
CYCLE/DRILL,1.0,IPM,2.0,0.2
GOTO/PTB
GOTO/PTC
CYCLE/OFF
where points PTB and PTC have different z coordinates.

and RETRCT), are the same as in the previous record. A GODLTA or GOHOME statement defines a point-to-point motion, just as a GOTO statement does. When the statement GODLTA, GOHOME, or GOCLER is specified, DATBUF(3) is the z coordinate specified in the statement. However, the value of DATBUF(3), corresponding to the RETRCT statement, is zero. The CLRSRF statement has no effect on the motion if it is specified within the cycle. This comparison indicates that different rules are used in the calculation routine of the DLWCYC for processing different statements and therefore should be considered during the design of the Machine Tool Module.

The CYCLE statement can also be used to define countersinking, peck-drilling, tapping, and milling cycles (Table 13-7). The tool motion in the tapping and countersinking cycles is similar to that in the drilling cycle. The DLWCYC routine calculates the value of DATBUF(29) on the basis of the given parameters d and a of the countersunk hole. The peck-drilling cycle is defined by the minor word DEEP (deep-hole drilling) or BRKCHP (chip-breakage drilling), together with the necessary parameters, z_i and f_i. A milling cycle does not involve up-and-down tool motion along the Z axis at one specific point. Although it can be defined through the use of a CYCLE statement, the programming saving is not significant.

When the first record of a cycle (i.e., statement CYCLE/...) is read, routine DLWPP2 is called to set up an auxiliary array where the parameters in the CYCLE statement can be stored; in addition, DATBUF(10) and (28) are set. Control then returns to the Reader Section. Then the GOTO data in the cycle is read and control goes to routine DLWLIN, wherein routine DLWCYC is called if the motion is defined at a feedrate other than RAPID. The canned-cycle routine DLWCYC sets the various z values at each point as variables DATBUF(3), (29), and (30), the feedrate and the dwelling time as the variables DATBUF(10) and (19), and the cycle status

TABLE 13-6 THE TOOL POSITION DATA CALCULATED BY THE DLWCYC ROUTINE (SEE ALSO FIG. 13-4). PTA: (0,0,1); PTB: (1,0,0); PTC: (2,0,0)

APT Program Segment Defining Cycle Motion	Tool Position Data for the Defined Cycle				
	DATBUF (1)	DATBUF (2)	DATBUF (3)	DATBUF (29)	DATBUF (30)
FROM/PTA					
CYCLE/DRILL,1.0,IPM,2.0,0.2					
GOTO/PTB	1.0	0	0.2	−1.0	0.2
GOTO/PTC	2.0	0	0.2	−1.0	0.2
CYCLE/OFF	*	*	*	−1.0	0.2
FROM/PTA					
CYCLE/DRILL,1.0,IPM,2.0,0.2					
GOTO/PTB	1.0	0	0.2	−1.0	0.7
GODLTA/0.5,0.0.5	1.5	0	0.7	−0.5	0.7
GOTO/PTC	2.0	0	0.2	−1.0	0.2
CYCLE/OFF	*	*	*	−1.0	0.2
(*Note:* Home position is point PTA.)					
FROM/PTA					
CYCLE/DRILL,1.0,IPM,2.0,0.2					
GOTO/PTB	1.0	0	0.2	−1.0	1.0
GOHOME	0	0	1.0	−1.0	1.0
GOTO/PTC	2.0	0	0.2	−1.0	0.2
CYCLE/OFF	*	*	*	−1.0	0.2
FROM/PTA					
CYCLE/DRILL,1.0,IPM,2.0,0.2					
GOTO/PTB	1.0	0	0.2	−1.0	0.2
CLEARP/XYPLAN,0.7					
GOCLER	1.0	0	0.7	−1.0	0.2
GOTO/PTC	2.0	0	0.2	−1.0	0.2
CYCLE/OFF	*	*	*	−1.0	0.2

(continued)

TABLE 13-6 THE TOOL POSITION DATA CALCULATED BY THE DLWCYC ROUTINE (SEE ALSO FIG. 13-4). PTA: (0,0,1); PTB: (1,0,0); PTC: (2,0,0) (continued)

APT Program Segment Defining Cycle Motion	Tool Position Data for the Defined Cycle				
	DATBUF (1)	DATBUF (2)	DATBUF (3)	DATBUF (29)	DATBUF (30)
....					
FROM/PTA					
CYCLE/DRILL,1.0,IPM,2.0,0.2					
GOTO/PTB	1.0	0	0.2	−1.0	0.2
CLRSRF/0.7					
RETRCT	1.0	0	0	−1.0	0.2
GOTO/PTC	2.0	0	0.2	−1.0	0.2
CYCLE/OFF	*	*	*	−1.0	0.2
....					
CLRSRF/0.7					
....					
FROM/PTA					
CYCLE/DRILL,1.0,IPM,2.0,0.2					
GOTO/PTB	1.0	0	0.2	−1.0	0.7
RETRCT	1.0	0	0	−1.0	0.7
GOTO/PTC	2.0	0	0.2	−1.0	0.2
CYCLE/OFF	*	*	*	−1.0	0.2

*ANULL.

TABLE 13-7   THE FORMAT OF THE CYCLE STATEMENT DEFINING COUNTERSINKING, PECK-DRILLING, AND MILLING

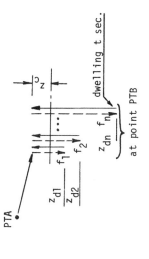

1. Countersinking cycle:
Statement format:

$$CYCLE/CSINK, d, a, \left\{ \begin{matrix} IPM \\ IPR \end{matrix} \right\}, f, d_c, DWL, t$$

Cutter motion shown in figure is defined by following program segment:

FROM/PTA
CYCLE/CSINK,0.5,90,IPM,1.0,0.6,DWL,1.0
GOTO/PTB
CYCLE/OFF

2. Peck-drilling (DEEP or BRKCHP) cycle:
Statement format:

$$CYCLE/ \left\{ \begin{matrix} DEEP \\ BRKCHP \end{matrix} \right\}, z_{d1}, \left\{ \begin{matrix} IPM \\ IPR \end{matrix} \right\}, f_1[, z_{d2}, \left\{ \begin{matrix} IPM \\ IPR \end{matrix} \right\}, f_2[, \ldots, [, z_{dn}, \left\{ \begin{matrix} IPM \\ IPR \end{matrix} \right\}, f_n], \ldots,]], z_c, DWL, t$$

Cutter motion shown in figure is defined by following program segment:

FROM/PTA
CYCLE/DEEP,z_{d1},IPM,f_1,z_{d2},IPM,f_2, . . . ,z_{dn},IPM,f_n,z_c,DWL,t
GOTO/PTB
CYCLE/OFF

3. Milling cycle:
Statement format:

$$CYCLE/MILL, z_d, IPM, f, IPM, f_1, z_c, DWL, t$$

Motion shown in figure is defined by following program segment:

FROM/PTA
CYCLE/MILL,0.2,IPM,1.0,IPM,3.0,0.5,DWL,1.0
GOTO/PTB
GOTO/PTC
CYCLE/OFF

Note: ---→ cutter motion at programmed feedrate; ⟶ cutter motion at rapid feedrate.

491

as variable DATBUF(31) (or array LOGCAL). Control then returns to routine DLWLIN, which sends control back to the DAPP main routine. Thereafter the MTM specified by the MACHIN statement in the APT program is called to process the data further. After the data have been output, control goes back to the Reader Section. If the cycle is a peck-drilling one (specified by the word DEEP or BRKCHP), routine DLWCYC is called again to set another set of z values and to repeat the process described above; otherwise, a new CLDATA record is read.

To summarize, the data transferred to the DLWCYC routine are as follows:

1. DATBUF(3), calculated by the routine DLWLIN
2. The clearance and depth values specified in the CYCLE statement
3. The corrections of the tool positions in the X, Y, and Z directions, calculated by the DLWROT and DLWORG routines
4. Class 5000 CLDATA record: GOTO/(point)
5. The feedrate specified in the previous FEDRAT statement or in the CYCLE statement
6. The RAPID feedrate: DATBUF(11)
7. The programmed spindle speed: DATBUF(14)
8. The dwelling time specified in the CYCLE statement

The variables calculated and set by the DLWCYC routine are as follows:

1. DATBUF(3), which is the sum of the input DATBUF(3) and the clearance height value
2. DATBUF(29) (the depth of drilling)
3. DATBUF(30) (the height of retreat motion)
4. DATBUF(31), which contains the variables LOGCAL(1) to (4), representing the cycle status
5. DATBUF(10) (the machining feedrate)
6. DATBUF(19) (the dwelling time)
7. TIME(1) (the machining time for the cycle motion)
8. DATBUF(32) and (33), if the machine uses drums to control the z coordinates of the tool

Error message 6106 is issued if any of the z coordinates DATBUF(3), (29), or (30) exceeds the motion limit in the Z direction. When the number of clearance or depth drums needed for controlling the cycle motion exceeds the number of drums that the machine has (see answers to questions M28 and M29), error message 6108 or 6109 is issued accordingly.

13.5 THE OUTPUT SECTION

The Output Section carries out the last postprocessing step in the DAPP. When a set of CLDATA has gone through the general and machine-dependent processing, the Output Section is called by the MTM. The functions of the Output Section are to convert floating-point numbers and characters to the required formats, such as

EBCDIC* characters, punched cards, and ISO or EIA punched tape image; to write the processed data and message as a formatted listing; and to punch the output program in the form of cards, magnetic tape or disk, or punched paper tape (in ISO or EIA code). The required format for punched output is selected through the first parameter in the MACHIN statement (see Section 6.12.2).

The Output Section consists of one main routine and several subroutines. The main routine directs control within the Output Section, selecting the pertinent subroutine that converts the input data to the required output format. Also included in this section is the routine DLWDER, which prints the postprocessing error messages.

There are three different situations in which the Output Section is called, and control flows accordingly. When the CLDATA has successfully gone through general and machine-dependent processing, the MTM calls the Output Section. After having output the data according to the requirement specified by the MTM (more specifically, by the variable OUTCTL in the MTM), the Output Section sends control back to the MTM.

If the read-in CLDATA is a postprocessor command INSERT or PPRINT, the Output Section is called by the DLWPP2 routine. Then the commentary text is written on the printed listing or, for the statement INSERT, punched on the card or tape. Control returns to routine DLWPP2, and then the Reader Section is called to read in a new set of CLDATA.

The Output Section is also invoked when an error occurs during processing by the General Processing Section and the MTM. The corresponding error messages are printed by the DLWDER routine, and then control returns to the calling routine.

The output from the Output Section usually consists of two types: the printed listing and the punched output in either card form or punched tape form. The printed output lists the processed NC codes, required machining time, length of the NC tape, and error messages if any. The values of the selected variables in the DATBUF array can also be printed, so that the processed result can be checked. The format of the DAPP printed listing is explained in Chapter 15.

The punched output begins with the following constituents in listed order:

1. The man-readable commentary text specified in the PARTNO statement, if the answer to question P02 is YES
2. The man-readable text regarding the serial number of, for example, the punched tape, the part, or the design modification, specified by the statement PLABEL
3. The tape leader, if specified in the answer to question P01
4. The tape rewind-stop code, if the answer to question C06 is YES
5. The end-of-block code, if the answer to question C07 is YES

These items are followed by the part machining program, in NC codes, generated by the postprocessor. No punched output is given if it is not required or if an error has been detected by the DAPP. As indicated in Chapter 6, the RESET statement can be specified to force the DAPP to punch the output even if an error occurs.

---

*The abbreviation for Extended Binary Coded Decimal Interchange Code.

PROBLEMS

13.1 Explain the difference between the functions of the Initialization Section and the Housekeeping Section.

13.2 Explain the reason for updating those elements of DATBUF array listed in Table 13-1.

13.3 Some of the elements in the DAPBUF array are not updated by the Housekeeping Section of the DAPP. How do we realize the revision of one of these elements if, according to the processing requirement of a particular postprocessor, they should be updated after the processing of a set of read-in CLDATA records is completed?

13.4 Postprocessor words (commands) without parameters are processed by the DLWPP1 routine. After completion of processing, control can be sent to three different sections of the DAPP. Give one example for each possible case, indicate the control flow, and explain the reason for sending control to the section indicated in your example.

13.5 The third and fourth parameters in the MACHIN statement can be used to transfer certain useful information from an APT program to the postprocessor. List the possible usages of these parameters.

13.6 Define the drilling motion specified in Problem 8.9, using the postprocessor command CYCLE.

13.7 Give the NC code(s) to be processed by the postprocessor for each postprocessor word (command) listed in Tables 13-4 and 13-5.

13.8 If an error is found in one of the machine-independent processing routines and the corresponding output — that is, DATBUF element(s) — is incorrect, suggest a method for correcting the DATBUF element(s) so that the correct NC code(s) can still be generated by the postprocessor.

chapter 14

The Machine-Dependent Processing Routine: Machine Tool Module

The machine tool module (MTM) of a DAPP-based postprocessor is a subroutine that performs further processing of the data, prepared by the General Processing Section, which is then passed to the Output Section and output as the NC codes.

After a set of CLDATA has gone through general processing, it is stored in the DATBUF array. When the MTM is called, each data element should first be examined to ascertain that it is within the corresponding limits of the NC machine. If the data element cannot be directly used to output the correct NC code(s), it should be recalculated or processed. The set of CLDATA that has been read in by the Reader Section usually contains more than one single data element; it might be output as one or more NC blocks. Thus the MTM should also determine the contents of an NC block (statement). The correct format of each NC code should also be set by the MTM before the data are passed to the Output Section. The MTM is also responsible for generating the special statements that are not defined in the APT program but are required by the NC machine. Therefore the functions of the MTM are as follows:

1. To examine the data regarding tool motion and machine operation, to ascertain that they are within the allowed limits of the NC machine
2. To convert or transform the data to the corresponding code(s) that is (are) acceptable to the controller system of the NC machine
3. To determine the content of an NC block

4. To develop motion commands that allow for requirements of the NC machine, such as safe acceleration or deceleration of slides and overshoot at a corner

5. To generate special statements required by the NC machine

The MTM is invoked after a set of CLDATA defining motion or machine tool operation has been processed by the General Processing Section of the DAPP. When the set of read-in CLDATA data, after processing, consists of codes that might not be output in one NC block (e.g., two M codes exist), a part of the CLDATA is output first, and then the MTM is called again to process the remaining data.

In this chapter, we first introduce the general structure of an MTM. Then we describe the method used to realize the desired functions and explain the structure of the various sections of the MTM. Finally, we describe the processing flow in the MTM and the DAPP-based postprocessor.

14.1 THE GENERAL STRUCTURE OF A MACHINE TOOL MODULE

The MTM for a DAPP-based postprocessor is written by the postprocessor designer and can be designed in a manner that he considers appropriate for the processing. However, the names and formats of the variables through which the MTM communicates with the DAPP should be followed. The variables that should usually be taken care of during design of an MTM consist of those described in Section 12.1. Generally, an MTM can be divided into five sections (Fig. 14-1):

1. The variable-definition section, for defining the COMMON variables through which the MTM communicates with the DAPP, and for defining the variables specific to the MTM

2. The initialization section, for setting the respective initial values of the variables, arrays, and parameters that are specific to the MTM (Another function of this section is to output some special NC statements at the beginning of an NC program.)

3. The processing section, for determining the format of NC code(s) and the content of an NC block (It consists of a number of subsections, each processing one kind of CLDATA.)

4. The preparation section, for finalizing the content of an NC block and for keeping the record of the NC codes that will be handed over to the Output Section of the DAPP

5. The control-directing section, for directing control after leaving the MTM

Since the MTM is a subroutine of the DAPP program, its first statement must be the FORTRAN statement that defines the name of the subroutine. In DAPP, the name that can be used for an MTM is MTMD*nn*, in which *nn* can be 01, 02, ..., or 10. Thus the first statement of an MTM must be

SUBROUTINE MTMD*nn*

where *nn* should be the same as the answer to question E02 (see Section 12.2.3). After the first statement, the five sections described above are listed consecutively.

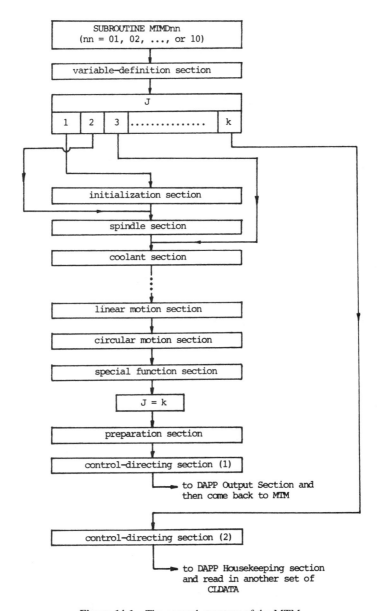

Figure 14-1   The general structure of the MTM.

The *variable-definition section* can be divided into two parts. The first part contains the statements defining the variables that serve as the communication linkage between the MTM and the DAPP (Fig. 14-2). They consist of one COMMOM field labeled DAPCOM, one labeled QUEST, which is used for communication with the corresponding BLOCK DATA program (see Section 12.3 and Fig. 12-4), and some other statements defining variables and common storage places that are used by both the DAPP and the MTM. Normally, these statements should not be

```
      IMPLICIT              REAL*8 (A-H,O-Z)
      COMMON   /DAPCOM/
     A      ANULL      ,CLEARS(3)  ,CSURF      ,CORREC(3)   ,FEDIPR
     B      ,FEDMAX    ,INNBUF(252),ORGX       ,ORGY        ,ORGZ
     C      ,PNO(6)    ,PTNOIP(11) ,RADI       ,RDRUM(20)   ,TABPOS
     D      ,THETA     ,TLNTH(4)   ,WORKR8     ,ZDRUM(20)
     E      ,AUXARR(6) ,CLKARG     ,CLKHRS     ,CLKMIN      ,CLKSEC
     F      ,DHFED(10) ,DPHOL(10)
     G      ,CLKSW6    ,CONTRL     ,DUMREC     ,ERRCOD      ,IDIREC
     H      ,IENDSW    ,IFED1      ,IFED2      ,IFIPR       ,IJSEQ
     I      ,ILIM      ,IERCNT     ,IERINC     ,IERROR(20)  ,INCBLK
     J      ,INSPUN    ,IPLAB      ,IPOINT     ,IRECNO      ,ISKIP
     K      ,ITAPL     ,IWORDS     ,JASGN      ,JEND        ,JLINE
     L      ,JP        ,LNCIRC     ,NBSTEP     ,NOSETP      ,NOWRD
     M      ,NSST      ,NSUBCL     ,OUTCTL     ,OUTRQD      ,PASS7
     N      ,PGCNT     ,PRTARG     ,PRTLIN(120),PUNBUF(144) ,PUNTAP
     O      ,RESET5    ,SEQNM7     ,SEQINC     ,ZPCTL1
      REAL*8             INNBUF
      REAL*4             AUXARR     ,CLKARG     ,CLKHRS      ,CLKMIN
     A      ,CLKSEC     ,DHFED      ,DPHOL
      INTEGER            CLKSW6     ,CONTRL     ,DUMREC      ,ERRCOD
     A      ,OUTCTL     ,OUTRQD     ,PASS7      ,PGCNT       ,PRTARG
     B      ,PUNTAP     ,RESET5     ,SEQNM7     ,SEQINC      ,ZPCTL1
      LOGICAL*1          PRTLIN     ,PUNBUF
      REAL*8             ORG(3)
      EQUIVALENCE        (ORG(1)    ,ORGX)
      REAL*4             DP         ,CL                      ,DW
      EQUIVALENCE        (DP,AUXARR(1))   ,(CL,AUXARR(2))   ,(DW,AUXARR(5))
      LOGICAL*1          PTNO(88)
      EQUIVALENCE        (PTNO(1)   ,PTNOIP(1))
      COMMON   /QUEST/
     A      CORTRS(3)  ,DATBUF(85) ,HOME(3)    ,LIMITS(11)  ,MAXRAD
     B      ,PPCODE    ,PPPARM     ,TPLIMT     ,ULINE0(15)  ,ULINE1(15)
     C      ,ULINE2(15),XCONVT     ,YCONVT     ,ZCONVT      ,XTRA(2)
     D      ,TIME(8)   ,ULINE3(15) ,ULINE4(15)
     E      ,CLDRMS    ,CNTOUR     ,DISKID     ,DPDRMS      ,DRUMZE
     F      ,IGOHOM    ,IGOTAB     ,IRPMOD     ,IZHOR       ,LDRLEN
     G      ,LEADSW    ,NLINE      ,OUTGRD(20,10)  ,R10      ,STRTSQ
     H      ,SIMOSW(4) ,TABLGO     ,ZMOD(2)    ,Z10
     I      ,CLKPRT    ,EOB        ,IEOB       ,IPART       ,IRSTO
     J      ,LCHAR(2)  ,TABCOD     ,TAPPRT     ,TAPMET
      REAL*8             LIMITS     ,MAXRAD
      REAL*4             TIME       ,ULINE3     ,ULINE4
      INTEGER*4          CLDRMS     ,CNTOUR     ,DISKID      ,DPDRMS
     A      ,DRUMZE     ,OUTGRD     ,R10        ,STRTSQ      ,SIMOSW
     B      ,TABLGO     ,ZMOD       ,Z10
      LOGICAL*1          CLKPRT     ,EOB        ,IEOB        ,IPART
     A      ,IRSTO      ,LCHAR      ,TABCOD     ,TAPPRT      ,TAPMET
      EQUIVALENCE        (ZDEPTH,DATBUF(29))   ,    (ZFINAL,DATBUF(30))
      REAL*8             CONVEN(3)
      EQUIVALENCE (CONVEN(1),XCDIG,XCONVT),(YCDIG,YCONVT),(ZCDIG,ZCONVT)
      INTEGER            ASSIGN(2)                 ,CYCTYP(2)
      EQUIVALENCE        (ASSIGN(1),DATBUF(20))    ,(CYCTYP(1),DATBUF(28))
      LOGICAL*1          LOGCAL(6)                 ,OUTSUX(40)
      EQUIVALENCE        (LOGCAL(1),DATBUF(31))    ,(OUTSUX(1),OUTGRD(1,1))
```

Figure 14-2 The statements specified in the variable-definition section of the MTM. These statements define the variables through which the MTM communicates with the DAPP.

changed. Therefore the first part of this section should be the same for every MTM, as shown in Figure 14-2.

The second part of the variable-definition section consists of statements defining the variables used within this particular MTM only. One type of variables usually needed for an MTM are those which keep the record of the current NC codes. These variables are used as references for processing, calculation, or both when the next NC block is being processed.

Those variables that should be reinitialized to the same values each time the MTM is called should also be initialized in this section. An example of the complete form of this section is given in Chapter 15.

After the variable-definition section, there should be a statement that directs the control flow. The FORTRAN statement normally used for this purpose is

```
      GOTO (n1,n2,...,nk),J
```

where $n1$, $n2$, . . . , nk are unsigned integer constants representing statement labels, and J is an unsubscripted integer variable.

The *initialization section* of the MTM has two functions. The first is to set, respectively, the initial values for those variables and arrays that should be initialized only once during the processing of a complete APT program. The second is to generate those special statements required by the NC machine at the beginning of an NC program (for example, the statement for returning the tool to the reference point for the lathe, as described in Chapter 2). Therefore this section should be entered only once, at the beginning of the processing of an APT program. This can be realized by setting the parameter J in the computed GOTO statement to a value other than 1 (if the value 1 corresponds to the initialization section) at the end of this section (see Fig. 14-1).

The *processing section* performs the major functions of an MTM, including the examination and conversion of the read-in data, the development of special motion commands to coordinate with the machine dynamics, and the determination of the content of an NC block. The method for realizing these functions is described in Section 14.3.

When a set of CLDATA has been passed by the General Processing Section of the DAPP to the MTM, it has been transformed and stored in the DATBUF array. The processing section of the MTM consists of several subsections (from spindle section to special function section in Fig. 14-1), each capable of processing only one kind of data. During processing, a set of CLDATA (or DATBUF variables) passes through the processing section from one subsection to another. Each subsection processes only those DATBUF positions that it is designed to handle. After passing through this section, all the DATBUF elements have been converted into the appropriate NC codes. However, control may leave this section before all the data have been processed (this case is not shown in Fig. 14-1) if it is discovered in a processing subsection that there are two codes that are not allowed in one NC block. In such a case, control is transferred to the *preparation section* to output the data. Then control reenters the MTM, which resumes processing the remaining data. After all the subsections have been executed and the complete set of read-in data has been processed, the MTM sets the variable J to k (see Fig. 14-1), which directs control to the second section of the *control-directing section* when reentering the MTM. As a result, control goes to the DAPP Housekeeping Section and the DAPP will read in another set of CLDATA.

The direction of control flow after leaving the MTM is determined by the value assumed by the variable OUTCTL. The control-directing section sets the proper value of OUTCTL so that control can be directed correctly. Another function of this section is to reset those variables that are not reset or reinitialized when the MTM is invoked again.

It is evident from the description presented above that the variable J in the control-branching statement is used to direct control flow *within the MTM*, whereas the variable OUTCTL directs control *in the DAPP*. Therefore a logical combination of these two variables determines the desired processing flow within the MTM and the DAPP.

14.2 THE INITIALIZATION SECTION OF THE MTM

As pointed out previously, one of the functions of the initialization section of the MTM is to set the initial values of those variables that must be initialized only once — the first time the MTM is called. The variables may include, for example, those defining the DATBUF positions to be printed on the output listing file (see Section 13.1 for the variables N1, ISTART, and LENGTH). The variables or array used to record the currently processed results (codes) should also be initialized in this section as ANULL (no input). Depending on the design of the MTM, there might also be other variables that should be initialized in this section.

A postprocessor designer may choose, instead, to use the nonexecutable DATA statement to initialize these variables in the variable-definition section of the MTM, if appropriate.

The other function of this section is to set and output the special statements required by the NC machine. For example, the NC block

G28U0W0

is required as the first statement in an NC program for the Cadillac NC-100 lathe/ FANUC 6T controller system. The FORTRAN program segment that will generate this NC block can be as follows:

```
DATBUF(73)=28
DATBUF(74)=0
DATBUF(75)=0
OUTGRD(2,10)=73
OUTGRD(11,10)=74
OUTGRD(12,10)=75
OUTCTL=6
J=2
RETURN
```

The first three statements are used to store the codes to be output. The fourth to sixth statements are related to G, U, and W codes, pointing to the DATBUF positions 73, 74, and 75, respectively. The control-directing variable OUTCTL is then set to 6, indicating that the NC codes should be output according to the format defined by the OUTGRD array. The variable J in the GOTO statement that directs the control flow in the MTM is set to 2, indicating that when the MTM is called again, control should go to the next section. The statement RETURN sends control back to the DAPP to output the defined NC block. Control then enters the MTM again; it goes directly from the variable-definition section to the section after this program segment. The variables used in this program segment should then be reinitialized.

The structure of a complete initialization section can be as shown in Fig. 14-3. After this section has completed its execution, the initial states of machine-related variables have been set and the NC statements required for the NC machine have been output. Hence the MTM is ready to process the read-in CLDATA.

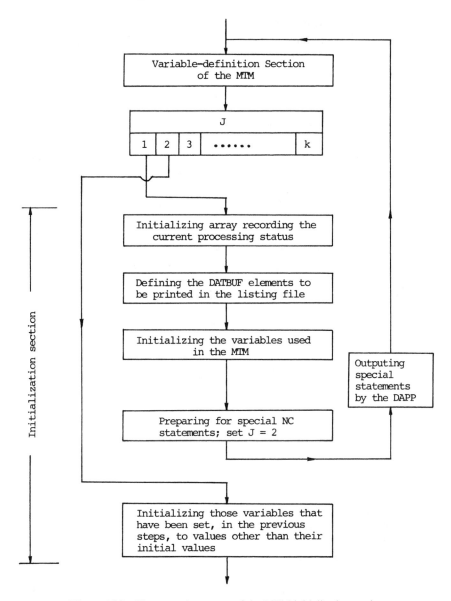

Figure 14-3   The general structure of the MTM initialization section.

14.3 THE PROCESSING SECTION OF THE MTM

The read-in CLDATA consists of various kinds of data. Generally, it can be divided into two types: the data regarding machine operation and that regarding cutter motion. These data should normally go through one or several types of processing, as described in the introduction to this chapter. Since each kind of data requires different processing routines, the processing section of the MTM should be further divided into several subsections, processing, respectively, various kinds of data, such as spindle rotation speed and direction, coolant, tool, linear and circular interpolation motion, and special machine operation functions.

14.3.1 The Method of Determining the Content of an NC Block

Before the processing section is designed, the rules for determining the kinds of codes that can be specified in one NC block should be established. For example, for the FANUC 6T and 6MB controllers, different kinds of codes can be specified in one NC block in an arbitrary order. However, codes of the same kind are not allowed to be in one block, with the exception of some of the G codes described in Chapter 2. There are two possible design rules in this case:

> **Rule 1:** No two codes of the same kind can be specified in one NC block, with the exception of those G codes that are allowed in one NC block.
>
> **Rule 2:** The codes in one NC block must be of different kinds.

The first rule reflects what we are allowed to do in NC programming for the FANUC system. To realize this, two or more sets of OUTGRD elements should be used to define the G code. For example, if the MTM is designed in such a way that two G codes can be output in one NC block, we might use $OUTGRD(1,i)$ and $OUTGRD(2,i)$, $i = 1, 2, \ldots, 9$, to define the same G code as shown in the footnote of Table 12-2. In the processing routine, we may set

    G1=OUTGRD(1,10)

and

    G2=OUTGRD(2,10)

When a data element to be output as a G code has been processed and stored in a DATBUF position, say DATBUF(73), then we set $G1 = 73$ so that the data in element DATBUF(73) can be output as a G code. If the next data element is stored in DATBUF(74) and is also to be output as another G code, it is necessary, first, to determine whether this G code can be output together with the previous one in one NC block. If so, then we can set $G2 = 74$ to output the second G code. It is not possible to use one set of OUTGRD positions, for example, $OUTGRD(1,i)$, $i = 1, \ldots, 10$, to output two G codes in the same block.

As can be seen readily, when the number of codes used for an NC machine is close to 20, the use of two or more sets of OUTGRD elements to define the same code might prevent the postprocessor from processing all the required codes. In addition, there is normally only a small percentage of the NC statements (blocks) that contain codes of the same kind in an NC program. Therefore the adoption of the second rule listed above does not cause a significant increase in the memory space required for storing the NC program. In the meantime, it does make the design of the MTM somewhat easier.

According to rule 2, each processing routine should first determine whether a code processed in the previous section is of the same type as the one to be processed in this routine. If so, the routine sets the parameter J in the control-directing state-ment GOTO to a value that points to the current section; it then sends control to the preparation section of the MTM to output the code processed previously. After the code is output, control returns to the MTM, and the control-directing statement GOTO will send control back to the routine to continue processing.

14.3.2 Processing of Machining Specification Codes

The processing logic of the subsection that processes the data concerning machining specifications (e.g., coolant, tool, spindle speed codes) can be as follows. Before processing is started, the routine should determine whether a code processed by the previous section is the same as the code to be processed in this section. If so, control is directed in the way described in Section 14.3.1. If there is no repeated code, the following processing begins.

The read-in data are first compared with the machine limits, if any, to ascertain whether the processed data are within the allowed range. If the machine limit has been exceeded, routine DLWDER should be called to print an error message, and then the corresponding action, such as terminating or continuing postprocessing, should be taken accordingly. Otherwise, the processing continues. The read-in data should be further transformed into codes that can be directly output by the DAPP Output Section. After the data have been transformed to the code(s) being readied for output, control goes to the next processing routine. The general structure of a subsection designed on the basis of the principles described above is shown in Fig. 14-4.

Consider the subsection that handles the spindle rotation speed and direction to be an example that explains its processing operation. A spindle speed is defined in revolutions per minute in APT and in CLDATA as well. The DATBUF elements that contain the information concerning spindle motion are DATBUF(13) and (14). If the NC machine uses a code other than rpm for spindle speed, then DATBUF(14) should be transformed into the corresponding code so that a correct S code can be generated. The processing routine should also generate — for example, for the Cadil-lac NC-100 lathe — the speed range code M41 or M42. The routine should then gen-erate the correct spindle rotation code M03 or M04 on the basis of the value of DATBUF(13) if the spindle has not yet been turned on. Otherwise, the routine should determine whether the spindle rotation is in the same direction as defined pre-viously. No M code should be generated when the two directions are the same; an M05 code and an M03 or M04 code to reverse the rotation direction are generated if the two rotation directions are different.

14.3.3 Processing of Codes Defining Cutter Motion

A simple cutter motion is defined by its path, its moving speed along the path, and its destination point, since the starting point of the motion is known as the end point of the previous motion. The function of a subsection that processes the motion codes is to convert the information stored in the DATBUF array to the correct motion

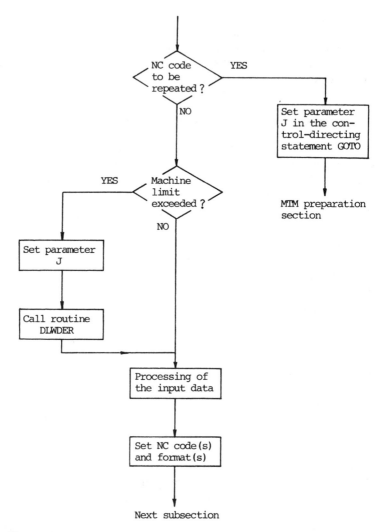

Figure 14-4 The general structure of an MTM processing subsection that processes the machine operation code. A postprocessor designer can choose to terminate the postprocessing after the error message is given if necessary.

codes, G, X, Y, Z, and F, for the milling machine—or G, X, Z, U, W, and F for
the lathe. The DATBUF elements that define a noncycle cutter motion are

DATBUF(i) i = 1-12, 19, 25 (for point-to-point and linear interpolation
motion)

i = 1-10, 36-71 (for circular interpolation motion)

Since the APT contouring motion statement has taken into account the tool radius
compensation through use of the word TLLFT, TLRGT, TLONPS, or TLOFPS, the
cutter position in the CLDATA normally need not be compensated for the tool
radius. DATBUF(27) can then be used to generate the code for defining the circular

interpolation plane (i.e., G17, G18, or G19). During cycle cutter motion, only point-to-point motion with nonrapid feedrate can be specified. Therefore the positions that define the cutter motion are

$$\text{DATBUF}(i) \qquad i = 1\text{-}11,\ 19,\ 25,\ 28\text{-}33$$

However, the retract motion in a cycle should still be a rapid motion, and for NC machines without the cycle function, a corresponding G00 code should be generated. As can be seen from Table 12-1, there is no element in the DATBUF array for recording the spatial orientation of the tool axis. Thus, if a postprocessor is also designed for processing multiaxis function, the tool axis data should be taken from the INNBUF array (see Section 13.3).

14.3.3.1 The Linear-Interpolation-Motion Subsection.

The data defining a linear or point-to-point motion are processed by the linear-interpolation-motion subsection (routine) whose major function is to convert the data stored in the DATBUF elements indicated previously into the correct codes, G, X, Y, Z, and F. In addition, functions should also be included to further transform or process these codes so that requirements resulting from machine dynamics and control can be met.

The NC control system can operate the machine in two different modes: it can execute each motion block as a discrete operation, or it can interpolate motion information smoothly and continuously from one block to the next, thus giving true continuous contouring during the cutter motion.

The first control method is used by NC machines with point-to-point motion only, such as NC drilling machines or those NC machines with NC controller of an older design, which can handle only straight-cut motion. Their control system is designed to provide both automatic acceleration up to a programmed feedrate at the beginning of each motion step and deceleration down to rest at the end. The motion defined by a motion block is not affected by that defined by a preceding or following motion block. In this case, the machine dynamics problem usually need not be considered during postprocessor design.[8] The second type of control is adopted in NC control systems that perform continuous contouring motion. Most of the NC controllers for milling machines and lathes are of this type. The situation for this type of control is different, and the machine dynamics problem should be considered accordingly.

For example, suppose that the motion defined in APT is $P0\text{-}B\text{-}C\text{-}D$, as shown in Fig. 14-5. The CLDATA corresponding to this motion consists of a series of cutter position coordinates, $x_{P0},\ y_{P0},\ z_{P0},\ x_b,\ y_b,\ z_b,\ x_c,\ y_c,\ z_c,\ x_{c1},\ y_{c1},\ z_{c1},\ \ldots,\ x_d,\ y_d,\ z_d$; the curved motion $C\text{-}D$ has been divided by the NC processor into many small linear motion segments, $C\text{-}C1,\ C1\text{-}C2,\ \ldots$, through linear interpolation according to the specified tolerance. The machine dynamics problem arises when the cutter changes its motion direction abruptly. For example, if the tool moves from point $P0$ through point B to point C, an overshoot results at point B since it takes time to decelerate the motion. The greater the feedrate of motion $P0\text{-}B$, the larger is the overshoot. The deviation, S_2, of the cutter from the defined profile depends on both the overshoot, S_1, and the angular change, A, in motion direction. Thus we have

$$S_1 = S_2/\sin A$$

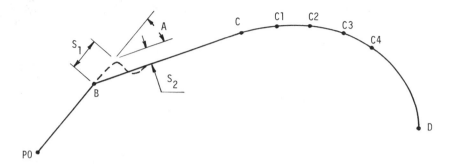

Figure 14-5 A cutter motion defined as P0-B-C-D in APT. It is redefined as
P0-B-C-C1-C2-C3-...-D in the CLDATA.
S_1: overshoot; S_2: deviation caused by overshoot;
- - - -: actual cutter path at the corner;
———: programmed cutter path.

where A is the included angle between the two motion vectors (Fig. 14-5). Substituting for S_2 the specified tolerance in this formula, the permissible overshoot S_1 can be
obtained. The deviation S_2 is insignificant if A is small. Thus, for motion C-$C1$-$C2$-
...-D, the overshoot has relatively little effect on the accuracy of the profile. To decrease the overshoot, NC controllers of older design require that each motion, such
as $P0$-B, be slowed down when the cutter is close to the destination point (i.e., point
B). Thus the segmentation of each motion step into two, the first one with the
programmed feedrate and the other with a lower feedrate, is required if the programmed feedrate causes excessive overshoot. Usually the data concerning the
feedrate-overshoot relationship and the relationship between the feedrate and
the minimum acceleration or deceleration step (Table 14-1[a and b]) are provided by

TABLE 14-1(a) THE OVERSHOOT S_1 (SEE FIG. 14-5) AS A FUNCTION OF THE
PROGRAMMED FEEDRATE[7]

Feedrate (in./min)	Overshoot S_1 (in.)
4.0	0
10.0	0.0004
20.0	0.0008
40.0	0.0020

TABLE 14-1(b) THE MINIMUM DISTANCE REQUIRED FOR THE MACHINE TO ACCELERATE
TO THE PROGRAMMED FEEDRATE OR DECELERATE FROM THE PROGRAMMED FEEDRATE
TO ZERO[7]

Feedrate (in./min)	Minimum Distance (in.)
10	0.02
20	0.02
40	0.04
80	0.08

the machine tool manufacturer. The allowed deviation is specified by the APT programmer, using the statement

MCHTOL/*n*

which is stored in DATBUF(26). The algorithm for processing the cutter motion can now be designed.

    With modern CNC machines, the situation is different. Nearly all the CNC milling machines and lathes have the automatic acceleration or deceleration (a/d) function, and their characteristic parameters — such as the exponential a/d time constant, the a/d distances at the inner corner, and the maximum position deviation in cutting movement — can be set according to the user's requirement. Therefore, for CNC machines with automatic a/d function, it is not necessary to include a routine that handles the deceleration at the end of a linear cutting motion. When a very high feedrate is used, it might still cause an excessive overshoot at the corner. However, this will happen only during a rough cut, and adequate allowance should be left over the surface so that the part profile can still be kept within the tolerance range after the finishing cut is made with a normal feedrate.

    Some NC controllers of older design also require the use of different G codes to define linear interpolation motion with different travel distances; besides, the controller resolution (the least incremental motion distance) for these codes also differs. For example, the Gorton Tapemaster 2-30 contouring mill with the Bunker-Ramo 3100 controller[10] requires that the X, Y, and Z codes be input as one to four digits without a decimal point. The decimal point is considered to precede the specified first digit if a G01 code is effective, or to be placed after the first digit and the second digit with code G10 and code G60, respectively. The X, Y, and Z codes that can be programmed with the G01, G10, and G60 codes should be incremental values. The minimum programmable motion distances or resolutions for G01, G10, and G60 codes are 0.0001, 0.001, and 0.01, respectively, whereas the respective maximum departure distances with these three codes are 0.9999, 9.999, and 99.99. Therefore a motion with an incremental motion of 14.2345 along the X axis cannot be programmed with one preparatory function code G01, G10, or G60. It should be divided into three motion steps; for example

$$14.2345 = 12.94 + 1.177 + 0.1175$$

each defined by an appropriate G code. If the motion is in the X direction, three blocks are needed:

G60X12.94
G10X1.177
G01X0.1175

For simultaneous three-axis motion, the incremental distances from the starting point to the end point in the X, Y, and Z axes should be proportionally segmented so that linearity of motion can be maintained. The postprocessor designer is assisted in segmenting the linear motion by two special routines, DLWSRR and DLWSEG, which have been incorporated into the DAPP. These two routines make use of the variables defined in the COMMON field labeled SEGCOM

COMMON/SEGCOM/RANGE1,RESLN1,RANGE2,RESLN2,RANGE3,RESLN3,NRANGE to communicate with the DAPP and MTM. RANGE1 and RESLN1 are used to define, respectively, the maximum incremental distance and the resolution for the G code with the largest definable incremental motion step (e.g., G60 in the example shown above); RANGE3 and RESLN3 are used similarly for the G code with the least definable incremental motion step (e.g., G01 in the example shown above). The variable NRANGE is used to specify the number of G codes used for defining the linear interpolation motion. The COMMON statement shown above should be specified in the variable-definition section of the MTM, and these variables should be defined before the two routines are called. For the example given above, they can be defined as follows:

```
NRANGE=3
RANGE1=99.99D0
RANGE2=9.999D0
RANGE3=0.9999D0
RESLN1=0.01D0
RESLN2=0.001D0
RESLN3=0.0001D0
```

Note that these are also the respective default values used in the DAPP for these variables. The ranges and resolutions should also be set and arranged in descending order. If, for example, an NC machine uses only two G codes, say G01 and G10, to define the linear interpolation motion, then these variables should be defined as follows:

```
NRANGE=2
RANGE1=9.999D0
RANGE2=0.9999D0
RESLN1=0.001D0
RESLN2=0.0001D0
```

These FORTRAN statements can be included in the initialization section of the MTM.

The function of routine DLWSRR is to set up the maximum segment lengths and resolutions according to the variables NRANGE, RANGE1, ..., RESLN3, defined as described above. It can be invoked through use of the statement

```
CALL DLWSRR
```

before calling the DLWSEG routine, which performs segmentation of the cutter path. The following statement calls the DLWSEG routine:

```
CALL DLWSEG (DATBUF(7),M,OUTPUT,IRET)
```

where

DATBUF(7) = a parameter that causes the three elements, DATBUF(7) through (9), that define the incremental motion to be stored in the array AINPUT, which is then used by routine DLWSEG to carry out the segmentation

M = an integer defining the number of elements available in the OUTPUT array

OUTPUT = the name of the array, with M elements, where the segmented data is stored (It can be replaced by a name given by the post-processor designer, e.g., D in the flowchart given in Fig. 14-6

IRET = the return code whose absolute value represents the number of elements, in the OUTPUT array, filled by the routine DLWSEG (The return code is positive if the data in the array AINPUT(1) through (3) have been completely segmented and reduced to zero. If the number of positions of the OUTPUT array, assigned by parameter M, is less than required, the return code IRET is negative.)

Consider again the Gorton Tapemaster 2-30 contouring mill as an example. A cutter motion is programmed with incremental X, Y, and Z values equal to 27.2543, 6.2543, and 0.2010, respectively. They are stored in DATBUF(7) through (9), respectively. The routine DLWSEG is called by means of the statement

CALL DLWSEG (DATBUF(7),9,OUTPUT,IRET)

As a result, the motion is divided into three segments and the following data are generated:

OUTPUT(1) = 24.78 ⎫
OUTPUT(2) = 5.69 ⎬ (incremental X, Y, and Z values for G60 code)
OUTPUT(3) = 0.18 ⎭
OUTPUT(4) = 2.249 ⎫
OUTPUT(5) = 0.513 ⎬ (incremental X, Y, and Z values for G10 code)
OUTPUT(6) = 0.019 ⎭
OUTPUT(7) = 0.2253 ⎫
OUTPUT(8) = 0.0513 ⎬ (incremental X, Y, and Z values for G01 code)
OUTPUT(9) = 0.0020 ⎭

If, in the calling statement, the parameter M is set to 3, the routine DLWSEG should be called three times to have the motion completely segmented. The data generated each time the DLWSEG routine is called are given in Table 14-2. These data can be

TABLE 14-2   THE RESULTANT DATA AFTER THE DLWSEG ROUTINE IS INVOKED

	CALL DLWSEG (DATBUF (7),3,OUTPUT,IRET)		
	First Call	Second Call	Third Call
OUTPUT (1)	24.78	2.249	0.2253
OUTPUT (2)	5.69	0.513	0.0513
OUTPUT (3)	0.18	0.019	0.0020
IRET	−3	−3	+3
DATBUF (7)*	2.4743	0.2253	0
DATBUF (8)*	0.5643	0.0513	0
DATBUF (9)*	0.0210	0.0020	0

*The initial values of these three variables are 27.2543, 6.2543, and 0.2010, respectively.

used to generate the X, Y, and Z codes for the linear interpolation motion segments defined by G60, G10, and G01 codes. As can be seen from this example, each data element, OUTPUT(1), (2), or (3), is generated by dividing the data in the DATBUF element by 1.1, and the value of the last segment, say OUTPUT(7), is the difference between the programmed distance, DATBUF(7), and the sum of the two corresponding segments generated previously, OUTPUT(1) and OUTPUT(4). Because the calculation is rounded off, the ratios among the incremental X, Y, and Z values in different motion segments cannot be kept the same, specifically

$$\text{OUTPUT(1)} : \text{OUTPUT(2)} : \text{OUTPUT(3)}$$
$$\neq \text{OUTPUT(4)} : \text{OUTPUT(5)} : \text{OUTPUT(6)}$$
$$\neq \text{OUTPUT(7)} : \text{OUTPUT(8)} : \text{OUTPUT(9)}$$

Thus the resultant cutter path is not exactly a straight line. The first three elements of the OUTPUT array should be output, respectively, as X, Y, and Z codes with the G60 code; the next two statements are the X, Y, and Z codes with the G10 and G01 codes, respectively. Since the formats of the X, Y, and Z codes are the same (four digits without a decimal point) for the G60, G10, and G01 codes, we are able to use only three sets of OUTGRD elements; for example

$$\text{OUTGRD}(i,j) \qquad i = 3, 4, 5 \text{ and } j = 1, \ldots, 9$$

to define the X, Y, and Z code formats. The program can be designed as shown in Fig. 14-6. After the COMMON field labeled SEGCOM and its variables have been defined, routine DLWSRR is called. Then the motion data stored in DATBUF(7) through (9) can be processed. First, routine DLWSEG, in which the parameter M is set to 3, is called to divide the motion into segments. As can be seen from the description presented above, the segmentation of the motion is carried out by DLWSEG according to the following principle: a linear motion is divided into

1. Three segments if
$$\text{Max.}(|\text{DATBUF(7)}|, |\text{DATBUF(8)}|, |\text{DATBUF(9)}|) > 9.999$$

2. Two segments if
$$0.9999 < \text{Max.}(|\text{DATBUF(7)}|, |\text{DATBUF(8)}|, |\text{DATBUF(9)}|) \leqslant 9.999$$

3. No segmentation is needed if
$$\text{Max.}(|\text{DATBUF(7)}|, |\text{DATBUF(8)}|, |\text{DATBUF(9)}|) \leqslant 0.9999$$

Thus execution branches to three different sections, depending on the value of the maximum motion component, D(1), D(2), or D(3). Sections A1, A2, and A3 are entered if the maximum motion component can be defined by G01, G10, and G60 codes, respectively. In each section, the processing routine first determines whether the G code in the previous block is the same as the one it must generate. If so, a new G01, G10, or G60 code need not be generated; otherwise, the routine examines whether there is a G code that was processed in the previous section but has not yet been output (we adopt the principle that no two codes of the same type can be specified in a block). If so, the G code processed previously should be output first; other-

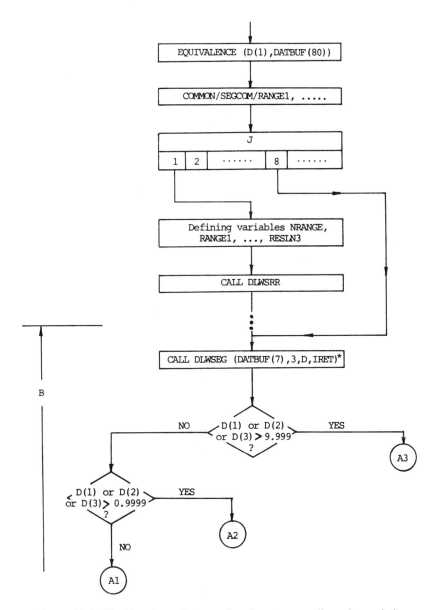

Figure 14-6 The flowchart of the section that processes linear interpolation motion for the Gorton Tapemaster 2-30 milling machine. The elements OUTGRD(2, i), OUTGRD(3, i), OUTGRD(4, i) and OUTGRD(5, i) ($i = 1, \ldots, 10$), are used for G, X, Y, and Z codes, respectively. *: The symbol D is used to name the OUTPUT array. A: preparation section of MTM; B: section processing linear interpolation motion.

wise, the appropriate X, Y, Z, and G codes are set according to the OUTPUT array (array D in Fig. 14-6). The feedrate should then be set according to either the programmed feedrate or a decreased feedrate to compensate for overshoot.

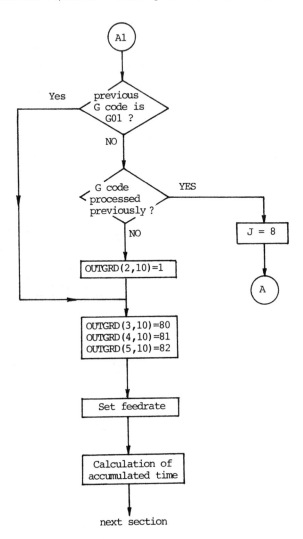

Figure 14-6 (continued)

For modern CNC machines, linear interpolation motion can be defined by a single code, namely, G01. It is not necessary to use routines DLWSRR and DLWSEG, and only one processing section is needed. Therefore the routine that processes linear interpolation motion can be simplified significantly.

Rapid point-to-point motion can also be processed by the linear interpolation motion section since the difference in the definitions of these two motions lies in their G codes only. The G code can be determined from the programmed feedrate. Thus it is necessary to check the feedrate before the G code is set. If the feedrate is RAPID, the G code should be set as G00; otherwise, it should be set as G01.

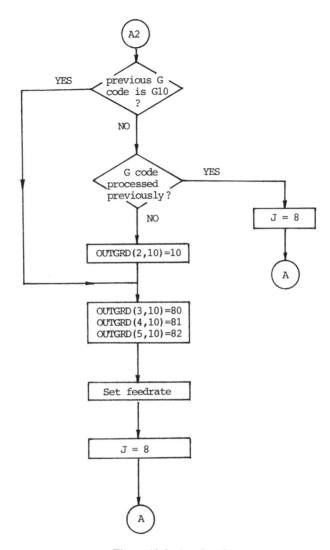

Figure 14-6   (continued)

The final processing step in this routine consists in calculating the accumulated time for execution of the NC program up to the current statement. This is realized through use of the FORTRAN statement

CLKARG=CLKARG+TIME(1)

For postprocessors designed for a lathe, the incremental x and y coordinates in the DATBUF array should be converted, respectively, into NC codes W and U (see Section 8.1); Furthermore, code U should be the diameteric value (i.e., twice the

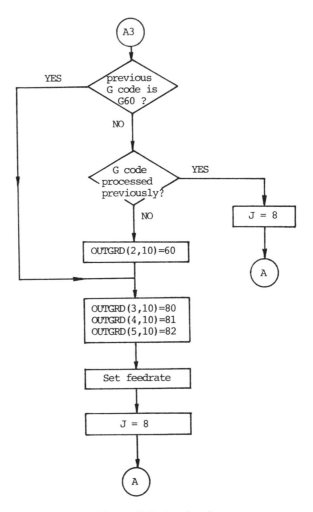

Figure 14-6 (continued)

value in the DATBUF array). The example given in Chapter 15 illustrates the details of this conversion.

14.3.3.2 The Circular-Interpolation-Motion Subsection.   As described in Section 13.4.2.2, a circular motion is divided into one to five segments by the DLWCIR routine of the DAPP, and the resultant data are stored in elements

$$\text{DATBUF}(i) \qquad i = 36\text{-}71$$

Segmentation of a circular path is required for some NC machines with controllers of older design. However, it is not necessary for modern CNC controllers, such as the FANUC 6MB and 6T.

The major function of this routine is to convert the data in

$$\text{DATBUF}(i) \qquad i = 1\text{-}10, \ 36\text{-}71$$

TABLE 14-3 THE RELATIONSHIP BETWEEN THE DATBUF VARIABLES AND THEIR
CORRESPONDING CODES

Elements in DATBUF Array	The Corresponding NC Codes
1–3	X, Y, Z (absolute value)
4–6	X (or W*), Y, Z (or U*) (incremental value)
10	F
36	G17, G18, or G19
37	G02 or G03
41	R
42–44, 48–50, 54–56, 60–62, 66–68	X (or W*), Y, Z (or U*) (incremental value)
45–47, 51–53, 57–59, 63–65, 69–71	I, J, K (see the following context)

*Codes U and W are for the lathe.

into correct NC codes according to the correspondence given in Table 14-3. Because of the error in the DLWCIR calculation routine, which was pointed out in Section 13.4.2.2, the DATBUF positions for center offsets

$$\text{DATBUF}(i) \qquad i = 45\text{-}47,\ 51\text{-}53,\ 57\text{-}59,\ 63\text{-}65,\ 69\text{-}71$$

cannot be used to output the I, J, and K codes. These values should be recalculated in the MTM as follows:

$$\text{DATBUF}(44 + i) = \text{DATBUF}(37 + i) - \text{DATBUF}(3 + i)$$
$$\text{DATBUF}(50 + i) = \text{DATBUF}(37 + i) - [\text{DATBUF}(3 + i) + \text{DATBUF}(41 + i)]$$
$$\text{DATBUF}(56 + i) = \text{DATBUF}(37 + i) - [\text{DATBUF}(3 + i) + \text{DATBUF}(41 + i) + \text{DATBUF}(47 + i)]$$
$$\text{DATBUF}(62 + i) = \text{DATBUF}(37 + i) - [\text{DATBUF}(3 + i) + \text{DATBUF}(41 + i) + \text{DATBUF}(47 + i) + \text{DATBUF}(53 + i)]$$
$$\text{DATBUF}(68 + i) = \text{DATBUF}(37 + i) - [\text{DATBUF}(3 + i) + \text{DATBUF}(41 + i) + \text{DATBUF}(47 + i) + \text{DATBUF}(53 + i) + \text{DATBUF}(59 + i)]$$

with $i = 1$, 2, and 3.

The general structure of the circular-interpolation-motion subsection is shown in Fig. 14-7. It is entered when circular-interpolation processing is requested by the statement ARCSLP/ON. The feedrate is then checked to ascertain whether the feedrate is other than RAPID. The routine might also determine whether the programmed feedrate will cause excessive deviation from the part profile (Fig. 14-8). The allowable feedrate for circular interpolation motion is determined by the characteristics of the NC control system and the allowable deviation from the defined profile (i.e., the tolerance). Usually the formula for calculating the maximum deviation in the radial direction during circular interpolation motion is provided by the ma-

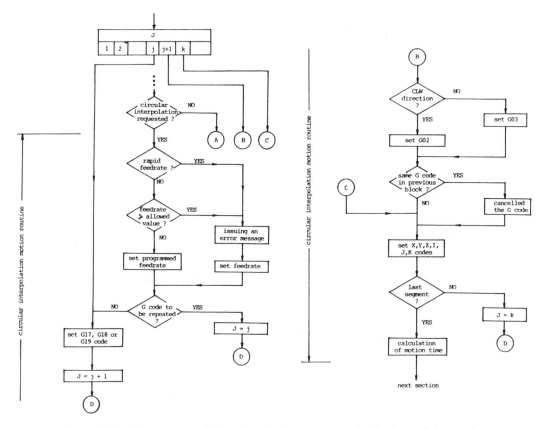

Figure 14-7   The structure of the subsection that processes circular-interpolation-motion data. A: linear-interpolation-motion section; D: preparation section of MTM.

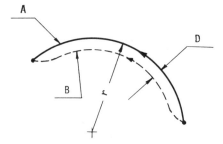

Figure 14-8   The deviation of the actual cutter path (B) from the programmed cutter path (A) because of the control system. D: maximum deviation.

chine tool builder or the NC controller manufacturer. For example, for the FANUC 6MB controller, the formula* is

$$D = [(T_1^2 + T_2^2) \cdot f^2]/(2 \cdot r)$$

*See reference 26 in the reference list for Part I of this text.

where

D = maximum deviation (in.) in the radial direction

T_1 = exponential acceleration or deceleration time constant (sec) of the NC controller (usually, set to zero)

T_2 = time constant (sec) for the positioning control system, which is the inverse of the position loop gain

f = feedrate (in./sec)

r = radius of the circular arc (in.)

If

$$T_1 = 0,$$
$$T_2 = 3.3 \times 10^{-2} \text{ sec}$$

and D is the tolerance D' (in.), then the allowable feedrate is

$$f = 2571.3 \cdot (D' \cdot r)^{-1/2}(\text{in./min})$$

Table 14-4 lists the allowable feedrates for various radii and tolerances calculated with this formula. It can be readily seen that, for the finishing and most roughing operations, the allowable feedrate will not be exceeded.

The next task of this routine is to convert the data in the DATBUF array to G, X, Y, Z, and F codes. Since a circular motion statement (block) will certainly contain a G code, the G code processed in the previous section, if there is one, should be output first by branching control to the preparation section of the MTM. Otherwise, circular-interpolation processing begins. The routine first sets the circular-interpolation plane code according to the value of DATBUF(36). Again, control is directed to the preparation section to output a G17, G18, or G19 code (because we follow the principle that no two codes of the same type can be specified in an NC block). Then the G02 or G03 code is set on the basis of the motion direction defined in DATBUF(37). If the circular motion should be divided into segments in different quadrants, then the X, Y, I, and J codes should be set for the first segment if the motion is in the X-Y plane. These codes are output by branching control to the preparation section. After outputting of these codes, control comes back and sets the X, Y, I, and J codes for the next motion segment. This process is repeated until all motion segments are defined. For those NC machines that do not require segmentation of the circular motion, the end coordinates given by

$$\text{DATBUF}(i) \qquad i = 1, 2, 3$$

TABLE 14-4 THE ALLOWABLE FEEDRATES, f (in./min), FOR DIFFERENT TOLERANCES, D' (in.), AND RADII, r (in.)

D' / f / r	0.005	0.001	0.0005	0.0001
5	406.6	181.8	128.6	57.5
1	181.8	81.3	57.5	25.7
0.5	128.6	57.5	40.7	18.2

can be used to process the X, Y, and Z codes. The center offset codes I, J, and K are
determined by DATBUF(45) through (47), which should be recalculated as described previously; the R code is determined by DATBUF(41).

The sign of the R code should be positive if the locus of the circular motion
has an included angle less than or equal to 180 degrees; it should be negative if the
angle is larger than 180 degrees (see Section 2.6.1). The method for determining the
sign of the R code can be as follows. Suppose that the starting point, S, and the end
point, E, which is in the third quadrant of a circular motion in the X-Y plane, are as
shown in Fig. 14-9. Point $S1$ is 180 degrees from the starting point, S. The sign of
the R code can be determined by comparing the distance between points $P2$ and $S1$
(i.e., D_0) and that between $P2$ and E (i.e., D_1). The R code should be positive if
$D_1 - D_0 < 0$; otherwise, it should be negative. The coordinates of points S, $P1$, $P2$,
and E can be determined from DATBUF(4) through (6), (42) through (44), (48)
through (50), (54) through (56), and (1) through (3). Thus distances D_1 and D_0 can
be computed as follows:

$$dx_1 = \text{DATBUF(42)}$$
$$dy_1 = \text{DATBUF(43)}$$
$$D_0 = [(r - dx_1)^2 + (r - dy_1)^2]^{\frac{1}{2}}$$
$$D_1 = \{[\text{DATBUF(54)}]^2 + [\text{DATBUF(55)}]^2\}^{\frac{1}{2}}$$

The sign of the R code can be determined easily if the end point of the circular motion is located in quadrants other than the third. It is positive if the end point is in the
first and second quadrants; it is negative if the end point is in the fourth and fifth
quadrants. No calculation is necessary in these cases. As a consequence, the routine
for determining the sign of the R code can be easily designed.

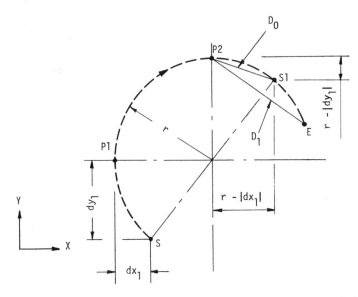

Figure 14-9   Determination of the sign of the R code for circular interpolation
motion.

The processing is completed after the accumulated times for execution of the NC program up to the current statement has been calculated.

14.3.3.3 The Machining Cycle Subsection.   Apart from the milling cycle, the cycle motion that can be processed by the DAPP consists of several motion steps in the Z direction. Before it is processed by the MTM, the motion cycle is defined by the DATBUF elements described in Section 13.4.2.3. There are two ways to process a cycle motion in the MTM: defining the cycle motion with noncycle motion codes, or defining it with cycle motion codes.

The first approach is to define the cycle motion through use of NC codes G00 and G01. Thus, for example, a simple drilling cycle can be divided into

1. Rapid motion (G00) to the point, defined by DATBUF(1) through (3), to be drilled
2. Drilling motion (G01) to the depth, defined by DATBUF(29), at the programmed feedrate
3. Rapid retract motion (G00), whose end point is defined by DATBUF(30)

Hence the processing is very similar to that in the linear interpolation motion subsection except that three NC motion blocks should be generated, instead of one, for each set of read-in CLDATA (or DATBUF elements). This method can be used for NC machines without the machining cycle function, and the general structure of this routine can be designed as shown in Fig. 14-10.

The routine sets the NC codes for defining the rapid motion to the point to be drilled, and the parameter J is set to $i + 1$. Then control is sent to the preparation section of the MTM to output the rapid motion block. The Output Section of the DAPP sends control back to the MTM, which starts generating the codes defining the drilling motion at the programmed feedrate. Control is then sent to the preparation section of the MTM after the parameter J is set to $i + 2$. After these codes have been output, the MTM is invoked again to generate the codes defining the retract motion. If the cycle is a simple drilling or boring cycle, control goes to the next section of the MTM. But if the cycle is a peck-drilling cycle (BRKCHP or DEEP), then the three motion steps described above define the first drilling cycle of the peck-drilling cycle, and more cycles follow. In this case, the codes defining the retract motion should be output first, and the cycle machining subsection should be invoked again to process the NC codes for the next cycle. Thus, for the BRKCHP or DEEP cycle, the routine sets J to i, which points to the beginning of this subsection, and OUTCTL to 1, which directs control to read in the next set of cycle data. A decision statement should then be inserted to determine whether the current cycle is the last one in the peck-drilling cycle. If so, control should go to the next processing section of the MTM.

The milling cycle can also be processed in a similar fashion.

If the postprocessor is designed for an NC machine that has cycle machining functions, then the MTM should be able to process the cycle machining codes — for example, codes G80 through G89 — for the FANUC 6MB controller. A G99 or G98 code can be specified by the statement PREFUN to define the level of retract motion; the values of DATBUF(1) through (3), (29), and (30) can be used to determine

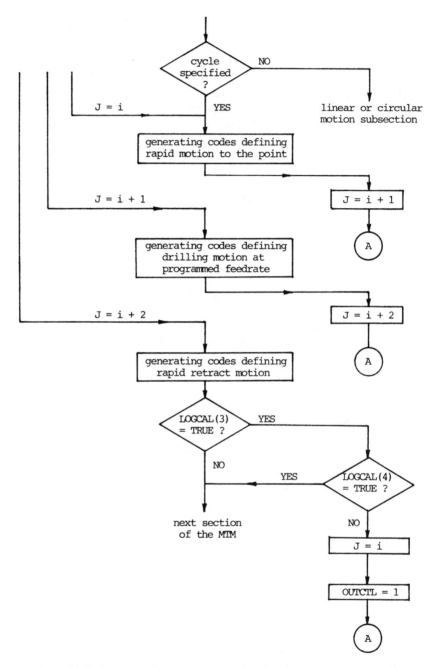

Figure 14-10 The general structure of the subsection that processes cycle motion codes. A: preparation section of MTM.

the X, Y, Z, and R codes for the motion defined by a G81, G82, G84, or G85 code. It is not possible to transfer all the parameters specified in a BRKCHP or DEEP cycle statement, representing different drilling depths, to one NC block. Thus the postprocessor designer should construct a rule that the postprocessor should follow for generating the Q code. For example, the depth of drilling, q (see Fig. 2-22), can be selected as

$$q = 1.5 \cdot \text{(diameter of the drill)}$$

or can be determined by a parameter specified in the BRKCHP or DEEP cycle statement.

14.3.4 Processing of Special Functions

In addition to codes concerning machining specification and cutter motion, NC codes specifying special functions, such as dwelling at the current position, end of a program, optional interruption of a program (or optional stop), or rotation of the rotary table, are also needed for successful operation. The data are stored in DATBUF(18), (19), (22), (25), and (26). The function of the subsection that handles the data is to transform them into the corresponding G or M code and some other appropriate codes, if needed (such as the X code after a G04 code).

As can be seen from Table 12-1, values have been assigned in DATBUF(25) to frequently used APT postprocessor commands, such as END, STOP, and DELAY. A special preparatory function specified by statement PREFUN/n in APT is stored as a negative value, $-n$, in DATBUF(25); it should be translated into a G code, namely, Gn. On the other hand, an auxiliary function defined by statement AUXFUN/n in APT is a positive scalar, n, in DATBUF(25); it should be converted into a corresponding M code, namely, Mn.

In the section that handles these special functions, the processing is very similar to that in the section that processes the machining specification codes. However, it should be noted that the processing routine should be designed in such a way that a logical sequence of NC codes is generated. For example, if a program termination code M02 or M30 is to be generated, the routine should be capable of examining the coolant and spindle status and should automatically generate M09 and M05 codes, if needed, to stop the spindle and coolant. This routine should also be able to issue error messages for those specified special functions that the controller does not have.

The general structure of this processing routine can be seen from the example in Chapter 15.

14.4 THE PREPARATION AND CONTROL-DIRECTING SECTIONS OF THE MTM

The data processed by the processing section of the MTM are sent to the preparation section before being output. One of the functions of this section is to compare a code to be output with the corresponding code in the previous block. If the two codes are identical, then the code need not be output in the current block. However, for those NC codes that are effective only in the block where they are specified (e.g., G04, M11), the above-mentioned routine should be skipped. After finalizing the content of an NC block, the routine updates the array that keeps a record of the output codes. Then control is sent to the control-directing section, which directs control after exiting the MTM. The control-directing section is divided into two subsections. The first subsection sets OUTCTL equal to 5, 6, 7, or 11, depending on the output

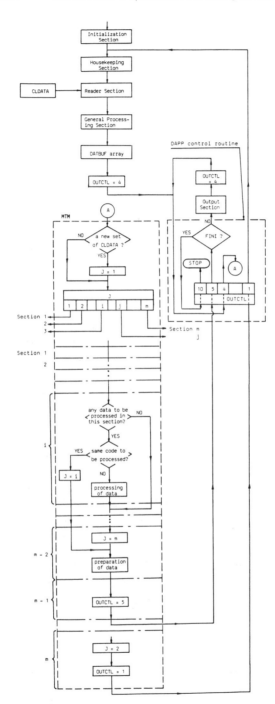

Figure 14-11 The MTM and the processing flow within a DAPP-based postprocessor. Note that the variable-definition section of the MTM is not shown.

requirement, thus sending control to the Output Section of the DAPP. The second subsection is entered only when $J = k$, which means that all the read-in data have been processed and control has gone through all the processing sections (see Fig.

14-1). As a result, control is directed to the DAPP Housekeeping Section and then to the Reader Section to read in another set of CLDATA.

Before control goes to the DAPP Housekeeping Section, the second subsection of the control-directing section should also reset those variables that are not reset in the variable-definition section but should be reset when a new set of CLDATA is to be processed. The structure of the control-directing section can be seen from the example given in Chapter 15.

14.5 THE PROCESSING FLOW IN THE DAPP-BASED POSTPROCESSOR

We shall now describe the processing flow in the MTM and the DAPP-based postprocessor. The structure of an MTM and its relationship with the various sections of the DAPP are shown in Fig. 14-11. After a set of CLDATA has been read, processed, and stored in the DATBUF array, the MTM is called. The variable-definition section (not shown) and initialization section of the MTM are executed. Then the data in the DATBUF array are processed, respectively, by different processing sections of the MTM. At the beginning of each section, a decision statement branches control to the next section if there are no data in the DATBUF array to be processed by the current section. Then another decision statement branches control to the preparation section (section $m - 2$ in Fig. 14-11) if the code to be processed by the current section will be repeated in an NC block. If so, parameter J is set and control goes to the preparation section, which carries out the needed preparation processing and sets the variable OUTCTL to 5, 6, 7, or 11. Control is then sent to the Output Section of the DAPP to output an NC block. The Output Section sets variable OUTCTL to 4 and sends control back to the MTM. Then the computed GOTO statement transfers control directly to the processing section exited previously. The processing continues from one section to another until all the data in the DATBUF array have been processed. In this case, the preparation section is entered from the beginning (statement $J = m$ in section $m - 2$). After the data are prepared and output, the MTM is called again, and control goes directly to the last section (section m), which sets the variable OUTCTL and J equal to 1 and 2, respectively. Thus control returns to the Housekeeping Section of the DAPP and a new set of CLDATA is retrieved. This processing cycle is repeated until the CLDATA of the complete APT program is processed.

PROBLEMS

14.1 The processing section of an MTM usually consists of several subsections arranged in an appropriate order, each of which processes the data regarding one specific function of the NC machine. For a three-axis milling machine (without a rotary table), into how many subsections can the processing section of the MTM be divided generally? Explain also the order of these subsections in the MTM.

14.2 The Cadillac NC-100 lathe described in Chapter 2 has a turret that can hold four turning tools. A T code followed by four digits is required to describe the tool used and its offset. The turret can rotate in the clockwise direction only. A clockwise rotation by 90 degrees is defined by the M11 code. The identification numbers for the consecutive tools on the turret are 1, 2, 3, and 4 in the counterclockwise direction (Fig. 6-11a). Thus one or several M11 codes may be needed to position the required tool to the working position. Assume that only the LOADTL/n (n = 1, 2, 3, or 4) statement is used in the APT program to change the tool. Design the corresponding processing routines (subsections) for processing the tool-changing codes (T and M codes) on the basis of the following assumptions:

a. This routine is the first subsection in the processing routine of the MTM.
b. This routine is not the first subsection in the processing routine of the MTM.
Give the flowcharts for these routines.

14.3 Suppose that the spindle speed of an NC machine is defined by a speed code, S, followed by four digits defining the spindle speed in revolutions per minute. Design the routine (subsection) for processing the spindle speed and direction codes, and draw the flowchart.

14.4 Suppose that an NC machine has only two coolant states, namely, COOLNT/ON and COOLNT/OFF. Design the routine for processing the coolant codes M08 and M09, and draw the flowchart.

14.5 For modern CNC machines, a linear interpolation motion usually need not be segmented. Design a routine that processes the linear and point-to-point motion codes (G01, G01, X, Y, and Z codes), and draw the flowchart.

14.6 Design a routine that processes the special functions of an NC machine, which include END, DELAY, OPSTOP, and STOP. The corresponding NC codes for these functions are M30, G04, M01, and M00, respectively.

14.7 Suppose that a three-axis CNC milling machine-controller system has only the limited functions listed below:

FUNCTION	CORRESPONDING CODE(S)
Spindle speed	S code followed by 4 digits representing rotation speed in rpm
Spindle stop	M05
Spindle on (CLW)	M03
Spindle on (CCLW)	M04
Coolant on	M08
Coolant off	M09
Linear interpolation motion in 3 axes (X, Y, and Z)	G01X_Y_Z_
Simultaneous rapid motion in 3 axes (X, Y, and Z)	G00X_Y_Z_
Dwelling	G04P_
Program end	M02
Program stop	M00
Feedrate	F code followed by 5 digits, of which the last 2 are fractional digits

There is no automatic tool changer on this machine, and thus no T code is needed. Assume (1) that the code formats, the machine and controller parameters, and the output requirements have been properly set by the answers to the DAPP Questionnaire and

(2) that the BLOCK DATA program has been generated. The OUTGRD elements that defined the code formats are as follows:

VALUE i IN OUTGRD(i,j), j = 1 TO 10	CODE
1	G
2	X
3	Y
4	Z
5	F
6	S
7	P
8	M

Assume that the output X, Y, and Z codes should be the absolute coordinates. The initial tool position is (x_1, y_1, z_1), where x_1 and y_1 are defined by the two parameters DATBUF(34) and (35) in the MACHIN statement, and z_1 is 1 in. Design an MTM for this machine that can handle all the functions listed above. Then draw the complete flowchart.

14.8 If different codes can be generated from a set of CLDATA, how does an MTM determine the contents of the output NC blocks?

14.9 If two G codes of different groups are allowed in one NC block, what measure should be taken in the design of the MTM?

chapter 15

The Creation of a DAPP-Based Postprocessor: An Example

In Chapters 11 to 14, we described the structure of the CLFILE (CLDATA) and the functions of various sections of the DAPP. The major variables used in the DAPP-based postprocessor, the method of initializing the machine-dependent variables, and the method of designing the machine tool module (MTM) were also explained in detail. Therefore the reader now has the information necessary for designing a DAPP-based postprocessor.

In this chapter, we first introduce the procedures used for creating a DAPP-based postprocessor. Then we use an NC lathe as an example to explain in detail the design and implementation of a postprocessor.

15.1 THE PROCEDURES USED FOR CREATING A DAPP-BASED POSTPROCESSOR

The procedure for creating a DAPP-based postprocessor is as follows:

1. Use the method described in Chapter 12 to generate the BLOCK DATA program in FORTRAN. For example, under the VM/CMS environment, it is created and saved as a file, with file name, type, and mode

   ANSWER FORTRAN A1

2. Write an MTM in FORTRAN based on the characteristics of the NC machine-controller system. For example, under the same environment, the designed MTM might have the file name, type, and mode

MTMD08 FORTRAN A1

3. Use the FORTRAN-H compiler to compile the two files listed above. The two resulting files are

ANSWER TEXT A1

and

MTMD08 TEXT A1

4. Link the two compiled files with the DAPP. Under the VM/CMS environment, the following job execution statements[11] (EXEC) can be used to generate the postprocessor named LATH2PP:

```
&CONTROL ERROR
GLOBAL TXTLIB CMSLIB FORTLIBX
LOAD DAPPPP DLWSSR DAPCOM ANSWER MTMD08 (RESET DAPPPP ORIGIN 30000)
GENMOD LATH2PP
PR LOAD MAP A5
&EXIT
```

The resulting file is

LATH2PP MODULE A1

which is the postprocessor load module and can be used to process the CLDATA and translate it into the NC codes. The JCL (Job Control Language) statements for generating the postprocessor under the OS/MVS environment are given in Appendix D.

   If other names are used for the BLOCK DATA program, the MTM, and the postprocessor, then the corresponding terms in the job execution statements shown above should be changed accordingly.

5. Test the postprocessor. Usually an APT program that defines a machining operation which consists of all the motions and operations that the postprocessor can handle should be processed by the NC processor and the postprocessor. The postprocessor can be verified by comparing the processed results (NC codes) with the designed machining sequence and by comparing the values of various DATBUF elements with the processed NC codes.

15.2 THE DESIGN AND IMPLEMENTATION OF A DAPP-BASED POSTPROCESSOR FOR THE CADILLAC NC-100 LATHE/FANUC 6T CONTROLLER SYSTEM

We shall now use the Cadillac NC-100 lathe/FANUC 6T controller system, introduced in Chapters 2 and 12, as an example to elaborate the design and creation of a postprocessor.

15.2.1 The Characteristics of the Machine-Controller System

The features offered by this machine are as follows:

1. Motion along the Z and X axes
2. Motion programmed in incremental or absolute coordinates
3. Programmable tool selection
4. Programmable spindle speed in two ranges and in two directions (clockwise and counterclockwise)
5. Programming in inches
6. Automatic reference point return
7. Programmable feedrate in inches per minute or per revolution
8. Miscellaneous functions

This controller accepts input in both ISO and EIA codes and provides the following preparatory and auxiliary functions:

M00	Program stop
M01	Optional stop
M02	End of program
M03	Spindle forward rotation (counterclockwise)
M04	Spindle reverse rotation (clockwise)
M05	Spindle stop
M08	Coolant on
M09	Coolant off
M11	Turret rotation by 90 degrees in clockwise direction
M30	End of program and memory rewind
M41	Low-gear range of spindle speed
M42	High-gear range of spindle speed
G00	Rapid positioning motion
G01	Linear interpolation motion
G02	Circular interpolation motion (clockwise)
G03	Circular interpolation motion (counterclockwise)
G04	Dwelling
G28	Return to reference point
G70-76	Multiple machining cycles as described in Chapter 2
G90,92,94	Turning, threading, and facing cycles, respectively
G98	Feedrate programmed in inches per minute
G99	Feedrate programmed in inches per revolution

This controller accepts the following NC codes in word-address formats given in Table 2-3: N, G, X, Z, U, W, I, K, F, S, T, M, L, P, Q, D, A. However, it does not accept the radius code R.

The postprocessor introduced below can handle all the G codes listed above except the cycle function codes (G90s and G70s), because the DAPP does not provide routines for processing the turning and multiple turning cycles. Hence codes L, P, Q , D, and A are not handled by this postprocessor.

The turret on this machine is pneumatically driven. It has four positions and can rotate in the clockwise direction only. It is assumed that the four tools on the turret are successively numbered 01, 02, 03, and 04 in the counterclockwise direction. Since the tool offset has been compensated for during the calculation of the tool cutter position by the geometric section of the DAPP, it is not necessary to generate the tool offset code for each tool. Thus the tool offset code is set to zero by this postprocessor, and the tool codes generated by this postprocessor are T0100, T0200, T0300, and T0400.

The dimension system used in APT is assumed to be in inches. According to measurements, the spindle speed code S can be calculated by using the following formulas:

$$s = (w - 12)/2.38 \text{ (low-gear range: } 12 < w < 250 \text{ rpm)}$$
$$s = (w - 50)/19.5 \text{ (high-gear range: } 250 \leqslant w < 2000 \text{ rpm)}$$

where w is the actual spindle speed in revolutions per minute.

This postprocessor is designed to output the feedrate in inches per revolution because it is the unit commonly used to define a feedrate for turning. However, the feedrate processed by the DAPP general processing routine, that is, DATBUF(10), is always expressed in inches per minute (see Table 12-1). For output of the feedrate in inches per revolution, the feedrate should be specified in an APT program by using the statement

FEDRAT/f

where f is expressed in inches per revolution, without the minor word IPR. In addition, the third parameter in the MACHIN statement should be specified as 2, which activates a special routine to calculate the machining time according to the feedrate specified in inches per revolution.

It is worth mentioning that the x and y coordinates in the APT program and the CLDATA should be transformed, respectively, into the z and x coordinates. Note also that the x coordinate in the NC program should be the value of the diameter (i.e., twice the value of the x coordinate in the APT program).

15.2.2 Preparation of the BLOCK DATA Program and the MTM

To start designing the postprocessor, one should prepare a list of the answers to the DAPP Questionnaire and then generate the BLOCK DATA program by using the program QUEST. These answers and the BLOCK DATA program are listed in Fig. 12-3 and 12-4, respectively. The proposed MTM subroutine (MTMD08) is given in Fig. 15-1, and the flowcharts for its various sections are given in Figs. 15-2 to 15-12.

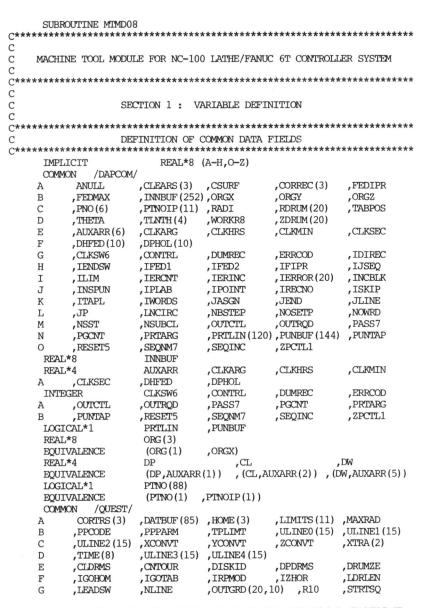

```
      SUBROUTINE MIMD08
C*************************************************************************
C
C     MACHINE TOOL MODULE FOR NC-100 LATHE/FANUC 6T CONTROLLER SYSTEM
C
C*************************************************************************
C
C                    SECTION 1 :  VARIABLE DEFINITION
C
C*************************************************************************
C                 DEFINITION OF COMMON DATA FIELDS
C*************************************************************************
      IMPLICIT             REAL*8  (A-H,O-Z)
      COMMON    /DAPCOM/
     A    ANULL       ,CLEARS(3)  ,CSURF      ,CORREC(3)    ,FEDIPR
     B    ,FEDMAX      ,INNBUF(252),ORGX       ,ORGY         ,ORGZ
     C    ,PNO(6)      ,PTNOIP(11) ,RADI       ,RDRUM(20)    ,TABPOS
     D    ,THETA       ,TLNTH(4)   ,WORKR8     ,ZDRUM(20)
     E    ,AUXARR(6)   ,CLKARG     ,CLKHRS     ,CLKMIN       ,CLKSEC
     F    ,DHFED(10)   ,DPHOL(10)
     G    ,CLKSW6      ,CONTRL     ,DUMREC     ,ERRCOD       ,IDIREC
     H    ,IENDSW      ,IFED1      ,IFED2      ,IFIPR        ,IJSEQ
     I    ,ILIM        ,IERCNT     ,IERINC     ,IERROR(20)   ,INCBLK
     J    ,INSPUN      ,IPLAB      ,IPOINT     ,IRECNO       ,ISKIP
     K    ,ITAPL       ,IWORDS     ,JASGN      ,JEND         ,JLINE
     L    ,JP          ,LNCIRC     ,NBSTEP     ,NOSETP       ,NOWRD
     M    ,NSST        ,NSUBCL     ,OUTCTL     ,OUTRQD       ,PASS7
     N    ,PGCNT       ,PRTARG     ,PRTLIN(120),PUNBUF(144)  ,PUNTAP
     O    ,RESET5      ,SEQNM7     ,SEQINC     ,ZPCTL1
      REAL*8            INNBUF
      REAL*4            AUXARR     ,CLKARG     ,CLKHRS       ,CLKMIN
     A    ,CLKSEC      ,DHFED      ,DPHOL
      INTEGER           CLKSW6     ,CONTRL     ,DUMREC       ,ERRCOD
     A    ,OUTCTL      ,OUTRQD     ,PASS7      ,PGCNT        ,PRTARG
     B    ,PUNTAP      ,RESET5     ,SEQNM7     ,SEQINC       ,ZPCTL1
      LOGICAL*1         PRTLIN     ,PUNBUF
      REAL*8            ORG(3)
      EQUIVALENCE       (ORG(1)    ,ORGX)
      REAL*4            DP                   ,CL                  ,DW
      EQUIVALENCE       (DP,AUXARR(1))   , (CL,AUXARR(2))  , (DW,AUXARR(5))
      LOGICAL*1         PTNO(88)
      EQUIVALENCE       (PTNO(1)   ,PTNOIP(1))
      COMMON    /QUEST/
     A    CORTRS(3)   ,DATBUF(85) ,HOME(3)    ,LIMITS(11)   ,MAXRAD
     B    ,PPCODE      ,PPPARM     ,TPLIMT     ,ULINE0(15)   ,ULINE1(15)
     C    ,ULINE2(15)  ,XCONVT     ,YCONVT     ,ZCONVT       ,XTRA(2)
     D    ,TIME(8)     ,ULINE3(15) ,ULINE4(15)
     E    ,CLDRMS      ,CNTOUR     ,DISKID     ,DPDRMS       ,DRUMZE
     F    ,IGOHOM      ,IGOTAB     ,IRPMOD     ,IZHOR        ,LDRLEN
     G    ,LEADSW      ,NLINE      ,OUTGRD(20,10)  ,R10       ,STRTSQ
```

Figure 15-1   The Machine Tool Module for the Cadillac NC-100 lathe/FANUC 6T
controller system.

The Variable-Definition Section   In the variable-definition section (Fig. 15-2), the
COMMON fields labeled DAPCOM and QUEST are defined, followed by the variables specific to this MTM. The meaning of these variables is as follows:

RCD1 = a REAL*8 array with 18 positions for keeping record of the output codes

```
H       ,SIMOSW(4)    ,TABLGO      ,ZMOD(2)      ,Z10
I       ,CLKPRT       ,EOB         ,IEOB         ,IPART       ,IRSTO
J       ,LCHAR(2)     ,TABCOD      ,TAPPRT       ,TAPMET
REAL*8               LIMITS       ,MAXRAD
REAL*4               TIME         ,ULINE3       ,ULINE4
INTEGER*4            CLDRMS       ,CNTOUR       ,DISKID      ,DPDRMS
A       ,DRUMZE       ,OUTGRD      ,R10          ,STRTSQ      ,SIMOSW
B       ,TABLGO       ,ZMOD        ,Z10
LOGICAL*1            CLKPRT       ,EOB          ,IEOB        ,IPART
A       ,IRSTO        ,LCHAR       ,TABCOD       ,TAPPRT      ,TAPMET
EQUIVALENCE          (ZDEPTH,DATBUF(29))    ,    (ZFINAL,DATBUF(30))
REAL*8               CONVEN(3)
EQUIVALENCE  (CONVEN(1),XCDIG,XCONVT),(YCDIG,YCONVT),(ZCDIG,ZCONVT)
INTEGER              ASSIGN(2)                ,CYCTYP(2)
EQUIVALENCE          (ASSIGN(1),DATBUF(20))   ,(CYCTYP(1),DATBUF(28))
LOGICAL*1            LOGCAL(6)                ,OUTSUX(40)
EQUIVALENCE          (LOGCAL(1),DATBUF(31))   ,(OUTSUX(1),OUTGRD(1,1))
C*********************************************************************
C
C        CURRENT STATUS OF MACHINE TOOL IS DEFINED BY ARRAY RCD1
C
C                    RCD1(1)  = G
C                    RCD1(2)  = X
C                    RCD1(3)  = Z
C                    RCD1(4)  = I
C                    RCD1(5)  = K
C                    RCD1(6)  = F
C                    RCD1(7)  = S
C                    RCD1(8)  = T
C                    RCD1(9)  = M
C                    RCD1(10) = U
C                    RCD1(11) = W
C                    RCD1(12) = RAPID FEEDRATE SWITCH
C                    RCD1(13) = SPINDLE DIRECTION
C                    RCD1(14) = SPINDLE RPM
C                    RCD1(15) = TOOL CODE
C                    RCD1(17) = COOLANT OUTPUT CODE
C                    RCD1(18) = SPINDLE SPEED RANGE
C
C*********************************************************************
C
C                    VARIABLE SPECIFIC TO THIS MTM
C
C*********************************************************************
      REAL *8 RCD1(18)
      REAL *4 LL1,SS,II1
      LOGICAL *1 NOTL, NOPRNT
      INTEGER *2 INDX11,J,IQ,GQES
```

Figure 15-1   (continued)

GQES = an INTEGER*2 variable that is used to output G, X, and Z codes whose value is zero

INDX11 = an INTEGER*2 variable that indicates the next available DAT-BUF element in the range DATBUF(73) to (85)

J = an INTEGER*2 variable that indicates to which section control is to be sent on reentry into the MTM, initialized to 1 to ensure execution of the initialization section

```
      INTEGER *4 G,X,Z,I,K,F,S,T,M,U,W
      DATA NOTL/T/,IQ/-10000/,INDX11/73/,J/1/,NOPRNT/T/
C********************************************************************
      COMMON /DBFPRT/ N1,ISTART(5),LENGTH(5)
      EQUIVALENCE (G,OUTGRD(2,10)),    (X,OUTGRD(3,10)),
     1            (Z,OUTGRD(4,10)),    (I,OUTGRD(5,10)),
     2            (K,OUTGRD(6,10)),    (F,OUTGRD(7,10)),
     3            (S,OUTGRD(8,10)),    (T,OUTGRD(9,10)),
     4            (M,OUTGRD(10,10)),   (U,OUTGRD(11,10)),
     5            (W,OUTGRD(12,10))
C********************************************************************
      GQES=0
C********************************************************************
50    GO TO (100,200,300,400,500,360,700,800,900,550,261,250,130,140,
     1 375,330), J
C********************************************************************
C
C
C            ARGUMENT "J" DETERMINES THE SECTION TO BE USED
C
C      J            LABEL               SECTION
C      1            100            INITIALIZATION
C      2            200            COOLANT AND TOOL
C      3            300            ROTATION DIRECTION OF SPINDLE
C      4            400            LINEAR INTERPOLATION
C      5            500            CIRCULAR INTERPOLATION
C      6            360            ROTATION SPEED OF SPINDLE
C      7            700            CONTROL FUNCTION AND FEEDRATE
C      8            800            PREPARATION SECTION
C      9            900            CONTROL DIRECTING SECTION (2)
C      10           550            SUBSEQUENT QUADRANTS OF CIRCULAR
C                                  INTERPOLATION
C      11           261            TOOL CHANGE
C      12           250            TOOL CODE
C      13           130            GENERATING G50 CODE
C      14           140            SPECIFYING DATBUF POSITIONS TO BE
C                                  PRINTED
C      15           375            GENERATING M03 OR M04 CODE
C      16           330            SET SPINDLE SPEED RANGE
C
C********************************************************************
C
C                  SECTION 2 :  INITIALIZATION
C
C********************************************************************
100   DO 110 J=1,18
110   RCD1(J)=ANULL
      DATBUF(73)=28
      DATBUF(74)=0
      DATBUF(75)=0
      OUTGRD(2,10)=73
      OUTGRD(11,10)=74
      OUTGRD(12,10)=75
      OUTCTL=6
      J=13
      RETURN
130   DATBUF(73)=50
      IF (DATBUF(27).NE.ANULL) GOTO 138
      DATBUF(27)=0.0D0
132   IF (DATBUF(34).NE.ANULL) GOTO 135
      DATBUF(34)=0.0D0
135   OUTGRD(2,10)=73
```

Figure 15-1 (continued)

```
            OUTGRD(3,10)=27
            OUTGRD(4,10)=34
            OUTCTL=6
            J=14
            RETURN
  138       DATBUF(27)=2*DATBUF(27)
            GOTO 132
  140       N1=4
            ISTART(1)=1
            ISTART(2)=24
            ISTART(3)=36
            ISTART(4)=72
            LENGTH(1)=17
            LENGTH(2)=2
            LENGTH(3)=12
            LENGTH(4)=9
            DATBUF(72)=0.0
            DATBUF(73)=ANULL
            DATBUF(74)=ANULL
            DATBUF(75)=ANULL
            DATBUF(27)=ANULL
            DATBUF(34)=ANULL
C**********************************************************************
C
C                    SECTION 3 :  COOLANT AND TOOL
C
C**********************************************************************
  200       IF (DATBUF(35).EQ.1.0D0.OR.DATBUF(35).EQ.2.0D0) CALL DLWDBF
            IF (DATBUF(17).EQ.ANULL) GOTO 250
            IF (DATBUF(17).EQ.RCD1(17)) GOTO 250
            IF (DATBUF(17).LE.4.0D0) DATBUF(17)=4.0D0
            DATBUF(17)=DATBUF(17)+4.0D0
            M=17
            RCD1(17)=DATBUF(17)
            J=12
            GO TO 801
C**********************************************************************
  250       IF (DATBUF(15).EQ.RCD1(15).OR.DATBUF(15).EQ.ANULL) GOTO 300
            DO 252 L=1,4
            IF (DATBUF(15).EQ.(1.0*L)) GO TO 260
  252       CONTINUE
            CALL DLWDER (6510)
            GOTO 300
  270       T=15
            NOTL=.FALSE.
            J=3
  272       CLKARG=CLKARG+TIME(5)
            GO TO 801
  260       IF (RCD1(15).EQ.ANULL) GO TO 275
            LL1=DATBUF(15)-RCD1(15)
            LLL=IFIX(LL1)
            IF (LLL.LT.0) LLL=4+LLL
            LLLL=1
  261       M=11*IQ
            IF (LLLL.EQ.LLL) GO TO 265
            J=11
            LLLL=LLLL+1
            GO TO 272
  265       RCD1(15)=DATBUF(15)
            DATBUF(15)=100*DATBUF(15)
            GO TO 270
```

Figure 15-1   (continued)

```
275    RCD1(15)=DATBUF(15)
       DATBUF(15)=100*DATBUF(15)
       T=15
       NOTL=.FALSE.
C*********************************************************************
C
C         SECTION 4 : DIRECTION AND SPEED OF SPINDLE ROTATION
C
C*********************************************************************
C
C                          SPEED AND RANGE
C
C*********************************************************************
300    VV=DATBUF(14)
       IF ((VV.EQ.RCD1(14)).OR.(VV.EQ.ANULL)) GO TO 360
       IF (RCD1(18).EQ.ANULL) GOTO 330
       IF ((DATBUF(13).NE.RCD1(13)).AND.(RCD1(13).NE.ANULL)) GOTO 350
       IF (((VV.GE.250).AND.(DABS(RCD1(18)-42.0).GT.0.5)).OR.
      1 ((VV.LT.250).AND.(DABS(RCD1(18)-41.0).GT.0.5))) GOTO 350
       IF (VV.GE.250.0) GO TO 319
       GOTO 315
330    IF (VV.GE.250.0) GO TO 318
       M=41*IQ
315    SS=((VV-12)/2.38)*IQ
       S=IFIX(SS)
       RCD1(18)=41
       GOTO 320
318    M=42*IQ
319    SS=((VV-50)/19.5)*IQ
       S=IFIX(SS)
       RCD1(18)=42
320    RCD1(14)=VV
       GOTO 360
350    M=5*IQ
       RCD1(13)=3
       RCD1(14)=0
       CLKARG=CLKARG+TIME(4)
       J=16
       GOTO 801
C*********************************************************************
C
C                          DIRECTION
C
C*********************************************************************
360    IF ((DATBUF(13).EQ.RCD1(13)).OR.(DATBUF(13).EQ.ANULL)) GOTO 400
       IF (M.EQ.0) GOTO 370
       J=6
       GO TO 801
370    IF (RCD1(13).EQ.ANULL) GOTO 375
       IF (DATBUF(13).NE.RCD1(13)) GOTO 375
       RCD1(13)=DATBUF(13)
       II1=DATBUF(13)
       III=IFIX(II1)
       GOTO (372,372,390),III
372    J=15
       GOTO 392
375    IF (DATBUF(13).EQ.2) GOTO 385
       IF (DATBUF(13).EQ.3) GOTO 388
380    M=(DATBUF(13)+3)*IQ
       GOTO 389
385    M=(DATBUF(13)+1)*IQ
       GOTO 389
```

Figure 15-1 (continued)

```
388     M=(DATBUF(13)+2)*IQ
389     RCD1(13)=DATBUF(13)
        J=4
        GOTO 396
390     J=4
392     M=5*IQ
        RCD1(13)=3
        RCD1(14)=0
396     CLKARG=CLKARG+TIME(4)
        GO TO 801
C*********************************************************************
C
C          SECTION 5 :   LINEAR INTERPOLATION AND QUICK POSITIONING
C
C*********************************************************************
400     IF ((DATBUF(7).EQ.0.0D0).AND.(DATBUF(8).EQ.0.0D0)) GO TO 700
450     IF ((DATBUF(36).NE.0.0).AND.(DATBUF(36).NE.ANULL)) GO TO 500
        IF (DATBUF(12).EQ.0.0D0) GO TO 452
        GQES=1
        G=0
        X=2
        Z=1
        GO TO 480
452     IF (RCD1(13).EQ.3.0) GOTO 490
455     IF (G.EQ.0) GO TO 460
        J=4
        GO TO 801
460     G=IQ
        IF (DATBUF(10).NE.RCD1(6)) F=10
470     X=2
        Z=1
        IF (DATBUF(1).NE.0.AND.DATBUF(2).NE.0) GO TO 472
        GQES=1
472     IF(DATBUF(35).NE.2.0) GOTO 480
        IF(DATBUF(14).EQ.0.0) GOTO 478
        TIME(1)=TIME(1)/DATBUF(14)
        GOTO 480
478     CALL DLWDER (6508)
480     CLKARG=CLKARG+TIME(1)
        J=7
        GO TO 801
490     CALL DLWDER (6501)
        DATBUF(24)=1.0
        GO TO 455
C*********************************************************************
C
C                    SECTION 6 :   CIRCULAR INTERPOLATION
C
C*********************************************************************
500     IF (DATBUF(12).NE.1.0) GO TO 510
        CALL DLWDER (6507)
        DATBUF(24)=1.0
510     F=10
        IDTB=42
        NUMQD=0
        IF (G.EQ.0) GO TO 520
        J=5
        GO TO 801
520     G=2*IQ
        IF (DATBUF(37).GT.1.0) G=3*IQ
550     U=IDTB+1
        W=IDTB
```

Figure 15-1   (continued)

```
         I=IDTB+4
         K=IDTB+3
         NUMQD=NUMQD+1
         GOTO (553,556,558),NUMQD
553      DATBUF(I)=DATBUF(39)-DATBUF(5)
         DATBUF(K)=DATBUF(38)-DATBUF(4)
         GOTO 559
556      DATBUF(I)=DATBUF(39)-(DATBUF(5)+DATBUF(43)/2)
         DATBUF(K)=DATBUF(38)-(DATBUF(4)+DATBUF(42))
         GOTO 559
558      DATBUF(I)=DATBUF(39)-(DATBUF(5)+DATBUF(43)/2+DATBUF(49)/2)
         DATBUF(K)=DATBUF(38)-(DATBUF(4)+DATBUF(42)+DATBUF(48))
559      IF (DATBUF(IDTB+6).EQ.ANULL.OR.NUMQD.EQ.3) GO TO 560
         J=10
         IDTB=IDTB+6
         GO TO 801
560      IF(DATBUF(35).NE.2.0) GOTO 570
         IF(DATBUF(14).EQ.0.0) GOTO 562
         TIME(1)=TIME(1)/DATBUF(14)
         GOTO 570
562      CALL DLWDER (6508)
570      CLKARG=CLKARG+TIME(1)
         J=7
         GO TO 801
C*******************************************************************
C
C        SECTION 7 :   SPECIAL CONTROL FUNCTIONS AND FEEDRATE
C
C                    THE FUNCTIONS HANDLED ARE:
C
C             FUNCTION                        DATBUF(25)
C        -------------------------------------------------------
C             END                             9001
C             OPSTOP                          9002
C             STOP                            9003
C             DELAY                           9004
C             PREFUN                          -1 TO -99
C             AUXFUN                          0 TO 99
C
C*******************************************************************
700      IF (DATBUF(25).EQ.ANULL) GO TO 790
         J=7
         IF (M.NE.0) GO TO 801
         IF (DATBUF(25).GE.0.0) GO TO 710
         DATBUF(25)=-DATBUF(25)
         G=25
         GO TO 800
710      JJ1=DATBUF(25)-9000.0
         IF (JJ1.GE.0.0) GO TO 720
         M=25
         GO TO 800
720      IF (JJ1.LT.1.OR.JJ1.GT.4) GO TO 770
         GO TO (730,740,750,760),JJ1
730      IF (RCD1(17).NE.8.0) GO TO 735
         M=9*IQ
         RCD1(17)=9.0
         GO TO 801
735      IF (RCD1(13).EQ.3.0) GO TO 738
         M=5*IQ
         RCD1(13)=3.0
         GO TO 801
```

Figure 15-1   (continued)

```
738    M=2*IQ
       GO TO 755
740    M=IQ
       GO TO 751
750    M=72
751    RCD1(12)=0.0
       RCD1(13)=3.0
       RCD1(17)=9.0
755    CLKARG=CLKARG+TIME(4)
       GO TO 800
760    G=4
       U=19
       GO TO 800
770    CALL DLWDER (6503)
       DATBUF(24)=0
790    IF (DATBUF(10).NE.RCD1(6)) GOTO 800
       IF (DATBUF(10).EQ.0) GOTO 800
       F=10
C*********************************************************************
C
C          SECTION 8 :  PREPARATION SECTION -- PREPARING DATA,
C                       UPDATING OUTGRD & RCD1 ARRAY
C
C*********************************************************************
800    J=9
801    DO 840 L=2,12
       IF (L.EQ.2.AND.G.EQ.4) GOTO 832
       IF ((L.EQ.2).AND.(GQES.EQ.1).AND.(G.EQ.0)) GO TO 830
       IF ((L.EQ.3).AND.(GQES.EQ.1).AND.(DATBUF(2).EQ.0)) GO TO 830
       IF ((L.EQ.4).AND.(GQES.EQ.1).AND.(DATBUF(1).EQ.0)) GOTO 830
       IF (OUTGRD(L,10)) 802,840,810
802    DATBUF(INDX11)=(OUTGRD(L,10))/IQ
       OUTGRD(L,10)=INDX11
       INDX11=INDX11+1
810    IF (L.EQ.5.OR.L.EQ.6.OR.L.EQ.11.OR.L.EQ.12) GOTO 812
       IF ((L.EQ.10).AND.(DATBUF(OUTGRD(L,10)).EQ.11)) GOTO 816
814    IF (DABS(DATBUF(OUTGRD(L,10))-RCD1(L-1)).LT.0.00001) GOTO 818
816    RCD1(L-1)=DATBUF(OUTGRD(L,10))
       IF (L.EQ.3.OR.L.EQ.11) GOTO 817
819    NOPRNT=.FALSE.
       GOTO 840
817    IF (L.EQ.11.AND.RCD1(1).EQ.4) GOTO 819
       DATBUF(OUTGRD(L,10))=2*RCD1(L-1)
       GOTO 819
812    RCD1(L-1)=0
       GOTO 814
818    OUTGRD(L,10)=0
       GOTO 840
830    DATBUF(INDX11)=0
       GOTO 835
832    DATBUF(INDX11)=4.0
835    OUTGRD(L,10)=INDX11
       INDX11=INDX11+1
       GOTO 816
840    CONTINUE
C*********************************************************************
C
C          SECTION 9 :  CONTROL-DIRECTING SECTION (1)
C
C*********************************************************************
       INDX11=73
```

Figure 15-1 (continued)

```
      IF (NOPRNT) GO TO 50
      OUTCTL=5
      NOPRNT=.TRUE.
      GQES=0
      RETURN
C***********************************************************************
C
C                        CONTROL DIRECTING SECTION (2)
C
C***********************************************************************
900   IF (DATBUF(15).LT.10) GOTO 901
      DATBUF(15)=DATBUF(15)/100
901   IF (DATBUF(24).LT.0.5) GO TO 905
      J=9
      DATBUF(24)=0.0
      OUTCTL=12
      RETURN
905   J=2
      OUTCTL=1
      DO 920 L=73,85
920   DATBUF(L)=ANULL
      RETURN
      END
```

Figure 15-1   (continued)

IQ = an INTEGER*2 variable initialized to a value of -10000; used to generate an output from elements DATBUF(73) through (85), based on the value of OUTGRD(i,10)

NOPRNT = a LOGICAL*1 variable, used to indicate whether an output is required (NOPRNT = .FALSE.) or not (NOPRNT = .TRUE.); initialized as .TRUE.

NOTL = a LOGICAL*1 variable, indicating whether a tool change is required (NOTL = .FALSE.) or not (NOTL = .TRUE.); initialized as .TRUE.

The common field labeled DBFPRT is also defined in this section so that the required DATBUF positions can be output in the listing output.

The Initialization Section   The function of the initialization section (Fig. 15-3) is twofold. It should generate the two NC statements

```
G28U0W0*
G50X_Z_*
```

as the first two statements in the NC output. The X and Z codes in the second statement are taken from the parameter in the CUTCOM statement and the second parameter in the MACHIN statement, respectively. Therefore, in the programming of an APT part machining program, an initial CUTCOM statement and the MACHIN statement should be used to define the origin of the machine coordinate system. The initialization section also sets the initial values of the array RCD1 and the DATBUF elements to be printed on the listing file. Finally, the DATBUF elements used in this section are also reset to "no input" status, represented by the symbol ANULL.

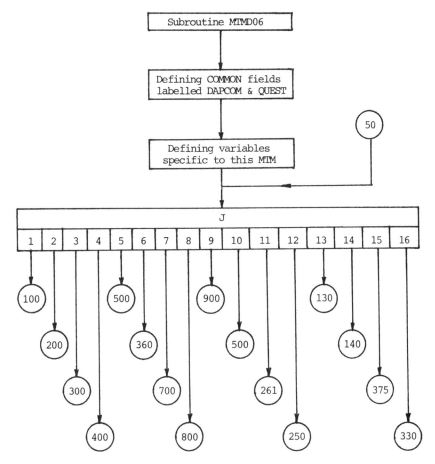

Figure 15-2   The flowchart for the variable-definition section of the MTM in Fig. 15-1.

The Coolant and Tool Sections   The coolant section (Fig. 15-4) processes the coolant data. It is the first processing section that control enters every time the MTM is called during processing of a set of CLDATA records. The DATBUF array printing routine DLWDBF is called to print the required DATBUF positions if the third parameter in the MACHIN statement is specified as 1 or 2. If coolant is requested, code M08 is set and output; control then goes to the tool section on reentry into the MTM. Otherwise, control goes directly to the tool section.

The function of the tool section (Fig. 15-5) is to generate the required tool code (T0100, T0200, T0300, or T0400) and the turret rotation code (M11), which orders the turret to rotate 90 degrees in the clockwise direction. A tool change is specified by the statement LOADTL/*n*, where *n* is the tool number of the requested tool, which might not be the tool at the next position on the turret. The first tool specified in an APT program is always assumed to be tool No. 1. The program calculates the

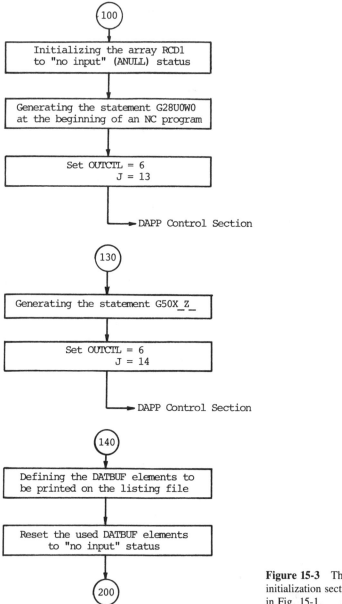

Figure 15-3   The flowchart for the initialization section of the MTM listed in Fig. 15-1.

required number of M11 codes and then generates the requested tool code and the M11 codes. For example, if the current tool is No. 2 and the statement LOADTL/1 is specified (i.e., tool No. 1 is requested), the tool processing routine generates the three NC blocks

M11*
M11*
M11T0100*

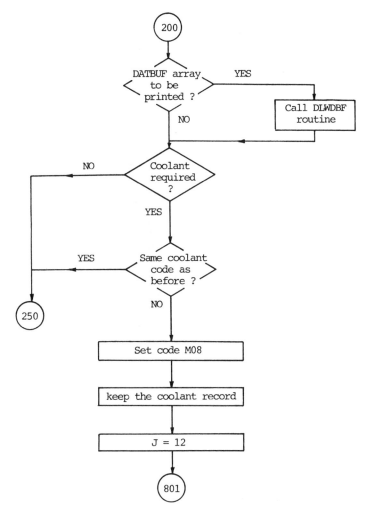

Figure 15-4   The flowchart for the coolant section of the MTM listed in Fig. 15-1.
250: The first statement in the tool section (Fig. 15-5). 801: The second statement
in the preparation section (Fig. 15-11).

As a result, the NC machine rotates the turret by 270 degrees and places tool No. 1
in position.

It is assumed that the tool number specified in the LOADTL statement is 1, 2,
3, or 4; the corresponding output code is T0100, T0200, T0300, or T0400. This
means that the value of DATBUF(15) has been multiplied by 100 before it is output.
Therefore the DATBUF(15) should be reset to its original value in the control-
directing section after the whole set of read-in CLDATA has been output.

At the end of this section, the machining time is updated by adding to it the
tool changing time.

The Spindle Rotation Section   The spindle rotation section (Figs. 15-6[a and b]
and 15-7) is the fourth section of the MTM; it processes the codes concerning

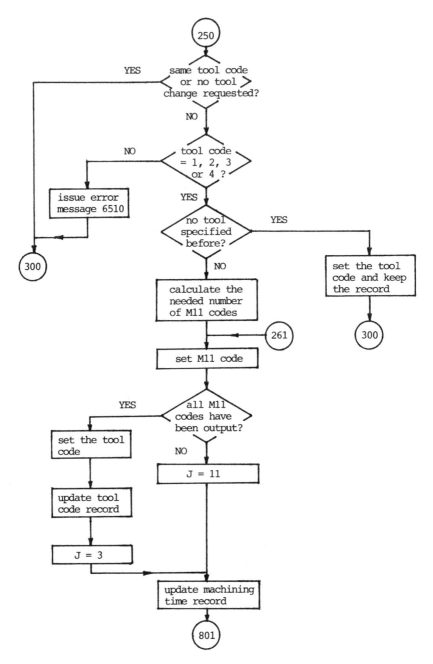

Figure 15-5   The flowchart for the tool section of the MTM listed in Fig. 15-1.

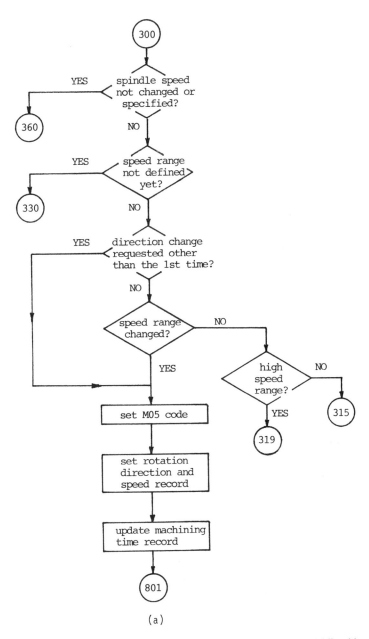

(a)

Figure 15-6(a)   The flowchart for the spindle rotation section of the MTM listed in Fig. 15-1.

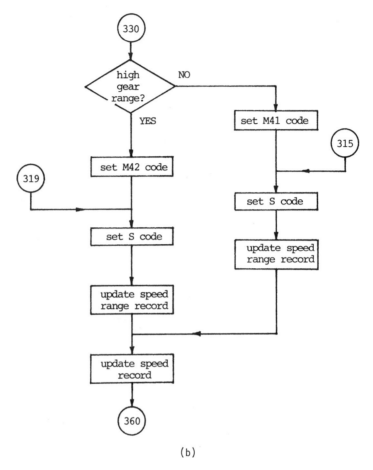

Figure 15-6(b)   The flowchart for the spindle speed and range section of the MTM listed in Fig. 15-1.

spindle rotation direction and speed. The NC codes generated in this section are the speed range code M41 or M42, speed code S, and spindle on or off code M03, M04, or M05. The program should generate the required codes for each specified SPINDL statement in APT. The following situations should be considered:

1. If the SPINDL statement is specified the first time in a program, the routine should generate a speed range code M41 or M42, a speed code S, and a spindle direction code M03 or M04.
2. If the SPINDL statement is other than the first one specified in a program, the routine should generate the following:
   a. An M05 code, an M41 or M42 code, an M03 or M04 code, and an S code, in that order, if both the speed range and the speed have been changed (This means that the machine should first stop the spindle, change gear, set the speed, and then turn on the spindle.)

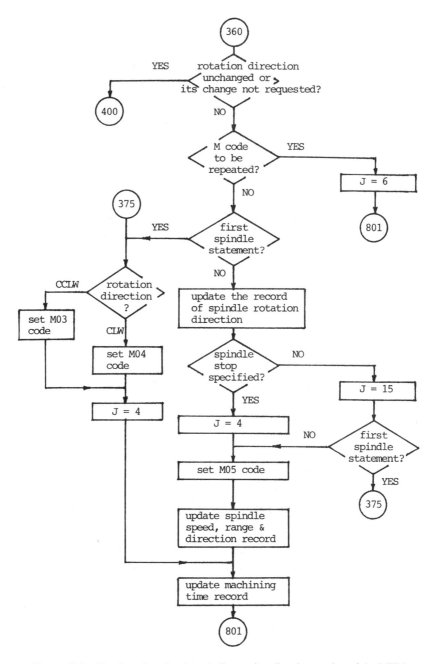

Figure 15-7 The flowchart for the spindle rotation direction section of the MTM listed in Fig. 15-1.

b. An M05 code, an M41 or M42 code, and an M03 or M04 code, in that order, if only the speed range has been changed

c. An M05 code and an M03 or M04 code if only the spindle direction is changed

d. A speed code S only, if the speed range and rotation direction have not been changed

e. An M05 code if the statement SPINDL/OFF is specified

Finally, the machining time record is updated.

The Linear-Interpolation-Motion Section The linear-interpolation-motion section (Fig. 15-8) processes point-to-point motion at rapid feedrate and linear interpolation motion. The codes generated in this section are the G00 or G01 code and the X, Z, and F codes. A G00 code is generated if the specified feedrate is RAPID; otherwise, code G01 is generated. There is no DATBUF element defining a G01 or G00 code. Thus the variable G, that is, OUTGRD(2,10), is first set to zero or IQ (= −10000), which is then transferred to and stored in one of the reserved elements, DAT-BUF(73) through (85), in the preparation section. Error message 6501 will be generated if linear interpolation motion is specified and the spindle speed is zero. When the feedrate is specified in inches per revolution, the routine issues error message 6508 if the spindle speed is zero. Since the routine that calculates motion time in the DAPP is designed for feedrate in inches per minute, the motion time should be recalculated when the feedrate is specified in inches per revolution. The third parameter of the MACHIN statement, namely, DATBUF(35), should be set to 2 to inform the MTM or postprocessor that the feedrate is specified in inches per revolution. The machining time record is updated accordingly before exiting this section.

The Circular-Interpolation-Motion Section The circular-interpolation-motion section (Fig. 15-9) processes the G02 or G03 code for the circular motion direction, the U and W codes for the destination point, the I and K codes for the center of the circle, and the feedrate code F. For the FANUC 6T controller, a circular motion can be defined without being segmented. However, in this program the circular motion is divided into several segments in different quadrants for examination of the value of the DATBUF elements that define the circular motion in different segments, that is, DATBUF(42) through (59). Because circular motion cannot be spread over three quadrants in turning, the maximum number of segments is set to three. The DAT-BUF elements used to define the end coordinates for the various segments are DAT-BUF(42), (43), (48), (49), (54), and (55). If the user wants to have unsegmented circular motion, he can simply use the end position defined by DATBUF(1) and (2) to generate Z and X codes, respectively.

A special statement calculates the motion time when the feedrate is specified in inches per revolution. The machining time is updated at the end of this routine. Error message 6508 is issued when the spindle speed is zero.

The Special Function Section The special function section (Fig. 15-10) processes codes F, M00, M01, M02, M08, M09, and G04 and the G codes defined by the APT statement PREFUN/*n*. It is divided into three sections processing, respectively, the F code, the G code defined by the PREFUN statement, and the M and G codes defined by the statements END, OPSTOP, STOP, and DELAY. A special process-

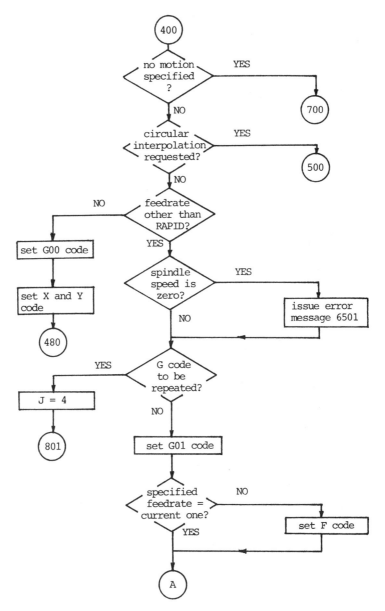

Figure 15-8   The flowchart for the linear interpolation motion section of the MTM listed in Fig. 15-1.

ing routine is incorporated into the program to ensure that the spindle and coolant have been turned off when the M02 code (end-of-program code) is generated. Error message 6503 is issued if an auxiliary function (AUXFUN) other than those listed above is programmed.

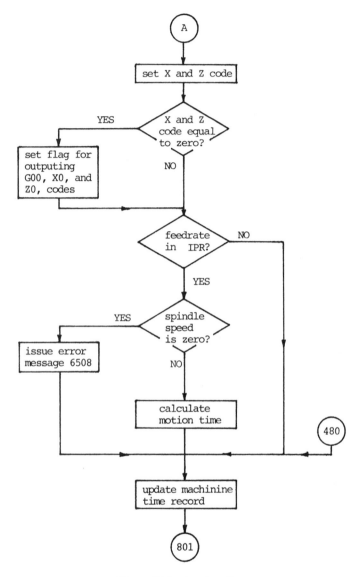

Figure 15-8 (continued)

The Preparation Section   Before one or more codes are output as a block after the processing by the previous sections, they are first sent to the preparation section (Fig. 15-11). There are several possible situations that must be handled by this section:

1. If a code to be output in the current block is the same as the one in the previous block, it should not be output unless it is one of the codes that are effective within a single block only (e.g., codes G04 and M11). The I, K, U, or W code

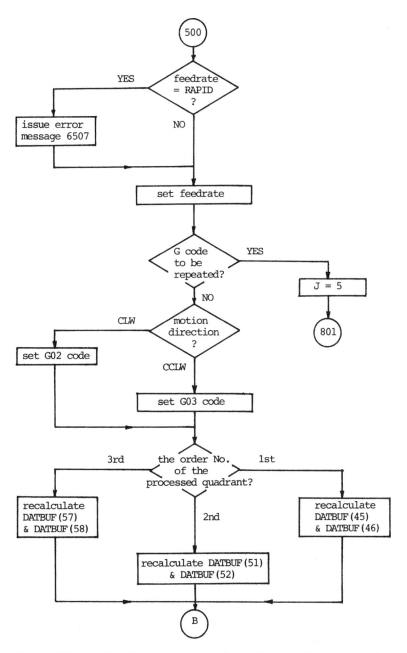

Figure 15-9 The flowchart for the circular interpolation motion section of the MTM listed in Fig. 15-1.

is zero if it is omitted in an NC block. Thus codes I, K, U, and W can be omitted only when their current and previous values are zero.

2. A code other than G is output on the basis of one of the DATBUF elements, DATBUF(1) through (71), and the processing is relatively straightforward.

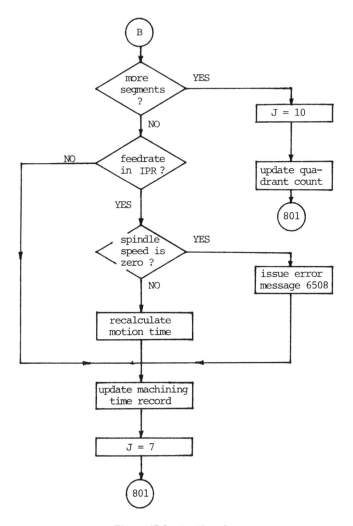

Figure 15-9 (continued)

However, a G code does not come from the CLDATA or the DATBUF array; it is generated during the processing by the MTM. In this MTM, a G code is set by first assigning a negative value to OUTGRD(2,10), which shares the same storage space as the variable G. It is then reset as a positive value and stored in one of the DATBUF elements, DATBUF(73) through (85), which is then output as the G code. For example, a G01 code is generated in the linear interpolation section by the statement

G=1*IQ

where IQ = −10000. Then it is transformed back to 1 in the preparation section and stored as a DATBUF element, DATBUF(i), where $73 \leqslant i \leqslant 85$. Thus we can specify, for example,

DATBUF(73)=1

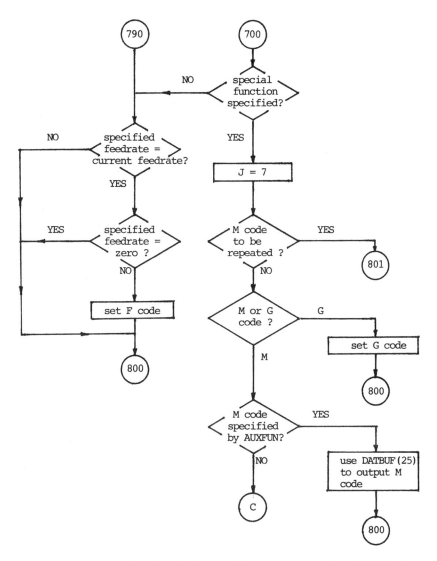

Figure 15-10   The flowchart for the special function section of the MTM listed in Fig. 15-1.

in the preparation section. Finally, we set

G=73

so that a G01 code can be output.

3. The x and y coordinates in APT should be transformed to z and x coordinates, respectively. With the exception of the U code, used together with the G04 code (representing the dwelling time), the X and U codes should be transformed to diameter values (i.e., twice the value of the absolute and incremental x coordinates, respectively) before they are output.

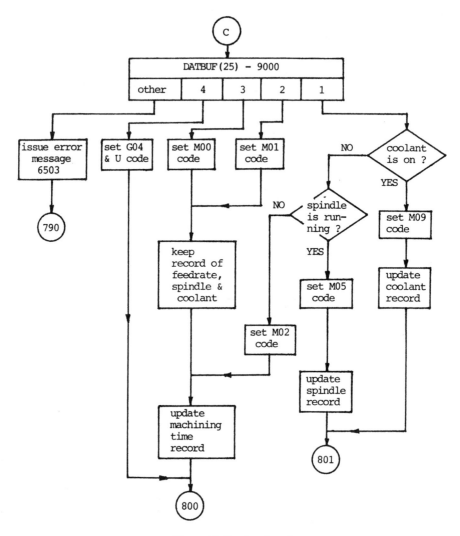

Figure 15-10   (continued)

The Control-Directing Section   Control enters the control-directing section (1) (Fig. 15-12; see also Fig. 15-1) when part of the CLDATA set has been prepared in the preparation section and is ready to be output. The control-directing variable OUTCTL is set to 5. Control is then sent to the DAPP Output Section. After the processed data have been output, control returns to the processing section of the MTM, directed by the value preset for variable J. This cycle repeats until the whole set of read-in CLDATA has been processed by the MTM processing section. Then the first statement of the preparation section (i.e., the statement labeled 800) can be executed, which sets the variable J to 9. After the processed data have been output, con-

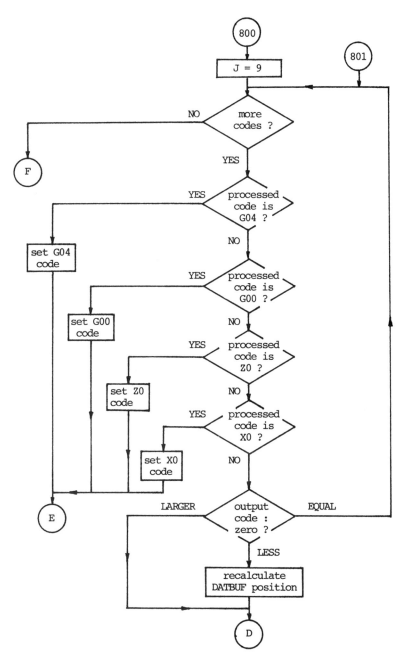

Figure 15-11 The flowchart for the preparation section of the MTM listed in Fig. 15-1. F: see Fig. 15-12.

trol branches directly to the control-directing section (2) (see also Fig. 15-1) when it reenters the MTM. In this section, DATBUF(15) is restored to its original value if it has been changed in the tool section. DATBUF elements (73) through (85) are also

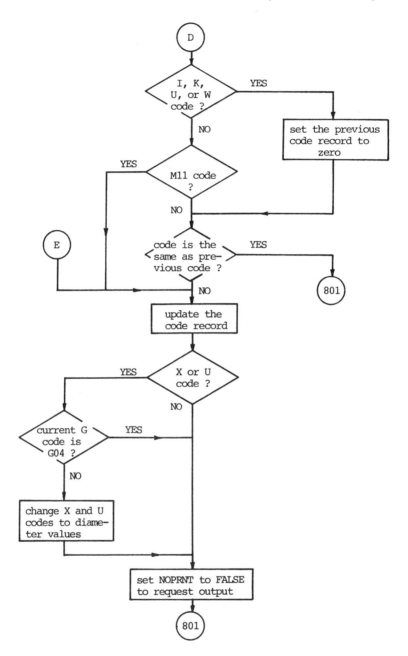

Figure 15-11   (continued)

reset. Control is sent to the Housekeeping Section of the DAPP to read in a new set of CLDATA if there is no processing error. Otherwise, a line of asterisks will be printed to indicate the existence of an error before control is sent to the DAPP Housekeeping Section.

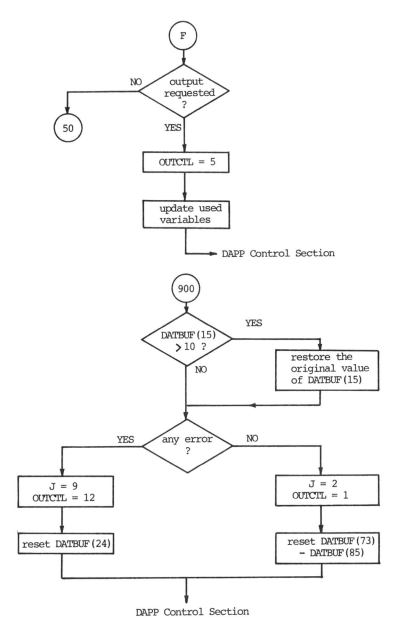

Figure 15-12   The control-directing section of the MTM listed in Fig. 15-1.

15.2.3 The Creation of the Postprocessor LTTH2PP and a Processing Example

The generated BLOCK DATA program and the designed MTM are compiled by the FORTRAN-H compiler. They are link-edited together with the DAPP by execution of the EXEC listed in Section 15-1. Thus the postprocessor named LATH2PP is created and is ready for processing of the CLDATA.

As an example, an APT program has been written for machining the part shown in Fig. 15-13. One side of the shaft has been machined to size in the previous operation. The APT program in Fig. 15-14 defines the machining operation for the right-hand side only. Four tools are used in this program for roughing (tool No. 1), finishing (tool No. 2), grooving (tool No. 3), and threading (tool No. 4), respectively. They are installed counterclockwise on the turret in that order. The origin of the coordinate system is the center of the finished end surface; the x and y coordinates of the machine reference point in this coordinate system are 12.5430 and 7.6400, respectively. An INSERT statement is used in this APT program to define the threading cycle in NC codes since the postprocessor LATH2PP cannot handle the threading operation. The postprocessor listing output and the processed NC program are listed in Figs. 15-15 and 15-16, respectively.

The use of the DAPP routine does simplify significantly the design of a postprocessor. The designer can make use of the machine-independent processing routines in the DAPP. He has only to provide the machine-dependent processing routine, the data regarding the characteristics of the NC machine-controller system, and the NC code format. However, if an NC machine has functions that cannot be handled by the DAPP, such as the threading cycle and the multiple turning cycle, some of the DAPP routines (e.g., DLWUP1 and DLWUP2) should be modified.[7] Special routines may be needed also in the MTM to realize the required processing function.

PROBLEMS

15.1 Select an NC milling machine in your workshop or laboratory, and design a DAPP-based postprocessor by following the procedures described in Section 15.1. A thorough understanding of the machine characteristics and its NC programming rules is necessary.

15.2 A three-axis CNC milling machine is used for reproducing the two-dimensional image (photograph) of a physical object on a transparent acrylic plate. A black-and-white image or photograph is a collection of a large number of black spots on a white background with varying densities in different areas. A black spot at a specific position can be represented by the status "TRUE" or the number "1," and a white one, by the status "FALSE" or the number "0." A series of coordinates representing the positions of the black spots can be obtained after the image of the object has been digitized. This image can be reproduced on a transparent acrylic plate by engraving on it a small spot at every position with a black spot. The engraved point on the plate becomes a white spot and is not transparent. Thus the image of the object can be shown clearly as a negative one by placing the transparent plate on a black paper (black background). Assume the following:

    a. After processing by the image-digitizer, the image of a physical object is represented by a number of coordinate-sets that represent the positions with black spots (Fig. P15-2). Each set of coordinates occupies one line. The minimum distance in the X or Y direction from one spot to another is 0.02 in.

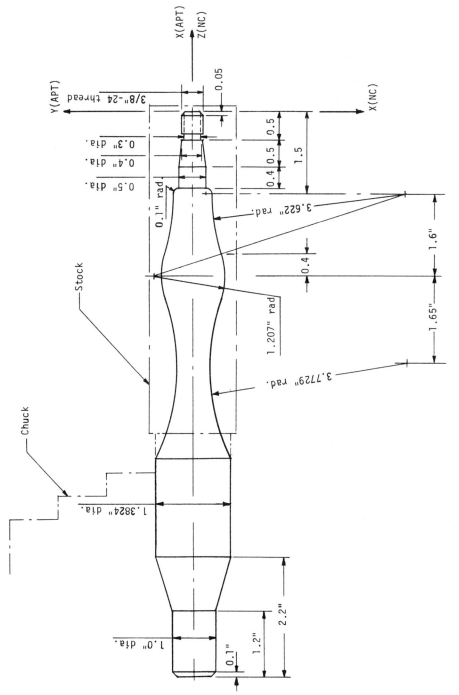

Figure 15-13 A profiled shaft with threads.

```
PARTNO SAMPLE PROGRAM FOR CUTTING THE PART IN FIG. 15-13
MACHIN/LATH2,8,OPTION,2,12.5430,2
CUTCOM/7.64
P0=POINT/4.0,2.0
P1=POINT/0.05,0.8
P2=POINT/0.05,0.65
P3=POINT/-5.9,0.65
P4=POINT/-3.1,0.65
P5=POINT/0.05,0.45
P6=POINT/-4.75,0.4
P7=POINT/0.05,0.25
L1=LINE/YAXIS
L2=LINE/XAXIS
L3=LINE/(P8=POINT/-0.05,0.1875),ATANGL,-45
L4=LINE/P8,(-0.5,0.1875)
L5=LINE/(-0.5,0.1875),(-1.0,0.25)
L6=LINE/(-1.0,0.25),PARLEL,L2
L7=LINE/PARLEL,L2,YLARGE,0.6912
C1=CIRCLE/-1.5,0.25,0.1
C2=CIRCLE/-1.5,(0.25+0.1+3.622),3.622
C3=CIRCLE/CENTER,(-3.1,-0.5843),SMALL,TANTO,C2
C4=CIRCLE/CENTER,(-4.75,4.1143),SMALL,TANTO,C3
SPINDL/800,CCLW
COOLNT/ON
CUTTER/0.03
LOADTL/1 $$ THE FIRST TOOL IS THE ROUGHING TOOL
FROM/(POINT/12.5430,7.64)
RAPID;    GOTO/P1
FEDRAT/0.005
GO/L1
TLLFT,GOLFT/L1,PAST,L2
RAPID;    GOTO/P2
GOTO/P3
RAPID;    GOTO/P1
RAPID;    GOTO/P5
GOTO/P4
GOTO/P6
GOTO/P3
RAPID;    GOTO/P1
RAPID;    GOTO/P7
GOTO/P4
RAPID;    GOTO/P1
THICK/0.02
A1=MACRO
GO/L2,(PLANE/0,0,1,0),L1
TLRGT,GOBACK/L1,PAST,L3
        GOLFT/L3,PAST,L4
        GOLFT/L4,L5
        GORGT/L5,PAST,L6
        GOLFT/L6,TO,C1
ARCSLP/ON;    GORGT/C1,TANTO,C2
ARCSLP/ON;    GOFWD/C2,TANTO,C3
ARCSLP/ON;    GOFWD/C3,TANTO,C4
ARCSLP/ON;    GOFWD/C4,PAST,L7
TERMAC
CALL/A1
RAPID;    GOTO/P0
LOADTL/2 $$ PUT THE FINICHING TOOL IN POSITION
CUTTER/0.005
FEDRAT/0.003
```

Figure 15-14   The APT program for machining the right-hand side of the part shown in Fig. 15-13.

```
THICK/0
SPINDL/1300,CCLW
CALL/A1
RAPID;    GOTO/P0
LOADTL/3 $$ PUT THE GROOVING TOOL IN POSITION
RAPID
GOTO/(POINT/-0.5,0.25)
FEDRAT/0.002
SPINDL/400,CCLW
GODLTA/0,-0.1,0
RAPID;    GODLTA/0,0.3,0
RAPID;    GOTO/P0
LOADTL/4 $$ PUT THE THREADING TOOL IN POSITION
RAPID;GOTO/P7
INSERT G76X0.3209Z-0.42K0.0271D0.01E0.041667A60
RAPID;GOTO/P0
SPINDL/OFF
COOLNT/OFF
END
FINI
```

Figure 15-14   (continued)

b. A tungsten carbide burr of a pointed tree shape is used to drill the small spot with a 0.01 in. depth on the acrylic plate.

A program is required to process the digitized image and to generate an NC program for reproducing the original image or an enlarged or reduced one on the acrylic plate. Describe the processing principles and draw the flowchart of the processing program, which can be written in any high-level language, such as FORTRAN or PASCAL.

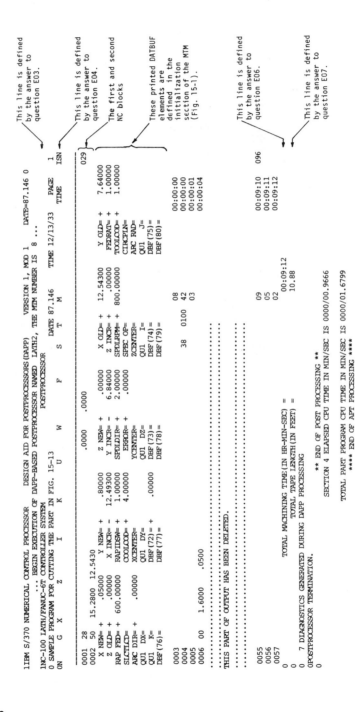

Figure 15-15 The postprocessor listing output for the APT program given in Fig. 15-14, with added explanation of its format.

560

```
%*N0001G28U.0000W.0000*
N0002G50X15.2800Z12.5430*
N0003M08*
N0004S38T0100M42*
N0005M03*
N0006G00X1.6000Z.0500*
N0007G01Z.0150F.0050*
N0008X-.0300*
N0009G00X1.3000Z.0500*
N0010G01Z-5.9000*
N0011G00X1.6000Z.0500*
N0012G00X.9000*
N0013G01X1.3000Z-3.1000*
N0014X.8000Z-4.7500*
N0015X1.3000Z-5.9000*
N0016G00X1.6000Z.0500*
N0017G00X.5000*
N0018G01X1.3000Z-3.1000*
N0019G00X1.6000Z.0500*
N0020G01X.0700Z.0350*
N0021X.3040*
N0022X.4450Z-.0355*
N0023Z-.4978*
N0024X.5700Z-.9978*
N0025Z-1.3696*
N0026G03I-.0350K-.1304U.2000W-.1304*
N0027G02I3.5870U.4052W-1.1885*
N0028G03I-1.1719K-.4115U.1402W-.4115*
N0029I-1.2420U-.1402W-.4115*
N0030G02I3.5267K-1.2385U-.4222W-1.2385*
N0031I3.7378U.6994W-1.5787*
N0032G00X4.0000Z4.0000*
N0033T0200M11*
N0034G01X.0050Z.0025F.0030S64*
N0035X.2770*
N0036X.3800Z-.0490*
N0037Z-.4998*
N0038X.5050Z-.9998*
N0039Z-1.3975*
N0040G03I-.0025K-.1025U.2000W-.1025*
N0041G02I3.6195U.4088W-1.1992*
N0042G03I-1.1412K-.4008U.1366W-.4008*
N0043I-1.2095U-.1366W-.4008*
N0044G02I3.5574K-1.2492U-.4260W-1.2492*
N0045I3.7704U.6996W-1.5858*
N0046G00X4.0000Z4.0000*
N0047T0300M11*
N0048G00X.5000Z-.5000*
N0049G01X.3000F.0020S17*
N0050G00X.9000*
N0051G00X4.0000Z4.0000*
N0052T0400M11*
N0053G00X.5000Z.0500*
G76X0.3209Z-0.42K0.0271D0.01E0.041667A60*
N0054G00X4.0000Z4.0000*
N0055M09*
N0056M05*
N0057M02*
```

Figure 15-16 The NC program generated from the APT program listed in Fig. 15-14.

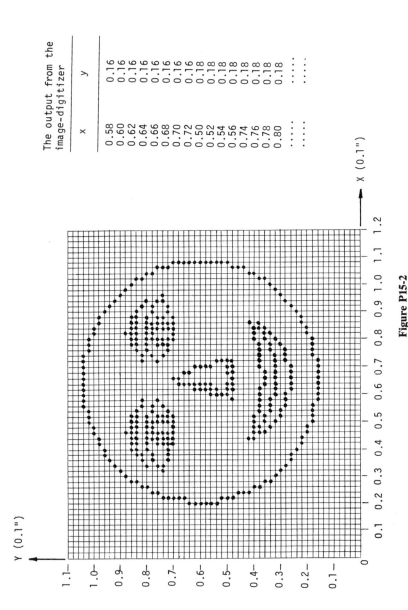

Figure P15-2

The output from the image-digitizer

x	y
0.58	0.16
0.60	0.16
0.62	0.16
0.64	0.16
0.66	0.16
0.68	0.16
0.70	0.16
0.72	0.16
0.50	0.18
0.52	0.18
0.54	0.18
0.56	0.18
0.74	0.18
0.76	0.18
0.78	0.18
0.80	0.18
....	....
....	....

X (0.1")

Y (0.1")

Epilogue to Part IV

The need for a postprocessor originates from the diversity of controller design, which prevents users from using the same NC program to cut the same part on NC machines of similar capability but of different design. A postprocessor is therefore needed for each machine to convert CLDATA into the required NC codes. A postprocessor-generated NC program is not portable; it can be used only with the machine for which that postprocessor was designed. Therefore the use of postprocessors cannot solve the problem of NC program portability. If the same part is to be cut on two NC machines of different design, the CLFILE must be processed twice, once by each postprocessor. Therefore this approach is passive in the sense that it is not intended to eliminate the existing problem but, rather, to accommodate to it.

The use of postprocessors increases the load on the computer in two aspects, namely, processing time and memory space. The cost (or time) for postprocessing an APT program is usually 30% to 60% of the total processing cost.[12] The number of postprocessors that should be kept in the computer library is determined by the number of NC machines. The more NC machines in a shop, the greater is the number of postprocessors (each having a large file) and the greater is the magnitude of related problems. Difficulty and confusion are created in production management, and production becomes less flexible and manageable.

The problems caused by a lack of exchangeability of NC programs become more critical in a computer-integrated manufacturing system (CIMS), such as a flexible manufacturing system (FMS), wherein a number of NC machines, robots, and movable vehicles are integrated under the control of a computer. Integration of CAD/CAM systems with factory equipment is often hampered at the NC postprocessing stage.[13] The hindrances confronting the users range from the time and expense needed to acquire and debug postprocessors to the frequent incompatibility of earlier-model postprocessors with newer CAD/CAM systems.

The ideal way to solve the problems is to standardize the input format (NC codes) for NC controllers. An NC processor can then be designed to yield its output in the standardized NC codes. As a result, no postprocessor is needed; an NC program is then portable among the same type of NC machines of different makes. This approach, although ideal in principle, is very difficult to realize because machine tool builders intend to keep their own design features and characteristics, and because there are already so many "nonstandard" NC machines in use. As long as the format of the input to the NC controllers is not standardized, postprocessors will always be needed. However, there are two ways to accomplish postprocessing:

1. The postprocessing is carried out by the computer.
2. The postprocessing is carried out by the NC controller.

The first is the currently adopted approach; it was discussed extensively in previous chapters. The second approach consists in using the CLDATA as the input to the NC machine-controller systems. The NC controllers should incorporate, or be retrofitted with, expanded functions so that they can accept and postprocess the CLDATA. This approach is somewhat similar to the idea mentioned above, but it is more realistic because existing NC machines can be retrofitted and used without fundamental changes.

The current NC processor output, CLDATA, defines the actions or operation relative to a part. The information contained in it is not machine related. The CLDATA is in a format that is easy for the computer to retrieve and process. Because of the limited processing capability of the CNC controller (excessive processing capability and functions are not desirable as the cost of the CNC controller would be too high), the CLDATA input to the NC controllers should be simplified. The basic concept of utilizing the CLDATA as input to NC controllers was originally defined by Rockwell International (El Segundo, Calif.) in 1974 and implemented in 1975 with the help of Vega Incorporated.[12,13] In view of the pressing needs caused by the rapid spread of FMSs, CAD/CAM systems, and CIMSs, the Electronic Industries Association has formulated a standard format for the CLDATA used as input to NC machines in 1983. According to the EIA RS-494 standard,[14] the contents of the CLDATA, including words, numbers, and characters, should be represented by a 32-bit binary exchange code, or BCL code for short, which is easier for the NC controller to process. The data format defined by the EIA RS-494 standard is the same for all machine types and all users. A BCL coded file as input to the NC controller defines actions relative to a part; thus EIA RS-494 is a part-oriented standard, in comparison with EIA RS-274-D, which is a machine-oriented standard

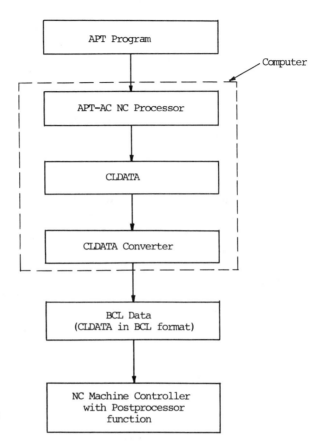

APT Program

Computer

APT-AC NC Processor

CLDATA

CLDATA Converter

BCL Data
(CLDATA in BCL format)

NC Machine Controller
with Postprocessor
function

A diagram showing the use of BCL data
as input to the NC machine.

defining machining operations for the NC machine. The figure shown here presents a diagram of the concept of using BCL as input to the NC controller.

It can be readily seen that the use of BCL data as input to NC machines requires a CLDATA converter (i.e., a computer program) to translate the CLDATA into a BCL standard format, since different NC processors have different physical representations of CLDATA records. Modifications of the NC controller are also needed because most NC machines cannot accept BCL data as input programs. Currently, very few NC machines are designed to EIA standard RS-494. Adopting the standard means a substantial investment in time, effort, and money. The advantages and disadvantages of using BCL data as input to the CNC controller must be explored because the EIA RS-494 standard has not been widely accepted by users and machine tool builders.[13,15] The use of BCL data input does make NC programs exchangeable (in the sense of statement and program format) between machines of the same type and similar capability. However, obstacles to the portability of NC programs result, not only from the diversity of NC codes acceptable to different NC machines, but also from different situations from one machine to the other, regarding, for example, the positions of fixtures and workpieces; the configuration of the NC machines; the attainable accuracy, stiffness, and horsepower; the limitations on spindle speed and feedrate; and the available tools. The fact that the BCL approach is beneficial to one company does not necessarily imply that it is also beneficial to others. As can be seen from the analysis presented above, the main purpose of the BCL standard is to provide portability of part programs among NC machines. Taking a step in this direction also means that the communication link between the computer and NC machines must be established. A computer-automated manufacturing system or a group of NC machines with direct or distributed numerical control can benefit from using BCL data as input. However, a company that has only one NC machine of each type, without linkage among computers and NC machines, will find it difficult to justify retrofitting just for the sake of using BCL. It is worth noting that the adoption of BCL data does not mean that we no longer need postprocessors. As a matter of fact, postprocessing is still required for the BCL data, but the postprocessor is removed from the computer and incorporated into the CNC controller.

In the meantime, many of the NC software vendors are trying to simplify the design of postprocessors by reducing their design to the answers to a series of questions regarding the NC machine.[16] If these answers are provided, a customized postprocessor for a specific NC machine can be generated. The realization of this approach requires that a number of processing routines be incorporated into the software for generating the postprocessor. The versatility of such software is determined by the number of processing routines included in it and the ease of answering these questions. One of the major problems in using such software is the difficulty in answering these questions correctly, because they are so numerous and interrelated and because some of them are difficult to understand or ambiguous. Another problem is that the processing routines cannot cover the functions of all types of machines. A user still must design one or more subroutines specifically for his own machine. In addition, if a processing routine in the postprocessor generator software does not meet the exact requirements for a function of a specific machine, an ordinary user usually has considerable difficulty in modifying it.

In summary, such software can be used by a person with little experience in postprocessor design to generate a customized postprocessor, provided he can understand and answer these questions correctly and the processing routines included in the software cover all the functions of his specific machine.

Is there another way to improve the portability of NC machining programs and the interface between CAD/CAM systems and NC machines? The computer numerical control (CNC) system, which has many more functions and is much more convenient to use than the hard-wired NC system, represents an evolution of the latter because of the decrease in the size and cost of computers with increasing capability. As the trend in the decreasing size and cost of computers continues, we can foresee that CNC controllers will incorporate more powerful dedicated computers. Interactive NC programming and graphic display of the cutter path have been realized on many CNC machines. The functions of a postprocessor can also be included as discussed above. The question, then, is, Can we also include the functions of the NC processor in a machine so that we may simply use the APT program as the input to the NC controller?* In view of the development of computer technology, we think

---

*This concept was proposed by Dr. David Grossman, of the IBM Corporation, during a private conversation.

that this goal can be reached and that reaching it is only a matter of time and effort. The competition among machine tool builders, and hence the great diversity in NC machine controller design, did promote the rapid development of NC technology in the past; however, they have become a hindrance to the development of computer-integrated manufacturing technology today. Standardization of design, as the history of industry has indicated in many fields, is the best way to save money, time, and effort in the realization of computer-integrated manufacturing systems.

REFERENCES FOR PART IV

1. ISO Standard (ISO 6983/1-1982): *Numerical Control of Machines: Program Format and Definition of Address Words. Part 1: Data Format for Positioning, Line Motion and Contouring Control Systems*, 1st ed. Geneva, Switzerland: International Organization for Standardization, 1982.

2. EIA Standard (RS-274-D): *Interchangeable Variable Block Data Format for Positioning, Contouring, and Contouring/Positioning Numerically Controlled Machines*. Washington, D.C.: Electronic Industries Association, 1979.

3. ISO Standard (ISO 3592-1978): *Numerical Control of Machines: NC Processor Output—Logical Structure and Major Words*. Geneva, Switzerland: International Organization for Standardization, 1978.

4. ISO Standard (ISO 4343-1978): *Numerical Control of Machine: NC Processor Output—Minor Elements of 2000 Type Records (Postprocessor Commands)*. Geneva, Switzerland: International Organization for Standardization, 1978.

5. Subramanian, M. L., "CLFILE Manipulator," *Advancing Manufacturing Technologies, Proceedings of the 21st Annual Meeting and Technical Conference*, Numerical Control Society, 1984, pp. 249–63.

6. IBM Manual (SH20-1414-2): *Automatically Programmed Tool—Advanced Contouring Numerical Control Processor: Program Reference Manual*, 3rd ed. Rye Brook, N.Y.: IBM Corp., 1985.

7. IBM Manual (SH20-2469-0): *System/370 Automatically Programmed Tool—Intermediate Contouring and Advanced Contouring Numerical Control Processor: Program Reference Manual*. Rye Brook, N.Y.: IBM Corp., 1980.

8. Sim, R. M., "Postprocessor," *Numerical Control User's Handbook*, W. H. P. Leslie, ed. New York: McGraw-Hill, 1970, pp. 299–344.

9. Chang, Chao-Hwa, "A Postprocessor for Computer-Aided Programming of the NC-100 Lathe/GN6T Controller systems based on the IBM System/370 APT-AC NC Processor," Master's degree thesis, University of California, Los Angeles, 1983.

10. Patton, W. J., *Numerical Control: Practice and Application*. Reston, Va.: Reston Publishing, 1972.

11. IBM Manual (SH20-1413-2): *System/370 Automatically Programmed Tool—Advanced Contouring Numerical Control Processor: Operations Guide*, 3rd ed. Rye Brook, N.Y.: IBM Corp., 1982.

12. Herndon, L. R., Jr., "CLdata Input for CNC: The End of the Line for Postprocessor," *NC—From the User's Point of View*, Numerical Control Society, 1977.

13. Ogorch, M., "CNC Standard Format," *Manufacturing Engineering*, Jan. 1985, pp. 43–5.

14. EIA Standard (RS-494): *32 Bit Binary CL Exchange (BCL) Input Format for Numerical-Controlled Machines.* Washington, D.C.: Electronic Industries Association, Aug. 1983.

15. Justice, R. K., "Capture the BCL Concept for Present Day Controllers," *Advancing Manufacturing Technologies, Proceedings of the 21st Annual Meeting and Technical Conference,* Numerical Control Society, 1984, pp. 241–8.

16. IBM Manual (GH20-6908-0): *Numerical Control Postprocessor Generator: General Information,* 1st ed. Rye Brook, N.Y.: IBM Corp., 1986.

Appendixes

APPENDIX A   THE VOCABULARY WORDS USED IN THE APT LANGUAGE

AAXIS	ABSF	ABSLTE	ACOSF
ADD	ADJUST	AIR	ALL
ALPNUM	ANGLE	ANGLF	ANGTOL
ANTSPI	APT360	ARC	ARCSLP
ASINF	ASLOPE	AT	ATANF
ATANGL	ATAN2F	ATTACH	AUTO
AUTOPS	AUTPOL	AUXFUN	AVOID
AXIS	BACK	BAXIS	BCD
BCHIP	BEVEL	BEVELS	BINARY
BISECT	BLACK	BLANK	BLUE
BORE	BOREOS	BOTH	BREAK
BRKCHP	CALL	CAM	CAMERA
CANON	CATLOG	CAXIS	CBORE
CBRTF	CCLW	CENTER	CHANGE
CHECK	CHORD	CHUCK	CIRCLE
CIRCUL	CIRLIN	CLAMP	CLDIST
CLEARP	CLEARV	CLFILE	CLPRNT
CLRSRF	CLTV	CLW	CM
CMIT	CNSINK	COLLET	COMBIN
CONE	CONSEC	CONST	CONT
CONTIN	CONTUR	COOLNT	COPY
CORNFD	COSF	COTANF	COUPLE
CROSS	CRSSPL	CS	CSINK
CTREAC	CURSEG	CUT	CUTANG
CUTCOM	CUTTER	CYCLE	CYLNDR
DAC	DARK	DASH	DATA
DATREF	DCOORD	DEBUG	DECR
DEEP	DEEPHL	DELAY	DELET
DELETE	DELTA	DEPTHV	DHOLE
DIAG1	DIAG2	DIAG3	DIAMTR
DISPLY	DISTF	DITTO	DMILL
DNTCUT	DNTLR	DNTLRP	DNTR
DO	DOTF	DOTTED	DOWN
DRAFT	DRAWLI	DRESS	DRILL
DS	DSTAN	DWELL	DWELLV
DWL	DYNDMP	EDIT	EDITND
ELLIPS	ELMSRF	END	ENDARC
ERCOND	EXEC	EXPF	FACE
FACEML	FAN	FEDRAT	FEDTAB
FEED	FEEDRT	FEET	FILE
FINI	FINISH	FIX	FLOOD
FLOW	FMATL	FORMAT	FOURPT
FREE	FROM	FRONT	FULL
FUNOFY	GAPLES	GCONIC	GENCUR
GO	GOBACK	GOCLER	GODLTA
GODOWN	GOFWD	GOHOME	GOLFT
GORGT	GOTO	GOUGCK	GOUP
GREEN	GRID	GROOVE	HEAD
HIGH	HOLDER	HYPERB	ICODEF

From IBM Manual SH20-1414-2: *Automatically Programmed Tool—Advanced Contouring (APT-AC) Numerical Control Processor: Program Reference Manual*, 3rd ed. Rye Brook, N.Y.: IBM Corp., 1985. (Courtesy of International Business Machines Corporation.)

IF	IFRO	IN	INCHES
INCR	INDEX	INDIRP	INDIRV
INDNRM	INDVEC	INSERT	INTCOD
INTEGR	INTENS	INTERC	INTGF
INTGRV	INTOF	INTOL	INVERS
INVOC	INVX	INVY	IPM
IPR	ISTOP	JUMPTO	KEYBOR
LARGE	LAST	LATER	LCONIC
LEADER	LEFT	LENGTH	LETTER
LIBRY	LIFTOF	LIGHT	LIMIT
LIMSRF	LINCIR	LINE	LINEAR
LINK	LINTOL	LIST	LITE
LNTHF	LOADTL	LOCAL	LOCK
LOCKX	LOFT	LOGF	LOG10F
LOOPND	LOOPST	LOW	LPRINT
LTV	MACH	MACHIN	MACRO
MAGTAP	MAIN	MAJOR	MANUAL
MATL	MATRIX	MAXDP	MAXDPM
MAXIPM	MAXRPM	MAXVEL	MAX1F
MCHFIN	MCHTOL	MDEND	MDWRIT
MED	MEDIUM	MESH	MILL
MINOR	MINUS	MIN1F	MIRROR
MIST	MIT	MM	MMPM
MMPR	MODE	MODF	MODIFY
MOTION	MOVETO	MULTAX	MULTRD
MXMMPM	NAUDIT	NCDB	NCTEST
NEGX	NEGY	NEGZ	NEXT
NIXIE	NOCS	NOMORE	NOPLOT
NOPOST	NOPS	NORMAL	NORMDS
NORMPS	NOSLP	NOW	NOX
NOY	NOZ	NUMBR	NUMF
NUMPTS	OBTAIN	OFF	OFFSET
OMIT	ON	OPEN	OPRINT
OPSKIP	OPSTOP	OPTION	ORIGIN
OUT	OUTTOL	OVCONT	OVPLOT
PARAB	PARAM	PARLEL	PART
PARTNO	PASS	PAST	PATCH
PATERN	PBS	PEN	PENDWN
PENUP	PERPTO	PERSP	PI
PICKUP	PILOTD	PITCH	PIVOTZ
PLABEL	PLANE	PLOT	PLUNGE
PLUS	PNTSON	PNTVCT	PNTVEC
POCKET	POINT	POLCON	POLYGN
POSMAP	POSTN	POSX	POSY
POSZ	POWER	PPLOT	PPRINT
PPWORD	PREFUN	PRINT	PROBX
PROBY	PROC	PROCND	PROPF
PS	PSIS	PSTAN	PTFORM
PTNORM	PTONLY	PTSLOP	PULBOR
PULFAC	PUNCH	QADRIC	QUILL
RADIUS	RAIL	RANDOM	RANGE
RAPID	READ	REAM	REAMA
REAR	RED	REDEF	REFSYS

(continued)

REGBRK	REMARK	REPLAC	RESERV
RESET	RETAIN	RETRCT	REV
REVERS	REVOLV	REWIND	RIGHT
RLDSRF	ROOT	ROTABL	ROTHED
ROTREF	ROUGH	ROUND	RPM
RTHETA	RULED	SADDLE	SAFETY
SAME	SCALE	SCRIBE	SCRUCT
SCULPT	SCURV	SECTN3	SEC1
SEC2	SEC3	SEG	SELCTL
SEQNO	SETANG	SETOOL	SFM
SIDE	SIGNF	SINF	SLOPE
SLOWDN	SMALL	SMESH	SOLID
SOURCE	SPALIN	SPDRL	SPDTAB
SPECDP	SPECFR	SPEED	SPHERE
SPINDL	SPINSP	SPLINE	SPMIL
SQRTF	SRFREV	SEFVCT	SSURF
START	STEP	STOP	SWITCH
SYN	SYSLIB	TABCYL	TABPRT
TANDS	TANF	TANON	TANSPL
TANTO	TAP	TAPKUL	TERMAC
TEXT	THETAR	THICK	THREAD
THRU	TIME	TIMES	TITLES
TLAXIS	TLLFT	TLNDON	TLOFPS
TLON	TLONPS	TLRGT	TMARK
TO	TOLER	TOOL	TOOLNO
TOOLST	TORS	TORUS	TP
TPI	TPMM	TRACUT	TRANPT
TRANS	TRANSL	TRANTO	TRAV
TRFORM	TRMCOD	TRYBOR	TRYBOS
TUNEUP	TURN	TURRET	TWOPT
TYPE	TYPEF	UAXIS	ULOCKX
UNIT	UNITS	UNLIKE	UNLOAD
UP	VAXIS	VECTOR	VTLAXS
WAXIS	WCORN	WDEFAC	WEIGHT
XAXIS	XCOORD	XLARGE	XREF
XSMALL	XYOR	XYPLAN	XYROT
XYVIEW	XYZ	YAXIS	YCOORD
YLARGE	YSMALL	YZPLAN	YZROT
YZVIEW	ZAXIS	ZCOORD	ZERO
ZIGZAG	ZLARGE	ZSMALL	ZSURF
ZXPLAN	ZXROT	ZXVIEW	2DCALC
3DCALC	3PT2SL	4PT1SL	5PT

APPENDIX B   THE CANONICAL FORMS OF SELECTED GEOMETRIC ENTITIES DEFINED IN APT

Type of Geometric Entity	Date Stored in Memory or Appearing in Listing File	Explanation
POINT	X, Y, Z	X, Y, and Z are the Cartesian coordinates of the point in the selected coordinate system.
LINE	A, B, C, D	A, B, C, and D are the coefficients of the normalized line equation $$AX + BY + CZ - D = 0$$ C is always zero, since, in APT, a line is treated as plane perpendicular to the X-Y plane.
CIRCLE	X, Y, Z, A, B, C, R	X, Y, and Z are the coordinates of the center point; R is the radius. A circle is treated as a cylinder parallel to the Z axis. A, B, and C are, respectively, the X, Y, and Z components of the unit vector parallel to the cylinder axis. Thus A, B, and C are 0, 0, and 1, respectively.
VECTOR	A, B, C	A, B, and C are, respectively, the X, Y, and Z components of the vector.
PLANE	A, B, C, D	A, B, C, and D are the coefficients of the plane equation: $$AX + BY + CZ - D = 0$$
CYLNDR	X, Y, Z, A, B, C, R	X, Y, and Z are the coordinates of a defined point on the axis of the cylinder. A, B, and C are, respectively, the X, Y, and Z components of the unit axial vector. R is the radius.
QADRIC	$A, B, C, D, F, G, H, P, Q, R$	These are the coefficients of the general quadric surface equation: $$AX^2 + BY^2 + CZ^2 + D + 2FYZ + 2GXZ + 2HXY + 2PX + 2QY + 2RZ = 0$$
GCONIC and LCONIC	$A_1, A_2, \ldots, A_{10}, C_1, C_2, \ldots, C_6,$ $M1, M2, \ldots, M12, N1, N2, \ldots, N12$	A_1 through A_{10} are the coefficients of the quadric surface equation: $$A_1x^2 + A_2y^2 + A_3z^2 + A_4 + 2A_5yz + 2A_6xz + 2A_7xy + 2A_8x + 2A_9y + 2A_{10}z = 0$$ C_1 through C_6 are the coefficients of the conic equation: $$C_1x^2 + C_2xy + C_3y^2 + C_4x + C_5y + C_6 = 0$$ $M1$ through $M12$ are the coefficients of the matrix M, which defines an inverse of the transformation that maps a point in the universal coordinate system where the conic is defined by the conic equation shown above. $N1$ through $N12$ are the coefficients of the inverse matrix of M.
MATRIX	$M1, M2, \ldots, M12$	These are the coefficients of the transformation matrix: $$\begin{bmatrix} M1 & M2 & M3 & M4 \\ M5 & M6 & M7 & M8 \\ M9 & M10 & M11 & M12 \\ 0 & 0 & 0 & 1 \end{bmatrix}$$

Modified from IBM Manual SH20-1414-2: *Automatically Programmed Tool — Advanced Contouring (APT-AC) Numerical Control Processor: Program Reference Manual*, 3rd ed., Rye Brook, N.Y.: IBM Corp., 1985. (Courtesy of International Business Machines Corporation.)

APPENDIX C **THE FORMAT AND CONTENTS OF THE PRINTOUT FOR A TABULATED CYLINDER**

Whenever a tabulated cylinder is defined, its data are printed in the listing output from the NC processor. The format and contents of this printout can be seen from the following example.

Let the defined tabulated cylinder (Fig. C-1) be

TABCYL/NOZ,SPLINE,2.5,3.5,3.5,4,7.5,3.5,11,4

then the corresponding data appearing in the listing is given in Fig. C-2.

The printout appearing after the TABCYL statement consists of the following three sections:

   I. A list of curvatures of the directrix at the given points together with a curvature plot. For example, the curvatures at the first and last defined points are −0.310099 and 0, respectively.

  II. The number of spaces needed to store the canonical form data. In this example, it is 48 (9 for the transformation matrix, 1 for the number of points, and 38 for the various parameters).

 III. The canonical form data.

Parameters U and V in section III are the coordinates of the defined points in the transformed coordinate system (U-V-W). Since no transformation matrix is specified in this statement, the coordinate system U-V-W is the same as the part coordinate system X-Y-Z. Two points are added as extensions of the defined directrix, namely, PT0 (−5.49871750,−2.50170961) and PT5 (19.9507012,8.45925412). As can be easily verified, the lengths of these extended lines, PT0-PT1 and PT4-PT5, are 10 units.

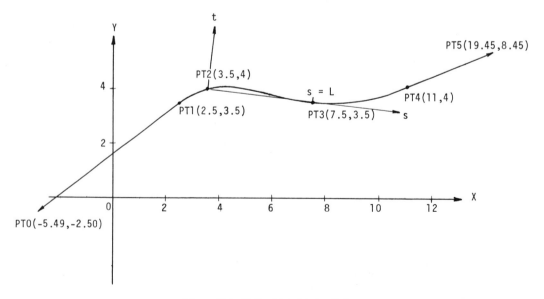

Figure C-1 Defined tabulated cylinder.

```
1                                                                                              PAGE 0001
 IBM S/370 APT-AC N/C PROGRAM VERSION=1.3.0000                                    DATE=09/29/87 TIME=09:54:29
0
                              ... BEGIN TRANSLATION PHASE... (SECTION 1) ...
0
 ISN 00001 PARTNO TEST TABCYL
 ISN 00002 T1 = TABCYL/NOZ,SPLINE,2.5,3.5,3.5,4.0,7.5,3.5,11,4
                     -.600000          -.350000          -.100000           .150000           .400000
      CURVATURE    .+...................+...................+...................+...................+.
  I    1 -.310099    .                   *                   .                                       .
       2 -.310932    .                   *                   .                                       .
       3  .160454    .                                       .                   *                   .
       4  0.0        .                                       *                                       .
      CURVATURE    .+...................+...................+...................+...................+.
0
 II -> DATA STORAGE =   48
  III ROTATION MATRIX
        1.00000000        0.0               -0.0
       -0.0               1.00000000        -0.0
        0.0               0.0                1.00000000
      NUMBER OF POINTS =    6
0
           U                V                A                  B               LENGTH          MAX             MIN
0
      -5.49871750      -2.50170961       0.0                0.0               10.0000000      0.0             0.0
       2.50000000       3.50000000      -.111022302 D-16  -.162820608        1.11803398      .045509743      0.0
       3.50000000       4.00000000       .023265677       -.200997302        4.03112887      .068226527      0.0
       7.50000000       3.50000000       .333066907 D-17   .093828662        3.53553390     -0.0            -.082933604
       11.0000000       4.00000000       0.0               0.0               10.0000000      0.0             0.0
       19.9507012       8.45925412
 ISN 00003 PRINT/3,T1

      T1               TABCYL               49 ITEMS        UNITS= INCHES
        48      6      0.0
       1.00000000      0.0               -0.0
      -0.0             1.00000000        -0.0
       0.0             0.0                1.00000000
      -5.49871750     -2.50170961         0.0                0.0               10.0000000      0.0             0.0
       2.50000000      3.50000000        -.111022302 D-16  -.162820608        1.11803398      .045509743      0.0
       3.50000000      4.00000000         .023265677       -.200997302        4.03112887      .068226527      0.0
       7.50000000      3.50000000         .333066907 D-17   .093828662        3.53553390     -0.0            -.082933604
       11.0000000      4.00000000         0.0               0.0               10.0000000      0.0             0.0
       19.9507012      8.45925412         1.00000000

 ISN 00004 END
 ISN 00005 FINI


    NO DIAGNOSTICS ELICITED DURING TRANSLATION PHASE
     5 N/C SOURCE RECORDS (SYSIN)
                                     **** END OF APT PROCESSING ****
```

Figure C-2

The directrix is divided into $n + 1$ sections, where n is the number of points specified in the TABCYL statement. The first and last sections are the extended lines. A parametric cubic equation with the following expression is used to define the curve sections between every two adjacent defined points:

$$t = As^3 + Bs^2 + Cs$$

where

A and B = the parameters listed, respectively, in columns A and B in section III

$C = -(AL^2 + BL)$, with L equal to the distance between the two end points of a curve section, which is listed in column LENGTH

s = the coordinate in the coordinate system t-s. Note that in each curve section, the s axis is in the direction from the starting point to the end point; thus $0 \leqslant s \leqslant L$.

For example, according to the output listed in Fig. C-2, the curve section PT2-PT3 is defined by the equation

$$t = 0.023265677s^3 - 0.200997302s^2 + 0.432178776s$$

The maximum and minimum of t for each curve section can be determined by multiplying L by the parameter listed in columns MAX and MIN, respectively. If there is a point of inflection within a curve section, then both a maximum and a minimum appear in the list; otherwise, only a maximum or a minimum is listed (the other is printed as zero).

APPENDIX D THE JOB CONTROL LANGUAGE (JCL) STATEMENTS FOR CREATION OF A
DAPP-BASED POSTPROCESSOR AND PROCESSING AN APT PROGRAM
UNDER OS/MVS ENVIRONMENT

1. The JCL statements for generating the BLOCK DATA program:

```
//IS79CHC1 JOB IS79CHC,MSGCLASS=A,TIME=(,20)
//          EXEC PGM=QUEST
//STEPLIB   DD DSN=SYS1.DKWLM.PP,UNIT=3330,VOL=SER=NCPACK,
           DISP=SHR
//FT06F001 DD SYSOUT=A
//FT07F001 DD SYSOUT=B
//FT08F001 DD UNIT=3330,SPACE=(TRK,4),
           DCB=(RECFM=FB,LRECL=80,BLKSIZE=400)
//FT05F001 DD *
.............................
.............................
.............................  (The anwswers to the DAPP Questionnaire)
.............................
.............................
/*
```

2. The JCL statements for creating a postprocessor:

```
//IS79CHC2 JOB IS79CHC
//          EXEC FORTHCL,PARM.FORT='OPT=2'
//FORT.SYSIN DD DSN=IS79CHC.COMBLO8,DISP=SHR
           DD DSN=IS79CHC.MTMD08,DISP=SHR
/*
//LKED.SYSLMOD DD DSN=IS79CHC.POSTP8,DISP=(NEW,KEEP),UNIT=3330,
//                VOL=SER=NCPACK,SPACE=(CYL,(5,1,5))
//LKED.OLDLIB  DD DSN=SYS1.DKWLM.PP,UNIT=3330,DISP=SHR,
              VOL=SER=NCPACK
//LKED.SYSIN DD *
         INCLUDE OLDLIB(ZDAPPPP)
         ENTRY DAPPPP
         NAME LATH2PP
/*
```

where the data-set names for the BLOCK DATA program and the Machine
Tool Module are COMBLO8 and MTMD08, respectively. The data-set name
for the created postprocessor is POSTP8. In the second last card (or line),
LATH2 is the postprocessor name defined by the answer to question E01.

Modified from references 2 and 8 for Part II and reference 9 for Part IV. In the JCL statements, IS79CHC is an assumed user's account name, and IS79CHCn, the job name. The reader should change the data-set name and parameters and the type of output according to the requirement of his or her computer system.

3. The JCL statements for processi   an APT program:

```
//IS79CHC3 JOB IS79CHC
//GO        EXEC PGM=DKW0,PARM='POOLSIZE=(10,5),LINECNT=76'
//STEPLIB   DD DSN=SYS1.DKWLM.PP,UNIT=SYSDA,VOL=SER=NCPACK,DISP=SHR
            DD DSN=IS79CHC.POSTP8,UNIT=SYSDA,VOL=SER=NCPACK,DISP=SHR
//ALINKLIB  DD DSN=SYS1.DKWLM.PP,UNIT=SYSDA,VOL=SER=NCPACK,DISP=SHR
//ERRFILE   DD SYSOUT=A
//FT02F001  DD UNIT=SYSDA,SPACE=(CYL,(2,2))
//FT06F001  DD SYSOUT=A,DCB=(RECFM=VBA,LRECL=133,BLKSIZE=1995)
//DPOOL     DD UNIT=SYSDA,SPACE=(CYL,(3,1))
//LIBDPOOL  DD DSN=APTAC.LIBDPOOL,UNIT=3330,VOL=SER=NCPACK,DISP=SHR
//SOURCE    DD DUMMY
//SYSPUNCH  DD SYSOUT=B
//POSPUNCH  DD SYSOUT=B
//FT07F001  DD SYSOUT=B
//FT05F001  DD DSN=IS79CHC.PTPGM1,DISP=SHR
//
```

where PTPGM1 is the APT program to be processed; POSTP8 is the data-set name of the used postprocessor; and LIBDPOOL is the data-set name of the NC processor diagnostic table.

subject index

name index

Aerospace Industries Association, 8
American National Standards Institute (ANSI), 418
Bendix Aviation Corporation, 5
Electronic Industries Association (EIA), 18, 564
EMI (British), 5
CADAM Inc., 380, 392
Cincinnati Milling Machine Company, 5
Computer Vision Corporation, 381
General Dynamic Corporation, 5
General Electric Company, 5
Giddings and Lewis Machine Tool Company, 5
Illinois Institute of Technology Research Institute (IITRI), 9
International Business Machine Corporation (IBM), 8, 9, 87, 354, 357, 380, 413
International Organization for Standardization (ISO), 18
Kearney and Tracker Corporation, 5
Massachusetts Institute of Technology (MIT), 5, 6, 7, 8
Milner, D. A., 12
Morey, 5
Parsons Corporation, 4, 5
Parsons, John, 4, 5
Rockwell International, 564
U.S. Air Force, 4, 5, 6
University of California, Los Angeles (UCLA), 63
Vasiliou, V. C., 12
Vega Incorporated, 564
Webster, William T., 5
Welch, A., 12